D0152850

Optical Effects in Amorphous Semiconductors

(Snowbird, Utah, 1984)

SC

AIP Conference Proceedings
Series Editor: Hugh C. Wolfe
Number 120

Optical Effects in Amorphous Semiconductors

(Snowbird, Utah, 1984)

Edited by
P. C. Taylor
University of Utah
and
S. G. Bishop
Naval Research Laboratory

American Institute of Physics
New York 1984

L.C. Catalog Card No. 84-72419
ISBN 0-88318-319-6
DOE CONF- 840881

PREFACE

An International Topical Conference on Optical Effects in Amorphous Semiconductors was held at the Snowbird Resort, Snowbird, Utah from the first to the fourth of August, 1984. This meeting was a satellite conference of the 17th International Conference on the Physics of Semiconductors, which was held on 6 through 10 August 1984 in San Francisco, California

The conference covered the basic optical processes which occur in amorphous semiconductors. A major area of emphasis concerned the electronic and structural metastabilities which occur in amorphous semiconductors as a result of optical excitation. The scope also included optically-induced degradation of devices which are based on hydrogenated amorphous silicon (a-Si:H).

The conference, which was sponsored by the American Physical Society and the University of Utah, is grateful to the following institutions for generous financial support:

Energy Conversion Devices, Inc.
Exxon Research and Engineering Company
IBM Corporation
3M Center
Solar Energy Research Institute
University of Utah
Xerox Corporation.

The Organizing Committee consisted of S.G. Bishop (Naval Research Laboratory), M.H. Brodsky (IBM, Yorktown), D.E. Carlson (Solarex Corp.), M.H. Cohen (Exxon, Linden), H. Fritzsche (U. of Chicago), M. Kastner (MIT), G. Lucovsky (North Carolina State U.), J. Mort (Xerox, Webster), W. Paul (Harvard U.), R.A. Street (Xerox, PARC), J. Tauc (Brown U.), and P.C. Taylor, Chairman (U. of Utah). Members of the International Advisory Committee included D. Adler (USA), M. Cardona (FRG), G.D. Cody (USA), E.A. Davis (UK), Y. Hamakawa (Japan), D. Kaplan (France), J.C. Knights (USA), K. Morigaki (Japan), I. Solomon (France), W.E. Spear (UK), J. Stuke (FRG), K. Tanaka (Japan), D.L. Weaire (Ireland), and K. Weiser (Israel).

These Proceedings have been organized into chapters which reflect the topics covered by the conference sessions:

Electronic States in the Gap
Photoluminescence
Light-Induced Effects
Urbach Edges and Band-Tail States
Amorphous Heterostructures and Superlattices
Structural and Photostructural Properties.

The Conference is grateful to the management at the Snowbird Resort for competent and courteous handling of the local arrangements. The Department of Conferences and Institutes at the University of Utah is gratefully acknowledged for skillful treatment of many of the administrative details. We also thank the Physics Department and the Salt Lake Valley Convention and Visitors Bureau for providing secretarial help. Mary Ann Woolf is gratefully acknowledged for invaluable assistance both with administrative details and with typing and editing of the proceedings. The Conference is indebted to several graduate and undergraduate students in the Physics Department for help with local arrangements. We also thank M. Abkowitz, D. Adler, D.K. Biegelsen, M.H. Brodsky, D.E. Carlson, B.C. Cavenett, G.D. Cody, H. Fritzsche, J.S. Lannin, G. Lucovsky, W. Paul, E. Sabisky, R.A. Street, U. Strom, J. Tauc and B. Weinstein for assistance in editing and in running the meeting.

Professor H. Fritzsche presented an excellent review of the light-induced changes in hydrogenated amorphous silicon and led a stimulating discussion on this topic. We are particularly grateful for these contributions.

P.C. Taylor
Salt Lake City, Utah

S.G. Bishop
Washington, D.C.

August, 1984

TABLE OF CONTENTS

PREFACE

CHAPTER 1: Electronic States in the Gap

Photoinduced Absorption Spectra in Amorphous Si:H and Ge:H and Microcrystalline Si:H
Z. Vardeny, D. Pfost, Hsian-na Liu, and J. Tauc.............. 1

Photoinduced Absorption and Photoinduced Absorption-Detected ESR in P-Doped a-Si:H
I. Hirabayashi, K. Morigaki, S. Yamasaki and K. Tanaka........ 8

Correlation of Optical and Thermal Emission Processes for Bound-to-Free Transitions from Mobility Gap States in Doped Hydrogenated Amorphous Silicon
A.V. Gelatos, J.D. Cohen and J.P. Harbison.................... 16

Photoconductivity of Intrinsic and Doped a-Si:H from 0.1 to 1.9 eV
T. Inushima, M.H. Brodsky, J. Kanicki and R.J. Serino......... 24

Photo-Depopulation-Induced ESR Measurement of Deep Gap States in a-Si:H
D.K. Biegelsen and N.M. Johnson............................... 32

Photoconductivity and Recombination in Amorphous Silicon Alloys
M. Hack, S. Guha and M. Shur.................................. 40

Photovoltaic Detection of Magnetic Resonance in a-Si:H Solar Cells
K.P. Homewood, B.C. Cavenett, C. van Berkel, W.E. Spear and P.G. LeComber.. 48

Systematic Trends in the energies of Dangling Bond Defect States in a-Si Alloys Containing C, N and O
G. Lucovsky and S.Y. Lin...................................... 55

Unified Theory of Bonding at Defects and Dopants in Amorphous Semiconductors
J. Robertson.. 63

Optical Determination of the Effective Correlation Energy of the Dangling Bond in Hydrogenated Amorphous Silicon
David Adler... 70

Photoinduced Paramagnetic Centers in a-SiO_2
J.H. Stathis and M.A. Kastner................................. 78

Paramagnetic States in Chalcogenide Glasses Induced by Near Mid-Gap Infrared Radiation
S.G. Bishop, J.A. Freitas, Jr. and U. Strom.................86

Transient Photo-Induced Absorption Spectroscopy in $a\text{-}As_2Se_3$ in the Presence of Strong Bias Illumination
Don Monroe and M.A. Kastner..................................94

Picosecond Optical Generation and Detection of Phonon Waves in $a\text{-}As_2Te_3$
C. Thomsen, J. Strait, Z. Vardeny, H.J. Maris, J. Tauc, and J.J. Hauser...102

Analysis of $a\text{-}As_2Se_3$ Transient Photocurrents in the Long-Time Regime
Guy J. Adriaenssens and Herman Michiel.......................110

Meastable Photoenhanced Thermal Generation in a-SeTe Alloys
M. Abkowitz, G.M.T. Foley, J.M. Markovics and A.C. Palumbo..117

CHAPTER 2: Photoluminescence

IR-Induced Transients of Photoluminescence and Photoconductivity in a-Si:H
R. Carius and W. Fuhs..125

Infrared Modulation of Photoluminescence in Glow Discharge Amorphous Silicon
C. Varmazis, M.D. Hirsch and P.E. Vanier.....................133

Time-Resolved Luminescence, ODMR and Light-Induced Effects in a-Si:H Films Prepared by the Glow-Discharge Decomposition of Disilane
M. Yoshida, K. Morigaki, I. Hirabayashi, H. Ohta, A. Amamou and S. Nitta...141

The Role of Dangling Bonds in Radiative and Nonradiative Processes in a-Si:H
B.A. Wilson, A.M. Sergent and J.P. Harbison..................149

Time Resolved Radiative and Nonradiative Recombination in a-Si:H
U. Strom, J.C. Culbertson, P.B. Klein and S.A. Wolf..........157

Picosecond Photoluminescence as a Probe of Band-Tail Thermalization in a-Si:H
T.E. Orlowski, B.A. Weinstein and H. Scher...................163

A Comparison of Photoinduced Absorption and Photoluminescence Measurements in a-Si:H
R.W. Collins and W.J. Biter.................................... 170

Metastable Carriers in a-Si:H: A Study by IR Stimulated Photoluminescence and Photoconductivity
F. Boulitrop.. 178

Triplet Photoluminescence in Crystalline Chalcogenide Semiconductors
L. Robins and M. Kastner.. 183

Luminescence and Photo-Effects in Obliquely Deposited a-Se-Ge Films
P.K. Bhat, T.M. Searle, I.G. Austin and K.L. Chopra.......... 189

Recombination Kinetics in Amorphous Semiconductors
Marvin Silver and David Adler................................... 197

CHAPTER 3: Light-Induced Effects

Dependence of the Metastable Light-Induced ESR in a-Si:H on Temperature and Power
C. Lee, W.D. Ohlsen, P.C. Taylor, H.S. Ullal and G.P. Ceasar... 205

The Kinetics of Formation and Annealing of Light-Induced Defects in Hydrogenated Amorphous Silicon
M. Stutzmann, W.B. Jackson and C.C. Tsai...................... 213

Light-Induced Creation of Defects and Related Phenomena in Silicon-Based Amorphous Semiconductors
A. Morimoto, H. Yokomichi, T. Atoji, M. Kumeda, I. Watanabe and T. Shimizu.................................... 221

Search for Reversible Light-Induced Changes in the Absorption Bands of a-Si:H
H. Fritzsche, J. Kakalios and D. Bernstein..................... 229

The Effects of Light Soaking on a-Si:H Films Containing Impurities
D.E. Carlson, A. Catalano, R.V. D'Aiello, C.R. Dickson and R.S. Oswald.. 234

The Staebler-Wronski Effect in Undoped a-Si:H: Its Intrinsic Nature and the Influence of Impurities
C.C. Tsai, M. Stutzmann and W.B. Jackson...................... 242

Gap States Dynamics Observed in Light-Soaked P-Doped a-Si:H
H. Okushi, M. Itoh, T. Okuno, Y. Hosokawa, S. Yamasaki and K. Tanaka.. 250

Photodarkening and Photobleaching in a-C:H Films
S. Iida, T. Ohtaki and T. Seki.................................. 258

Light-Induced Effect in a-Si_xC_{1-x}:H Films
Chen Guanghua, Zhang Fangqing, Wang Yinyue, Wang
Huisheng, Xu Xixiang and T. Shimizu.............................. 266

Photoconductivity Detected S-W Related Defect Levels in
a-Si:H
A. Tříska, M. Vaněčeck. O. Stika, J. Stuchlík, A.
Kosarev and J. Kočka.. 272

Light Induced Change in Photoluminescence Intensity of
Hydrogenated Amorphous Silicon
Jin Jang and Choochon Lee... 280

Optically Induced Changes in Photoluminescence in
Amorphous Si:H
P.K. Bhat, T.M. Searle and I.G. Austin............................ 288

The Effect of Extended Light Exposure on the Urbach Tail
and the Subbandgap Absorption in
a-Si:H
Daxing Han and H. Fritzsche....................................... 296

Difference in the Influence of Oxygen and Nitrogen on
the Light Induced Effect
N. Nakamura, S. Tsuda, T. Takahama, M. Nishikuni,
K. Watanabe, M. Ohnishi and Y. Kuwano............................. 303

Influence of Bias and Photo Stress on a-Si:H-Diodes with
nip- and pip-Structures
W. Krühler, H. Pfleiderer, R. Plättner and W. Stetter......... 311

The Effect of Dopants on the Stability of a-Si Solar
Cells
H. Sakai, A. Asano, M. Nishiura, M. Kamiyama, Y. Uchida
and H. Haruki... 318

Optically Detected ESR of Luminescence Centers in
a-As_2S_3
T. Tada, H. Suzuki, K. Murayama and T. Ninomiya................... 326

Optically-Induced ESR and Absorption Edge Shift in GeS_x
Glasses
J. Shirafuji, N. Kumagai and Y. Inuishi........................... 333

CHAPTER 4: Urbach Edges and Band-Tail States

The Conduction Band of Hydrogenated Amorphous Silicon
W.B. Jackson, S.-J. Oh, C.C. Tsai and J.W. Allen.................. 341

Optical Properties of a-Ge:H--Structural Disorder and Hydrogen Alloying
P.D. Persans, A.F. Ruppert, G.D. Cody, B.G. Brooks and W. Lanford..349

Electro-optical Effects on the Absorption Edge of a-Si:H
U. Mescheder and G. Weiser....................................356

Comparative Study of the Photoconductivity and the Photoabsorption of Amorphous Si:F and Si:H Films
M.Janai, R. Weil, B. Pratt, Z. Vardeny and M. Olshaker........364

Interband Optical Absorption in Amorphous Semiconductors
Morrel H. Cohen, Costas M. Soukoulis and E.N. Economou........371

Electronic Properties of Strained Bonds in Amorphous Silicon: The Origin of the Band-Tail States?
L. Schweitzer and M. Scheffler................................379

Structural Order and the Urbach Slope in Amorphous Phosphorus
L.J. Pilione, R.J. Pomian and J.S. Lannin.....................386

CHAPTER 5: Amorphous Heterostructures and Layer Structures

Electronic Structure of Amorphous Semiconductor Heterojunctions by Photoemission and Photoabsorption Spectroscopy
B. Abeles, I. Wagner, W. Eberhardt, J. Stöhr, H. Stasiewski and F. Sette....................................394

Photoemission Studies of Amorphous Silicon Heterostructures
F. Patella, F. Evangelisti, P. Fiorini, P. Perfetti, C. Quaresima, M.K. Ketty, R.A. Riedel and G. Margaritondo..402

Amorphous Silicon Heterojunctions Studied by Transient Photoconductivity
R.A. Street and M.J. Thompson.................................410

Photoluminescence in Amorphous Silicon/Amorphous Silicon Nitride Double Heterostructures
T. Tiedje, B. Abeles and B.G. Brooks..........................417

Doping Modulated Amorphous Semiconductors
K. Kakalios, H. Fritzsche and K.L. Narasimhan.................425

Electroabsorption Measurements of Interface Charges in a-Si:H/a-SiN_x:H Superlattices
C.B. Roxlo, T. Tiedje and B. Abeles...........................433

CHAPTER 6: Structural and Photostructural Properties

Optical and Raman Investigation of Amorphous Polyphosphides
D. Olego, R. Schachter, J. Baumann, M. Kuck and S. Gersten.......... 441

Pressure Induced Structural Change of Clusters in Chalcogenide Glasses
Kazuo Murase and Toshiaki Fukunaga.......... 449

Reversible and Metastable Changes in the Raman Spectrum of GeS_2 Glass Induced by Compression
B.A. Weinstein and M.L. Slade.......... 457

Raman Scattering, Laser Annealing and Pressure-Optical Studies of Ion Beam Deposited Amorphous Carbon Films
S.K. Hark, M.A. Machonkin, F. Jansen, M.L. Slade and B.A. Weinstein.......... 465

Spin Polarized Photoemission as a Sensitive Tool to Study Structural Surface Phase Transitions
F. Meier and D. Pescia.......... 473

CHAPTER 7: Metastable Defects

Review of Metastable Defects in a-Si:H
H. Fritzsche.......... 478

PHOTOINDUCED ABSORPTION SPECTRA IN AMORPHOUS Si:H and Ge:H AND MICROCRYSTALLINE Si:H

Z. Vardeny*, D. Pfost, Hsian-na Liu** and J. Tauc
Division of Engineering and Department of Physics
Brown University, Providence, Rhode Island 02912

ABSTRACT

The steady state subgap photoinduced absorpotion bands in a-Ge:H and a-Si:H are interpreted in terms of two kinds of optical transitions of photoexcited carriers from traps in the gap into the bands of which one produces absorption and the other produces bleaching. The photoinduced absorption in microcrystalline Si:H is due to free carriers in the crystalline grains whose recombination is dominated by the surrounding amorphous matrix.

INTRODUCTION

We have applied the steady state (ss) photoinduced absorption technique to hydrogenated a-Si, a-Ge and microcrystalline (μc) silicon. We show that the PA bands observed in both a-Si:H and a-Ge:H can be interpreted in terms of photoionization of trapped carriers, provided a bleaching process is added that sharpens the high energy side of the spectrum and makes the spectrum more symmetrical. We show that the inclusion of the bleaching process provides a unique opportunity to determine the effective electron correlation energy in the localized state involved in the induced optical transitions. The PA spectrum in the μc-Si:H samples is very different: the absorption increases monotonously with increasing wavelength. We interpret it as due to free carrier absorption which occurs in the crystalline grains. The temperature and pump intensity dependences indicate that the recombination in μc materials is governed by the amorphous matrix.

EXPERIMENTAL

The experimental set-up for the ss PA is described elsewhere[1]. For excitation (pump) we used a CW Ar^+ laser at 2.4 eV, chopped at 150 Hz. The variable wavelength probe beam was an incandescent light source (glow-bar or tungsten lamp) followed by a monochromator. The probe beam transmission (T) and its photoinduced changes (ΔT) were measured with detectors (Si, PbS or PbSe) and a lock-in amplifier. The system response was accounted for by calculating the ratio $\Delta T/T$ which is proportional to the change $\Delta\alpha$ of the absorption coefficient α. The samples were held in an open-cycle cryostat and measurements were taken from 10 to 300 K.

All our samples were thin films prepared by the glow-discharge process. The a-Si:H films were prepared at Exxon

Research (Linden) and the a-Ge:H sample was prepared at IBM (San Jose). The μc samples were made at Nanjing University; different grain sizes were obtained by varying the rf power from 13 to 145 W, while keeping the substrate temperature at 300°C. The substrates were either fused silica or crystalline Si.

PA SPECTRA in a-Ge:H and a-Si:H

The PA spectra of a-Ge:H and a-Si:H are shown in Figs. 1 and 2. The dominant feature is the subgap PA band which is sharper in a-Ge:H. The spectrum of the PA band and its strength do not change much with temperature in a-Ge:H while in a-Si:H the spectrum shifts to higher energies and its strength decreases as the temperature increases[1]. The additional absorption at high energies which is stronger at high temperatures is associated with the temperature modulation of the energy gap[2] and will not be discussed in this work.

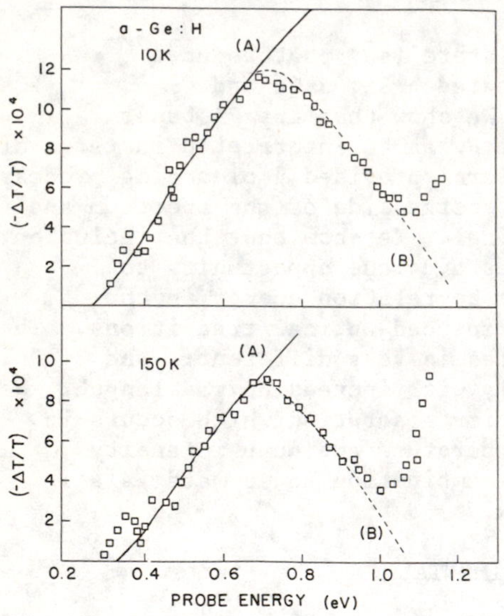

Fig. 1: PA spectra in a-Ge:H at 10 and 150 K. The theoretical curves are based on the ITO model. Curve A absorption only, curve B includes the contribution of the bleaching process.

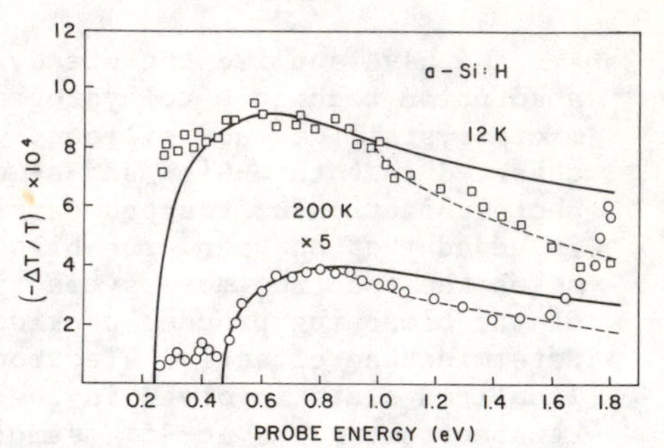

Fig. 2: PA in a-Si:H at 12 and 200K. Curves A and B are theoretical fits as in Fig. 1 but using the OT model.

The PA band in a-Si:H has been ascribed to photoionization of holes trapped in the tail of the valence band into the valence band[1,3]. These transitions are schematically shown in Fig. 3(a) with an onset t_1. If the distribution of tail states (DOS) is sufficiently sharp, most photoexcited holes are concentrated in a narrow energy interval around the quasi-Fermi level F_p. Assuming that the optical matrix elements are independent of energy (O'Connor-Tauc model[4] - OT) and the DOS in the valence band is proportional to $(E_v-E)^{1/2}$, $\Delta\alpha$ near the onset of the PA band is

$$\Delta\alpha = N_{ss}B(\hbar\omega - E_o)^{1/2}/\hbar\omega \qquad (1)$$

where the threshold $E_o = F_p - E_v$ (Fig. 3(a)), N_{ss} is the ss carrier density and B is a constant[4].

A slightly different equation for $\Delta\alpha$ can be deduced in a similar way as Eq. (1), if the DOS above E_v is not sharp in the region of F_p. Assuming the DOS to be constant near F_p, the initial states can be integrated in Eq. (1) to yield[2]

$$\Delta\alpha = N_{ss}C(\hbar\omega - E_o)^{3/2}/\hbar\omega \qquad (2)$$

where C is a constant. We refer to this model as integrated O'Connor-Tauc model (IOT).

These formulas fit reasonably well the onsets of the PA bands in a-Si:H and in a-Ge:H as shown in Figs. 1, 2 (curves A). For well prepared a-Si:H films, Eq. (1) was shown to fit the PA onset data at all temperatures[1,4]. For a-Ge:H and not so well prepared a-Si:H films with a higher DOS in the gap, Eq. (2) fits better the PA onsets. None of the proposed equations can account for the PA band decay at high energies as seen in Figs. 1 and 2. This discrepancy could be removed by making suitable assumptions about the optical matrix elements which depend on the wavefunction of the trapping state, as was recently proposed by Hirabayashi and Morigaki[5] for a-Si:H. However these choices are arbitrary unless they can be justified on theoretical grounds. Instead, we have recently proposed[2] a straight-forward explanation to account for the systematic deviations at high energies from the curves expressed by Eqs. (1) and (2) (Figs. 1,2). This explanation is based on considering a bleaching process associated with photoinduced reduction of the optical transitions which contribute to the absorption below the absorption edge.

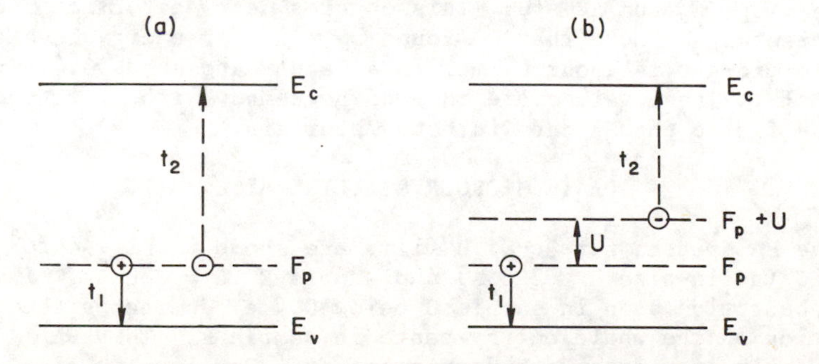

Fig. 3: (a) Schematic representation of the optical transitions onsets: t_1 for absorption, t_2 for bleaching, F_p is the hole quasi-Fermi level. (b) The effective correlation energy U is included.

MODEL FOR PA SPECTRA

In the dark, the valence-band tail is filled with electrons. These electrons produce ordinary absorption associated with their transitions into the conduction band[6] whose onset is represented in Fig. 3(a) by transition t_2. When photoexcited holes are trapped at F_p, the density of initial states of electrons is reduced, therefore transitions into E_c are bleached. If the hole density at F_p is p, the total photoinduced change in α is

$$\Delta\alpha = [\sigma_1(\omega) - \sigma_2(\omega)]p \quad (3)$$

where $\sigma_1(\omega)$ is the absorption cross-section for hole transitions into the valence band and $\sigma_2(\omega)$ is the absorption corss-section for electron transitions into the conduction band. For their frequency dependencies we have used Eq. (1) (OT model) or Eq. (2) (IOT model). We have obtained good fits to the data in Figs. 1,2 (curve B) using the IOT model for a-Ge:H and the OT model for a-Si:H, applying the same model for absorption (σ_1) and bleaching (σ_2). We note that the ratio C_1/C_2 of the factors C defined in Eq. (2) (where 1 represents absorption and 2 represents bleaching) is 0.6 in a-Ge:H and the ratio B_1/B_2 in Eq. (1) for a-Si:H is 1.5.

As apparent from Fig. 3(a) the threshold energies for absorption E_1 and for bleaching E_2 should satisfy the relation $E_1 + E_2 = E_g$, where E_g is the energy-gap. However, the values E_1, E_2 extracted from the fits in Figs. 1 and 2, do not satisfy this relation[2]. The deviation ΔE, $\Delta E = E_g - (E_1 + E_2)$, in a-Ge:H is of order 0.1 eV, while in a-Si:H ΔE is of order 0.3 eV. This deviation can be explained if electron correlation energy is considered[7].

The states which produce PA are empty under steady state illumination conditions and are concentrated near F_p. The corresponding states in the dark are doubly occupied and therefore have energies close to $F_p + U$, where U is the effective electron correlation energy. This is shown in a one-electron energy diagram[8] in Fig. 3(b) for a positive U. In this case the threshold energies observed in PA satisfy the relation $E_1 + E_2 + U = E_g$ and $\Delta E = U$. This enables us to determine U by PA measurements. We note that U found by ESR[9] for energy levels close to mid-gap is about 0.1 eV in a-Ge:H[10] and about 0.4 eV in a-Si:H[9,10]. These values are in good agreement with ΔE extracted from the fit to the PA data in both materials.

PA in MICROCRYSTALLINE Si:H

The PA spectra for μc-Si:H films are shown in Fig. 4 for sample C (grain size L ≃ 220 Å) and sample E (L ≃ 830 Å). The extra absorption seen in sample C below 0.4 eV dominates the absorption in the whole energy range in sample E. This PA monotonously increases with decreasing $\hbar\omega$. The log-log plot in Fig. 5 shows that the wavelength dependence is a power law, $-\Delta T/T \sim \lambda^{\gamma}$, where γ is between 1.5 to 1.8, similar to that observed

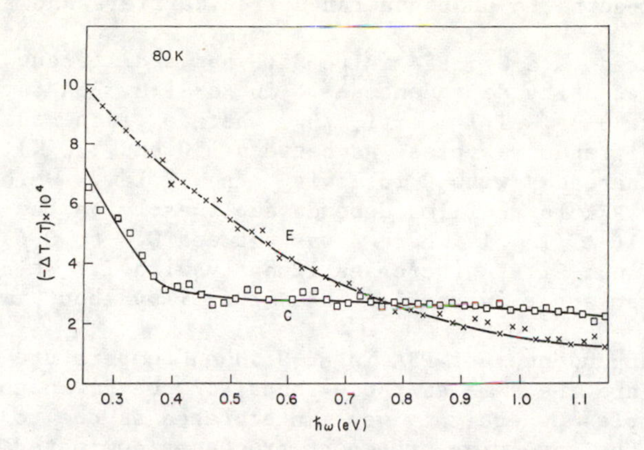

Fig. 4: PA spectra of μc-Si:H with different grain size L. Sample C: L ≃ 220 Å, sample E: L ≃ 830 Å.

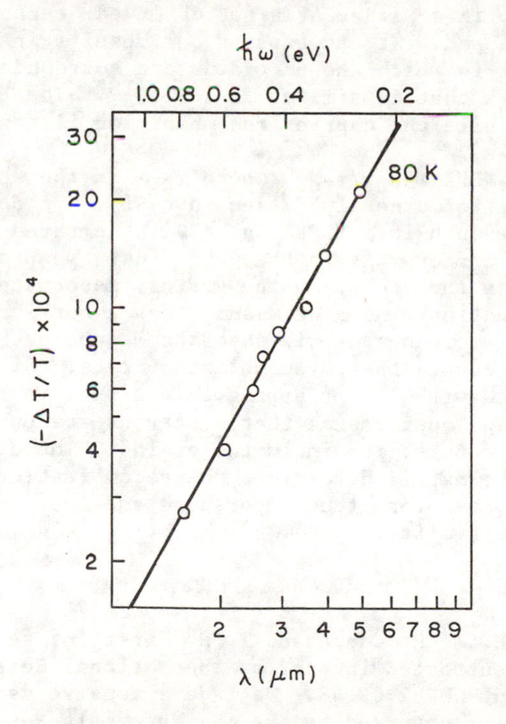

Fig. 5: Wavelength dependence of PA in μc-Si:H, sample E. The slope of the straight line is γ = 1.8.

for free-carrier absorption in crystals[11]. We therefore ascribed these PA spectra to photogenerated free carriers and refer to them as PFA.

Compared to c-Si, PFA in μc-Si:H has a different temperature and pump intensity dependences[12]. In μc-Si:H, PFA decreases with temperature θ as $-\Delta T/T \sim \exp(-\theta/\theta_o)$ where θ_o is a constant (whose value in different samples was between 150 and 270 K), while in c-Si PFA increases with θ following a power law. We have also found that PFA in μc films depends sublinearly on the laser intensity I($-\Delta T/T \sim I^{\delta}$ where δ was between 0.4 to 0.6). On the contrary, in c-Si PFA increases linearly with I. We also note that PFA is larger in μc-Si:H than in c-Si by about two orders of magnitude.

The dependences of PFA in μc-Si:H on temperature and laser intensity are the same as those found for the PA in good quality a-Si:H samples[1] where they were interpreted as due to recombination involving transport processes dominated by exponential distributions of trap densities below the band edges[3]. It is therefore plausible to assume that the μc-Si:H films have large densities of traps comparable to those in a-Si:H. However, in a-Si:H PFA is not observed[3] because carriers are quickly trapped. We can explain our results in μc-Si:H by a two phase model in which the μc grains are surrounded by an amorphous matrix that is similar to a-Si:H[13]. The PFA occurs in the μc grains while the carrier recombination is dominated by the amorphous matrix.

Generally, PFA $\sim N_{ss}/\tau_{rel}$, where τ_{rel} is the momentum relaxation time (of order 10^{-13} sec in c-Si). It does not seem likely that the much larger PFA in μc-Si:H compared to c-Si is due to a much shorter τ_{rel} in the μc grains. We propose that the main reason is a larger N_{ss} in the grains, associated with a longer recombination time. We assume that a barrier between the grains and the amorphous matrix separates electrons and holes[12] and therefore reduces their recombination rate. Since the PA band, associated with hole trapping in a-Si:H[3], is not observed in μc samples, we must assume that electrons are pushed into the matrix while the holes remain in the grains. The dynamics of electrons in the matrix determines the recombination process as in a-Si:H, and therefore the temperature and intensity dependences are similar.

ACKNOWLEDGEMENTS

We thank H. A. Stoddart and T. R. Kirst for technical help. This work was supported in part by the National Science Foundation grant DMR82-09148. We made extensive use of the Optical Facility supported by the NSF Materials Research Laboratory program at Brown University.

* Permanent address: Physics Department, Technion, Haifa, Israel.

** Permanent address: Physics Department, Nanging University, Nanging, The People's Republic of China.

1. P. O'Connor and J. Tauc, Phys. Rev. B25, 2748 (1982).
2. D. Pfost, Hsiang-na Liu, Z. Vardeny and J. Tauc, Phys. Rev. B30 (1984).
3. J. Tauc, in Festkoerperprobleme (Advances in Solid State Physics), edited by P. Gross (Vieweg, Braunschweig, 1982), Vol. XXII, p. 85.
4. P. O'Connor and J. Tauc, Sol. State Commun. 36, 947 (1980).
5. I. Hirabayashi and K. Morigaki, J. Non-Cryst. Solids 59-60, 433 (1983).
6. C. R. Wronski, B. Abeles, T. Tiedje and G. D. Cody, Sol. State Commun. 44, 1423 (1982).
7. D. Adler and E. T. Joffa, Phys. Rev. Lett. 36, 1197 (1976).
8. L. Schweitzer, M. Grunewald and H. Dersch, Sol. State Commun. 39, 355 (1981).
9. H. Dersch, J. Stuke and J. Beichler, Phys. Status. Solidi (b) 105, 265 (1981).
10. M. Stutzman and J. Stuke, Sol. State Commun. 47, 635 (1983).
11. J. P. Woerdman, Philips Res. Rep. Suppl. No. 7 (1971).
12. Hsiang-na Liu, D. Pfost and J. Tauc, Sol. State Commun. 50, 987 (1984).
13. H. Richter and L. Ley, J. Phys. (Paris) 42, C4-261 (1981).

PHOTOINDUCED ABSORPTION AND PHOTOINDUCED ABSORPTION-DETECTED ESR IN P-DOPED a-Si:H

I. Hirabayashi and K. Morigaki
Institute for Solid State Physics, University of Tokyo
Roppongi, Tokyo 106, Japan

S. Yamasaki and K. Tanaka
Electrotechnical Laboratory
Sakuramura, Ibaraki 305, Japan

ABSTRACT

Photoinduced absorption (PA) and photoinduced absorption-detected ESR (PADESR) measurements have been carried out on P-doped a-Si:H. We have observed two PA components in the PA spectra at low temperatures for highly doped samples, which correspond to excitation of trapped electrons at P_4^+ and excitation of trapped holes at P_4^+-D^- pairs. For these samples, the observed PADESR spectrum consists of a narrow line and a broad line (hyperfine doublet), which are identified as being due to trapped holes at P_4^+-D^- pairs and phosphorus dangling bonds (P_2^0), respectively.

INTRODUCTION

Photoinduced absorption (PA) has been known as a powerful optical means to investigate relaxation and recombination processes of photocreated carriers in amorphous materials.[1,9,10] Furthermore, optical cross sections obtained from PA measurements provide information about wavefunctions of the initial and final states of optical transition responsible for PA, impurity (defect) potentials, charged states of localized states involved in PA, etc[2]. Thus, the PA measurement seems to us very useful for characterization of gap states of hydrogenated amorphous silicon (a-Si:H) we are concerned with. We have already reported PA measurements carried out along this line on undoped a-Si:H, by which the position of the energy levels of doubly occupied dangling bond centers (D^-) and A centers (trapped hole centers) were estimated to be 0.6 eV below the edge of conduction band and 0.25 eV above the edge of valence band, respectively[3]. Furthermore, we have observed spin-dependent effects on PA in undoped a-Si:H, which are used to detect ESR signals[4,5]. Such PA-detected ESR (PADESR) measurement monitors the PA intensity to detect ESR signals of metastable states involved in the PA processes under optical excitation. By combining the spectral dependences of such PADESR signals with the PA spectra, we can determine the optical absorption cross section associated with a particular defect center identified by PADESR, taking into account differences in spin-dependences of each defect center.

In this paper we report the results of PA and PADESR measurements on P-doped a-Si:H, in which two localized levels, i.e., a shallow electron trap and a deep hole trap are identified in PA spectra. The trapped hole centers are also found to give rise to a main PADESR signal. Additional PADESR signals are identified as being due to twofold coordinated phosphorous centers (dangling bonds). We discuss the nature of defect centers responsible for PA and PADESR and also the creation mechanism for those defects.

EXPERIMENTAL

The PA and PADESR measurements were carried out, using the same apparatus as that reported in ref. (2)-(5). Changes in the transmitted probe light intensity by samples, ΔT, associated with optical excitation by unfocused laser light (Ar^+ laser at 514.5 nm or Kr^+ laser at 530.9 nm) were measured, i.e., -ΔT/T corresponds to the PA intensity. The PADESR signal intensity was proportional to a change in ΔT at resonance, i.e., δ(ΔT). A cooled Ge detector or a PbS detector were used for the PA measurement, while a cooled Ge detector was used for the PADESR measurement. Thus the detection limit for the low energy was about 0.5 eV for PA and 0.7 eV for PADESR. all data were corrected for slight overlapping of emitted light from the samples in the same way as described in ref. (3)-(5).

P doped a-Si:H samples were prepared by glow discharge decomposition of the gas mixture of SiH_4 and PH_3, using a inductively coupled system. The gas ratio in volume, $[PH_3]/[SiH_4]$ was from 10^{-2} to 10^{-5}. The sample thickness ranged from 5 μm to 10 μm. The substrate temperature was 300°C for all samples.

RESULTS

A. PA Measurements

Figure 1 shows log-log plots of PA spectra for three P-doped a-Si:H samples. The characteristics of PA spectra are highly dependent on the doping level. Lightly-doped sample No.EP-5 exhibits a similar PA spectrum to that for undoped samples which can be interpreted in terms of optical transition of holes at the A centers into the valence band, using Lucovsky's model that takes δ-potential as an impurity potential.[6] In this case, asymptotic spectral dependence in higher energy region can be expressed as $(h\omega)^{-1.5}$, while the PA spectrum for sample No.EP-5 shown in Fig. 1 is fitted with $(h\omega)^{-1.0\sim-1.2}$. For sample No.EP-4, the slope of the PA spectrum is fitted with $(h\omega)^{-3}$ which is close to that expected from a localized center in Coulomb potential, i.e., $(h\omega)^{-3.5}$.[7,8]

Highly-doped sample No.EP-2 shows additional PA component besides the PA observed for sample No.EP-4. The presence of two components in this PA spectrum is also inferred from the temperature dependence of the PA intensity measured at 0.8 eV for this sample, as shown in Fig. 2. As seen from this figure, the temperature dependence of the PA intensity for sample No.EP-2 is very different from those of the PA intensity observed for the D^- center and the A center in undoped a-Si:H and has two components, i.e., a sharp decrease and a slow decrease with temperature. The sharp decrease and the slow decrease may correspond to the PA component observed for low photon energies and the additional PA component appearing with a hump around 0.85 eV in Fig. 1, respectively. A detailed analysis of the PA spectra can be done by combining them with spectral dependences of the PADESR signals, as will be described below.

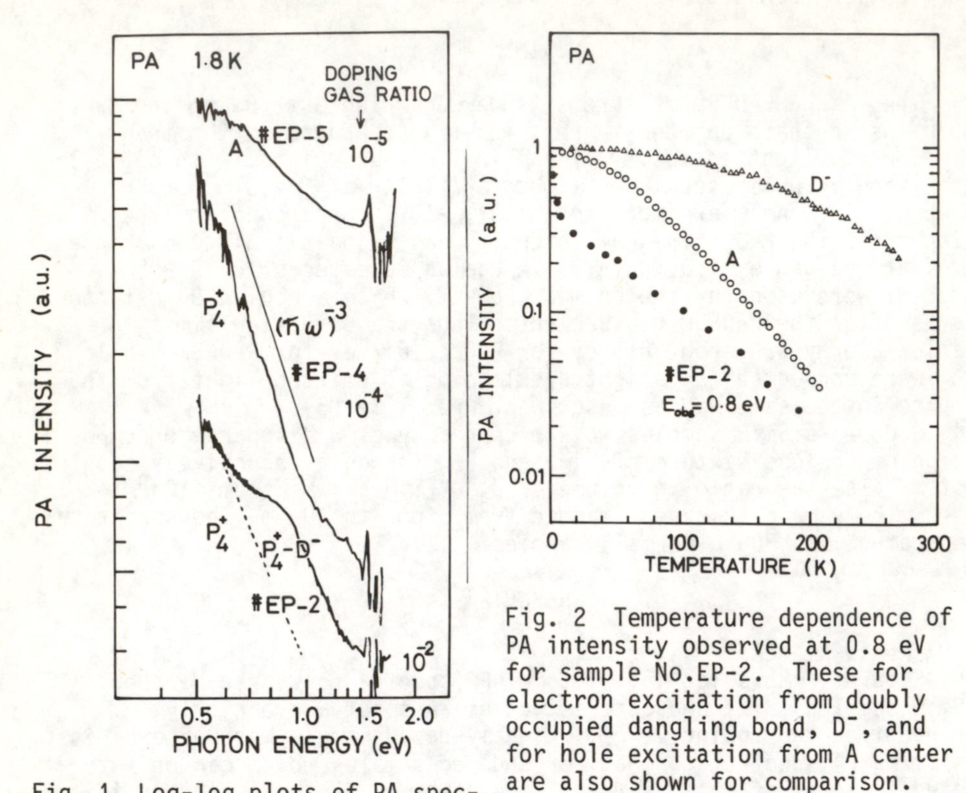

Fig. 1 Log-log plots of PA spectra for P-doped samples obseved at 1.8K.

Fig. 2 Temperature dependence of PA intensity observed at 0.8 eV for sample No.EP-2. These for electron excitation from doubly occupied dangling bond, D^-, and for hole excitation from A center are also shown for comparison.

B. PADESR Measurements

The PADESR spectrum observed at 2K for sample No.EP-2 is shown in Fig. 3, where transmission intensity of probe light corresponding to PA, ΔT, increases at resonance. It consists of a narrow line with g = 2.0038 and a broad one superposed over the narrow line. This broad line seems to be composed of a doublet with splitting of 220 G arising from hyperfine interaction, as will be discussed later. We measured the PADESR intensity, $\delta(\Delta T)/T$, as a function of photon energy of probe light, namely the spectral dependence of PADESR signals. The spectral dependence of $\delta(\Delta T)/T$ measured at the peak position of the PADESR spectrum, i.e., that of the narrow line corresponding to g = 2.0038 is shown in Fig. 4 as well as the PA spectrum, i.e., $-\Delta T/T$ vs. $h\omega$ plot. This figure also shows the spectral dependence of $\delta(\Delta T)/T$ measured at various magnetic fields where the broad line, i.e., the hyperfine doublet predominantly contributes to the PADESR spectrum. The spectral dependence of $\delta(\Delta T)/T$ measured at the peak position of the narrow line shows a peak, corresponding to a similar hump in the PA

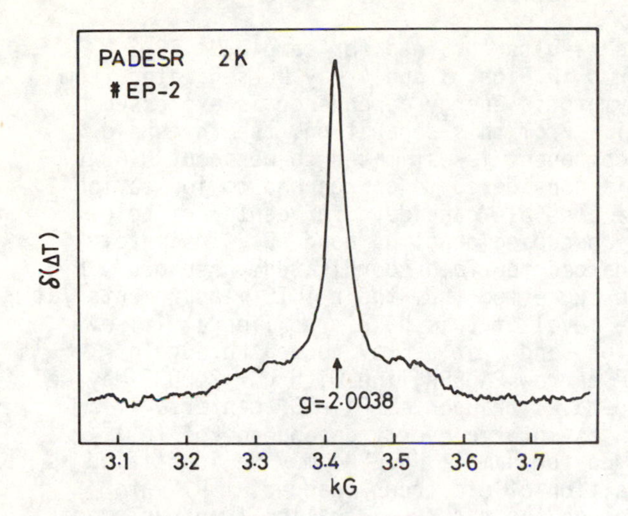

Fig. 3
PADESR spectrum for highly P-doped sample No.EP-2 (doping ratio $\cong 10^{-2}$). Microwave frequency = 9.58 GHz.

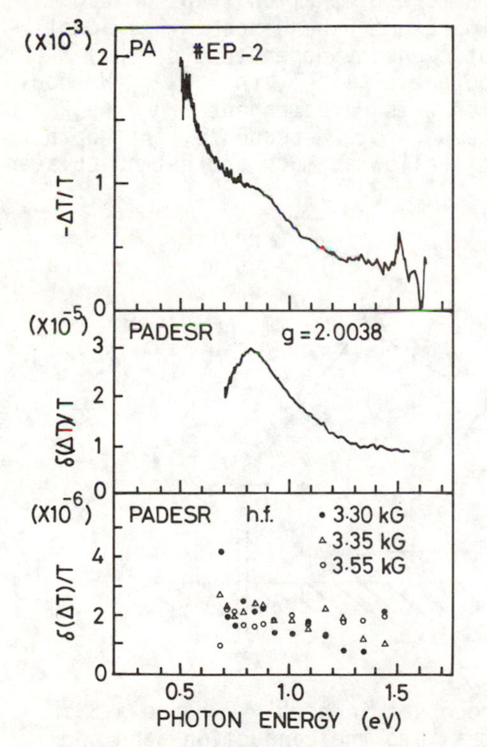

Fig. 4
Dependences of PA and PADESR on probe photon energy for highly P-doped sample No.EP-2.

spectrum. Thus this narrow line is closely correlated with the PA component with a hump.

For sample No.EP-4, only narrow line with g = 2.0046 was observed in the PADESR measurement. Thus the broad line may be associated with high doping of phosphorous.

DISCUSSION

First, we discuss the results obtained for sample No.EP-2. The PA component with a hump in Figs. 1 and 4 may be separated from the PA component whose asymptotic energy dependence is expressed by $(h\omega)^{-3}$, as shown in Fig. 1. From this separation, the threshold energy for the former PA component is estimated to be about 0.6 eV. If this threshold energy is considered to correspond to ionization energy of trapped holes, a possible candidate for center of hole trapping would be a doubly occupied dangling bond, D^-, intimately coupled with positively charged fourfold coordinated phosphorus, P_4^+, because Okushi et al.[12,13] have suggested from their ICTS measurements at room temperature, that the level of this hole trap lies at 1.1 eV below the edge of conduction band (at 0.6 eV above the edge of valence band). Thus, the narrow PADESR line with g = 2.0038 may be identified as being due to holes trapped at P_4^+ - D^- centers.

The PA component with asymptotic energy dependence of $(h\omega)^{-3}$ which has also been observed for sample No.EP-4 may be identified as being due to optical transition of electrons trapped at P_4^+ into conduction band by considering the nature of shallow traps as suggested from the PA spectrum and its temperature dependence. This identification is consistent with asymptotic energy dependence of $(h\omega)^{-3}$ that is close to that for Coulomb potential, i.e., $(h\omega)^{-3.5}$. We could not estimate the threshold energy for this PA component from the PA spectra, but as inferred from its temperature dependence, the depth of trapped electrons at P_4^+ would be shallow as much as that calculated by Robertson, i.e., 0.1 eV.[14]

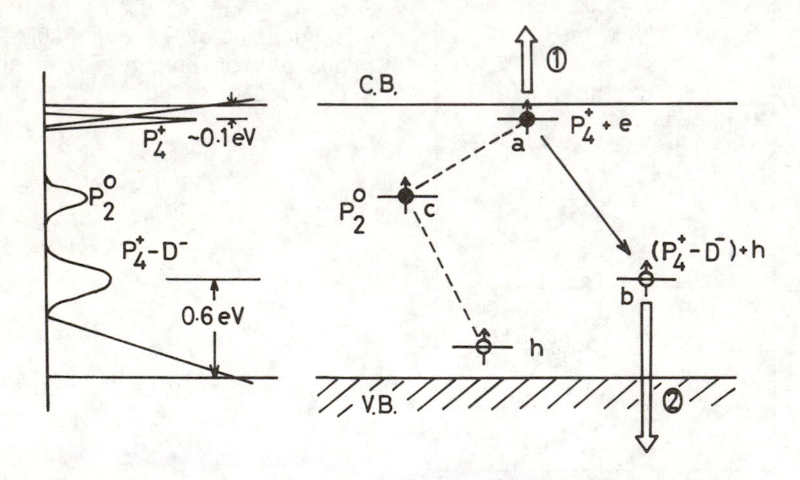

Fig. 5 Schematic diagram of PA processes in highly P-doped a-Si:H. 1 electron excitation from P_4^+ states into the conduction band and 2 hole excitation from P_4^+ - D^- pair states into the valence band. A dashed line shows nonradiative recombination channels. The recombination channel shown by a solid line is either radiative or nonradiative process, depending on doping ratio. The D^- level is omitted in this figure.

A model for PA and PADESR results discussed above is shown in Fig. 5. In the following we consider the reason why the PA intensity changes at the resonance of trapped holes at the $P_4^+ - D^-$ centers. Associated with ESR of trapped holes, their recombination rate with trapped electrons at the P_4^+ state and other tail- and gap-states is enhanced, as has been pointed out by several authors on the basis of an electron-hole pair model[15-17]. This causes the life time of trapped holes to be decreased, so that the number of trapped holes in the steady state decreases and eventually the intensity of PA involving these traps decreases, as has already been discussed .

Secondly, we discuss the nature of magnetic centers responsible for the broad line (doublet) of PADESR spectrum observed for highly P-doped a-Si:H sample No.EP-2. From the fact that this PADESR line was observed only for highly P-doped a-Si:H samples, we consider the origin of this doublet as hyperfine structure with ^{31}P nucleus (nuclear spin of 1/2)[18]. A possible candidate for this center is neutral twofold coordinated phosphorus, P_2^0. For this center, an unpaired electron giving rise to ESR lies in the 3p-orbital of phosphorus atom.[20] However, admixture of the 3s-orbital into the unpaired electron orbital must be expected. Such normalized admixture factors are given by α^2 and β^2 for 3s- and 3p-orbitals, respectively. Furthermore we assume that the unpaired electron occupies the site of phosphorus atom with a factor of η^2 and is located at its surrounding atom with $1 - \eta^2$. The factors, α^2, β^2 and η^2 can be estimated from isotropic and anisotropic hyperfine interaction with ^{31}P nucleus. Anisotropic hyperfine interaction gives rise to a line broadening in amorphous materials. So, the splitting of doublet, i.e., 220 G and the line width, i.e., 70 G leads us to estimate α^2, β^2 and η^2 as 0.1, 0.9 and 0.74, respectively, in a similar way to the case of phosphorus centers in radiation-damaged crystalline silicon[19]. Dominant p-character of the unpaired electron of P_2^0 with a slight admixture of the s-character is consistent with the nature of p-bonding of phosphorus atom. The unpaired electron spends 74% at the central phosphorus site and 26% at its neighboring sites.

Very recently, the doublet structure of the ESR line has independently been observed by Stutzmann and Stuke[21] in the dark, using a conventional ESR spectromater. However, they considered the three center bond as a possible candidate for this line.

In the following, we consider a creation mechanism for P_2^0 that is a phosphorus-dangling bond. The normal coordination of phosphorus atom is threefold, but valence alternation pairs are created with the following reaction:

$$2\,P_3^0 \rightleftarrows P_4^+ + P_2^- \quad .$$

If we assume that P_2^- reacts a neutral silicon-dangling bond, D^0, negatively charged twofold coordinated phosphorus would become neutral ones in the following way:

$$P_2^- + D^0 \rightleftarrows P_2^0 + D^- .$$

As was considered above, it is quite reasonable that the P_2^0 centers are created in highly P-doped a-Si:H samples during preparation of the samples. As regards the role of these centers as recombination centers, the spectral dependence of the PADESR doublet signal measured at three magnetic fields shown in Fig. 4 provides the following information: This spectral dependence appears not to exhibit such a hump as shown in the spectral dependence of the PADESR signal with g = 2.0038. This fact implies that the spectral dependence of the PADESR doublet signal almost corresponds to the PA spectrum by trapped electrons at P_4^+. Spin-dependent effect on this PA arises from an enhancement of the transfer rate of an electron at P_4^+ into a P_2^0 center, associated with ESR of the P_2^0 center, as shown in Fig. 5. This is similar to the case of dangling bond centers in undoped a-Si:H. Preliminary ODMR measurements suggest that the P_2^0 centers act as nonradiative centers. This is also consistent with our model shown in Fig. 5.

Finally, we would like to comment about isolated D^- centers. Creation of P_4^+ is associated with that of D^-,[11] so that the presence of isolated P_4^+ leads to that of isolated D^-. The degree of association of P_4^+ with D^- is approximately estimated as a function of doping ratio in a similar way to ref.(22). The calculation shows that the density of the associated pairs is proportional to the doping ratio, while that of isolated ones is proportional to the square root of the doping ratio, as has already been pointed out by Street.[11] So we can expect the number of isolated P_4^+ and D^- relative to that of pairs becomes ten times larger in sample No.EP-4 than in highly doped sample No.EP-2. Considering the level of D^- (0.6 eV below conduction band edge) and the Fermi level of sample No.EP-4 (~0.25 eV below conduction band edge), the PA spectrum associated with D^- is expected to consist of the photoinduced transmission (negative PA)[23] associated with the decrease of the electron excitation by probe light from D^- into the conduction band (threshold energy ≅ 0.6 eV) and photoinduced absorption due to the hole excitation into the valence band. The excitation energy of the latter process is estimated from the energy level of neutral dangling bond states in undoped a-Si:H, namely if lattice distortion is not so large, associated with hole trapping, it would range between 0.8 eV and 1.0 eV. In fact, a PA component with about 1.0 eV can be seen in the PA spectrum for sample No. EP-4, as shown in Fig. 1. On the other hand, the photoinduced transmission (a negative component of PA) is not observed in our experiments, probably owing to the overlapping of PA of trapped electrons at P_4^+. Preliminary ODMR measurements[25] show that trapped holes at D^- give rise to a quenching ODMR signal that indicates the nonradiative nature of this center, as has seen in undoped a-Si:H. Further study is required to clarify the nature of trapped holes at D^-.

CONCLUSION

We have observed the PA spectra associated with phosphorus doping that are attributed to optical transition of electrons at P_4^+ into the conduction band and that of holes at P_4^+ - D^- pairs into the valence band. The hole ionization energy from P_4^+ - D^- pairs was 0.6 eV. As the trapped electron level at P_4^+ was shallow as much as 0.1 eV predicted by Robertson, the threshold energy for this PA was beyond our detection limit for PA. The PADESR measurements allow us to identify two types of defects, i.e., trapped holes at $P_4^+ - D^-$ pairs and phosphorus dangling bonds, P_2^0. Our preliminary ODMR [25] measurements suggest the role of the P_2^0 centers as nonradiative centers and radiative and nonradiative natures of trapped holes at $P_4^+ - D^-$ pairs, depending on the doping level, namely radiative nature for sample with doping ratio of 10^{-3} and nonradiative nature for sample with doping ratio of 10^{-2}. This result suggests that the low energy luminescence [24] associated with P doping arises from recombination of electrons at P_4^+ and/or at tail states with holes at $P_4^+ - D^-$ centers.

REFERENCE

1. J. Tauc, Festkoerperprobleme Vol.XXII p.85 (Vieweg, Braunschweig, 1982).
2. I. Hirabayashi and K. Morigaki, to be published.
3. I. Hirabayashi and K. Morigaki, J. Non-Cryst.Solids 59&60, 433 (1983).
4. I. Hirabayashi and K. Morigaki, Solid State Commun. 47, 469 (1983)
5. I. Hirabayashi and K. Morigaki, J. Non-Cryst. Solids 59&60, 133 (1983).
6. G. Lucovsky, Solid State Commun. 3, 299 (1965).
7. H. B. Bebb, Phys. Rev. 185, 1116 (1969).
8. A. M. Stoneham, Theory of Defects in Solids (Clarendon, Oxford, 1975).
9. Z. Vardeny, J.Strait, D. Pfost, J. Tauc and G. D. Cody, Phys. Rev. Lett. 48, 195 (1983).
10. D. Pfost and J. Tauc, Solid State Commun. 48, 195 (1983).
11. R. A. Street, Phys. Rev. Lett. 49, 1187 (1982).
12. H. Okushi, Y. Tokumaru, S. Yamasaki, H. Oheda and K. Tanaka, Phys. Rev. B25, 4313 (1982).
13. K. Tanaka and H. Okushi, Proc. Int'l Conf. Transport and Defects in Amorphous Semiconductors (Michigan, 1984).
14. J. Robertson, Phys. Rev. B28, 4647 (1983).
15. K. Morigaki, J. Phys. Soc. Japan 50, 2279 (1981).
16. B. Movaghar, B. Ries and L. Schweitzer, Phil. Mag. B41, 141 (1980).
17. S. Depinna, B. C. Cavenett, T. M. Searle and I. G. Austin, Phil. Mag. B46, 501 (1982).
18. B. A. Goodman and J. B. Raynor, Electron Spin Resonance of Transition Metal Complexes, Adv. Inorg. Chem. and Radio Chem. Vol.13 (Academic Press, NY, 1970).
19. G. D. Watkins and J. W. Corbett, Phys. Rev. 134, A1359 (1964).
20. S. R. Elliot and E. A. Davis, J. Phys. C12, 2577 (1979).
21. M. Stutzmann and J. Stuke, Proc. Int'l Conf. Transport and Defects in Amorphous Semiconductors (Michigan, 1984).
22. A. B. Lidiard, Phys. Rev. 94, 29 (1954).
23. D. Pfost, Hsiang-na Liu, Z. Vardeny and J. Tauc, Phys. Rev. to be published.
24. D. K. Biegelsen, R. A. Street, W. A. Jackson and R. L. Weisfield, Proc. Int'l Conf. Transport and Defects in Amorphous Semiconductors (Michigan, 1984).
25. I. Hirabayashi and K. Morigaki, to be published.

CORRELATION OF OPTICAL AND THERMAL EMISSION PROCESSES FOR BOUND-TO-FREE TRANSITIONS FROM MOBILITY GAP STATES IN DOPED HYDROGENATED AMORPHOUS SILICON

A.V. Gelatos and J.D. Cohen
University of Oregon, Eugene, OR 97403

J.P. Harbison
Bell Communications Research Inc., Murray Hill, NJ 07974

ABSTRACT

We have employed photocapacitance techniques to n-type doped a-Si:H Schottky diode samples and have deduced the density of states, g(E), in the upper half of the mobility gap versus the optical detrapping energy. This photocapacitance g(E) is found to be in good agreement with the density of states deduced by deep-level transient spectroscopy measurements performed on the same samples. Small differences in the energy scale are observed in some samples which can be interpreted in terms of lattice relaxation effects. We have also applied sub-band gap optical excitation in conjunction with the observation of capacitance transient measurements following voltage pulse excitation. This has allowed us to observe the direct competition between the thermal detrapping of electrons from gap states with optical detrapping from the same states. These measurements have allowed us to observe the thermal enhancement of the optical detrapping processes in a-Si:H. They have also allowed us to explicitly measure for the first time the optical cross sections for the bound-to-free transitions over a limited region of the mobility gap.

INTRODUCTION

A great variety of experimental techniques have emerged over recent years to deduce the energy distribution of mobility gap states in hydrogenated amorphous silicon (a-Si:H). Among these, the two techniques that have allowed this distribution to be deduced over the largest energy range are deep-level transient spectroscopy (DLTS)[1-3], and sensitive optical absorption measurements using photo deflection spectroscopy[4,5].Photoconductivity measurements in several cases have also been calibrated to deduce optical absorption in the sub-band gap regime and thus deduce g(E)[6].

The DLTS deduced spectra give the distribution of gap states in terms of a thermal emission energy of the gap state electrons or holes to the nearest mobility edge. The absolute number of such emitted charges can be determined by analyzing the change of the barrier capacitance that is observed. Absorption measurements give g(E) in terms of an optical energy scale; the number of states is determined by an assumed (constant) optical cross section. The energy scales determined by these two kinds of measurements are not necessarily identical; in fact, they may differ substantially if some degree of lattice relaxation exists at a particular localized site between filled and empty states[7]. However, in spite of possible differences between energy scales and uncertain optical cross-sections, results obtained by the two methods seem to be in fairly close agreement.

In the study reported here we investigate these issues in detail by comparing the DLTS density of states in n-type doped a-Si:H with that deduced by photocapacitance measurements in the same samples. We also study the direct competition between the thermal and optical induced emission of gap state electrons. These latter measurements allow us to observe the thermal enhancement of the optical detrapping process and to directly measure the optical cross sections for the bound-to-free transitions over a limited range of mobility gap energies.

SAMPLE PREPARATION

Two n-type a-Si:H samples (100 vol. ppm Ph_3 in SiH_4) were prepared by the standard rf decomposition of silane as described in detail eleswhere[2]. Sample 1 was grown with a ratio SiH_4/Ar of 30% and a substrate temperature of 250°C; sample 2 was grown with a 50% ratio and a substrate temperature of 258°C. Both samples were grown with identical rf power density (250 mW/cm^2), gas pressure (0.36 torr) and were deposited with equal thickness (2.6μm) on p^+ crystalline Si substrates. All measurements reported below were taken with reverse bias applied to the substrate-amorphous silicon interface which allows the p^+n depletion region to be probed.

EXPERIMENTAL PROCEDURE

We performed three distinct kinds of measurements on each sample. The first, steady state photocapacitance, was made at 10kHz at a temperature just above dielectric turn-on for each sample (170 K for sample 1 and 200 K for sample 2). At these temperatures the thermal detrapping of gap state charge was determined to be negligible. Measurements were made at several values of applied bias using weak sub-band gap light of 1.0 - 2.0 μm obtained by a scanning monochrometor with a tungsten-halogen source. Before light exposure at each wavelength, the sample bias was reduced to zero and then restored to fill gap states in the depletion region. Exposing the sample then emptied states from E_c down to the corresponding optical energy. Light exposure at each energy continued until no further capacitance change was observed (typically 1-10 minutes). Measurements of the capacitance, C, were taken using two nearby values of applied bias, V_A so that, in addition, a value of

$$N_{cv} = \frac{-C^3}{\varepsilon q A^2}\left(\frac{dC}{dV_A}\right)^{-1} \qquad (1)$$

could be determined. Here ε denotes the dielectric constant, A is the sample area, and q is the electronic charge.

In carrying out these measurements we determined that the pre-

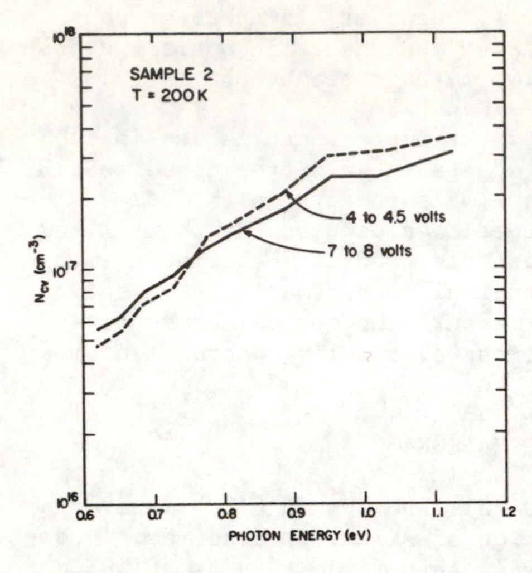

Fig. 1. Photocapacitance N_{cv} as determined by Eq. (1) vs. photon energy in the vicinity of two values of applied sample bias.

sence of light could shift the quasi-Fermi level in the near depletion region of the p^+n barrier which caused the asymptotic value of the capacitance to change with light intensity and also vary in an anomalous manner with applied bias. However, the value obtained for N_{cv} was largely insensitive to these problems and gave consistent values to within a factor of 1.5 over a range of applied bias of 4 to 8 volts. Although the actual capacitance behavior can probably also be understood and analyzed in detail with additional modeling of the a-Si:H barrier, for the present time we believe the value of N_{cv} to give a reasonably reliable measurement of the photodetrapped charge. Representative data for the variation of N_{cv} versus optical energy obtained for sample 2 in the vicinity of both 4 and 7.5 volts applied bias are displayed in Fig. 1.

In addition to steady state photocapacitance we made measurements of capacitance transients induced by a combination of thermal and optical detrapping of gap state charge. Here we employed two methods as shown schematically in Fig. 2. For both methods we recorded the capacitance transient using a DLTS correlator function following a voltage filling pulse in the standard manner[2] except that the measurement of dark (purely thermal) DLTS was interleaved with a measurement of the transient affected in some manner by the presence of sub-band gap light. In method 1, the optical excitation

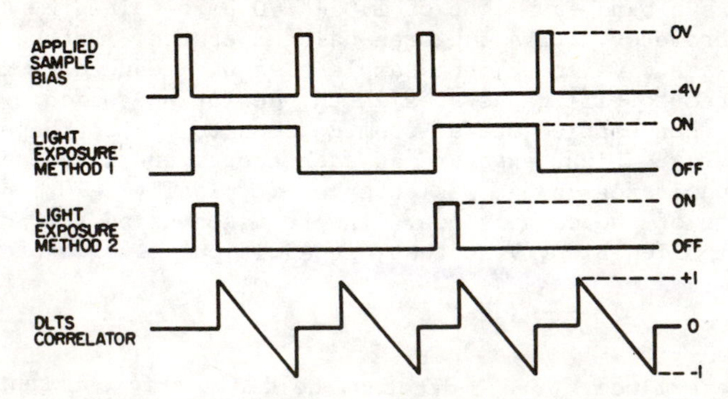

Fig. 2. Voltage and light pulse sequence for two methods of transient capacitance measurement as described in the text.

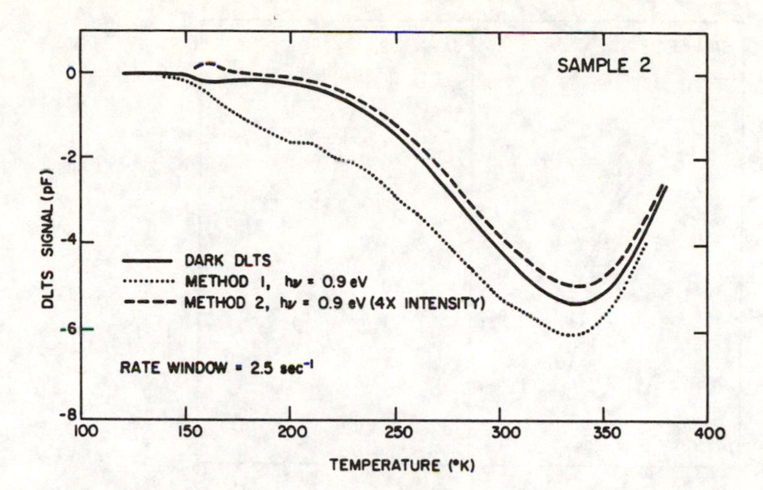

Fig. 3. Dark DLTS spectrum along with transients recorded during "light-on" phase for two methods illustrated in Fig. 2.

was present during the entire recording interval. Thus the transient observed was due to thermal plus optical detrapping. Subtracting the dark phase signal from the "light-on" phase signal gave the component of the capacitance transient due to the optical emission alone.

In method 2 the optical excitation was present only for a relatively short time prior to the recording interval for the light-on phase. With this method we may measure how many states have been optically depopulated for each specific thermal emission energy as determined in dark DLTS. Thus we may directly determine an optical transition rate versus optical energy for states corresponding to a particular thermal emission energy.

The raw dark DLTS spectrum for sample 2 is shown in Fig. 3 along with the signals obtained by the two methods during the light-on phase. Note that method 1 which is "thermal plus optical" enhances the transient signal while method 2 which is "thermal minus optical" decreases it. Although the dark signal phase gives identical spectra for both methods, it is taken concurrently in the manner of Fig. 1 to discriminate against instrumental drift.

We have verified that for the low light intensities employed, the difference signals (light-on minus dark) are strictly proportional to light intensity. Thus we normalize all data to the actual photon flux for each optical energy as determined by a thermo-electric photodetector. Since we have also measured the transmission coefficient through the semitransparent contact of each sample, we may also calculate the absolute optical cross sections for the transition rates obtained using method 2.

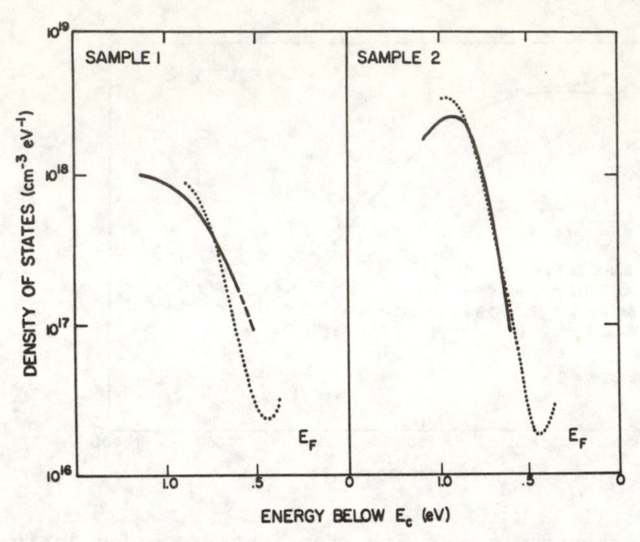

Fig. 4. Densities of states vs. optical energy derived from photocapacitance (solid lines) compared to the densities of states vs. thermal emergy derived from DLTS (dotted lines) for the same samples.

RESULTS

A. Steady-State Photocapacitance

The densities of states deduced by the low temperature photocapacitance measurements for samples 1 and 2 are displayed in Fig. 4 along with densities of states determined by voltage pulse DLTS. For the two samples there is good qualitative and quantitive agreement considering that sources of uncertainty can easily account for a factor of two in the absolute magnitude of g(E) deduced by the two kinds of measurements. The photocapacitance deduced g(E) for sample 1 displays a shift to deeper energy plus a broadening of the midgap defect band. Such a shift of roughly 0.2eV suggests a significant degree of lattice relaxation associated with the bound-to-free transition of the gap states involved[7]. For sample 2, however, the thermal and optical energies are the same to within experimental error.

B. Photocapacitance Transients

The results of combining optical and thermal detrapping in the measurement of capacitance transients are given for sample 1 in Fig. 5. This figure displays the difference signal using method 1 (e.g., between the dotted and solid curves in Fig. 3) for a series of different wavelengths of sub-bandgap illumination. These data have been normalized to the photon flux at each wavelength and also corrected for area spreading of the barrier junction as discussed previously[8].

The rapid falloff of all the curves below 170K is due to the dielectric "freeze-out" of the ac response of the sample. Between 180K and 250K, the more modest increase is due to an actual increase in the optical bound-to-free transition rate for gap state electrons.

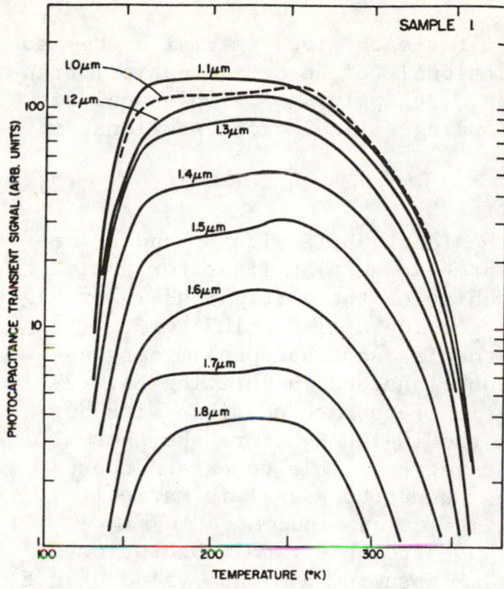

Fig. 5. Photocapacitance transient signal vs. temperature for different wavelengths of sub-bandgap illumination.

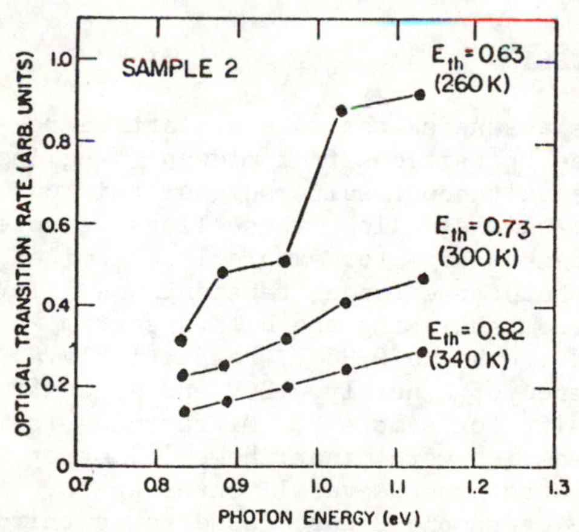

Fig. 6. Optical transition rate vs. photon energy for gap states at three different thermal energy depths. The value of unity on the vertical scale corresponds to an optical cross section of roughly $1 \times 10^{-15} cm^2$.

Since the purely thermal transient component has been subtracted out, this increase corresponds to a thermal enhancement of the optical transition rate. Note that at the highest optical energies (at $\lambda \lesssim 1.1\mu m$), however, there is almost no significant thermal enhancement component.

For wavelengths shorter than 1.1μm, the capacitance transient signal begins to decrease. This is expected for optical energies significantly above $E_g/2$ since gap states may then be filled from the valence band as well as emptied to the conductive band. This argument suggests that the optical energy corresponding to $E_g/2$ for sample 1 lies between 1.1 and 1.2 eV.

Above 260K the photocapacitance signal decreases because of the loss of a substantial fraction of gap state electrons due to thermal emission prior to the transient recording interval. Above approximately 360K the difference signals become zero. For $\lambda \gtrsim 1.6\mu m$ this difference signal reaches zero at lower temperatures which indicates that optical energies below 0.75eV are not able to depopulate states near midgap (even with thermal assistance).

C. Optical Cross Section Versus Optical Energy

The difference signal obtained by method 2 (between the solid and dashed curves in Fig. 3) indicate how many electrons, normally emitted at a particular thermal energy, have been removed by prior exposure to sub-bandgap light. This allows us to deduce the optical transition rate for states at that thermal energy depth. The results displayed in Fig. 6 are normalized to the dark DLTS signal for each given thermal energy so that the values plotted are proportional to the actual optical transition rate per gap state electron. By applying a simple analysis we can also determine the corresponding optical cross sections, σ:

$$\sigma = \frac{\delta S}{S}\frac{1}{t\phi} \tag{2}$$

Here ϕ is the photon flux, S is the (dark) DLTS signal, and δS denotes the change in S due to the pre-exposure of light for a time t. We thus estimate that a value of unity on the vertical scale of Fig. 6 corresponds to an optical cross section of $1.0 \times 10^{-15} cm^2$.

These results should be regarded as somewhat preliminary because of the simplifying assumptions inherent in deriving Eq. (2). Also, because of the low total optical exposure possible with our monochromator light source during the 10-100mS before the thermal transient is recorded, we have not yet been able to extend these measurements to longer wavelengths where the signal is markedly weaker. However, the optical cross sections indicated in Fig. 6 are generally in good agreement with sub-bandgap optical absorption lengths measured by other groups [4-6] assuming a midgap value of g(E) between 10^{17} and 10^{18} $cm^{-3}eV^{-1}$ for their doped a-Si:H samples. The <u>increase</u> in optical cross section with optical energy at fixed thermal energy and the <u>decrease</u> for deeper states is also reasonable assuming that the density of final states in the conduction band is increasing in energy. Perhaps a more suprising feature of these data is the very modest increase with optical energy of σ for the (thermally) deepest states.

DISCUSSION

A key issue in this study is assessing the role of lattice relaxation involved in bound-to-free transitions from midgap (dangling band) defect states. The photocapacitance results suggest this can vary significantly from sample to sample. Along these lines we note that the value of the optical midgap energy for sample 1 determined by the onset of the decrease in photocapacitance transient was 1.1eV while the midgap energy deduced from analyzing the DLTS spectrum was 0.95eV. For sample 2 we obtained 1.0eV and 0.98eV, respectively. This gives some independent evidence of a nearly 0.2eV energy shift for sample 1 with a negligible shift for sample 2. Differences in growth parameters for these two samples were minor; however, other differences may exist since they were grown several months apart. Results nearly identical to those of sample 2 were found for a third sample grown under identical conditions and immediately subsequent to sample 2.

The thermal enhancement behavior observed in the photocapacitance transient measurements for sample 1 is also consistant with a significant degree of lattice relaxation. Similar data for sample 2 indicate enhancement effects also, but only at optical energies below about 0.85eV.

Finally, while the optical cross sections appear quantitatively correct and vary in a general manner as expected with optical and thermal energies, the details are at present somewhat difficult to interpret. Hopefully we will be able to better resolve these issues by further study on a wider variety of samples and also by extending these measurements to states in the lower half of the gap using bandgap laser pulse excitation.

ACKNOWLEDGEMENTS

Research at Oregon was supported by NSF grant DMR-8207437. Additional support for this project came from a Northwest Area Foundation Grant of Research Corporation.

REFERENCES

1. J.D. Cohen, D.V. Lang and J.P. Harbison, Phys. Rev. Lett. 45, 197 (1980).
2. D.V. Lang, J.D. Cohen and J.P. Harbison, Phys. Rev. B 25, 5285 (1982).
3. J.D. Cohen and D.V. Lang, Phys. Rev. B 25, 5321 (1982).
4. W.B. Jackson and N.M. Amer, Phys. Rev. B 25, 5559 (1982).
5. W.B. Jackson, Sol. State Comm. 44, 477 (1982).
6. C.R. Wronski, B. Abeles, T. Tiedje and G.D. Cody, Sol. State Comm. 44, 1423 (1982).
7. See, for example, the discussion by D.V. Lang, H.G. Grimmeiss, E. Meijer and M. Jaros, Phys. Rev. B 22, 3917 (1980).
8. D.V. Lang, J.D. Cohen, J.P. Harbison and A.M. Sergent, Appl. Phys. Lett. 40, 474 (1982).

PHOTOCONDUCTIVITY OF INTRINSIC AND DOPED a-Si:H FROM 0.1 TO 1.9 eV

T. Inushima, M. H. Brodsky, J. Kanicki, and R. J. Serino
IBM Thomas J. Watson Research Center
Yorktown Heights, New York 10958, U.S.A.

ABSTRACT

The photoconductivity of intrinsic and doped a-Si:H was investigated from 0.1 to 1.9 eV by the use of a fourier-transform infrared spectrometer (FTIR) with a-Si:H as the detector. The photoconductivity spectra of intrinsic a-Si:H consist of two broad peaks. One, at 1.8 eV, is the band-to-band transition and the other is the subband at 0.9 eV. The peak position of the subband does not show a temperature dependence. Doping with PH_3 increases the intensity of the subband photoconductivity and shifts the peak to 1.2 eV, which is the only peak observed in the photoconductivity spectra for highly doped a-Si:H. For B_2H_6 doping, the subband spectrum becomes broader and merges into the tail part of the band-to-band transitions with a bump at 1.3 eV. From these results the impurity levels in a-Si:H are discussed.

INTRODUCTION

It is well established that hydrogenated amorphous silicon (a-Si:H) has midgap states due to dangling bonds, defects and impurities.[1] These states make a significant contribution to the electronic properties of a-Si:H. To detect these levels there are many methods which include deep-level-trap spectroscopy[2] (DLTS), photoacoustic spectroscopy[3] (PAS), photoluminescence[4], photothermal deflection spectroscopy[5] (PDS) and photoconductivity[6] measurements. Each of these methods has advantages for the understanding of the midgap states. The data obtained from these methods, however, have little information about the temperature dependence of the density-of-states of a-Si:H or the photoconductive mechanism for impurity doped a-Si:H.

In this paper we present the photoconductivity spectra of a-Si:H from 0.1 to 1.9 eV by the use of a fourier-transform infrared spectrometer (FTIR) with a-Si:H as a detector. Michelson interferometer optics provides a method for increasing the amount of light on the sample over that of dispersive spectroscopy, hence increasing the signal to noise at the detector. This method is an a.c. photoconductivity measurement and has a high efficiency to detect the weak photoconductivity of heavily doped a-Si:H where the dark current is comparable to the photocurrent. By the use of this method we present the photoconductivity spectra of a-Si:H as a function of temperature and dopant concentration.

There is no clear information about the impurity levels in a-Si:H, although it is well established[7] that the introduction of PH_3 or B_2H_6 reduces the activation energy of dark conductivity (E_d) toward a value similar to the activation energy of photoconductivity (E_p). From this experiment and the published data about photoluminescence of doped a-Si:H, we provide information on the energy positions of P and B impurity levels in a-Si:H.

EXPERIMENT

The a-Si:H films, which we used in this experiments, are listed in Table I. These samples were deposited on the anode of a capacitance reactor at a rate of 0.8 Å/sec. The flow rate of pure SiH_4 was 10 standard cm^3/min at 150 mtorr. The substrate temperatures are given in Table I. Dopants in the form of PH_3 and B_2H_6 were introduced and in this paper the impurity concentration (Table I) is indicated by the gas volume ratio.

TABLE I

Sample #	Dopant	Ratio(1)	Thickness (μm)	Substrate Temperature (°C)	σ_D(2) ($\Omega^{-1}cm^{-1}$)	σ_L(3) ($\Omega^{-1}cm^{-1}$)
84-1141I	-----	-----	7.2	250	2.5×10^{-9}	6.1×10^{-6}
84-1164N	PH_3	2×10^{-6}	7.8	250	2.0×10^{-8}	8.4×10^{-6}
84-1159N	PH_3	2×10^{-4}	5.1	250	3.8×10^{-3}	4.1×10^{-3}
83-1134N	PH_3	1×10^{-2}	14.5	250	4.11×10^{-3}	4.19×10^{-3}
84-1181P	B_2H_6	2×10^{-5}	8.2	250	2.8×10^{-9}	3.2×10^{-7}
84-1158P	B_2H_6	2×10^{-4}	7.2	250	2.3×10^{-7}	5.2×10^{-6}
82-1051P	B_2H_6	1×10^{-2}	1.14	275	8.93×10^{-6}	1.13×10^{-5}

(1) Doping is indicated by the ratio of impurity gas to SiH_4.
(2) σ_D is the dark conductivity with the electric field of 18V/mm.
(3) σ_L is the conductivity with the electric field of 18V/mm and under 50 mW/cm^2 illumination.

For the measurement of the photoconductivity we used an IBM FTIR Model 98 as a spectrometer and the sample as a detector. The light source was a 150 W tungsten-halogen lamp and a quartz beamsplitter was used for the measurement from 4000 to 15800 cm^{-1}. For the infrared region from 400 to 8000 cm^{-1}, a Ge-coated KBr beamsplitter was used. In both cases the size of the beam spot on the surface of the sample was 1 cm in diameter. When we used the quartz beamsplitter white light, about 50 mW/cm^2, illuminated the sample constantly and the interferogram light superposed on it had an intensity of 15% of the bias light. In case of the Ge-coated KBr beamsplitter the visible light was filtered out and the interferogram of the infrared light was 30% of the bias light.

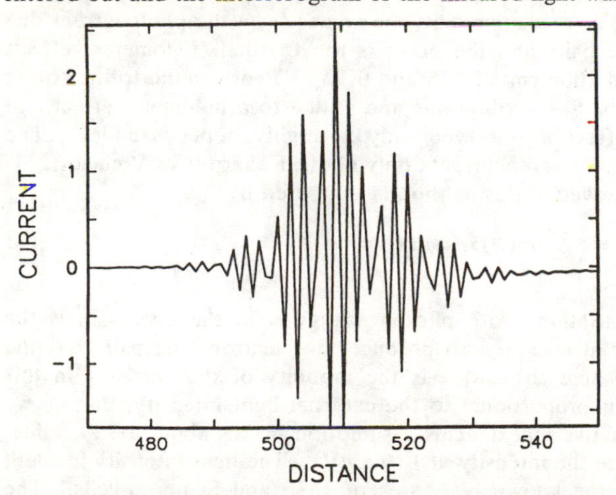

Fig. 1: Typical interferogram of a-Si:H photoconductance.

The photoconductivity was measured between the Mo electrodes 2200 Å thick, 2.5 cm long and 0.5 mm apart. A d.c. voltage of from 3 to 15 V was applied to the sample through a load resistor with a resistance similar to the sample resistance at room temperature. The photoconductivity was detected as a voltage response across the load resistor. The output signal was amplified and then extraneous frequencies were removed by electronic filtering. After that the signal was digitized.

In Fig. 1 we show an example of the interferogram of the photocurrent before fourier conversion to get an actual spectrum. In this figure the interferogram has frequency components from 1.1 to 17.7 kHz, and these frequencies are related to the spectral wavenumber by $\Omega = 4V\omega$, where Ω is called heterodyne frequency of spectral wavenumber ω (in units of cm^{-1}) and V is the velocity of mirror. The V of Fig. 1 is 0.28 cm/sec (VEL=9 in the notation of the IBM FTIR). Therefore, Fig. 1 has the photoconductivity information from 1000 to 15800 cm^{-1}. In this experiment the wavenumber ω dependence of the photoconductivity has the same meaning as the intensity response of the photocurrent at the frequency Ω.

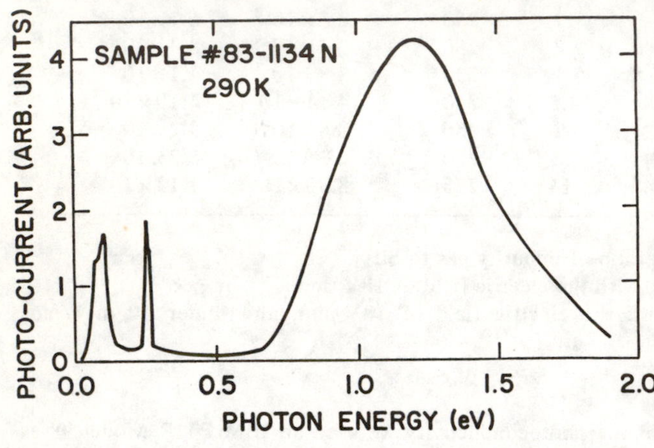

Fig. 2: Photoresponse spectrum of $PH_3/SiH_4 = 1\times10^{-2}$.

The data shown in Fig. 1 is a function of the mirror displacement. The wavenumber dependence of the photoconductivity was obtained by the fourier transformation of Fig. 1, where we used the Happ-Genzel function as an apodization function and a phase correction was made. The number of scans for each data collection varied from 128 to 4192 according to signal to noise conditions. The resolution throughout this experiment was 4 cm^{-1}. For the temperature dependence measurements, the sample was attached to a copper holder in a cryostat with a quartz window.

An example of the photoconductivity spectrum is shown in Fig. 2, which was obtained for the 1×10^{-2} P-doped a-Si:H. This spectrum was constructed from two spectra, which were obtained separately using quartz and Ge-coated KBr beamsplitters. In this spectrum the photocurrent has two components, one comes from the photocurrent due to the production of electron-hole pairs and the other is the thermal (bolometer) effect. The peaks observed near 2000 and 1000 cm^{-1} (0.25 and 0.12 eV) correspond to the strong absorptions of the Si-H (and maybe Si-O) vibrations and is due to a bolometer effect. In this experiment the bolometer effect is observed only in highly doped samples. The intrinsic and lightly doped a-Si:H have photocurrents only above a sharp 0.4 eV cut-off.

The photoconductivity observed in this method is expressed by

$$\sigma_p(\omega) = \sum_{i=e,h} q\eta(\omega)\tau_i(\omega)\mu_i(\omega)L(\omega), \tag{1}$$

where $L(\omega)$ is the photon concentration with photon energy ω in the crystal, η is the photo-generation efficiency factor at energy ω to produce the electron-hole pair, τ is the life time of the electron (e) and hole (h) and μ is the mobility of the carrier. In this experiment we assume that $L(\omega)$ is proportional to the external light intensity, that is, we neglect the dispersion of the refractive index.[8] This assumption makes about a 7% reduction at L(2.0 eV) when we compare the intensity at L(0.5 eV). The light intensity incident on the sample was calibrated by the known responses of InSb and Si photo-cells. The maximum incident intensity at 0.5 eV was about an order of magnitude greater than that incident at 1.9 eV. The parameters in the right side of Eq. (1) depend on photon energy.[9]

The mirror-velocity dependence of the photoconductivity spectra of intrinsic a-Si:H is shown in Fig. 3. In Fig. 3 VEL= 0 is V= 0.059 cm/sec and VEL= 9 is V=0.28

cm/sec. The decrease of σ_p due to the increase of V means that $\mu\tau$ is sensitive to the frequency response.

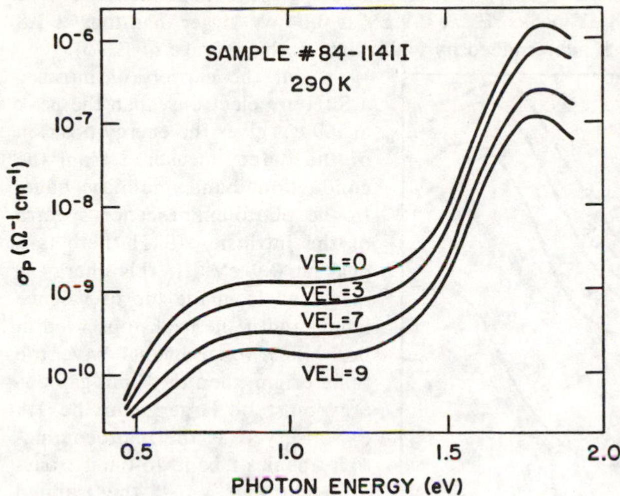

Fig. 3: Mirror velocity dependence of the photoconductivity of intrinsic a-Si:H.

The spectral shape, however, does not depend on the velocity. This result suggests that μ has only a small energy dependence in the energy region investigated and the peak observed at 1.8 eV in the band-to-band transition region is due to the energy dependence of τ (surface recombination).

As is shown in Fig. 1 the detected photocurrent is an a.c. photocurrent under d.c. bias illumination. In our experiment, the photoconductivity spectra were taken under the condition where the photocurrent is proportional to the external light intensity. Then the spectrum can be normalized by the d.c. photoconductivity of the bias light. The y-scale is, therefore, written as follows,

$$\sigma_p(\omega_i) = \frac{\sigma_p^{meas}(\omega_i) \times (\sigma_L - \sigma_D)}{\sum_{i=0.4}^{1.9} \sigma_p^{meas}(\omega_i)}, \tag{2}$$

where σ_L and σ_D are the d.c. conductivities under bias light and in the dark, respectively, and $\sigma_p{}^{meas}$ is the observed a.c. photoconductivity. The assumption used in Eq. (2) is that the integrated photocurrent from 0.4 to 1.9 eV is equal to the d.c. photocurrent under bias light. This is a good approximation when there is no photocurrent above 1.9 eV. The y-scale of Fig. 3 was calculated by Eq. (2) with VEL=3. The values of σ_L and σ_D of the samples used in this experiment are listed in Table I.

RESULTS AND DISCUSSION

Temperature dependence of the photoconductivity of a-Si:H

The temperature dependence of the photoconductivity of intrinsic a-Si:H is shown in Fig. 4. The scale of the abscissa of the figure is obtained from (2). The d.c. photoconductivity spectrum of intrinsic a-Si:H obtained at 300 K by Fuhs et al.[6] is similar to that shown in Fig. 4. Our sample has the activation energy of dark conductivity (E_d) of 0.7 eV and that of photoconductivity (E_p) of 0.2 eV. Above 480 K it was difficult to detect the photoresponse where the dark current exceeded the photocurrent. Photocurrent was observed from 0.5 to 1.9 eV; there was no energy region between 0.5 and 1.9 eV where the photoconductivity was zero. There are two peaks in the photoconductivity spectra; one is at 1.8 and the other is at 0.9 eV. The peak at 1.8 eV shifts to lower energies with increasing temperature. The magnitude of the shift is 3.5×10^{-4} eV/K. This value is the same as that reported for the temperature dependence of the band gap of intrinsic a-Si:H.[10] On the other hand the peak at 0.9 eV does not show the temperature dependence. This result shows that the origin of the peak in the midgap is due to the defect or the dangling bond of the a-Si:H. From the temperature dependence of the intensity ratios of the 0.9 and 1.8 eV peaks, E_p at 0.9 eV is 0.04 eV larger than that at 1.8 eV. This result suggests that the E_p determined by white light is the average of $E_p(\omega)$.

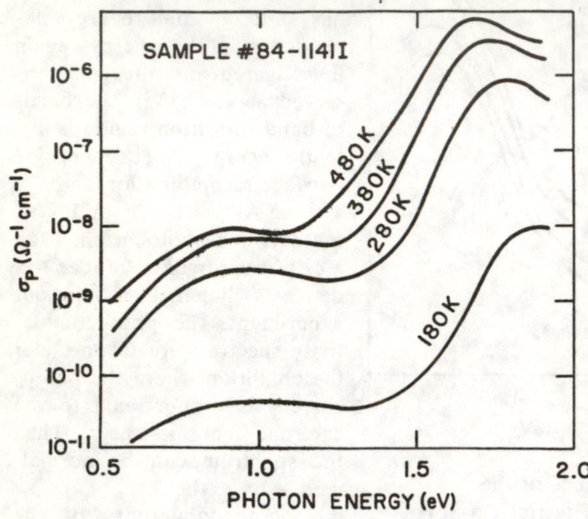

Fig. 4: Temperature dependence of the photoconductivity of intrinsic a-Si:H.

If the carriers of intrinsic a-Si:H are electrons, then the peak at 0.9 eV gives the energy position of the defect measured from the conduction band mobility edge. In the photoluminescence spectra of the intrinsic a-Si:H there is a peak at 0.9 eV. If this energy is measured from the top of valence band[4] and if the peaks observed in both these experiments have the same origin, then the band gap energy must be larger than the 1.8 eV observed as the photoconductivity peak of band-to-band transitions in Fig. 4. If the trapped hole state is 0.2 or 0.3 eV as proposed by Street, the mobility gap energy of a-Si:H is 2.0 or 2.1 eV. This is consistent with the literature.[4,11] Then the peak at 1.8 eV is the result of the photon energy dependence of the parameters given in Eq. (1) and the fact that the temperature dependence of the peak position is the same as that of band gap suggests that these parameters are connected to the band structure of a-Si:H.

When the temperature increases the tail of band-to-band transitions becomes broader and at 480 K, where σ_d is comparable to the σ_p, the tail part extends to 1.0 eV. When we apply the theory[12] of the Urbach tail to Fig. 4, the steepness parameter, which is obtained from the straight part of the photoconductivity spectra, is ~0.3. According to Toyozawa,[12] this value is large enough to produce a self-trap state of the electron or hole at the band edge.

P-doped a-Si:H

In Fig. 5 we show the PH_3 concentration dependence of the photoconductivity of a-Si:H. These spectra were all obtained at the same velocity of VEL=3. In the 2×10^{-6} case the peak position due to band-to-band transitions has the same energy as that of intrinsic a-Si:H. The excitation from midgap states, however, increases due to doping and the subband peak shifts to the higher energy side. In the 2×10^{-4} case the spectrum still shows band-to-band transitions but the peak is broadened. It is obvious that the photoconductivity spectra consists of two peaks (about 1.2 and 1.9 eV) with similar intensities. In the 1×10^{-2} case, the part due to band-to-band transitions is no longer observed and only midgap photoconductivity is seen with a peak at 1.2 eV. When the concentration of PH_3 increases, the velocity dependence of the photoconductivity decreases to point where, for the 1×10^{-2} case, it is difficult to see the velocity dependence.

The density-of-states of P-doped a-Si:H measured by the PAS method was reported by Yamasaki et al.[3] They suggested that there is a trap center due to P atoms in a-Si:H around 1.0 eV below the top of the conduction band. In PDS measurements[5] the absorption coefficient from midgap states increased continuously with doping level and it was not clear whether there was a peak in the midgap or not.

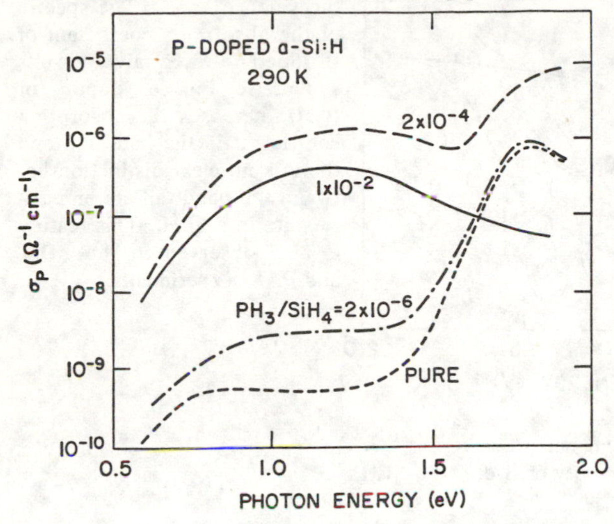

Fig. 5: Concentration dependence of the photoconductivity of doped a-Si:H.

In Fig. 5 it is clearly observed that the P in a-Si:H has a peak at 1.2 eV. Then the level due to P impurities is 1.2 eV below the conduction band. We observed that midgap response is independent of additional visible light bias. This result suggests that the midgap states have their own photo-response and this response is stronger than that from band-to-band transitions.

It is well established that the E_d of P-doped a-Si:H decreases from 0.7 eV to 0.2 eV with doping. On the contrary the E_p only has a concentration dependence of about 0.1 eV. The E_p is related to the trap levels (band tail) of the electron which is 0.2 eV below the conduction band. The spectra observed in Fig. 5 show that for the P-doped a-Si:H the main part of the photoconductivity comes from the donor level.

B-doped a-Si:H

In Fig. 6 we show the photoconductivity spectra of B-doped a-Si:H. All of the data were taken at VEL=3. The velocity dependence of the photoconductivity of B-doped a-Si:H is stronger than that of intrinsic a-Si:H and it is difficult to take a spectra at higher velocities. In the 2×10^{-5} case, the peak due to the band-to-band transitions becomes broad and the total photocurrent is greatly reduced. It is well established[7] that a-Si:H lightly doped with B_2H_6 has low photoconductivity and has a compensation effect as in our 2×10^{-5} case. This sample has no clear peak in the midgap, the photoconductivity at 0.8 eV is less than for intrinsic a-Si:H. The photoconductivity peak due to B_2H_6 is observed around 1.3 eV, which is observed as a bump at a concentration of 2×10^{-4}. This energy is 0.1 eV higher than that observed for PH_3 doping. At high concentrations, there is photoconductivity due to band-to-band transitions and the midgap photoconductivity increases its intensity. Finally the photoconductivity in the midgap merges into the tail of the band-to-band transitions.

The photoconductivity spectra observed in Fig. 6 gives the joint density-of-states of the B-doped a-Si:H under the assumption that Eq. (1) does not depend on the energy. It is well established that the concentration dependence of E_p is very small and is around 0.2 eV and that of E_d is also 0.2 eV in the highly B doped a-Si:H. As we showed in Fig. 6 the B in a-Si:H has no clear peak and modifies the band edge profile completely. In this case it is

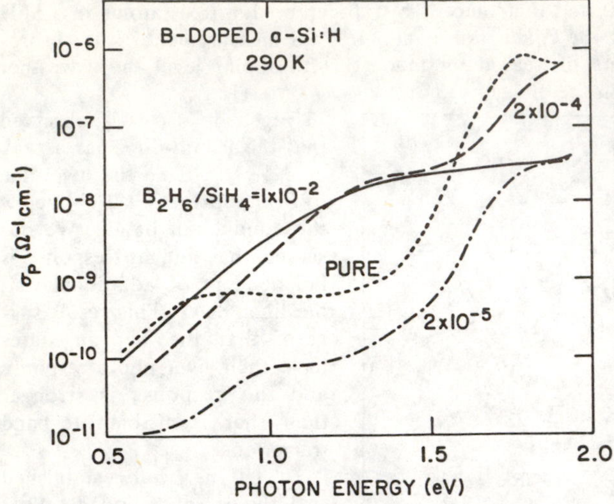

Fig. 6: Concentration dependence of the photoconductivity of B-doped a-Si:H.

not clear whether the optically excited electron or hole produces the localized state below the conduction band or above the valence band. The spectra of the absorption coefficient of B-doped a-Si:H give results similar to Fig. 6; doping of B_2H_6 increases the absorption coefficient in the band tail and there is no clear distinction between the band tail and impurity state in a-Si:H. This feature is also observed in the PDS[5] and PAS[3] experiments.

Consideration of the impurity state in a-Si:H

Many models of the impurity states in a-Si:H have been proposed. Street[13] proposes that the existence of P or B levels is related to the shift of the Fermi level and that the impurity does not by itself make a clear midgap level. Excited donor or acceptor states lie above the Fermi level. The experimental results given in Fig. 5 and 6, however, are difficult to explain by the shift of Fermi level or the increase of compensating defects. The role of P in a-Si:H, though qualitatively similar, is quantitatively different to that of B. The peak at 1.2 eV is related to the P impurity and the bump at 1.3 eV is related to the B impurity.

The photoluminescence process is complementary to photoconductivity. Biegelsen et al.[14] reported that intrinsic a-Si:H has a defect emission peak at 0.8 eV. In the lightly P-doped a-Si:H, the emission consists of two peaks, one is from band-to-band transitions and the other from the midgap emission. When the concentration increases, the relative intensity of the midgap peak increases and finally only the midgap peak at 0.8 eV remains. In B-doped a-Si:H, the peak due to band-to-band transitions shifts to the lower energy side with increasing impurity concentration until it finally reaches 0.9 eV. These features are quite similar to what we see in Fig. 5 and 6. The energy difference of the peak positions of the photoconductivity and the luminescence can be understood by taking account of the relaxation of the carriers before radiative recombination. This 0.4 eV excess energy is also present for the band-to-band photoconductivity relative to the band-to-band luminescent peak at 1.4 eV.

SUMMARY

As we have seen, the density of states of a-Si:H is sensitive to P and B doping. The photoconductivity measurement by the use of FTIR spectroscopy with a-Si:H as a detector is highly sensitive to the midgap states. The impurity level of P is 1.2 eV below the conduction band and that of B is 1.3 eV above the valence band.

REFERENCES

1. C. C. Tsai, H. Fritzsche, M. H. Tanielian, P. J. Gaczi, P. D. Persans, and M. A. Vesaghi, in Proceedings of the 7th Int. Conf. on Amorphous and Liquid semiconductors, edited by W. E. Spear (Univ. of Edinburgh, Edingurgh, Scotland, 1977), p.339.
2. J. D. Cohen, D. V. Lang and J. P. Harbison, Phys. Rev. Letters **45**, 197 (1980).
3. S. Yamasaki, H. Oheda, A. Matsuda, H. Okushi and K. Tanaka, Jpn. J. Appl. Phys. **21**, L539 (1982); S. Yamasaki, A. Matsuda, and K. Tanaka, Jpn. J. Appl. Phys. **21**, L789 (1982).
4. R. A. Street, Phys. Rev. **B21**, 5775 (1980).
5. W. B. Jackson and N. M. Amer, J. Physique **42**, C4-293 (1981).
6. W. Fuhs, H. M. Welsch and D. G. Booth, Phys. Stat. Sol. (b) **120**, 197 (1983).
7. W. Rehm, R. Fischer, J. Stuke and H. Wagner, Phys. Stat. Sol. (b) **79**, 539 (1977).
8. M. H. Brodsky and P. A. Leary, J. Non-Cryst. Solids, **35-36** 487 (1980).
9. C. R. Wronski, B. Abeles, T. Tiedje and G. D. Cody, Solid State Commun. **44** 1423 (1982).
10. J. M. Berger, F. de Chelle, J. P. Ferraton, A. Donnadieu, J. Beichler and G. Weiser, Solar Energy Mat. **9**, 301 (1983).
11. M. H. Brodsky, Solid State Commun. **36**, 55 (1980).
12. Y. Toyozawa, Physica **117B&118**, 23 (1983).
13. R. A. Street, Phys. Rev. Letters **49**, 1187 (1982).
14. D. K. Biegelsen, R. A. Street, W. A. Jackson and R. L. Weisfield, (in Press).

PHOTO-DEPOPULATION-INDUCED ESR MEASUREMENT OF DEEP GAP STATES IN a-Si:H

D. K. Biegelsen and N. M. Johnson
Xerox Palo Alto Research Center, Palo Alto, CA 94304

ABSTRACT

We describe here the technique of photo-depopulation and photo-depopulation-induced ESR and apply them to n-type a-Si:H. From the optical energy dependence of the induced spin and charge densities we conclude (1) that deep states are predominantly, if not entirely, dangling bonds, and (2) that the D^- band has a peak ~0.9±0.1 eV below E_c.

INTRODUCTION

A variety of techniques have been used to ascertain the density of states (DOS) of deep gap levels in a-Si:H[1]. Two principal areas of remaining uncertainty are (1) the energy scale of the DOS spectra and (2) the atomic nature of the deep states. In this paper we introduce a novel technique which was designed to address these two issues directly.

The method consists of photo-depopulation of deep states in a strongly reverse-biased abrupt junction or MOS device which is held at low temperature. Photo-excited majority carriers are swept away from their origin leaving behind states of different charge and magnetic character. Ideally the experiment works as follows: an n-type sample, for example, is prepared with one contact blocking for majority carriers and one Ohmic. Initially the structure is placed under zero bias or flat band conditions so that most deep states are filled. A strong reverse bias voltage, V_R, is then applied. For sufficiently low temperatures, T, states deeper than $E_D = E_C - kT\ln(\nu_o t)$ will not thermally emit in the time, t, of the experiment. (Here ν_o is the attempt-to-escape frequency and E_C is the conduction band mobility edge.) T is however sufficiently high so that zero bias filling can occur and 100 Hz capacitance measurements can be made. We now irradiate the sample with uniformly absorbed light starting with energy $h\nu = |E_C - E_D|$ and increase $h\nu$ incrementally. At each step we photo-excite majority carriers to a narrow band of states near E_C and sweep the charge from the sample. There is no retrapping in deep states if $\mu\tau V_R/L$, the average carrier range, is greater than the sample thickness, L. If CV_R (where C is the depletion width capacitance) is greater than the integrated density of states between approximately midgap and E_D, then all the states in the upper half of the gap can be swept out.

In Figure 1 we show schematically a simplified DOS for a-Si:H. The zero of electron energy is fixed at E_C and a rectangular conduction band DOS is assumed. We neglect localized band tail states at this point. In Fig. 1A a band of deep states is shown occupied after a zero bias filling pulse. At t=0, strong reverse bias is applied and E_D moves down into the band logarithmically in time (Fig. 1B.) For the time scale of these experiments, $|E_D| < 0.35$eV. In Fig. 1C the effect of optical excitation is shown. States down to ~E_c - $h\nu$ are depopulated. In fact, for long times, depopulation can progress even further from E_C because excitation to tail states, although much less

0094-243X/84/1200032-08 $3.00

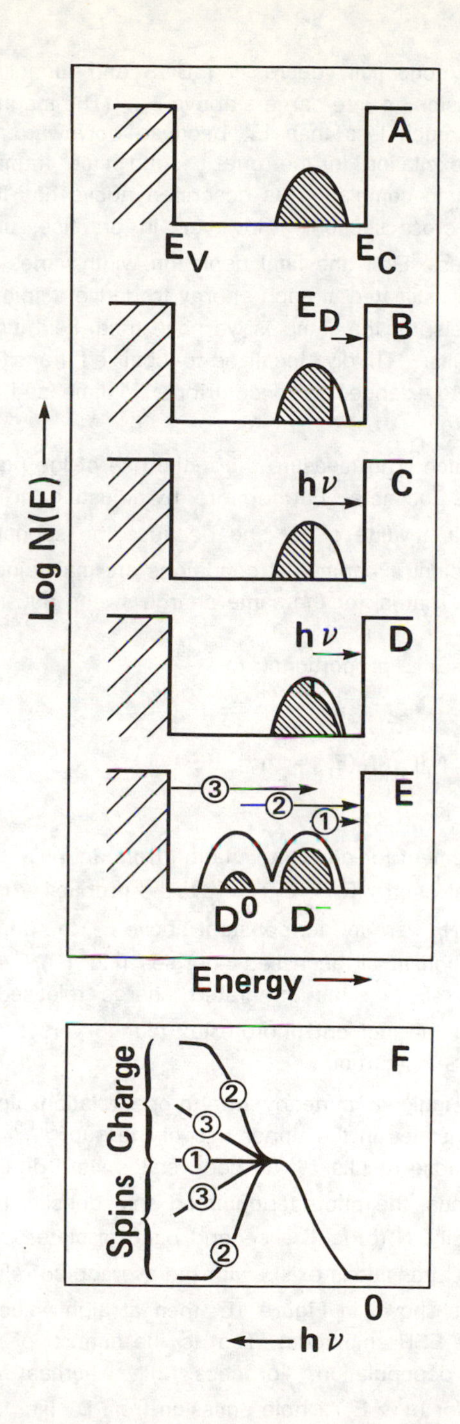

Fig. 1 Photodepopulation effects on DOS of a-Si:H. A)-E), model DOS spectra and partial occupation; F), charge and spin response for several possible processes.

probable (due to the exponentially decreasing DOS and tunneling overlap factors) leads, by thermal re-emission, to free carriers above E_C. (The magnitude of this shift in the photogeneration is much less than E_D because downward transitions strongly compete with thermal excitation in the one second time frame.)

Photo-depopulation to completion as described above has the beauty of being independent of optical cross section. However, in practice, unfortunately, it has undesirable consequences. First, the final depletion width varies with $h\nu$ so that an effective volume must be estimated at each energy to derive a spin density. Secondly, at high depopulation levels, for the samples we have tried, field induced injection sets in and distorts the results. Thirdly, localized-to-localized transitions, although less probable than localized-to-extended, also contribute in time and, in effect, shift the spectrum to lower energy.

An alternate approach is to measure the initial rise of the photo-excitation. The most probable transitions dominate. Furthermore, by adjusting the photon-flux to keep the slope constant at each value of $h\nu$ and because the subgap light is absorbed uniformly for all $h\nu$, constant experimental conditions are maintained and the effective sample volume is fixed. That is, for the same charge sweep out, the band bending is the same.

The initial response is proportional to

$$\int_{E_C - h\nu}^{E_D} M(E_i, E_f)\; N_i(E_i)\; N_f(E_i + h\nu)\; dE_i$$

where M is the matrix element for optical excitation from an initial state at energy E_i to an extended final state at energy $E_f = E_i + h\nu$. If the probability for optical excitation at each initial energy, $M \cdot N_f$, were in fact constant above E_c, the populated states would drop by a constant fraction at all accessed energies (Fig 1D). As $h\nu$ is varied, the photon flux is adjusted so that the integrated charge released is held constant. Therefore, the inverse of the incident photon flux plotted versus $h\nu$ is approximately proportional to the DOS spectrum.

Many experiments indicate a nearly one-to-one relationship between dangling bond states, D, and gap states in the upper half of the gap[2,3,4,5]. The results in this paper strongly confirm those results. If all deep states were due to doubly-occupied spin-paired dangling bonds, the photodepopulated spin density, $N_s(h\nu)$, would equal the induced charge density, $N_e(h\nu)$. If a second band of states corresponding to the energy level for $D^+ \leftrightarrow D^o$ transitions exists with the average correlation energy greater than the peak widths, as shown in Figure 1E, then at high values of $h\nu$ one would expect a decrease in the ESR amplitude. That is, the number of paramagnetic states is reduced by second depopulation, "process (2)", whereas N_e will continue to increase. Furthermore, for $h\nu > E_g/2$ hole emission from D^o (i.e., photoexcitation of a valence band electron into D^o) can also be expected ("process (3)" in Fig. 1E) as well as hole emission from D^+. The latter process can occur, of course, only after second depopulation.

In Fig. 1F we show possible characteristic responses for the depopulation induced charge and spin densities. If the DOS consists of only a single band of dangling bond states, then both N_s and N_e would follow curve (1). If a second charge state of the dangling bond exists, resolved from the first, as sketched in Fig. 1E, then N_e will follow the upper curve (2) and N_s will follow the lower curve (2). If the D^o and D^- bands overlap, N_s will be much less than N_e. Hole emission competes with second depopulation and depending on relative cross sections, tends to drive curves (2) back toward (1); that is, curve (3).

EXPERIMENTAL DETAILS

In the work reported here an a-Si:H p^+n diode was used. An MOS capacitor was also used but the ability to neglect interface state contributions in the former led us to focus our effort there. The active area, approximately 5mm x 9 mm had the following structure. On a fused silica substrate were deposited a 20nm Cr layer, a 50nm p^+ layer ($[B_2H_6]/[SiH_4] = 10^{-2}$), a 7$\mu$m n layer ($[PH_3]/[SiH_4] = 1\text{-}2\times10^{-5}$) a 50nm n^+ layer and a semi-transparant Al top electrode 14nm thick. The sample was held at 130K and irradiated through appropriate filters using a Xenon arc lamp and double monochromator.

The thick sample used here allows us to demonstrate directly the difference between light induced ESR, a dynamic process, and the depopulation induced ESR, a persistent signal. In n-type material subgap excitation results in an ESR composite signal from a neutralized dangling bond at g = 2.0055 and a band tail electron at g = 2.0043. At 130K, after the light is removed, the LESR decays in times less than 100ms[4]. In Fig. 2 the stronger signal arises from the neutral region under 1.1eV illumination several orders of magnitude larger than in the depopulation experiments. The weaker signal is the persistent signal remaining in the dark and located in the volume of sample which was depleted during irradiation. The shifts in zero crossing and line width are consistent with the picture above.

The technique used for the ESR initial slope measurements was a second-derivative method. The magnetic field was swept at 33 Hz between the peaks in the first derivative dangling bond ESR response and lock-in amplified with a 300ms integration time. The peak position was set using a known dangling bond resonance in an undoped a-Si:H sample. The system gain was determined using a measured amount of spin-calibrated pitch configured into the standard sample geometry. A zero bias pulse ~1.5 seconds long filled ~50% of the states in steady state cycling. A strong reverse bias of -39 volts was applied for 8 seconds. This cycle was iterated every ten seconds and the data averaged ~3000 times per point. The capacitance transient was monitored similarly in parallel with the ESR measurements. At -39 volts bias, the minimum depletion width (~3.0μm) was always much larger than the zero bias depletion width (~0.7μm) so that the latter could be neglected in assessing the effective volume of depopulation.

Initial measurements were made at each energy to demonstrate the linear

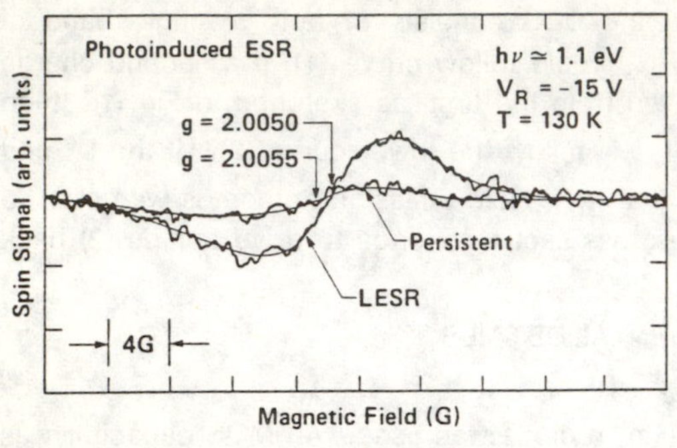

Fig. 2 LESR and persistent photodepopulation ESR spectra.

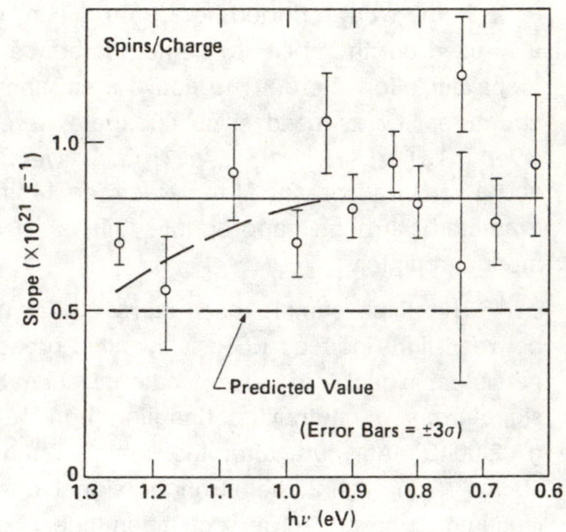

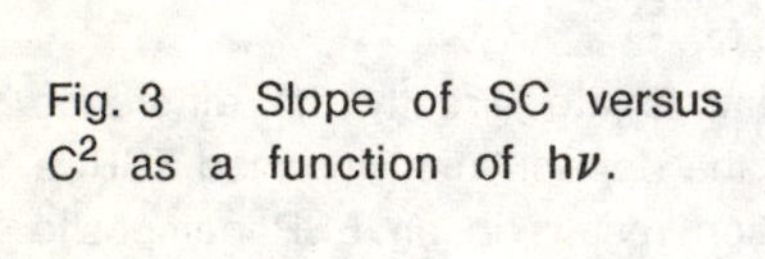

Fig. 3 Slope of SC versus C^2 as a function of $h\nu$.

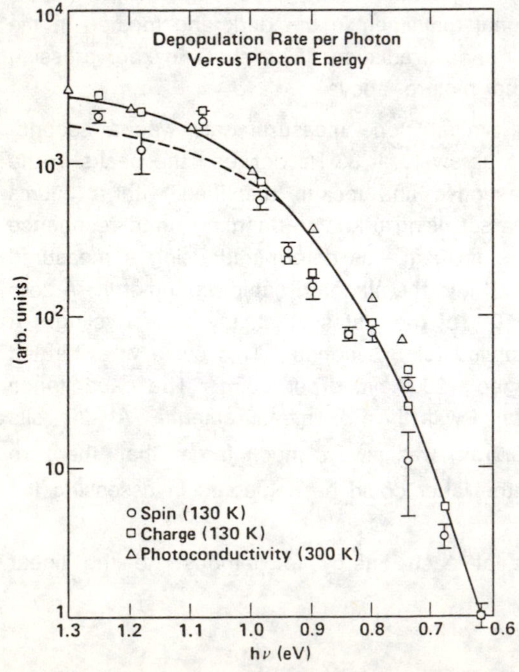

Fig. 4 Reciprocal of photon flux for constant spin and charge depopulation rate as a function of $h\nu$. Triangles are light-biased photoconductivity results.

dependence of ESR slope on illumination intensity. Subsequently the capacitance transient was used to set the incident photon flux to give approximately constant initial slope. The slopes were later scaled to a uniform value and the scaling factor applied to obtain the photon flux (after correction for electrode transmission) for constant slope.

To test the relation between N_s and N_e we plotted S(t)C(t), the number of induced spins within the depletion width, versus $C^2(t)$-$C^2(0)$, the induced change in space charge density, where S and C are the ESR and capacitance signals. In Figure 3 we show the slope of the linear regression fit of SC versus C^2 at each energy. The large error bars are due to the very weak ESR response, made more extreme by the necessity of working with fractional depopulation. We have drawn two curves through the data. Neither behavior can be ruled out. The dashed line is the predicted value of the slope in the uniform space charge, depletion approximation. Assuming $N_s = N_e$, the slope is given by $2(V_R + V_{bi})/q = 5\times10^{20}\ F^{-1}$ where V_{bi} is the built in potential. The accuracy of the estimate is limited primarily by the absolute calibration of the ESR and cannot, without great difficulty, be determined to better than a factor ~2. The observed deviation from the predicted value is therefore within the likely systematic error of the calibration. The linearity of SC versus C^2, the constancy of the slope with energy for $h\nu < 1$eV and the near agreement of experiment and prediction (within the accumulated uncertainties) all indicate that the gap states consist only of dangling bond states, and that the Hubbard bands of the dangling bond are well-resolved. Above $h\nu \sim 1$ eV we seem to observe a drop in the number of spins/charge. This could correlate with the second depopulation of dangling bonds (or the presence of other states which are non-spin-active).

Finally in Fig. 4 we show the depopulation rate per photon of spins and charge, normalized to a value of 1 at the lowest energy measured here of 0.62 eV. (In fact, we have found that the curve continues to drop to below 0.5 eV). (The small rise at 1.1eV and dip near .9eV we believe arise from weak interference artifacts). The triangles, shown for later comparison, are room temperature primary photoconductivity data measured in the presence of bias light.[6] These measurements used n^+-n-n^+ samples prepared under conditions similar to those for our p^+n sample.

DISCUSSION

The finding that $N_s \sim N_e$ above midgap simplifies the interpretation of the gap states in a-Si:H. Firstly we may infer that the dangling bond bands are reasonably separate. Further, as we will attempt to show below, the photoexcitation is dominated by first depopulation. We will then conclude that the D^- band lies at ~ 1.0 eV below E_c at low temperatures. Figure 1F shows three characteristic behaviors for the charge and spin depopulation signals. If the DOS consists in effect of only one band of states which are dangling bonds, then curves (1) result. In the initial slope method this can occur either because the D^o peak is well resolved from the D^- band (correlation energy > peak width) or because the net probability for second depopulation, P_2 is much smaller than that for first depopulation, P_1. (Second depopulation must follow the first emission and in the initial slope technique only 10% of all states are depopulated.) Curves (2) result if $P_2 \gg P_1$ i.e. as soon as one dangling bond electron is photoemitted, the second is removed rapidly, leading to no spin. An intermediate state is achieved if the rates are comparable. We note again that hole emission from D^o at energy E_{D^-} (which can occur for $h\nu > |E_v - E_{D^-}|$) would tend to reduce the effect of P_2 in the ESR data, but not in the capacitance data. We also point out that in initial slope measurements hole emission can only modify the P_1 process by at most a fraction equal to the depopulation fraction.

We make the following observations about Fig. 4. We note that the photodepopulated charge increases approximately exponentially up to $h\nu \sim 1.1$eV then rises less than a factor of two out to 1.25eV. The half-height below the plateau occurs at $h\nu \sim 1.0$ eV. The spin density also increases up to ~1.0eV then stays constant or drops slightly. (cf. Fig. 3). This behavior is similar to curves (1) and (3) in Fig. 1F (plotted there on a linear scale). The solid line in Fig. 4 is the curve calculated for the photogeneration from a Gaussian DOS peak centered at 1.0 eV with FWHH = 0.3 eV to a conduction band DOS given by Jackson *et al.*[7] in this conference and assuming constant matrix elements. The data are thus consistent with a dangling bond band at 130K located at ~1.0 eV with FWHH of 0.3-0.4 eV and a second depopulation band ~0.3 eV deeper. In the initial slope technique the gap state occupation is hardly perturbed. Therefore the first depopulation transition is strongly weighted relative to other transitions requiring emptied gap states. However, to the extent that N_e and N_s do diverge at high $h\nu$ we would infer that the relative magnitudes of the optical matrix elements $M_2 \gg M_1$.

The room temperature photoconductivity data normalized to constant photon flux are remarkably similar to the present 130K capacitance data. The energy shift noticeable at low energies is, as pointed out by Moddel[8], partly (<0.1 eV) due to a temperature dependent shift of D^- relative to the conduction band. Another part is due to high temperature contributions to conduction from localized tail states deeper than E_c which are in quasi-thermal-equilibrium with the conduction band states at room temperature but are "frozen out" at 130K. We thus find that the capacitance data and photoconductivity data are in close agreement. For completeness we mention that the low energy onset observed in optical absorption data is shifted to still lower energies.[9] This is most likely due to photoexcited localized to localized transitions which do not generate free carriers through thermal excitation before re-deep-trapping.

We conclude by discussing the apparant discrepancy between energy scales in experiments involving thermal versus optical emission. In agreement with refs. 10 and 11, we believe the relative displacement of scales arises from the different band of excited states which predominate in each experiment, and which represent "reference" energies different from E_c. Photoexcitation occurs on average into states ~0.1 eV above E_c due to the linearly increasing conduction band DOS.[7] Transport of thermally excited electrons, in contrast, is dominated by states at, and possibly just below E_c. The direction and magnitude of these final state shifts seem adequate to bring into agreement the various energy scales surmised. Moreover, uncertainties in the energy dependence of optical and thermal emission cross sections can lead to spectral distortions also of order 0.1eV. The variations between energy scales thus can be explained by known, quantitative uncertainties and not necessarily by qualitative differences.

In summary we have demonstrated that the deep gap states above midgap in n-type material are predominantly dangling bond states. Furthermore, the D^- band lies at 1.0 ± 0.1 eV below E_c at 130K. We then estimate that the room temperature peak of the D^- distribution lies at E_c-0.9 ± 0.1 eV with a FWHH ~0.3eV.

ACKNOWLEDGEMENTS

We are pleased to thank R. A Street, W. B. Jackson and M. Stutzmann for enlightening discussions and M. D. Moyer and R. Thompson for help in sample fabrication. This work was supported in part by Solar Energy Research Institute Contract No. XB-3-03112-1.

REFERENCES

1. R. A. Street and D. K. Biegelsen, The Spectroscopy of Localized States, in The Physics of Hydrogenated Amorphous Silicon II, J. D. Joannopoulos and G. Lucovsky, eds. (Springer-Verlag, Berlin, 1984), p.195.
2. J. D. Cohen, J. P. Harbison and K. W. Wecht, Phys. Rev. Letters 48, 109 (1982).
3. R. A. Street, J. C. Zesch and M. J. Thompson, Appl. Phys. Lett. 43, 672 (1983).
4. D. K. Biegelsen, R. A. Street and R. L. Weisfield, paper presented at the International Topical Conference on Transport and Defects in Amorphous Semiconductors, Bloomfield Hills, MI, March 1984, to be published.
5. W. B. Jackson and N. M. Amer, Phys. Rev. B25, 5559 (1982).
6. W. B. Jackson, private communication.
7. W. B. Jackson, S.-J. Oh, C. C. Tsai and J. W.Allen, this conference, to be published.
8. G. Moddel, D. A. Anderson and W. Paul, Phys. Rev. B22, 1918 (1980).
9. W. B. Jackson, R. J. Nemanich and N. M. Amer, Phys. Rev. B27, 4861 (1983).
10. W. B. Jackson, Solid State Commun. 44, 477 (1982).
11. N. M. Johnson and W. B. Jackson, paper presented at the International Topical Conference on Transport and Defects in Amorphous Semiconductors, Bloomfield Hills, MI, March 1984, to be published.

PHOTOCONDUCTIVITY AND RECOMBINATION IN AMORPHOUS SILICON ALLOYS

M. Hack and S. Guha
Energy Conversion Devices, Inc., Troy, Michigan 48084

M. Shur
University of Minnesota, Minneapolis, Minnesota 55455

ABSTRACT

We present results of a new model to realistically describe the steady-state photoconductivity in amorphous silicon alloys. The room temperature photoconductivity is very dependent on the position of the dark Fermi level and the sensitization is a consequence of both a change in the recombination path and dopant created gap states.

We also demonstrate the relationship between the power dependence of photoconductivity and dark Fermi level position and show that as a result of space charge neutrality, this dependence can be related to a characteristic energy slope of the density of states only in the absence of injected charge or dopants. Moreover, in agreement with recent experimental data, we show that our model predicts a power dependence of less than 0.5 for high intensity illumination on n-type amorphous silicon. Finally we examine the temperature dependence of photoconductivity and find good agreement between our theory and experimental results.

INTRODUCTION

There is now much experimental data[1-4] in the literature showing the relationship between the photoconductivity and dark conductivity in amorphous silicon alloys. In particular, experimental results show that the photoconductivity increases as the Fermi level is moved closer to the conduction band edge. This was first reported on by Anderson and Spear[1] and more recently by Beyer and Hoheisel[2] who have shown the dependence of photoconductivity on Fermi level position for both undoped and lightly doped samples from different laboratories. Simultaneously, there is a reduction in γ, the power dependence of the photoconductivity with intensity.[1,3,5] It is interesting to note that there appears to be a general trend relating these properties for films grown under different deposition conditions.

Rose[6] proposed a model for photoconductivity based on the assumption of an exponential distribution of traps, and directly related the power dependence to the characteristic energy slope of this distribution. This model is appropriate for undoped samples but a more detailed approach is required in the analysis of doped samples, or films into which charge has been injected. We propose a new model based on two exponential distributions of both acceptor-like and donor-like traps that adequately explains both the change in photoconductivity and its power dependency with dark Fermi level position, E_{FO}. The sensitization of the photoconductivity is interpreted in terms of a change in the occupation of these recombination centres, which is also responsible for the variation of the power dependence.

In the case of Fermi level shifts caused by the addition of dopants we also investigate the effects of dopant created gap states and show that in some cases they can actually increase the sensitization with respect to dark Fermi level position. We show that, as a result of space charge neutrality conditions, the power dependency of n-type samples can be less than 0.5.

Finally, we have also examined the temperature dependence of the photoconductivity of undoped samples as predicted by our model and find it to be in good agreement with experimental results. At low temperatures, the photoconductivity increases with increasing temperature with a power law dependence which appears activated over specific temperature ranges. Around room temperature or above the photoconductivity peaks, and this peak is associated with the dark conductivity becoming larger than the photoconductivity and the dominant role played by thermally generated carriers at high temperatures.

In our model, we assume that there are both acceptor-like states and donor-like states with the former predominantly in the upper half of the gap and the latter dominating below mid-gap. Various experimental measurements[7-9] made on undoped amorphous silicon alloys suggest that the acceptor-like states consist of two different distributions both exponential in energy. Firstly, there are tail states near to the band edge, whose characteristic energy slope E_1 is comparable with the thermal voltage at room temperature, and deep localized states whose characteristic energy, E_2 is approximately 86 meV. For the lower half of the gap the tail states[9] have a characteristic energy E_3 of 43 meV and we have assumed that the deep donor-like states have an exponential distribution with a characteristic energy E_4 equal to 129 meV.

There have now been various approaches used to model the photoconductivity in amorphous materials, based on a continuous distribution of localized states in the mobility gap[10-12]. For uniform absorption of light in the absence of electric fields, the material must be neutral and the generation rate, G, equal the recombination rate R. Hence the space charge density ρ is given by $\rho = q\,[p_t - n_t + N_D^+ - N_A^- + p - n] = 0$ where p_t, n_t are the densities of trapped charge; N_D^+ , N_A^- are the densities of ionized dopants and n,p the free carrier densities. To determine the densities of trapped charge we have used the approach first proposed by Taylor and Simmons[10].

A more complete description of our photoconductivity model is given in ref. 12. However, in this present paper we have added two important new features. First, we have modeled the gap state spectrum by four exponential distributions (as opposed to two) so that we can now account for both deep states and tail states as separate distributions. Second, our present claculations of space charge densities have been made using the full Fermi-Dirac distribution functions without using zero temperature approximations.

Finally, we have also attempted to include the effects of dopant created gap states into our model. As we shall see later, these states drastically alter the sensitization of the photoconductivity to dark Fermi level position. Recent experimental evidence [13-14] suggests that the created defect density is proportional to the square-root of the dopant density and so in our model for the density of states we have put

$$g_{min}(N) = g_{min}(N{=}0) + K\,[N/g_{min}(N{=}0)]^{1/2} \qquad (1)$$

where $g_{min}(N{=}0) = 10^{16}\,cm^{-3}\,eV^{-1}$ and N is the dopant concentration per cm^3. We also have assumed that the tail state distributions are unaffected by dopants and hence an increase in g_{min} with dopants will consequently alter E_2 and E_4 accordingly.

RESULTS AND DISCUSSION

A. SENSITIZATION AND DEFECT CREATED GAP STATES

In Figure 1, we show the sensitization of the photoconductivity as a function of E_c-E_{FO} for varying values of the parameter K as defined by equation (1), i.e., varying dependency of defect creation on the dopant concentration. Also plotted on Figure 1 are experimental results from Anderson et al [1] showing the sensitization of photoconductivity in gap cell samples where E_c-E_{FO} was shifted by the introduction of dopants.

As expected, we can see that the addition of dopant created gap states (K > 0) does indeed reduce the photoconductivity of samples with $E_c - E_{FO} > 650$meV. However, it is at first surprising to see from Fig. 1 that for $E_c-E_{FO} \sim 600$ meV the photoconductivity increases as K increases i.e., we introduce more defect created states. This is a consequence of the relationship between the density of states spectrum, dark Fermi level position and photoconductivity. Movement of the dark Fermi level is determined by states around itself whereas photoconductivity is controlled by the states around the trap quasi-Fermi levels. Now the sensitization of the photoconductivity is a result of the induced or ionized charge altering the occupation probability of the traps and hence changing the recombination mechanism. This will be discussed in more detail in a later publication [15].

Briefly one may note that as a result of the asymmetry of the localized state spectrum in amorphous silicon alloys, in undoped samples most of the recombination takes place around the hole trap quasi-Fermi of the donor-like states. However, the ratio of the recombination rate at donor-like states to acceptor-like states is very dependent on the ratio of free electrons to holes. As this ratio is increased (i.e., as we make the sample more n-type) the main recombination path changes from electrons to charged donor-like states to holes to charged acceptor-like states. This is because as n/p increases, the donor-like states become full of electrons and therefore neutral, lowering their effectiveness as recombination centres. This is indeed the basic mechanism responsible for the sensitization of photoconductivity for E_c-E_{FO} less than 650 meV. Thus, if we can pin the dark Fermi level but also introduce charge into the material, we will increase the sensitization with respect to Fermi level shift. The photoconductivity will be increased by the addition of negative charge to the system. Defect created mid-gap states will result in more charge being added to the system for any given shift of the dark Fermi level, increasing the photoconductivity with respect to E_{FO}.

Hence, defect created gap-states, will increase the ratio of the photoconductivity to dark conductivity of lightly n-type samples as shown in Figure 1. Obviously a large increase in the density of the states around the trap quasi-Fermi levels would also suppress the photoconductivity, reducing the sensitization dependency on E_{FO}. However, as the trap quasi-Fermi levels are in a much higher density of states region than the dark Fermi level, it is reasonable to expect that dopant created states will suppress movement of the dark Fermi level more than increasing the recombination around the trap quasi-Fermi levels. Finally, we may note that a value of $K=3x10^{16} cm^{-3} eV^{-1}$ yields a very good fit to the experimental data of Anderson[6] et al.

B. INTENSITY DEPENDENCE

In Figure 2, we compare the high intensity γ dependence on E_c-E_{FO} for K=0 and $K=3x10^{16}$ cm^{-3} eV^{-1} corresponding to Fermi level shifts by injected charge and ionized dopants respectively. It appears that dopant created gap states are responsible for a steeper variation of γ with E_c-E_{FO}. Also shown in Figure 2 is our own experimental result of a gap-cell phosphorous doped amorphous silicon sample with an activation energy of 0.2eV whose γ factor was 0.37. This also agrees with published results by Kagawa[3] et al, who found γ less than 0.5 in field effect structures.

From previous models for photoconductivity in undoped amorphous silicon alloys[6,11,12] it can be seen that the power dependence of photoconductivity, γ, can be simply related to the energy slope of the acceptor-like states. However in deriving these equations, it has been assumed that $n_t = p_t$. If charge is introduced into a sample, this condition no longer holds and γ no longer equals $T_2/(T+T_2)$ where $E_2 = kT_2$. The addition of dopants changes the space-charge neutrality condition to $p_t = n_t - N$ where $N = N_D^+ - N_A^-$ and this change invalidates simplified models for this γ dependence.

For low light levels where the trap quasi-Fermi levels do not enter the tail states, we find that the addition of negative charge to the system reduces γ, even though for E_c-E_{FO} < 650meV this will actually increase the value of T_2, due to an increase in g_{min}. Hence, from the simple formula $\gamma = T_2/ [T + T_2]$ we would expect γ to increase.

For n-type samples, as the illumination intensity is increased, the ratio of free electrons to holes will decrease and the main recombination path will once again be electrons to charged donor-like states.

At high intensities, for n-type films, we see γ reduce to less than 0.5, which cannot be explained by conventional models of monomolecular or bimolecular recombination. The recombination rate R may therefore be written as $R = G = K_1 n p_t$ where K_1 is a constant of proportionality. Thus, from the neutrality condition it can be seen that $R = G = K_1 n(n_t - N)$. It may be noted that at high intensities, the trap quasi-Fermi levels will enter the tail states, and hence most of the negative space charge n_t lies in the tail states. It can then be easily shown[15] that n_t will be proportional to n. Thus we may write $n_t = K_3 n$ and therefore, $R=G=K_1 n(K_3 n-N)$. Hence

$$n = \frac{K_1 N + (K_1^2 N^2 + 4 G K_1 K_3)^{1/2}}{2 K_1 K_3} \quad (2)$$

Looking at equation (2) we can see therefore that at high fluxes, in the absence of dopants, n is proportional to the square-root of intensity but that the addition of negative charge ($N > 0$) will reduce the apparent power dependence to less than 0.5. This is clearly shown in Figure 2 and confirmed by our own experimental result.

C. CARRIER LIFETIMES

Figure 3 gives plots of the computed electron and hole lifetimes as a function of the dark Fermi level position. As a consequence of the asymmetrical localized state distribution in amorphous silicon alloys, the electron lifetime is greater than the hole lifetime in undoped samples.

In interpreting the results of Figure 3, we first consider the electron lifetime. As the dark Fermi level moves from its intrinsic position towards the conduction band edge, there is a change in the main recombination path which causes sensitization of the photoconductivity. This then is the origin of the rapid increase in the electron lifetime with decreasing E_c-E_{FO}. Increasing E_c-E_{FO} to greater than 630 meV leads to a large reduction in the electron lifetime for two reasons. First, the addition of acceptor created defect gap states will increase recombination. Secondly, as E_c-E_{FO} increases p_t will also increase as a result of space charge neutrality considerations, and this will also increase the recombination rate of electrons, reducing the carrier lifetime.

The reduction of hole lifetime for E_c-E_{FO} less than 630 meV is analogous to the reduction in electron lifetime for $E_c-E_{FO} > 630$ meV.

However, there is no large sensitization of the hole lifetime for $E_c - E_{FO} > 630$ meV, i.e., as the sample becomes p-type as there is for the electron lifetime in n-type samples. This is because movement of the Fermi level away from the intrinsic position towards the valence band does not alter the basic recombination path and does not lower the overall recombination rate.

D. TEMPERATURE DEPENDENCE OF PHOTOCONDUCTIVITY

In figure 4, we show both experimental[16] and computed plots of the temperature dependence of the photoconductivity of undoped amorphous silicon alloys, together with their corresponding dark conductivities. These plots can be divided into three regimes, as indicated. In regime one, the photoconductivity appears to be weakly activated with an activation energy ~ 0.1 eV. In regime two, the activation energy increases to ~ 0.2 eV and the photoconductivity reaches a peak at a temperature just above where the photoconductivity equals the dark conductivity. Finally, at high temperatures, the photoconductivity decreases with increasing temperature.

At low temperatures, region 1, both n and p are photogenerated and thermal generation of carriers can be ignored. Recombination predominantly takes place at charged donor-like states[15] and it can also be shown that the trapped charge decreases with increasing temperature (as it lies in deep localized states). From equation (3) we can see that the photoconductivity will therefore increase with increasing temperature, although the exact relationship between σ_{ph} and temperature is a power and not an exponential dependence. At moderate temperatures, where σ_{ph} is still greater than the dark conductivity n/p will increase with increasing temperature causing recombination to shift from charged donor-like states to charged acceptor-like states. This is a result of the asymmetry in the density of states spectrum. This sensitization will therefore increase the temperature dependence of the photoconductivity as compared to regime I. At high temperatures, regime 3, where σ_{ph} is now less than the dark conductivity, thermally generated carriers cannot be neglected and the carrier quasi-Fermi levels will converge on the equilibrium Fermi level, E_{FO}.

We note from Fig. 4 that although the general shape of the photoconductivity vs. temperature plot as obtained theoretically is in excellent agreement with the experimental data, the curves are shifted with respect to each other in the temperature axis. As we have discussed earlier, the temperature regime, where σ_{ph} starts decreasing with increasing temperature takes place when $\sigma_d > \sigma_{ph}$ where σ_d is the dark conductivity. In our model, we have calculated σ_d by using $\sigma_d = \sigma_o \exp -\left(\frac{E_c - E_F}{kT}\right)$ where σ_o, as determined by mobility and the density of states at the mobility edge, was taken to be 32 $(\text{ohm cm})^{-1}$. Experimentally, it has been observed that the magnitude of σ_o depends on $(E_c - E_F)$, the so-called Meyer-Neldel rule. Since the physical basis of this phenomenon is still not known, we have not attempted to introduce this in our theory. However, if we choose $\sigma_0 \sim 5,000(\Omega\text{-cm})^{-1}$ as predicted from Meyer-Neldel rule for samples of this activation energy, the discrepancy between our theory and experimental results is considerably minimised.

We acknowledge the expert assistance of J. Call in preparation of samples and thank C.-Y. Huang, W. Czubatyj, C.R. Wronski and S.J. Hudgens for valuable discussions. We also gratefully acknowledge partial financial support from The Standard Oil Company Ohio (SOHIO) and the constant encouragement of S.R. Ovshinsky.

REFERENCES

1. D A Anderson and W E Spear, Phil Mag, 36, 695 (1977).
2. W Beyer and B Hoheisel, Solid State Comm, 47, 573 (1983).
3. T Kagawa, N Matsumoto and K Kumabe, Physical Review B, 28, 4570 (1983).
4. P E Vanier, Solar Cells, 9, 85, (1983).
5. C R Wronski and R E Daniel, Physical Review B, 23, 794 (1981).
6. A Rose, Concepts in Photoconductivity and Allied Problems, Krieger, New York, 1978).
7. C Y Huang, S Guha and S J Hudgens, Phys Rev B, 27, 7460 (1983).
8. M Shur and M Hack, Journal of Applied Physics, 55, 3831 (1984).
9. T Tiedje, J M Cebulka, D L Morel and B Abeles, Phys Rev Lett 46, 1425 (1981).
10. G W Taylor and J G Simmons, J Non-Cryst Solids, 8-10, 940 (1972).
11. T Tiedje, Applied Physics Letters, 40, 627 (1982).
12. M Hack and M Shur, Journal of Applied Physics, 54, 5858 (1983).
13. C R Wronski, B Abeles, T Tiedje and G D Cody, Solid State Comm 44, 1423 (1982).
14. R A Street, J Zesch, M J Thompson, Appl Phys Lett 43, 672 (1983).
15. M Hack, S Guha, M Shur, unpublished.
16. W E Spear, R J Loveland and A Al-Sharbaty, J Non-Cryst Solids, 15, 410 (1974).

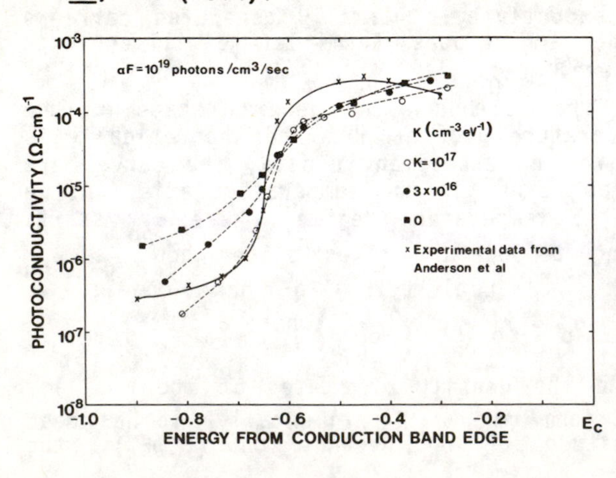

Figure 1. - Computed and experimental photoconductivity as a function of dark Fermi level position for various values of K, i.e. dependency of dopant created gap states on dopant density.

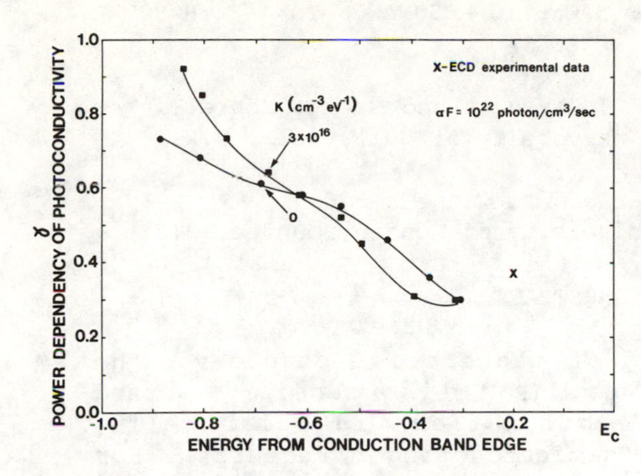

Figure 2. - Power dependence of dark Fermi level position for samples where this has been moved by doping ($K=3x10^{16}$) and injected charge (K=0). Also shown is experimental data showing $\Upsilon=0.37$

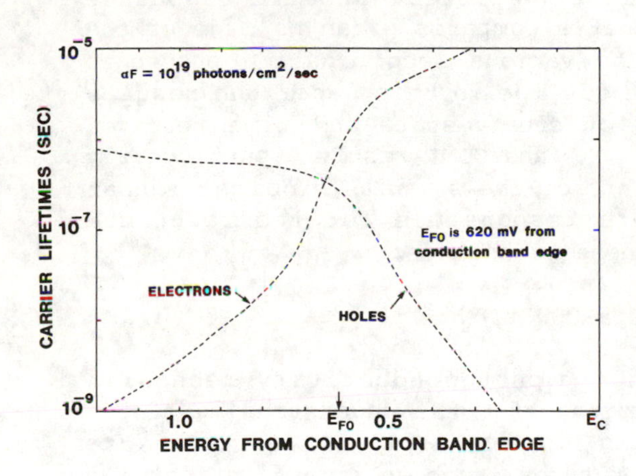

Figure 3. - Computed free carrier lifetimes as a function of dark Fermi level position.

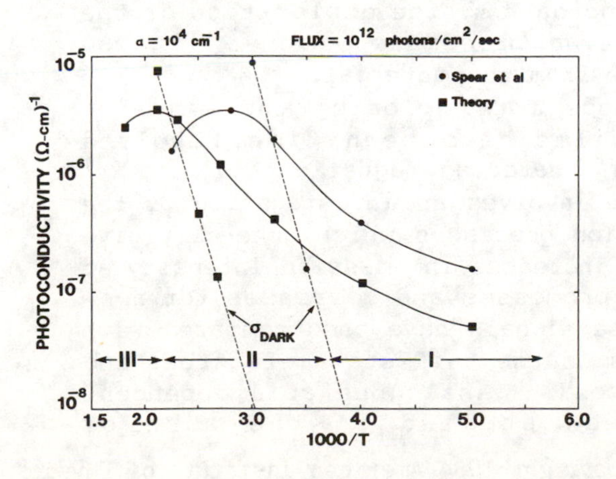

Figure 4. - Computed temperature dependence of photoconductivity together with experimental results from Spear et al. Also shown are the corresponding dark conductivities.

PHOTOVOLTAIC DETECTION OF MAGNETIC RESONANCE IN a-Si:H SOLAR CELLS

K. P. Homewood, B. C. Cavenett and C. van Berkel
Physics Department, University of Hull, Hull, U.K.

W. E. Spear and P. G. LeComber
Physics Department, University of Dundee, Dundee, U.K.

ABSTRACT

The detection of magnetic resonance via the photovoltage or photocurrent (PDMR) is a sensitive method for investigating recombination in solar cells, particularly amorphous semiconducting structures. For samples which also show luminescence the results can be directly related to optically detected magnetic resonance (ODMR) measurements so that a comparison can be made of the relative importance of the various recombination processes on the two effects. We outline in this paper the results of comprehensive studies of steady-state and time resolved PDMR and a comparison of PDMR and ODMR results, particularly the temperature dependence of the signals, indicates that different mechanisms are responsible for the resonance signals observed using these techniques.

INTRODUCTION

An investigation of the photoconductivity mechanism amorphous silicon is important for the understanding of the photovoltaic effect in solar cells [1]. In fact, particularly at low temperatures, such an investigation will provide an alternative approach to the exploration of the electron-hole recombination processes which will also determine the luminescence from the material.

The principal methods for exploring the emission processes in a-Si:H films have been time resolved spectroscopy [2] and optically detected magnetic resonance [3,4,5]. This latter technique involves an investigation of the spin dependent recombination processes which generally give rise to microwave induced increases in emission intensity at resonance for radiative processes and decreases for non-radiative processes. These signals have been compared with the known EPR resonances of the tail state electrons and holes and the dangling bond. Analogous spin dependent measurements have been studied in the case of

0094-243X/84/1200048-07 $3.00

photoconductivity by Street [6] and Dersch et al [7], but these measurements have been principally at high temperatures, i.e. greater than 100K. We have recently explored an alternative approach by investigating the spin dependent recombination processes in p^+-i-n^+ structures where either the photo-voltage or photocurrent can be monitored [8]. This is an exceedingly sensitive technique which we have explored in detail for both CW laser excitation of the cell and for pulsed laser excitation which has allowed time-resolved measurements of the photovoltaic or photocurrent detected magnetic resonance (PDMR) to be carried out. In this paper we discuss the results, including temperature dependent measurements, in terms of a model for the low temperature photoconductivity.

EXPERIMENTAL RESULTS

The p^+-i-n^+ cells were prepared at Dundee by the glow discharge method. The intrinsic region was 500 nm thick and details of the doped contact regions are given in figure 1 which also illustrates the method used for monitoring the PDMR. For lifetime measurements and time resolved PDMR measurements the photocurrent was monitoredwith a 100 load. The cell was placed at the centre of an X-band cavity and cooled by a gas flow cryostat. The microwaves

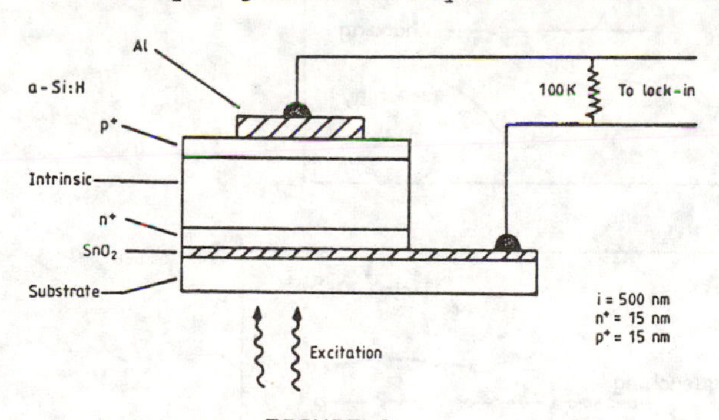

FIGURE 1
p^+-i-n^+ cell configuration for PDMR measurements

were modulated by a microwave switch and the photovoltage or photocurrent signals were monitored with a lock-in detector. This lock-in (Brookdeal 9505 SC) was operated in the gated mode for the time resolved measurements when the laser was switched by an acousto-optic modulator. This method was also used to explore the resonances present at different times within the microwave pulses in the condition of CW

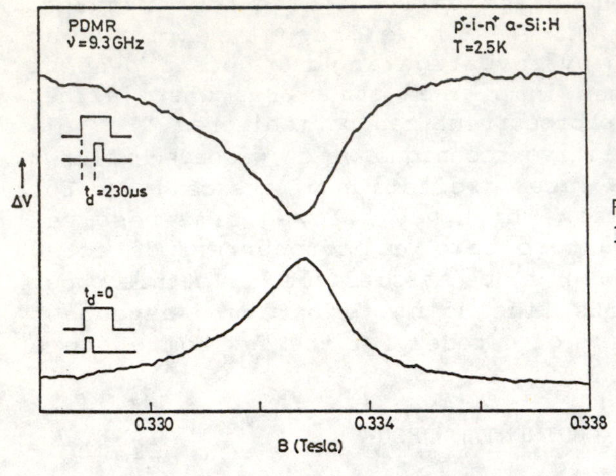

FIGURE 2
PDMR signals using gated lock-in

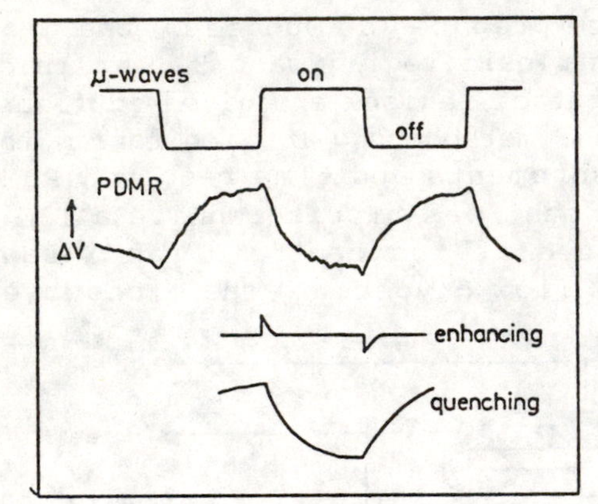

FIGURE 3
PDMR waveform showing enhancing and quenching components

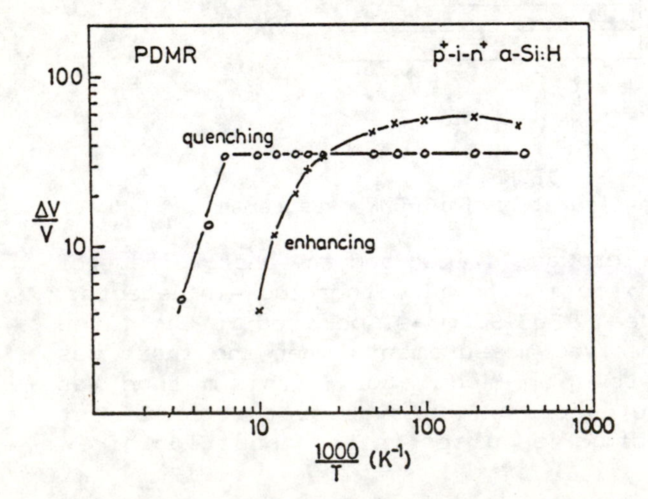

FIGURE 4
Temperature dependence of the PDMR signals

laser excitation.

In fact, the results of this latter experiment are shown in figure 2. The insets show the position of the lock-in gate (width 1Ø usec) with respect to the microwave pulse (square wave at 1 kHz). For the gate at the beginning of the pulse the enhancing signal, (lower curve) is observed with g = 2.ØØ59 and ΔB =19.5 mT. When the lock-in gate is moved to the centre of the microwave pulse the quenching PDMR signal is observed as shown in the lower curve with g = 2.ØØ65 and ΔB = 22.5 mT. These results are consistent with the microwave induced waveform of the photo-voltage change which is shown in figure 3. The signal which is at 47Ø Hz is made up of two components as shown in the lower part of figure 3. The enhancing signal has the transient response typically observed for radiative spin dependent emission processes in ODMR measurements while the quenching signal is typical of non-radiative spin dependent processes.

We note that the g-values given here differ from those in our previous letter [8] and this is attributed to the fact that the phase shift measurements did not separate the two signals adequately.

The temperature dependences of the enhancing and quenching PDMR resonances are shown in figure 4. The results for the enhancing PDMR and ODMR [9] signals are very similar suggesting that the same centres are involved in both the photoconductivity and the luminescence. The quenching PDMR signal has a very different behaviour, being independent of temperature up to 2ØØ K and then falling rapidly in magnitude to room temperature.

DISCUSSION

Previous spin dependent photoconductivity measurements cited above have been carried out at temperatures greater than ~ 1ØØ K where only the quenching signal has been observed. For example, Street [6] observed the dangling bond resonance and interpreted this as a spin dependent capture process which competed with the photoconductivity. Dersch et al [7] interpreted their resonances in terms of mixed electron, hole and dangling bond resonances. In both of these investigations the maximum signal-to-noise was obtained at room temperature, though $-\Delta\sigma/\sigma$ increased at lower temperatures due to the larger degree of unthermalization for coupled pairs as the tunnelling or recombination time becomes greater than the spin lattice relaxation time for the coupled system.

The PDMR results, in contrast, show maximum sensitivity at low temperatures but $\Delta V/V$ decreases as the temperature is raised as in the case of the photoconductivity. In fact, the PDMR results at first sight appear similar to the ODMR but in detail there are many differences and these will be discussed below.

The photovoltage falls rapidly with temperature and becomes effectively independent of T below approximately 50 K. This behaviour has been reported for photo-conductivity measurements by Hoheisel et al [10] who suggested that this independence on T is the result of the drift of the photo-carriers during the thermalization from the extended states down through the localized states. Their report supports the view that non-geminate recombination occurs at low temperatures simply because transport still exists at these temperatures.

In our investigation we have been concerned with the comparison of the luminescence and photovoltage or photo-current measurements. The spin dependency of the luminescence shows enhancing and quenching resonances and, although the results from different groups vary considerably in detail and interpretation, it is generally agreed that the signal is due to electron-hole tail state recombination and that the large g-value (~ 2.007) is the result of exchange interaction between coupled distant pairs. In fact, as pointed out above, the temperature dependence of the enhancing PDMR and ODMR signals are very similar suggesting that the same centres are involved in both processes.

In order to explore this further we have compared the decay of the photocurrent (PC) and photoluminescence (PL)

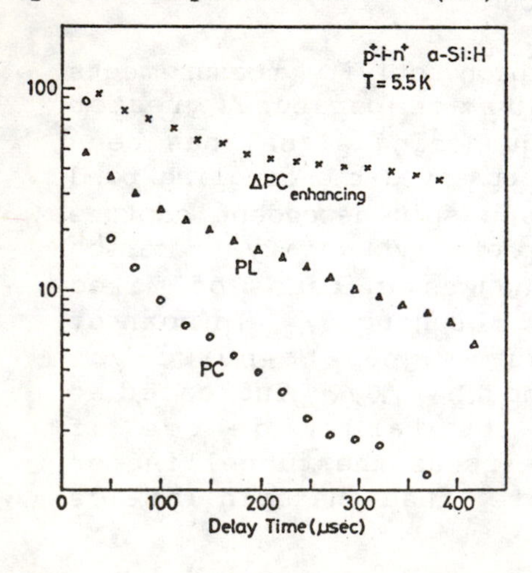

FIGURE 5
Lifetime dependence of photoconductivity, luminescence and enhancing signal.

and also measured the time resolved PDMR in the photoconductivity detection mode. The results are shown in figure 5 where it can be seen that the PL contains longer lived components than the PC. The enhancing PDMR signal is also very long lived (almost constant lifetime of ~ 400 μsec) while the quencing PDMR (not illustrated) shows a t^{-1} dependence which is almost as rapid in decay as the PC.

In our previous paper on the PDMR we suggested that the geminate model would explain the enhancing PDMR. Although, one cannot rule out such a possibility the observed slow decay of this signal with time is perhaps longer than one would expect for such a process. Alternatively one must consider a tunnelling model within tail states. In fact, the lifetime result rules out the possibility that the effect is occurring during the thermalization time (~ 10^{-12} usec) but the p^{+}-i-n^{+} cell has a built in electric field and so carrier tunnelling to states with lower energy could account for the temperature independence of the photovoltage or photocurrent.

The quenching signal in the ODMR always has a lower g-value than the enhancing signal; it also shows asymmetry associated with the presence of the hole tail state and so we have suggested that it results from recombination via a shunt non-radiative path, in contrast to the Street model where electron capture at a dangling bond competes with the luminescence. In the PDMR measurements, it is clear that the quenching signal has a very different temperature dependence from the ODMR and, in fact, the g-value is greater than the enhancing signal. The quenching signal also has a non-exponential dependence on delay time in the time resolved measurements implying a distribution of lifetimes. In fact, many processes can compete with the photoconductivity. The non radiative luminescence processes are long lived and so may not be a major contributor but the short lived radiative processes are expected to compete to some extent and this view is supported by the similarity of the quenching PDMR and enhancing ODMR g-values. However, since the PDMR temperature dependence is considerably different from the ODMR one cannot rule out the proposal of Street [6] and Dersch et al [7] that dangling bond capture is a very likely competing process.

Finally, it should be noted that for both the enhancing and quenching PDMR signals, the hole resonance produces an asymmetric line shape to low field. One can assume that similar spin dependent hopping processes occur for the holes but one might expect for different exerimental conditions that these separate processes would contribute varying amounts to the resonant lineshape. However, no such variation of the lineshape with temperature, time delay or microwave power was observed.

In conclusion, the results suggest that at low temperatures (<50 K) the temperature independent photoconductivity is due to electric field assisted tunnelling to singly occupied tail states and the principal competing processes are the radiative recombination and the capture of carriers at dangling bonds.

REFERENCES

1. R.A. Gibson, E.E. Spear, P.G. Le Comber and A.J. Snell, J. Non Cryst. Solids 36/36 725 (1980)
2. C. Tsang and R.A. Street, Phys. Rev. B19 3027 (1979)
3. B.C. Cavenett, S.P. Depinna, I.G. Austin and T.M. Searle, Phil. Mag. B48 169 (1983)
4. R.A. Street, Phys. Rev. 26 3588 (1982)
5. K. Morigaki, Jap. J. Applied Physics 22 375 (1983)
6. R.A. Street, Phil. Mag. B46 273 (1982)
7. H. Dersch, L. Schweitzer and J. Stuke, Phys. Rev. B28 4678 (1983)
8. K.P. Homewood, B.C. Cavenett, W.E. Spear and P.G. LeComber, J. Phys. C. 16 L427 (1983)
9. S.P. Depinna, B.C. Cavenett, T.M. Searle and I.G. Austin Sol. State. Comm. 43 79 (1982)
10. M. Hoheisel, R. Casius and W. Fuhs, J. Non Cryst. Solids 6 313 (1984)

SYSTEMATIC TRENDS IN THE ENERGIES OF DANGLING BOND DEFECT STATES IN a-Si ALLOYS CONTAINING C, N AND O

G. Lucovsky and S.Y. Lin
North Carolina State University, Raleigh NC 27695-8202, USA

ABSTRACT

We have used a tight-binding formalism to calculate the energies of Si-atom dangling bond defects within the pseudo-gap of a-Si alloys containing C, N and O. We have determined the energy of the dangling bond state relative to the width of the pseudo-gap as a function of the chemical nature of the atoms that are back-bonded to the Si-atom with the dangling bond. We have considered local geometries in which all bonded neighbors are Si-atoms, or where one or more of these is replaced by an alloy (or impurity) atom such as C, N or O. We find that the relative energy scales with the average electronegativity of these atoms, being near mid-gap for three Si neighbors, and about three-quarters of the way to the conduction band for two Si, and one O neighbor. We have found that this trend is independent of the way the tight-binding parameters are chosen, in particular that the inclusion of the s* state formalism of Dow and his coworkers yields the same trends as other parameterizations used by Chadi and his coworkers.

INTRODUCTION

The dominant defect in a-Si alloys has been identified as a Si-atom dangling bond which can also be described as a tricovalent Si-atom.[1] When this state is singly occupied it gives rise to a characteristic electron spin resonance (ESR) signal. It also plays a role as an intermediate capture state in the defect luminescence band in the vicinity of 0.8 to 0.9 eV.[1,2] It has been reported that the density of electronically active defect centers increases in a-Si:H alloys containing C, N or O impurities at levels of about $10^{19}/cm^3$ or higher.[1] Increased concentrations of dangling bond defects can also be produced in a-Si:H alloys by long term exposure to light that is sufficiently energetic to produce excitations across the forbidden gap.[3,4] The efficiency for the generation of such light induced defects is found to be greater in materials containing relatively high concentrations of impurity atoms, in particular oxygen.

This paper addresses the question of the dangling bond defects in a-Si alloys, and emphasizes the effects of impurity atoms on the energy of Si-atom dangling bond states. In studies of the local bonding in the ternary alloys a-Si:H:O[5,6] and a-Si:H:N[7], it was established that there were spatial correlations between the H and O, and H and N alloy atoms; specifically that they were bonded to common Si-atoms. In an earlier paper, we showed that this spatial correlation did not introduce additional states into the pseudo-gap, and further, that gap states could generally only be generated through dangling bonds.[8] This paper represents a continuation of the theoretical investigation of the properties of gap states in a-Si alloys. We employ a tight-binding approach in order to identify the effects of local chemistry. The approach we have taken reveals the chemical trends in positions of Si-atom dangling bond states, even though the absolute energy scale may have some uncertainty.[8] We have established this using different ways of assigning tight-binding parameters. Specifically two representative approaches to this problem yield

0094-243X/84/1200055-08 $3.00

different energy gaps for the a-Si host material, but give the same trends in the relative energies of the defect states as function of the near-neighbor chemistry.

TIGHT-BINDING METHOD

The structural basis for our calculations is the cluster Bethe Lattice which has been applied successfully for both vibrational and electronic calculations in disordered solids.[9,10] Our approach will be to calculate the electronic density of states (DOS) for a Si Bethe Lattice using two different sets of tight-binding parameters, and then to study the properties of dangling bond defect states as a function of the near-neighbor chemical environment. We will consider impurity atom effects in a dilute limit, basing our calculations on local bonding clusters which include the Si-atom with the dangling bond, and its three nearest neighbors, one of which may be an impurity atom such as C, N or O. We will terminate the clusters with Si Bethe Lattices. We are primarily concerned with defects in hydrogenated alloys, a-Si:H, in which there may be additional impurity atom species. The results of Ref. 8 have shown that near-neighbor H-atoms do not have a significant effect on the the positions of the dangling bond states, and in particular do not change any of the trends that result from changes in the local chemistry. Figure 1 gives a schematic representation the local bonding configurations that we consider.

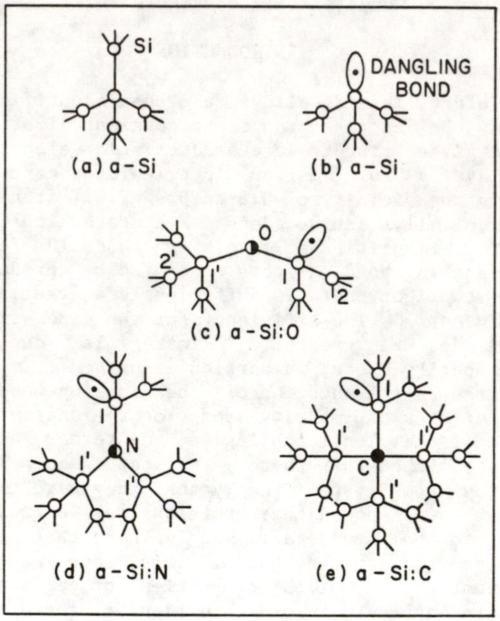

Fig.1 Schematic Representation of Local Bonding Environemnts: (a) Si Bethe Lattice; (b)-(e) Si-atom Dangling Bonds in a-Si, a-Si:O, a-Si:N and a-Si:C.

We employ two different tight-binding Hamiltonians; one using the conventional sp^3 representation[11,12], and a second using an additional non-bonding s-state in the basis set, sp^3s^*.[13] When applied to crystalline Si (and other group IV semiconductors), the first approach gives an excellent representation of the valence bands, but overestimates the band gap and gives a poor representation of the conduction bands. The overestimate stems from the fact that it yields direct, rather than indirect band gaps for the elemental group IV materials, C, Si and Ge.[13] The inclusion of the s* state remedies these defects. The calculations that follow represent the first use of the s* formalism for disordered solids.

The tight-binding formalism requires parameterization that includes self-energies for the constituent atoms and matrix elements for the interactions between the various orbitals of different atoms. The self-energies represent interactions between orbitals on the same atom and can be expressed in the form of a diagonal matrix, E(b). For a basis set including s and p functions the matrix is given by:

$$E(b) = \begin{vmatrix} Ep & 0 & 0 & 0 \\ 0 & Ep & 0 & 0 \\ 0 & 0 & Ep & 0 \\ 0 & 0 & 0 & Es \end{vmatrix} \qquad (1)$$

The differences between the self-energies of the p and s states of Si and the impurity atoms are proportional to the differences in their orbital energies, where the appropriate factors are 0.8 for s-orbitals and 0.6 for p-orbitals. The interaction Hamiltonian H(int) is given by the following matrix:

$$H(int) = \begin{vmatrix} V(x,x) & V(x,y) & V(x,z) & V(x,s) \\ V(y,x) & V(y,y) & V(y,z) & V(y,s) \\ V(z,x) & V(z,y) & V(z,z) & V(z,s) \\ V(s,x) & V(s,y) & V(s,z) & V(s,s) \end{vmatrix} \qquad (2)$$

where x,y and z refer to the p-orbitals. These matrix elements show chemical bonding trends that are reflected in a d^{-2} dependence, where d is the bond-length. Introducing the s* state increases this matrix from 4 X 4 to 5 X 5, and also introduces an additional diagonal element in the self-energy matrix. The self-energy Es* is approximated by scaling with respect to states of the free atom, in particular by adjusting its value relative to the p-state self-energy. The terms in the Hamiltonian matrix have been estimated from pseudo-potential calculations, where available[12], or via scaling using empirically determined rules.

TABLE I TIGHT-BINDING PARAMETERS--sp^3 MODEL

	SELF-ENERGIES(eV) Si,C,N,O		Si		INTERACTION PARAMETERS(eV)				
	Es	Ep	Es'	Ep'	Vs,s	Vx,x	Vx,y	Vs,p'	Vs',p
Si,Si	-5.05	1.20	-5.05	1.20	-1.94	0.30	1.38	1.01	1.01
Si,C	-7.60	0.97	-4.00	3.20	-3.10	0.76	1.73	2.37	2.30
Si,N	-14.2	-5.17	-2.18	4.92	-2.47	0.75	1.96	4.75	2.70
Si,O	-19.6	-6.80	-0.55	6.55	-2.85	0.87	2.27	5.48	3.12

TABLE II TIGHT-BINDING PARAMETERS--sp^3s^* MODEL

A. SELF-ENERGIES(eV)

	Si,C,N,O			Si		
	Es	Ep	Es*	Es'	Ep'	Es*'
Si,Si	-4.20	1.72	6.69	-4.20	1.72	6.69
Si,C	-7.60	0.97	8.50	-4.00	3.30	8.17
Si,N	-14.2	-5.17	7.72	-2.18	4.92	9.60
Si,O	-19.6	-6.80	6.40	-0.55	6.55	11.2

B. INTERACTION PARAMETERS(eV)

	Vs,s	Vx,x	Vx,y	Vs,p'	Vs',p	Vs*,p'	Vs*',p
Si,Si	-8.30	1.72	4.58	5.73	5.73	5.37	5.37
Si,C	-3.10	0.76	1.73	2.37	2.30	2.18	1.10
Si,N	-2.47	0.75	1.96	4.75	2.70	2.60	1.31
Si,O	-2.85	0.87	2.27	5.48	3.12	3.00	1.52

Consider first the sp^3 Hamiltonian. The relevant parameters for the configurations shown in Fig. 1 are obtained from calculations for c-Si, and for defect configurations containing C, N or O impurities from SiC, SiO_2, and Si_3N_4, respectively. The parameters for Si and O (both self-energies and interaction terms) are obtained from the calculations of Laughlin et al.[II] who employed a tight-binding/CBL approach to study defects at the c-Si/SiO_2 interface. The self-energies for N and Si, and the matirx elements are between Si and N are scaled from those of Si and O using the empirically determined rules. The parameters relative to C are obtained from Ref. 13, with the exception of the self-energy terms, which have been adjusted to maintain values that are proportional to the free atom values. The expansion of the basis set to sp^3s^* uses parameters for Si and C from Ref. 13, with values of self-energies and interaction matrix elements being scaled in an appropriate manner.

CALCULATION OF DEFECT STATE ENERGIES

The Green's function method, and transfer matrix or effective field techniques are employed, and the details of the mathematics are described elsewhere.[I,II,12] We will present DOS functions for the Si atom with the dangling bond defect, and its immediate neighbors within the defect cluster. Space does not permit presenting the p and s projections for all of the DOS functions. We first present results for the sp^3 Hamiltonian, and then for the calculation that includes the s* states in the expanded basis set.

Figure 2 gives the total DOS as calculated for the Si Bethe Latttice using the sp^3 representation. The figure also includes the projections on the s and p orbitals. The s-states contribute most strongly to the lower portion of the valence band, and to the peak at the bottom of the conduction band, while the p-states contribute to the dominant feature at the top of the valence band and the dominant conduction band states. This particular parameterization overestimates the width of the pseudo-gap giving a value of 2.8 eV as compared to the measured values of about 1.5-1.8 eV. Figure 3 gives the local density of state (LDOS) functions for the Bethe

Lattice configuration that includes a dangling bond defect (see Fig. 1(a) and the insert in Fig. 3). The dangling bond state is the relatively sharp feature just above mid-gap. The projections on the s and p orbitals establish that the defect state has the expected sp^3 character. The LDOS functions for successive shells of near-neighbors establish the local character of this defect state.

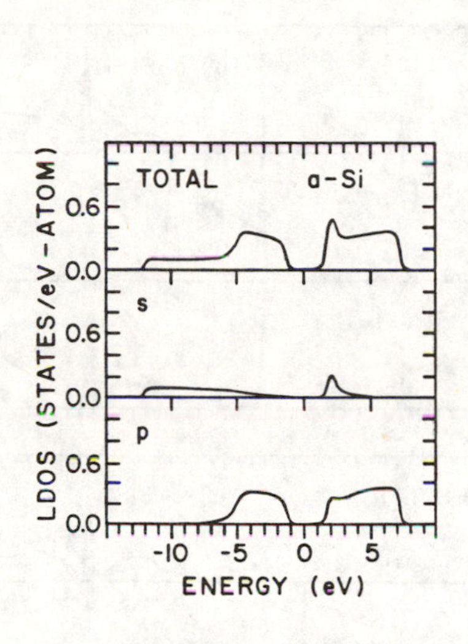

Fig.2 DOS Functions: Si Bethe Lattice

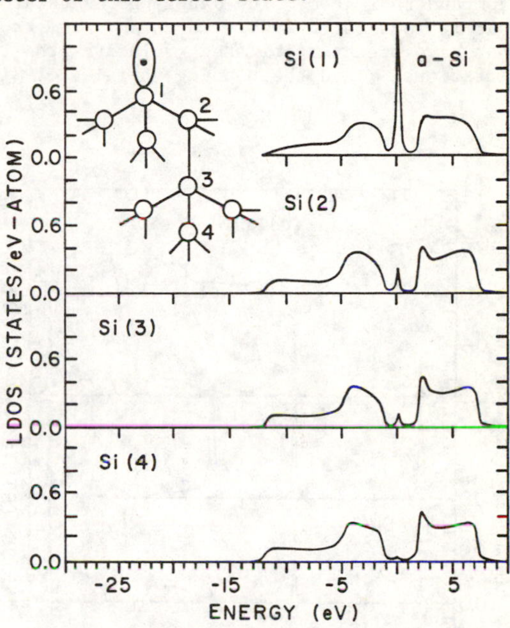

Fig.3 DOS Functions: Dangling Bond

Figure 4 gives the LDOS functions for the defect cluster which includes an O impurity (see Fig. 1(b) for the bonding geometry). The energy gap is also calculated to be 2.8 eV. The dangling bond state is the sharp feature just above mid-gap for the atom designated as Si(1). Note that as in the case of the Si Bethe Lattice, the defect is localized. The relatively sharp features for the LDOS functions for the O-atom are, in order of increasing energy, associated with bonding s-states, and bonding and non-bonding p-states. The bonding character of the O states is evidenced by the contributions to the LDOS functions of the two neighboring Si-atoms, Si(1) and Si'(1). Figure 5 gives the LDOS functions for three atoms in each of the defect clusters for N and C, Figs. 5(a) and 5(b), respectively. The N 2s and 2p states have bonding and non-bonding character that is essentially the same as the corresponding states of the O-atoms, while the 2s and 2p states of C are both involved in bonding as expected by the four fold-coordination of that atom. As in Figs. 3 and 4, the dangling bond states are the relatively sharp features near mid-gap.

Figure 6 summarizes the trends in the relative position of the dangling bond defect state as a function of the local bonding chemistry. We have plotted the relative energy of the dangling bond state, defined here as $(E_D - E_{VB})/E_G$, where E_D is the energy of the defect state, E_{VB} is the energy of the top of the valence band and E_G is the width of the pseudo-gap. We have plotted this relative energy as a function of the Pauling electronegativity of the impurity atom (this plot is completely equivalent to using the sum of the electronegativities of all three atoms that are back-bonded to the Si-atom with the dangling bond). Note that a linear relationship prevails. This derives from the fact that increasing the electronegativity of one the atoms strenthens the bond between the tricovalent Si-atom and that atom. The dangling bond state is essentially an anti-bonding state so that it is shifted upward in energy. The trend shown here parallels two other trends that have been established relative to the frequencies of Si-H bond-stretching vibrations[14], and the binding energies of Si 2p core states.[15]

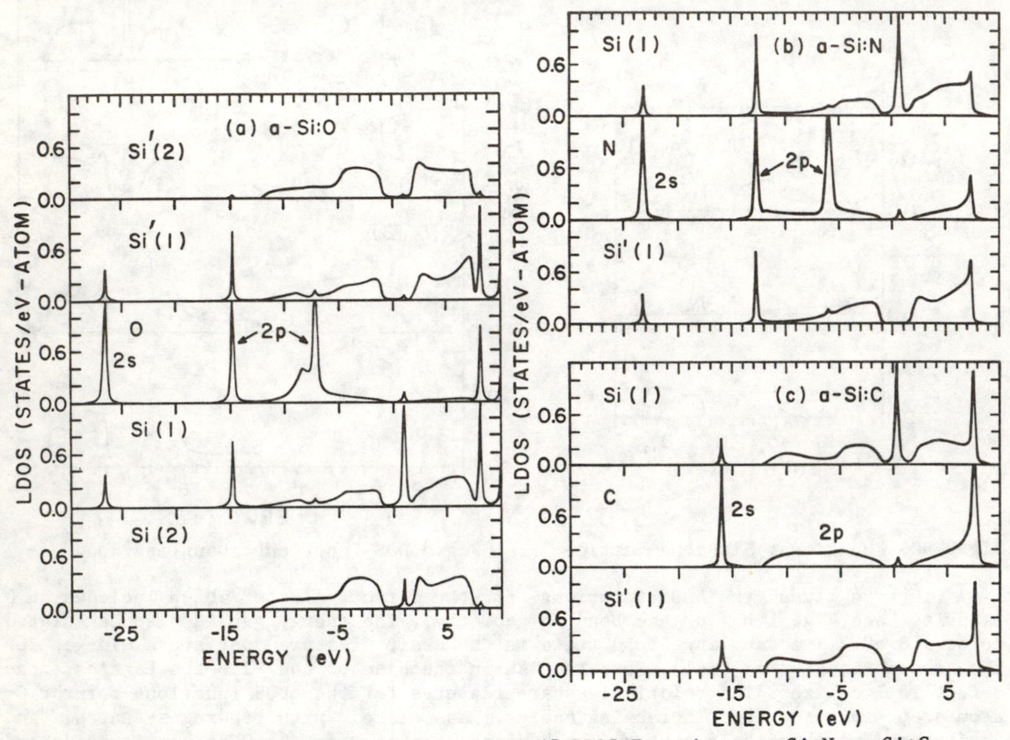

Fig.4 LDOS Functions: a-Si:O.

Fig.5 LDOS Functions: a-Si:N, a-Si:C.

Figure 7 presents the results of the calculations employing the s* states for the Si Bethe Lattice and the dangling bond defect. There are two important things to note: (1) the use of the same parameters that give an excellent representation of the valence and conduction bands of c-Si gives an energy gap for the Si Bethe Lattice of 1.8 eV that is very close to the observed value, and a valence band DOS that replicates all of the features observed via photoelectron spectroscopy[16]; and (2) a localized state for the dangling bond defect that is above the mid-gap energy. We have calculted the properties of the dangling bond defect state for the bonding configurations shown in Fig. 1 with the inclusion of the s* state parameters. We find: (1) the width of the pseudo-gap is maintained at 1.8 eV; and (2) that the relative energy of the dangling bond state shows the same trends with local chemistry that are displayed in Fig. 6. The calculation using the s* state formalism gives appoximately the same energy difference for $E_D - E_{VB}$ as the calculations using the sp^3 Hamiltonian, but since the s* state calculations yield a lower gap energy, they give a higher relative dangling bond energy.

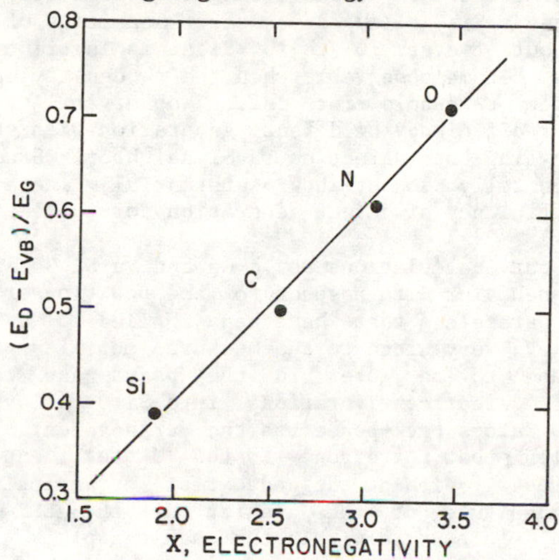

Fig.6 Chemical Trends in Dangling Bond State Energies.

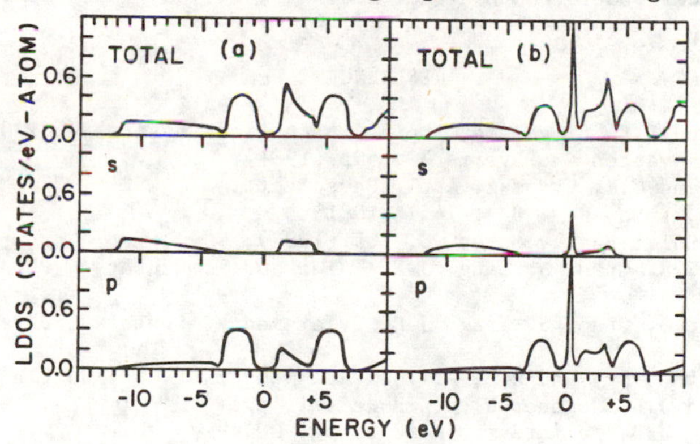

Fig.7 LDOS Functions sp^3s* Basis: Si Bethe Lattice; Dangling Bond Defect.

DISCUSSION

There have been measurements the defect luminesence bands in a-Si alloys with and without O doping.[1,2] Analysis of these spectra in terms of a band structure model yields $E_D - E_{VB}$ values of 0.8 eV for the Si dangling bond state in non-oxygen containing a-Si, and 1.1 eV in a-Si with O. These values are in accord with the trends found in this study, assuming that the defect luminescence in a-Si:O alloys proceeds through Si dangling bond states which have an O-atom neighbor. An increase in trapping and recombination states is observed in a-Si:H alloys for O concentrations in excess of about $10^{19}/cm^3$. This is consistent with the calculations given above if we assume that O "induces" dangling bond defects on Si-atoms. The best semiconductor grade a-Si:H alloys have defect densities of the order of 10^{15} to $10^{16}/cm^3$, so that one out of every 10^6 to 10^7 Si atoms is tricovalent. If we assume that O induced defects become observable when their density approaches that of the a-Si:H host, then the threshold for defect observability of about $10^{19}/cm^3$ corresponds to a higher efficiency of defect generation with about one in every 10^4 O-atoms inducing a dangling bond defect on a Si neighbor. Studies of a-Si:H alloys with added C and N generally do not show substantially increased defect densities indicating that the efficiency of defect generation for these atomic species is less than that of O.

We have extended our calculations to a-Ge and a-Si,Ge alloys and have found qualitatively similar behavior with respect to the position of dangling bond defect states. Using the s* parameters that have been applied to c-Ge gives a pseudo-gap for a-Ge of about 1.3 eV, very close to the measured mobility gap. The dangling bond defect in a-Ge is found to be lower in the pseudo-gap than in a-Si. We have calculated energies of defect configurations in a-Si:O alloys in which there are different numbers of O atoms back-bonded to the tricovalent Si-atom. We find four distinct defect energies that correspond to the states identified by the Marburg group in their ESR study.[17] Within the context of our calculations, an important contribution to the magnitude of the g-shift of the ESR lines comes from the relative energy of the defect state within the pseudo-gap.

Supported under under ONR Contract N00014-79-C-0133, and SERI Contract XB-2-02065-1.

REFERENCES

1. R.A. Street and D.K. Biegelsen, Topics in Applied Physics 56, 195 (1984).
2. R.A. Street, Advances in Physics 30, 593 (1981).
3. D.L. Stabler and C.R. Wronski, Appl. Phys. Lett. 31, 292 (1977).
4. R.S. Crandall, Phys. Rev. B24, 7457 (1981).
5. G. Lucovsky, Solar Energy Mater. 8, 165 (1982).
6. G. Lucovsky, S.S. Chao, J. Yang, J.E. Tyler and W. Czubatyj, Phys. Rev. B28, 3225 (1983).
7. G. Lucovsky, S.S. Chao, J. Yang, J.E. Tyler and W. Czubatyj, Phys. Rev. B28, 3234 (1983).
8. S.Y. Lin, G. Lucovsky and W.B. Pollard, J. Non-Cryst. Solids (in press).
9. D.C. Allan and J.D. Joannopoulos, in Ref. 1, pg. 5.
10. G.Lucovsky and W.B. Pollard, in Ref. 1, pg. 301.
11. R.B. Laughlin, J.D. Joannopoulos and D.J. Chadi, Phys. Rev. B21, 5733 (1980).
12. D.J. Chadi and M.L. Cohen, Phys. Stat. Solidi B6, 405 (1975).
13. P. Vogl, P.H. Hjamarson and J.D. Dow, J. Phys. Chem. Solids 44, 365 (1983).
14. G. Lucovsky, Solid State Commun. 29, 571 (1979).
15. G. Lucovsky, J. Phys. (Paris) C4 42, 741 (1981).
16. L. Ley. in Ref. 1, pg. 61.
17. E. Holzenkampfer, F.W. Richter, J. Stuke and U. Voget-Grote, J. Non-Cryst. Solids 32, 327 (1979).

UNIFIED THEORY OF BONDING AT DEFECTS AND DOPANTS IN AMORPHOUS SEMICONDUCTORS

J. Robertson
Central Electricity Research Laboratories, Leatherhead, Surrey, UK

ABSTRACT

A framework to describe the bonding at defect and impurity states is developed, which applies equally to defects of amorphous semiconductors of group four to six elements, and also accounts for the novel doping mechanism in a-Si:H.

The gap states due to defects control may properties of amorphous semiconductors. The properties of amorphous (a-) chalcogenides such as a-As_2Se_3 are frequently contrasted with those of hydrogenated amorphous silicon (a-Si:H) and attributed to differences in the effective correlation energy (U) of the defects, $U < 0$ for chalcogenides and $U > 0$ for Si[1,2]. The bonding of normal sites in amorphous semiconductors is now well understood[3] in that the coordination of each site generally follows the "8-N rule". The substitutional doping of a-Si:H breaks this rule and has only recently been explained in terms of doping by charged dopant-defect site pairs[4,5]. We show here that it is possible to describe the bonding at defects and dopants within a common framework which for the first time applies equally to both chalcogenides and Si, negative and positive U systems, and which also gives insight into metastable processes. We concluded by discussing the least understood defects and metastable process - the Staebler-Wronski effect.

Bulk sites in an amorphous semiconductor satisfy their valence requirements locally in the absence of constraints due to periodicity: electrons are paired and coordinations N_c obey the 8-N rule, viz $N_c = 8-N$ for $N > 4$ and $N_c = N$ for $N < 4$. Defects can be defined as sites with different coordinations. The defects in amorphous chalcogenides are now well understood. They are valence alternation pairs (VAPs) consisting for a-Se of charged Se_3^+ and Se_1^- sites (where the subscript indicates coordination) formed from the ground state by valence alteration:

$$2Se_2^{\bullet} \rightleftharpoons Se_3^+ + Se_1^- \tag{1}$$

A defect has three charge states, empty (D^+), singly occupied (D°) and full and spin-paired (D^-). The additional energy to fully occupy D^- is U, which includes both the direct electronic repulsion and a term due to lattice relaxation in (1). If electron repulsion dominates, then $U > 0$ and D^+, D° and D^- occur sequentially with rising E_f. If lattice relaxation dominates $U < 0$, and

$$2D^\circ \rightarrow D^+ + D^- \tag{2}$$

is exothermic. For Se this is

$$2Se_1^{\bullet} \rightarrow Se_3^{+} + Se_1^{-} \quad (3)$$

and is exothermic because of the extra dative bond formed by Se_1^{+}:

$$Se_1^{+} + Se_2^{\bullet} \rightarrow Se_2^{\bullet} + Se_3^{+} \quad (4)$$

For $U < 0$, D° is unstable and only created by excitation. The principal conclusions are
(1) negative U centers explain most properties of gap sites in chalcogenides (except perhaps aspects of impurities[6]),
(2) the number of bonds in a VAP is the same as in the ground state, so they have a low creation energy[2],
(3) non-bonded states are needed for overcoordination of D^{+}; pnictides like As use their s^2 electrons for this purpose but Si has none so a negative U is not possible in a-Si:H by this mechanism[2]. This is confirmed by experiment, the major observed defect in a-Si:H is the simple dangling bond Si_3 for which[7-9] $U \sim 0.3$ eV.

Substitutional doping is observed[10] in a-Si:H. For phosphorus $P_4^{\bullet}$ sites are paramagnetic and clearly break the 8-N rule. Doping of a-Si:H differs from that of crystalline (c-) Si in a number of ways:
(1) electrically inactive three-fold coordinated sites, $P_3^{\bullet}$, coexist with substitutional sites,
(2) the conductivity increase saturates and E_f does not enter extended states at high doping levels, and
(3) the density of dangling bonds increases with doping level[8,11,12] and depends on the position of E_f as it decreases again with compensation.
A theory must also account for
(4) the absence of any ESR signal associated with deep or shallow dopant sites, and should recognise that
(5) while the presence of hydrogen is advantageous in reducing the density of gap states it is not essential[13] nor central to the doping mechanism and the breaking of the 8-N rule.
The explanation is that substitutional sites are always ionised, P_4^{+} not P_4°, and the carriers are trapped at dangling bonds which must then be present in equal numbers[4]. The doping species is a charged pair P_4^{+}-Si_3^{-}, a dopant VAP or DVAP, rather than a single atom[4,5,14], and is diamagnetic:

$$P_3^{\bullet} + Si_4^{\circ} \rightleftharpoons P_4^{+} + Si_3^{-} \quad (5)$$

The valence alternation character accounts for its greater stability compared to P_4°. Clearly, the 8-N site $P_3^{\bullet}$ is most stable in a network, followed by the DVAP, while the P_4° site is the least stable (fig. 1). P_3° is the ground state and the DVAP the first excited state, but both are topologically forbidden in c-Si, which P must then enter as $P_4^{\bullet}$.

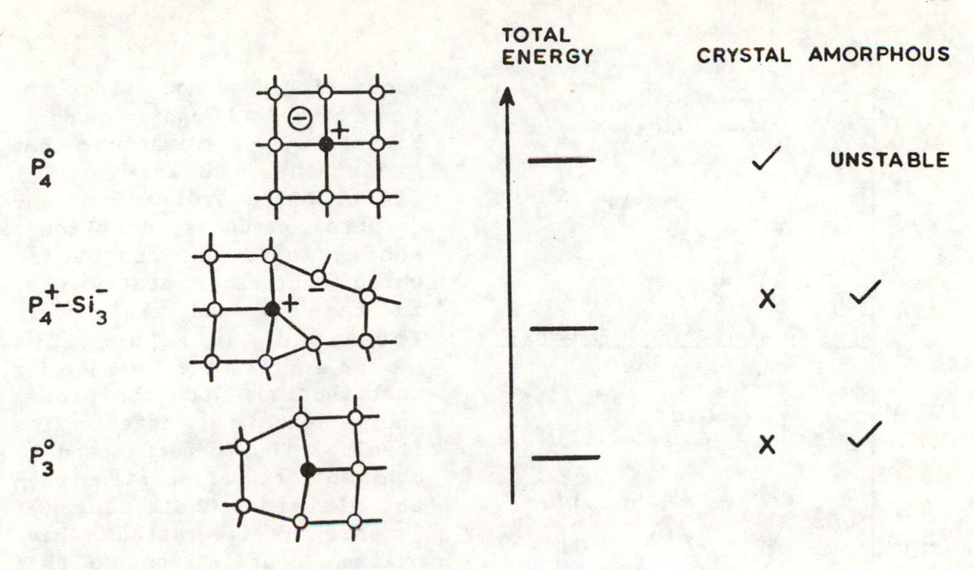

Fig. 1. Configurations of phosphorus in a-Si:H and c-Si.

DVAPs produce a more complicated form of doping because of their double site nature and because they have $U < 0$. The latter is seen by considering reaction (2) for Si:

$$2Si_3^\circ + P_3^\circ \rightarrow P_4^+ + Si_3^- + Si_4^\circ \qquad (6)$$

$U < 0$ because while in undoped a-Si:H Si_3^+ cannot overcoordinate, in doped a-Si:H it can:

$$Si_3^+ + P_3^\circ \rightarrow P_4^+ + Si_4^\circ \qquad (7)$$

Thus, $U > 0$ in intrinsic a-Si:H but $U < 0$ for dopants in doped a-Si:H. The P_4 levels lie just below E_c (fig. 2) and the Si_3 levels around midgap[4]. The requirement that P_4° does not form is equivalent to E_f always lying below the P_4 levels which in turn is equivalent to the familiar pinning of E_f between the D^+ (P_4) and D^- (Si_3) states of the negative U centers - i.e. the negative U is a general description of the saturation of doping efficiency of a-Si:H at high doping levels. Note, however, that the valence alternation occurs only during deposition in a-Si:H rather than at all times as in chalcogenides.

The defects of amorphous semiconductors and the novel doping mechanism of a-Si:H can now be rationalised in a single framework by concentrating on the wavefunctions of gap states. We recall that the basis of the 8-N rule is that bonding energy is maximised by occupying only states of bonding character while keeping empty those of antibonding character. Sites generally also possess non-bonding states which do not change energy during bonding and so their

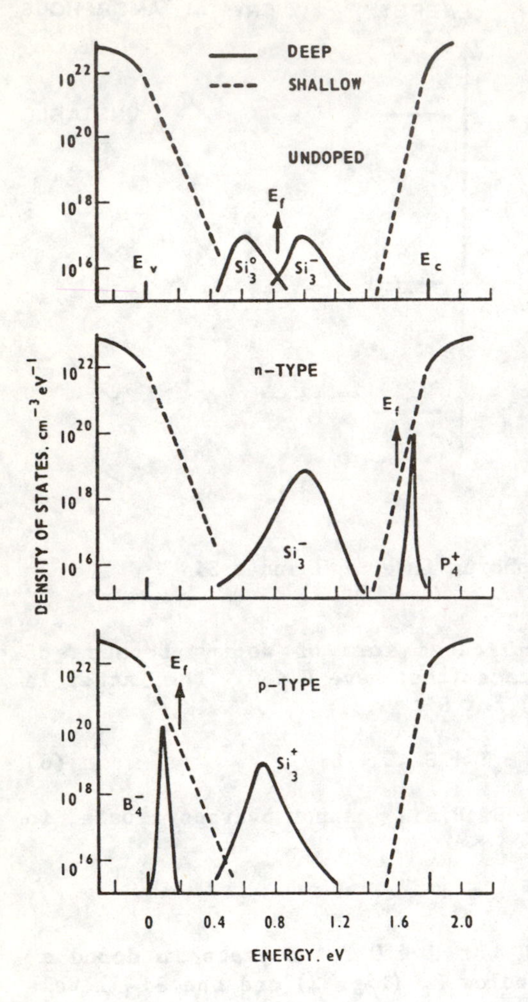

Fig. 2. Gap states in undoped and doped a-Si:H according to the DVAP model.

occupation does not affect the total bonding energy. Maximising coordination then gives the 8-N rule. The different coordinations of defects produces a strong, short-ranged perturbation which introduces states deep in the gap and their very localized wavefunctions concentrate their contributions to the total energy onto the defect site itself. Thus defects tend to obey an "occupation principle" to minimise their energy: "stable configurations have maximum coordination so that any deep gap states of bonding character are full, any of antibonding character are empty while non-bonding states can have any occupation". We see that deep states are usually diamagnetic and unpaired electrons are only permitted in non-bonding states, i.e. dangling bonds or lone pairs. Thus networks are predominantly diamagnetic as proposed by Anderson[15], with paramagnetic states appearing only in unusual circumstances such as by excitation or sometimes intrinsically[16] to relieve strain as in a-Si:H.

The principle can also be expressed as a "modified 8-N rule" where N is now the actual number of electrons at the site[4].

In chalcogens, for Se (Fig. 3), Se_1^- possesses a full non-bonding (π) gap state, while D^+ must be Se_3^+ to maximise coordination and thus possess an empty σ^* gap state. Adding electrons to Se_3^+ would cause occupation of the σ^* state. Instead the bond breaks, giving Se_1°, converting the σ^* state to π which is allowed to be partially occupied. Thus the principle requires D° to be Se_1° rather than Se_3° as found by ESR[17] and total energy calculations[18]. Similarly for pnictides, P_4^+ has an empty σ^* state, P_2^- a full π state while D° is P_2° rather than P_4° as the partially occupied state is now π-like. In a-Si:H the dangling bonds are non-bonding and so can have any occupation. The origin of the stability of DVAPs like P_4^+-Si_3^- compared to neutral substitutional sites P_4° is now much clearer. P_4° sites involve partial occupancy of σ*-like states, while in DVAPs these are empty. The only occupied states are σ and non-bonding states. Similarly for boron, B_4° has a partially occupied σ-like state while the B_4^--Si_3^+ DVAP has full σ states and an empty non-bonding state (Fig. 4). Thus, bonding at defects and impurities in amorphous semiconductors of groups IV, V and VI are for the first time expressed in a single conceptual framework.

We suggest that this framework may provide a basis to explain photostructural changes and related metastable effects in amorphous chalcogenides[19]. Many mechanisms have been suggested to account for the changes in structure factor at low q - defects, bond angle changes, and two-level systems[19]. However, these mechanisms do not readily account for the changes at high q, seen recently, indicating changes in the first neighbour environment[20]. We see that if gap states of σ or σ*-like character exist, then sub-gap illumination will excite electrons out of or into these states. As partially occupied σ and σ* states are forbidden according to our occupation principle, the associated bond will break, thus giving a very general mechanism for photostructural changes.

The reversible photo-induced decrease in dark- and photoconductivity, the Staebler-Wronski effect[21], is the most important defect related phenomenon in a-Si:H with has as yet no completely satisfactory explanation. An associated increase in ESR dangling bond signal has been seen [22,23] and a related increase in trapping and decrease in μτ product and conductivity was seen, suggesting that dangling bonds are created and pin E_f nearer midgap[23]. The precursor states have a very small cross-section for conversion, as persistent illumination is needed[24]. The photocreated dangling bonds may differ slightly from the intrinsic centers as two different annealing responses are observed in photoconductivity[25]. An increase in density of other occupied deep states has been detected by DLTS in n-type samples[26,27] and may be related to the unexplained new gap states found in compensated films[11,28]. A mechanism involving just the breaking of weak Si-Si bonds is inconsistent with ESR as the resulting centers are concluded to be widely separated[22]. The present theory and that of Street[4] neatly accounts for the effect in doped films - the photocreation of P_4° or B_4° sites causes the dopant-Si bond to break, creating single Si_3° centers - but these theories need additional centers to account for the effect in intrinsic films, unless the effect is totally impurity-related[29].

The lower spin density than total gap state density suggests to Adler[14,24] the presence in a-Si:H of additional spinless and negative U defects. There are several attractive features to this proposal, but there are difficulties once specific configurations and their gap states are tried, such as nearest neighbour Si_3^+-Si_3^- sites or Si_2° sites. The stable configuration of adjacent Si_3 sites is contentious; early calculations[30] favored charge separation while surface studies[31] with better methods disagree and favour slight π bonding between neutral cis dangling bonds. Interestingly, the

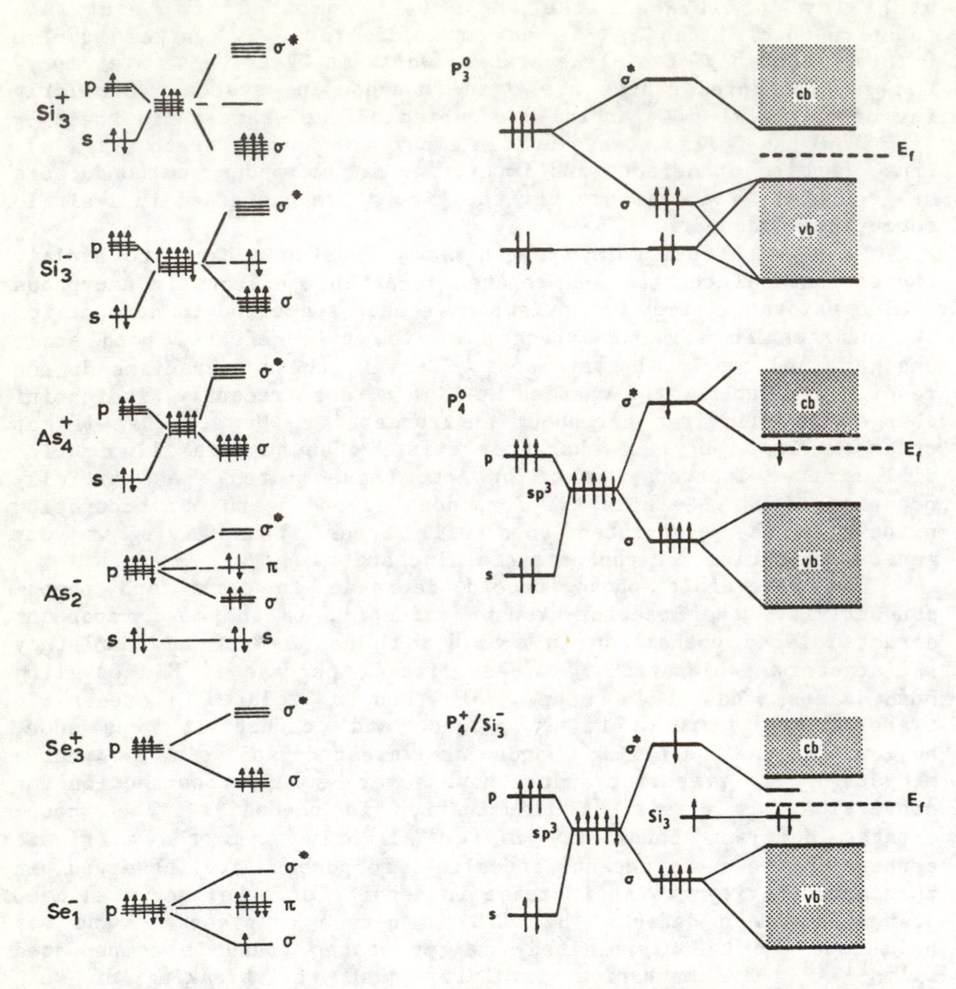

Fig. 3. Occupancies for diamagnetic defects in Se, As and Si.

Fig. 4. Orbitals at the three configurations of P in Si.

molecular analogue[32] Si_2H_4 is stable in the trans $SiH_2.SiH_2$ configuration with trans not cis dangling bonds, and both configurations have similar energies to the Si_2° analogue $SiH.SiH_3$. These "cluster" calculations should be representative of a-Si:H because they compare stabilities of neutral molecules, whereas comparisons of different charge states using clusters[33] greatly overestimate coulombic effects in the solid state[34]. Thus, the Staebler-Wronski effect could involve charge-induced rearrangement and valence alternation between these configurations. Further studies on the electronic structure of these defects and on impurity-related centers is continuing.

REFERENCES

1. R.A. Street, N.F. Mott, Phys. Rev. Lett, 35 1293 (1975).
2. M. Kastner, D. Adler, H. Fritzsche, Phys. Rev. Lett, 37 1504 (1976).
3. J. Robertson, Adv. Phys. 32 361 (1983).
4. R.A. Street, Phys. Rev. Let, 49 1187 (1982).
5. J. Robertson, J. Phys. C. 17 L 349 (1984).
6. G. Pfister, P.C. Taylor, J. Non-Cryst Solids 35 793 (1980).
7. R.A. Street, D.K. Biegelsen, Solid State Commun. 33 1159 (1980).
8. H. Dersch, J. Stuke, J. Biechler, Phys Stak Solidi b, 105 265 (1981).
9. J.D. Cohen, J.P. Harbison, K.W. Wecht, Phys. Rev. Lett, 48 109 (1982).
10. W.E. Spear, Adv. Phys., 26 811 (1977).
11. R.A. Street, D.K. Biegelsen, J.C. Knights, Phys. Rev. B, 24 969 (1981).
12. R.A. Street, J. Zesch, M.J. Thompson, App. Phys. Let, 43 672 (1983).
13. E.G. Harbeke, et al., App. Phys. Lett, 42 249 (1983)
14. D. Adler, Phys. Rev. Let, 41 1755 (1980).
15. P.W. Anderson, Phys. Rev. Let., 34 953 (1976).
16. J.C. Phillips, Phys. Rev. Let., 42 1151 (1979).
17. S.G. Bishop, P.C. Taylor, U. Strom, Phys. Rev. B., 15 2278 (1977).
18. D. Vanderbilt, J.D. Joannopoulos, Solid State Commun., 35 535 (1980).
19. K. Tanaka, A. Odajima, J. Non-Cryst Solids, 46 259 (1981).
20. S.R. Elliott, J. Non-Cryst Solids, 59 899 (1983).
21. D.L. Staebler, C.R. Wronski, App. Phys. Lett. 31 292 (1977).
22. H. Dersch, J. Stuke, J. Biechler, App. Phys. Let, 38 456 (1981).
23. R.A. Street, App. Phys. Lett, 42 507 (1983); N.M. Amer, A. Skumanich, W.B. Jackson, J. Non-Cryst Solids, 59 409 (1983).
24. D. Adler, Solar Cells, 9 133 (1983).
25. D. Han, H. Fritzsche, J. Non-Cryst Solids, 59 397 (1983).
26. J.D. Cohen, D.V. Lang, J.P. Harbison, A.M. Sergent, Solar Cells, 9 119 (1983).
27. H. Okushi, et al., J. Non-Cryst Solids, 59 393 (1983).
28. P. Cullen, J.P. Harbison, D.V. Lang, D. Adler, Solid State Commun., 50 991 (1984).
29. R.S. Crandall, D.E. Carlson, A. Catalano, H.A. Weakliem, App. Phys. Let, 44 200 (1984).
30. D.C. Allan, J.D. Joannopoulos, Phys. Rev. Lett, 44 43 (1980).
31. K. Pandey, Phys. Rev. Lett, 49 223 (1982).
32. H. Lischka, H. Kohler, Chem. Phys. Lett, 85 467 (1982); L.C. Synder, Z.R. Wasserman, J. Am. Chem. Soc., 107 5222 (1979); T. Fjeldberg et al., J. Chem. Soc., Chem. Commun., 1407 (1982).
33. G.T. Surratt, W.A. Goddard, Solid State Commun., 22 413 (1977).
34. M. Lannoo, J. Phys. C, 17 3137 (1984).

OPTICAL DETERMINATION OF THE EFFECTIVE CORRELATION ENERGY OF THE DANGLING BOND IN HYDROGENATED AMORPHOUS SILICON

DAVID ADLER
Massachusetts Institute of Technology Cambridge, MA 02139

ABSTRACT

It is generally agreed that the dangling bond, T_3, is the major defect in a-Si:H films; however, the sign and magnitude of the effective correlation energy, U_{eff}, for T_3 centers remains a subject of some controversy. Several recent measurements have purported to be direct measurements of U_{eff}, but they are non-self-consistent. A correct interpretation of optical data involving states for which electronic correlations are significant (i.e., $U_{eff} \neq 0$) requires taking into consideration the necessary shifts in energy with changes in occupancy. Furthermore, since the difference between U and U_{eff} is entirely due to atomic relaxation effects, careful consideration of the Franck-Condon principle is required. In this paper, it is pointed out that unambiguous optical determination of U_{eff} is possible, in principle. If the absorption spectra for both n-type and p-type samples can provide the values of the energy required for the transition between the defect level and the appropriate band edge, E_n and E_p, respectively, then $U_{eff}=E_{opt}-E_n-E_p$, where E_{opt} is the optical gap. Indications to date are that the sign of U_{eff} is negative for the T_3 defect in a-Si:H.

INTRODUCTION

Silicon is a tathogen atom, from Column IV in the Periodic Table; its lowest-energy chemical configuration is one in which it forms tetrahedrally coordinated bonds with four different neighboring atoms. Since this is the maximum number of bonds that can be formed without d-electron hybridization, valence alternation is impossible and it is likely that no simple low-energy defects characterize amorphous silicon-based alloys such as hydrogenated amorphous silicon (a-Si:H). However, these alloys are generally quite overconstrained and both large-scale structural defects such as cracks and microvoids and local flaws such as distorted bonds and defect centers are introduced during processing. Most attention has been focused on the three-fold-coordinated Si atom, or dangling bond (T_3 in the the conventional notation[1]), since in its neutral state (T_3^o) it can be detected even in relatively small concentrations by ESR experiments. The most common ESR signal in undoped a-Si:H, with a g-value of 2.0055, is believed to be the signature of the T_3^o defect, and typically appears with concentrations in the 10^{15}-10^{20} cm^{-3} range.

The only other hypothesis with regard to atomic-scale flaws that is generally agreed upon is the existence of strained bonds, usually believed to take the form of distorted bond angles. If sufficiently distorted, these should lead to a distribution of localized band-tail states, and there is evidence for ~10^{21} cm^{-3} of these in a-Si:H.

On a larger scale, in addition to the high probability of microvoids in typical films, there is also strong evidence for the existence of heterogeneities, which take the form of alternating hydrogen-rich and hydrogen-poor regions.[2]

The bulk of the electronic states in a-Si:H fall into a filled valence band and an empty conduction band that are separated by about 1.8 eV. However, the electrical transport properties of the material are controlled by localized states near the Fermi energy, E_F, which arise primarily from defect centers. The nature and energy distribution of these states has been the subject of much controversy over the last few years. It is the purpose of this paper to try to elucidate this problem and summarize the experimental situation to date. In particular, special attention is paid to a simple optical technique for measuring the effective correlation energy of the predominant defect in any material that can be substitutionally doped both p-type and n-type.

LOCALIZED DEFECT STATES

The conventional band theory of solids is based upon two major assumptions, the validity of the adiabatic and the one-electron approximations; the former neglects electron-phonon interactions, the latter neglects electronic correlations. It is an unfortunate fact that neither assumption is applicable to the localized electronic states arising from defect centers. The breakdown of either approximation is sufficient to invalidate the existence of a well-defined density of states function, g(E), for the solid. If this is ignored, basic experimental results can be misinterpreted completely and incorrect conclusions drawn about the nature of the material.

When the concentration of defects is sufficiently small that the localized wave functions do not overlap significantly, a technique exists for maintaining the concept of a density of states.[3] However, the effective g(E) necessarily changes with variations in the occupancy, n(E). Consequently, g(E) is a function of electron concentration, of temperature, and after a non-equilbrium disturbance even of time.

The important parameter in understanding the effects of electronic correlations is the intraionic correlation energy, U, defined as:

$$U = \iint d\underline{r}\, d\underline{r}'\ \psi^* (\underline{r})\ \psi^* (\underline{r}')\ \frac{e^2}{4\pi\varepsilon\,|\underline{r}-\underline{r}'|}\ \psi (\underline{r})\ \psi (\underline{r}') \qquad (1)$$

where $\psi(\underline{r})$ is the wave function for an electron localized near the defect center. U is a positive definite parameter whose upper limit is given by the difference between the ionization potential and the electron affinity of the isolated defect center. In the solid, of course, electronic screening considerably reduces the magnitude of U below this upper limit. The net effect of U is to increase the total energy when two electrons are simulataneously present near the same center. In principle, U can be measured optically, since it represents the energy required for the transition:

$$2D^o \longrightarrow D^+ + D^- \quad (2)$$

when no atomic relaxations take place around the charged defect centers, D^+ and D^-. According to the Franck-Condon principle, no such relaxations occur at optical frequencies. However, unless the centers are spatially close, the optical matrix element for (2) should be extremely small, particularly when the defect concentration is small.

At lower frequencies, breakdown of the adiabatic approximation introduces another serious problem. It is clear that in a primarily covalent solid, significant local relaxations must take place around both D^+ and D^- centers, tending to reduce the total energies of both. When these relaxations occur, the energy necessary to induce transition (2) is thus considerably lower than U; the reduced value is usually called U_{eff}. It is even possible for U_{eff} to be negative, resulting in stabilization of the charged rather than the neutral centers.

Whenever the adiabatic approximation is invalid, different experiments measure different energies, and great care must be taken to avoid inconsistent conclusions. Transport experiments carried out under near-equilibrium conditions are ordinarily sensitive to the energy differences between centers after atomic relaxations have occurred, and thus they measure U_{eff} rather than U. However, such relaxations may be retarded in strained regions and, in any event, may involve overcoming of potential barriers, thus introducing time and temperature dependent processes into the experiments. A complete analysis would require a knowledge of the many-body energy levels as a function of local configuration coordinates, an enormously difficult calculation even in crystals, given the techniques currently available. In disordered solids, the existence of regions with different local environments introduces still further complexity. An array of values for U_{eff} could characterize the same defect center in different regions of the material, depending on the extent of the local strains.

The major consequences of U_{eff} follow not so much from its magnitude than from its sign. If $U_{eff} > 0$, neutral defects are energetically favorable compared to positively and negatively charged pairs, and E_F is unpinned. If D^o centers contain an unpaired spin, they should be observable by ESR whenever E_F is located between the filled and empty defect levels. D^+ and D^- states would not be simultaneously present at equilibrium, and no spatial correlation would ordinarily be expected between the defect centers. In contrast, when $U_{eff} < 0$, D^+-D^- pairs would be stable and would be further stabilized when they are spatially close. E_F would be pinned to the extent of the concentration of distant pairs. No unpaired spins from D^o centers would be present at equilibrium at low temperatures, although these could be excited optically or thermally.

TIGHT-BINDING ANALYSIS

Accurate calculations of defect energies in a-Si:H are difficult for many reasons, not the least of which is the fact that large clusters are essential to properly understand the overconstrained nature of the network. Nevertheless, some insight can be gained form the opposite extreme, an analysis in which only nearest neighbors are considered and complete relaxations are allowed. This should provide a lower limit on U_{eff}, since any strain-induced retardation of the relaxations must increase U_{eff}. Such a calculation[1] indicates that U_{eff} could well be negative for the T_3 defect in a-Si:H. The physical origin of this result can be extracted from Fig. 1. If an electron is removed from a T_3^o center, it converts to T_3^+ [Fig. 1 (c)], which optimally has bond angles of 120°. Thus, the local distortion involved tends to move the central Si atoms into a position within the plane of the three neighbors to which it bonds. Analogously, a T_3^- center optimally has bond angles in the range 95°-100° [Fig. 1 (d)], induced by a relaxation in which the central Si

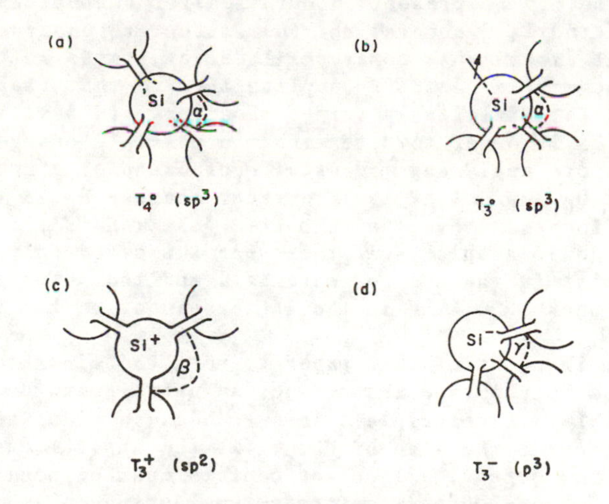

Fig.1 Sketches of the optimal local coordinations of a Si atom in several different configurations: (a) ground state, T_4^o; (b) neutral dangling bond, T_3^o (c) positively charged dangling bond, T_3^o; (d) negatively charged dangling bond, T_3^-.

atom moves away from the plane of the three neighbors to which it bonds. Since three-fold coordination represents the ground-state configurations of both Si^+ and Si^- ions but not of neutral Si atoms, it is not unreasonable to conclude that U_{eff} could be negative. Somewhat more sophisticated calculations[4] have indicated that U_{eff} is positive, but complete relaxations were not taken into account. In any event, since there are always some unpaired spins in a-Si:H films, and their concentration is not a strong function of tempera- we must conclude that U_{eff} is positive for at least some of the T_3 defects. On the other hand, there is some evidence[1] that large concentrations of T_3^+ and T_3^- centers are also present in a-Si:H films. A possible explanation[1] is that $U_{eff} < 0$ when complete relaxations around both the T_3^+ and T_3^- centers can take place, but that such relaxations are retarded in particularly strained regions of the film. In that case, isolated T_3^o centers would coexitst with $T_3^+ - T_3^-$ pairs. We shall return to this possibility later.

OPTICAL DETERMINATION OF U_{eff}

Many experiments have claimed to have measured U_{eff} for the T_3 defect in a-Si:H. However, because of the problems discussed previously, it is dangerous to accept them uncritically. For example, any measurement that depends on ESR observations is sensitive only to T_3^o centers; these may be present in much smaller concentrations then the spinless T_3^+ and T_3^- centers, but the latter are undetected. In contrast, optical experiments observe all defect states in the gap, provided that the matrix elements coupling them to the valence or conduction bands is sufficiently large. However, it is vital to know the position of E_F in order to determine the possible charge states of the defect before analyzing the data. For example, a recent conclusion[5] that $U_{eff} = 0.4$ eV is incorrect, because E_F in the undoped film is located above the purported position of the T_3^- levels in the P-doped samples. If this were the case, only T_3^- centers could exist in the undoped material, and the downward shift of the observed peak by 0.4 eV in the latter cannot be the result of U_{eff}.

It is the main point of this paper to note that in materials that can be doped both p-type and n-type, an unambiguous determination of U_{eff} is possible, in principle. This is because p doping depresses E_F, independent of the sign of U_{eff}. If the dopant concentration exceeds that of the defect, all defect centers must be positively charged, e.g. T_3^+. Analogously, after n-type doping, only negatively charged centers, e.g. T_3^-, can appear. If electrons from the T_3^- centers are optically excited into the conduction band in p-doped a-Si:H, then a deconvolution of the resulting shoulder should yield $E_n = E_c - E_-$, where E_c is the conduction band edge and E_- is the effective energy of the T_3^+ level. Similarly, in B-doped a-Si:H, $E_p = E_+$

- E_n, where E_n is the valence band edge and E_+ is the effective energy of the T_3^+ level, should be experimentally accessible. Since $U_{eff} = E_- - E_+$ and $E_{opt} - E_c - E_n$, it is clear that:

$$U_{eff} = E_{opt} - (E_n + E_p) \qquad (3)$$

This analysis is sketched in Figs. 2 and 3 for the cases $U_{eff} > 0$ and $U_{eff} < 0$, respectively. When $U_{eff} > 0$, we expect relatively small values for E_n and E_p; in any event, their sum cannot exceed 1.8 eV. Alternatively, if $U_{eff} < 0$, $E_n + E_p$ must be greater than 1.8 eV.

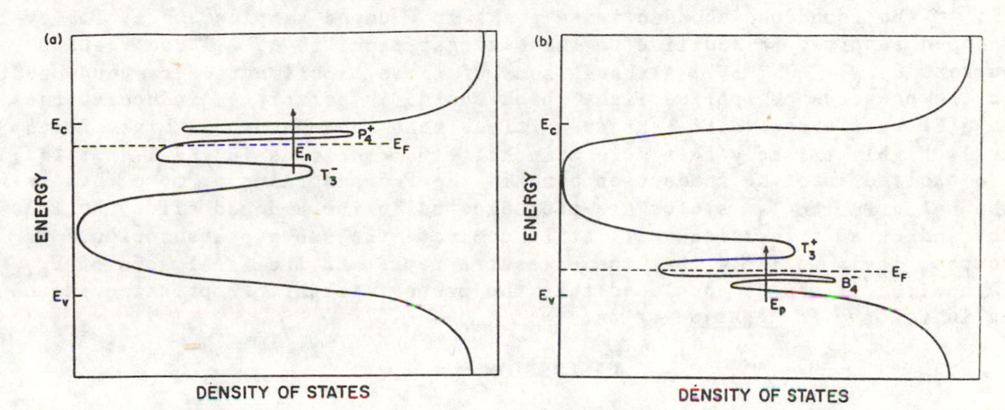

Fig.2 Optical determination of U_{eff} for the T_3 defect in a-Si:H for $U_{eff} > 0$: (a) measurement of E_n in P-doped sample; (b) measurement of E_p in B-doped sample.

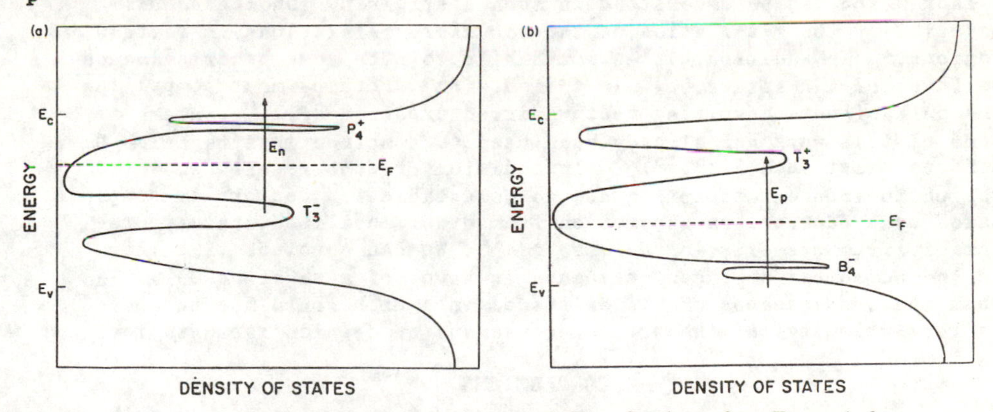

Fig.3 Analogous experiments to those of Fig.2, but for $U_{eff} < 0$.

EXPERIMENTAL EVIDENCE

There have been many careful transmission experiments performed on P-doped a-Si:H films. Although deconvolution of the absorption shoulder has some degree of error associated with it, the best fit for a series of well-characterized films observed by Cody[6] yielded E_n = 1.1 eV. Almost identical results were obtained by Yamasaki et al.[7] and by Grasso et al.[8] from photoacoustic spectra. Triska et al.[9] recently studied both P-doped and B-doped samples, and found E_n = 1.25 eV, E_p = 1.03 eV, and E_{opt} = 1.74 eV, with a stated accuracy of ±0.02 eV. Use of (3) for these results indicate U_{eff} = −0.5 eV. Again, the absorption spectrum in B-doped films was essentially the same as those obtained by other groups.[10,11] Perhaps the most convincing evidence is the recent work of the Inushima et al. [12] who found photoconductivity peaks in P-doped samples and at 1.3 eV in B-doped samples, in addition to the band-gap peak 1.8 eV. These results suggest U_{eff} = −0.7 eV. Although some of these experiments were conducted in the presence of applied light which could, in principle, induce changes in g(E) that expose different transitions than those shown in Figs. 2 and 3, it is highly unlikely that more than half the electrons in the T_3^- states are excited into the conduction band in the P-doped films or more than half the holes in the T_3^- states are photoexcited in the B-doped films, so that the indicated transitions will still dominate the sub-gap absorption. Of course, it is unlikely that these results represent the final word on U_{eff} in a-Si:H, it can be concluded that the present weight of optical evidence is in favor of a <u>negative</u> value.

DISCUSSION

Clearly, the idea of a negative U_{eff} for the T_3 defect in a-Si:H is, at best, controversial. However, as previously noted, the major reason for this is simply the presence of T_3^o centers in all samples, a fact which can be understood in such a strained, inhomogeneous material by the retardation of the necessary relaxations in certain regions.[1] In addition, it is worthwhile to note some other evidence in favor of a negative U_{eff} in a-Si:H.[13,14] Perhaps most compelling is the bulk of data involving photo-induced creation of T_3^o centers.[1] Some of this work has already been used to conclude that negative U_{eff} defects exist in a-Si:H.[13,14] In addition, recent results show that the photo-induced absorption due to metastable T_3^o centers is compensated by a decrease in absorption from other localized states, presumably from pre-existing defects.[15,16] We can conclude that there is indeed some independent evidence in favor of a negative U_{eff}, and that the consequences of its existence in a-Si:H would not be quite as revolutionary as the present consensus of opinion seems to feel.

CONCLUSIONS

It has been demonstrated that an optical determination of U_{eff} for the dangling bond in a-Si:H is possible, and that the present weight of evidence is in favor of a <u>negative</u> valve, at least in those regions in which local relaxations can freely occur.

ACKNOWDLEDGEMENTS

I should like to thank G. D. Cody and P. C. Taylor for useful discussions. This work was supported by the National Science Foundation Materials Research Laboratory Grant No. DMR 81-19295.

REFERENCES

1. D. Adler, in Semiconductors and Semimetals, Vol. 21A, R. K. Willardson and A. C. Beer, eds. (Academic Press, NY, 1984), p. 291.
2. J. A. Reimer, J. Phys. (Paris) 42, C4-715 (1981).
3. D. Adler, Solid State Phys. 21, 1 (1968).
4. D. C. Allen and J. D. Joannopoulos, Phys. Rev. Lett. 44, 43 (1980).
5. N. M. Amer and W. B. Jackson, in Semiconductors and Semimetals, Vol. 21B, R. K. Willardson and A. C. Beer, eds. (Academic Press, NY, 1984), in press.
6. G. D. Cody, in Semiconductors and Semimetals, ibid.
7. S. Yamasaki, K. Nakagawa, H. Yamamoto, A. Matsuda, H. Okushi, and K. Tanaka, AIP Conf. Proc. 73, 258 (1981).
8. V. Grasso, G. Morabito, F. Neri, G. Saitta, L. Baldasarre, and A. Cingolani, to be published.
9. A. Triska, I. Shimizu, J. Kocka, L. Tichy, and M. Vanecek, J. Non-Crystall. Solids 59-60, 493 (1983).
10. W. B. Jackson and N. M. Amer, Phys. Rev. B 25, 5559 (1982).
11. S. Yamasaki, A. Matsuda, and K. Tanaka, J. Appl. Phys. 21, 1789 (1982).
12. T. Inushima, M. H. Brodsky, J. Kanicki, and R. J.. Serino, these proceedings
13. A. Morimoto, H. Yokomichi, T. Atoji, M. Kumeda, I. Watanabe, and T. Shimizu, these proceedings.
14. C. Lee, W. D. Ohlsen, P. C. Taylor, H. S. Ullal, and G. P. Ceasar, these proceedings
15. H. Okushi, M. Itoh, T. Okuno, Y. Hosokawa, S. Yamasaki, and K. Tanaka, these proceedings
16. S. Guha, C-Y. Huang, and S. J. Hudgens, Phys. Rev. B 29, 5995

PHOTOINDUCED PARAMAGNETIC CENTERS IN a-SiO_2

J. H. Stathis and M. A. Kastner
Department of Physics and Research Laboratory of Electronics
Massachusetts Institute of Technology, Cambridge, Ma. 02139

ABSTRACT

Paramagnetic defects induced by sub-bandgap light have been observed in vitreous SiO_2. These defects are stable at room temperature, but anneal well below T_g. The observed EPR spectrum changes as the excitation photon energy is varied, and is sensitive to the hydrogen content of the material. Other photoinduced effects include a mid-gap absorption band and photogenerated luminescence centers. The behavior of a-SiO_2 is compared to that of the narrow-gap chalcogenide glasses.

INTRODUCTION

The fact that a metastable spin resonance signal can be photoinduced in the semiconducting chalcogenide glasses has been known for almost ten years[1]. After exposure to Urbach-tail light, at low temperature, a density of $\sim 10^{17}$ cm^{-3} spins is observed. Accompanying the photoinduced spins is a broad mid-gap absorption band, and a reduction (fatiguing) of the photoluminescence. These effects are attributed to the trapping of carriers at preexisting diamagnetic defects. No spin is observed in the annealed glass, in spite of evidence for a large density of defect states near the fermi level. This is explained by a model[2-4] in which all gap states are either doubly occupied (negatively charged) or unoccupied (positively charged). The negative effective correlation energy necessary for this spin-pairing to occur is thought to be a result of the special chemistry of the chalcogenides[4]. The valence band of these materials is composed of the non-bonding, or lone-pair, electrons on each chalcogen atom[5]. These electrons are available to take part in bond-rearrangements accompanying changes in the charge state (occupation) of defects in the glass network. This provides the mechanism for the large electron-phonon coupling that is necessary[2] for the existence of a negative effective correlation energy. The lowest energy, and hence most numerous defects are therefore believed to be charged over- and under-coordinated sites.

Although the equilibrium state of the native defects in the glass is therefore diamagnetic, the paramagnetic states of the defects are accessible by optical excitation, as was first shown by Bishop, Strom, and Taylor. Moreover, the paramagnetic states are metastable. They remain indefinitely if the sample is held at sufficiently low temperature, but they anneal at temperatures well below the glass softening point.

Amorphous SiO_2, like the semiconducting chalcogenides, shows no unpaired spins in the as-quenched state, in spite of the existence of defects with states in the gap that give rise to photoluminescence[6]. SiO_2 is a 3-dimensional network glass composed of well defined SiO_4

0094-243X/84/1200078-08 $3.00

tetrahedra joined at their oxygen corners. Each (two-fold coordinated) oxygen is therefore said to be "bridging" two silicons. The O-Si-O angles are quite rigidly constrained[7] at the tetrahedral angle (109.5°), while the Si-O-Si angles range[8,9] from 120° to 180° with the peak of the distribution near 144° which is the value it takes in crystalline quartz. The presence of heteropolar bonding introduces considerable complications[10-12] which are at the root of the theoretical dispute over the nature of the intrinsic defects in a-SiO_2. The top of the SiO_2 valence band is composed of oxygen lone-pair electrons[13] so that its defect chemistry may be similar to that of the narrower gap chalcogenides. In fact, however, the role that the oxygen lone pairs play in defects in SiO_2 is not agreed upon. The complication is due to the large electronegativity of oxygen, which causes the bonding in SiO_2 to have a large ionic character, and to fairly rigid steric constraints. Greaves[14] has therefore suggested that the charged defects, rather than being negative-U defects (valence-alternation pairs) are simply charged dangling silicon and oxygen bonds, the charge transfer being driven purely by the electronegativity difference rather than by electron-phonon coupling. The dangling bonds will nonetheless interact weakly with nearby oxygen lone pairs, so that the energy levels depend on occupation as in the model of reference 3. Lucovsky[15], on the other hand, argues that Greaves' positively charged 3-fold coordinated silicon should completely bond to the lone-pair electrons of a nearby bridging oxygen to form a positively charged 3-fold coordinated oxygen, and that it is this reconstruction which stabilizes the charge transfer. Broadly speaking, therefore, both Greaves and Lucovsky propose that the positively charged defect involves an oxygen interacting with three silicons. Although ionicity effects tend to favor heteropolar bonds[10] other defects are possible if homopolar bonds are allowed. Mott[16] has proposed that positive defects may exist in the form of 3-fold coordinated oxygens that are bonded to two silicons and one oxygen, resulting from the re-bonding of an oxygen dangling bond in direct analogy to the chalcogenides. In all cases the negative defect is argued to be a singly-coordinated oxygen. Mott's picture also requires the existence of oxygen vacancies or neutral silicon-silicon bonds if the glass is stoichiometric. These might prefer to undergo valence-alternation to give three-fold coordinated negative silicon and three-fold coordinated positive oxygen.

The aim of the present work is to learn more about the intrinsic defects by investigating the response of a-SiO_2 to optical excitation, in hopes of helping to resolve these disputes. We would like to know to what extent SiO_2 behaves like a chalcogenide, and to what extent its properties are unique. Conversely, if certain properties could be shown to be a consequence of the lone-pair chemistry, then it is possible that a study of a-SiO_2 may help to shed light on the semiconducting chalcogenides.

An additional motivation for studying SiO_2 lies in its considerable technological importance. The most important applications are in optical fibers, whose core and cladding regions are made of SiO_2 either in its pure state or doped with various modifiers, and in integrated circuit technology, where thermally oxidized silicon serves as an insulating or passivating layer. In both of these areas knowledge of the structure and properties of native defects is crucial. Defects (and impurities) give rise to optical absorption which limits the performance of optical fibers, and to charge traps which affect the behavior of semiconductor integrated circuits. Additional knowledge may make possible further improvements in these technologies.

PHOTOINDUCED EPR

Recently, we reported[17] the discovery that metastable paramagnetic defects can be photoinduced in a-SiO_2 by sub-bandgap light, providing a compelling analogy to the sulfide and selenide glasses. In this paper we shall first summarize and discuss the observations reported previously. Then we will present some additional results, some of which serve to emphasize the similarity between SiO_2 and the narrower gap chalcogenides, and others which may be indicative of properties unique to SiO_2. All of the results will be germane to eventually attaining a complete understanding of the defects in SiO_2.

Figure 1, reproduced from reference 17, shows the electron paramagnetic resonance (EPR) spectra observed after exposure of high purity bulk amorphous SiO_2 to ultraviolet photons of various energies, as labeled. Photon energies of 7.9 eV, 6.4 eV, and 5.0 eV were obtained from a multigas laser (Lumonics) using F_2, ArF, and KrF, respectively, as the lasing medium. All of these energies are below the optical gap of SiO_2, which is around 10 eV[18,19]. In all cases the absorbed dose of photons (estimated using published[14] absorption coefficient data) is a few times 10^{19} in a volume of approximately 0.15 cm^3. Two types of commercially available[20] a-SiO_2 have been investigated: Suprasil 1, which contains a large amount of hydrogen impurity in the form of 1200 ppm (by weight) OH groups, and Suprasil W1 ("water-free"), which contains only ~5 ppm OH. Both of these synthetic fused silicas have a total metallic impurity level on the order of 1 ppm. All EPR measurements were made at ~9.4 GHz and at room temperature. The microwave power was 0.2 mW.

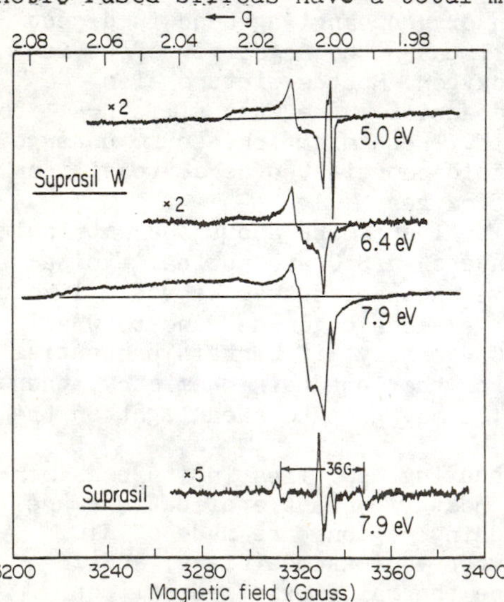

Fig. 1. Photoinduced EPR in a-SiO_2.

The variation of the photoinduced EPR spectrum with excitation energy is an important aspect of this work. The most dramatic feature of the data in figure 1 is the difference between the 7.9 eV-induced spectrum in Suprasil W and the spectrum induced in the same material by the lower photon energies. The 7.9 eV-induced spectrum shows a low-field (g_3) shoulder extending to g≃2.065, which disappears completely upon lowering the excitation energy to 6.4 eV. Clearly it would be extremely useful to extend these measurements to other photon energies, to map out the excitation spectrum of the EPR between 6.4 eV and 7.9 eV and to extend the measurements above 7.9 eV.

All of the spectra exhibit the well-known E' resonance, which is due to a dangling silicon bond. This is the narrow resonance, 2.4 gauss wide, very close to g=2. In all of the spectra displayed the E' resonance is in microwave saturation, so that absolute number densities cannot be directly extracted. Nonetheless, trends in this density as a function of excitation photon energy are apparent. Most striking is the larger number of E' centers produced by 5.0 eV light than by 6.4 eV light.

Also shown in figure 1 is the EPR spectrum obtained after 7.9 eV irradiation of Suprasil. The three-component resonance, consisting of the central g=2.004 resonance and the two hyperfine lines split off by 18 gauss, has been identified[17] as due to nitrogen impurity; we will not discuss it further in this paper. None of the broad resonances seen in Suprasil W are observed in Suprasil, although we do see the E' resonance. It is remarkable that the behavior of Suprasil is so different from that of Suprasil W, since the only difference between these materials is purported to be their hydrogen content.

DISCUSSION OF THE EPR SPECTRA

Paramagnetic defects produced by irradiating SiO_2 with γ-rays, x-rays, or neutrons have been extensively investigated in the past. These radiation-induced defects have been identified as consisting of three types: the E' center[7], which is a silicon dangling bond (≡Si•); the non-bridging oxygen[21], consisting of an oxygen bonded to a single silicon (≡Si-O•); and the peroxy radical[22] (≡Si-O-O•). It is important to keep in mind that these identifications were all made in samples which had been exposed to high-energy radiation, which is certainly capable of creating additional structural defects in the glass, either by direct knock-ons or by radiolysis[23]. Therefore one gains little information about the intrinsic defects. In fact, the radiation-induced EPR is generally interpreted[24,25] in terms of oxygen vacancies (≡Si-Si≡) and interstitials (≡Si-O-O-Si≡), i.e. Frenkel defect pairs, which could result from the displacement of an oxygen. Sub-bandgap light, on the other hand, is more likely to excite only existing defects, although there is evidence in the chalcogenides[26] for the creation of additional defects. In either case EPR only gives information about the excited state of the defects, so that one can only speculate about the structure of the diamagnetic ground state from which the paramagnetic state was created. It is here that the investigation of optically-induced defects has potential, since by varying the photon energy one obtains an excitation spectrum of the defects in question. The work described here represents the first step in this direction.

Some insight may be gained by comparing the results of figure 1 with what is known about the optical absorption spectrum of a-SiO_2. It is known, for example, that the optical absorption edges of Suprasil and Suprasil W are different[27]. In particular, in Suprasil W there is evidence of an absorption band at 7.6 eV. This band is known to be defect related[28], and its strength even in annealed OH-free silica depends on the method of preparation[27,29]. Since the 7.9 eV photons used in the present work are within this absorption band, this might explain the difference between the Suprasil W EPR induced by 7.9 eV photons and by the lower energy photons, as well as the difference between the Suprasil and Suprasil W 7.9 eV-induced EPR. O'Reilly and Robertson[30] have suggested that absorption near 7-8 eV could arise from a variety of different centers, since all positively charged defects possess a shallow state 1-2 eV below the conduction band.

We have already described the narrow E' resonance as being due to a dangling silicon bond. The broader resonances, seen in all of the Suprasil W spectra, are generally associated with oxygen-related defects, specifically oxygen dangling bonds. This accounts for the relatively large g_3 value since this is determined by the (small) splitting between the two non-bonding p orbitals, and since this splitting should be very sensitive to the local environment one can account for the broadness of the low-field shoulder. The broad resonance seen in 7.9 eV-irradiated Suprasil W most closely resembles the spectrum ascribed to the peroxy radical (≡Si-O-O•) in γ-irradiated Suprasil W[21]. This identification is somewhat tentative, since it is based principally on the position of the low-field shoulder, and the spectrum of the non-bridging oxygen (≡Si-O•) is not easily distinguished from that of the peroxy since the shoulder is quite broad in both cases. Although at first glance the sharp structure between g=2.00 and g=2.01 seems to lend support to this identification (see the data in reference 21), we will see in the next section that the sharp features are not associated with the low-field shoulder.

Gaczi[31] has done a careful analysis of the photoinduced EPR spectra in sulfide glasses, and finds evidence for disulfide radicals (R-S-S•) analagous to the peroxy radical discussed here.

The fact that no oxygen-related defects are observed in 7.9 eV-exposed Suprasil is a very significant result since it says that the intrinsic defects are strongly modified by hydrogen, which serves to tie up dangling bonds and relieve strains in the glass network. This has been discussed by Greaves[14] in the context of his defect model. Other manifestations of the effect of hydrogen will be seen when we discuss the photoinduced absorption and luminescence in a later section.

The density of photoinduced paramagnetic centers can be obtained by twice integrating the derivative spectra of figure 1 and comparing with a known standard. For the 7.9 eV-induced EPR in Suprasil W we obtain a density of 2.5×10^{17} spins/cm^3 if we use the penetration depth of the exciting light to calculate the sample volume. However, it is quite likely, based on what we know[32] about the growth and saturation behavior of photogenerated absorption and luminescence centers in this material, that the photon dose used to obtain this spectrum was more than sufficient to saturate the photoinduced EPR, so that the sample was exposed throughout its 3 mm thickness. This reduces the density estimate to $\sim 4 \times 10^{16}/cm^3$. For the spectra induced by the lower photon energies we estimate $\sim 2 \times 10^{15}$ spins/cm^3, Although this is close to the density of alkali metal impurities (~0.1 ppm) in this material, these defects cannot be impurity related because the alkali impurity level is the same in Suprasil in which these resonances are not seen. The photoinduced E' density (measured at lower microwave power) is surprisingly low: only ~$5x10^{13}/cm^3$ result from 5.0 eV excitation.

ANNEALING OF THE EPR

From the shape of the EPR spectra, and from the variation of the spectra with excitation energy, it is clear that these spectra consist of several components arising from different defects. If the different defects have different activation energies for annealing, it is possible to empirically separate the observed spectra into their components. The annealing behavior is also important in attempting to correlate the photoinduced EPR with the photoinduced optical effects to be described in the next section.

Since the largest effects result from 7.9 eV excitation of Suprasil W we have begun by investigating the annealing behavior of this system. The results of a series of 15 minute anneals at successively higher temperatures, in increments of 100°C, are shown in figures 2 and 3.

The solid curve in figure 2 is the total integrated EPR intensity (proportional to the number of paramagnetic centers). It can be seen that there is no single activation energy, since the annealing is fairly gradual. However, all of the paramagnetic centers are annealed out by 700°C. The annealing temperature of the photoinduced EPR in a-SiO_2 is higher than that in the chalcogenide semiconductors[1,26], consistent with the larger band gap and bond strengths.

In figure 3 we show the 7.9 eV-induced EPR spectrum at different stages of the annealing process. The dotted curve is the same data shown in figure 1; the solid and dashed curves are the spectra observed after 15 minute anneals at 300°C and 500°C, respectively. This figure makes it apparent that the gradual annealing seen in figure 2 is due, at least in part, to the separate annealing of several components. For example, the amplitude of the low-field shoulder

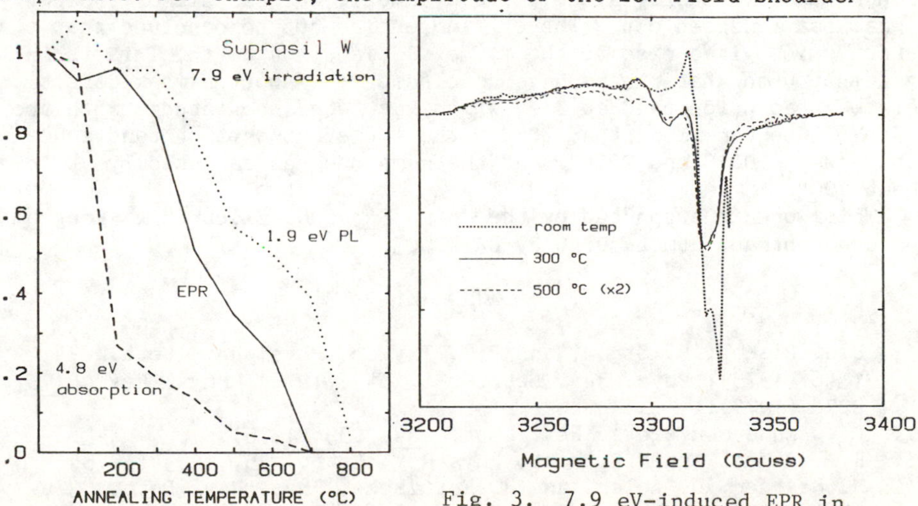

Fig. 2.

Fig. 3. 7.9 eV-induced EPR in Suprasil W after various anneals.

(g≈2.065) is not changed by the 300°C anneal, while all of the sharp features between g=2.00 and g=2.02 have annealed out. Therefore the shoulder and the sharp features must arise from different defects.

LIGHT-INDUCED LUMINESCENCE AND ABSORPTION

In addition to the EPR, there are other metastable effects which result from exposure of a-SiO_2 to sub-bandgap photons. Recall that in the sulfides and selenides[2] one observes a photoinduced mid-gap absorption band and a decrease in the luminescence quantum efficiency. The situation seems to be somewhat different in SiO_2. Here one also observes[32,33] a mid-gap absorption at 4.8 eV, but the spectrum is a relatively narrow (~1 eV FWHM) gaussian, in contrast to the broad featureless "ionization-type" spectrum observed in As_2S_3 and As_2Se_3. The absorption therefore seems to arise from transitions between localized defect levels rather than between a defect level and the continuum. The number of 4.8 eV absorption centers saturates at ~$2x10^{16}/cm^3$ assuming an oscillator strength of unity. The photoinduced absorption is only seen in Suprasil W, not in Suprasil.

We have also discovered[32] a photogenerated luminescence center, which gives rise to photoluminescence (PL) at 1.9 eV. Although originally observed to result from exposure to 7.9 eV photons, the 1.9 eV PL centers have since been found to be created by photons as low as 5.0 eV. The generation spectrum must fall off below this, however, since 3.7 eV photons, although capable of exciting the 1.9 eV PL once the defects are in place, are unable to create more of these defects. The PL centers are created in both Suprasil and Suprasil W, although only a portion of them are stable at room temperature in Suprasil.

The annealing behavior of the 4.8 eV absorption band and of the 1.9 eV PL band (both resulting from 7.9 eV irradiation; the PL was excited at 3.7 eV) are shown in figure 2. An important thing to note about the 1.9 eV PL is that it still has 40% of its initial (room temperature) intensity after a 700°C anneal, whereas all of the EPR signal has annealed out. There is no choice but to conclude that the defect which gives rise to the 1.9 eV PL is diamagnetic. This is in agreement with what others have concluded[21,25] about a weak defect-related absorption band at 2 eV which has been associated[34] with the 1.9 eV PL center. More than 70% of the 4.8 eV absorption band anneals out between 100°C and 200°C, and the remainder anneals slowly up to about 700°C.

This work is supported by the Joint Services Electronics Program through contract number DAAG-29-83-K0003.

REFERENCES

1. S.G. Bishop, U. Strom, and P.C. Taylor, Phys. Rev. Lett. 34, 1346 (1975); Phys. Rev. Lett. 36, 543 (1976); Phys. Rev. B 15, 2278 (1977).
2. P.W. Anderson, Phys. Rev. Lett. 34, 953 (1975).
3. R.A. Street and N.F. Mott, Phys Rev. Lett. 35, 1293 (1975).
4. M. Kastner, D. Adler, and H. Fritzsche, Phys. Rev. Lett. 37, 1504 (1976).
5. M. Kastner, Phys. Rev. Lett. 28, 355 (1972).

6. C.M. Gee and M. Kastner, Phys. Rev. Lett. 42, 1765 (1979); J. Non-Cryst. Solids 40, 577 (1980).
7. D.L. Griscom, E.J. Friebele, and G.H. Sigel, Solid State Commun. 15, 479 (1974).
8. R.L. Mozzi and B.E. Warren, J. Appl. Crystallogr. 2, 164 (1969).
9. R.J. Bell and P. Dean, Philos. Mag. 25, 138 (1972).
10. R.A. Street and G. Lucovsky, Solid State Commun. 31, 289 (1979).
11. D. Vanderbilt and J.D. Joannopoulos, Phys. Rev. B 23, 2596 (1981).
12. J. Robertson, Phys. Chem. Glasses 23, 1 (1982).
13. T.H. DiStefano and D.E. Eastman, Phys. Rev. Lett. 27, 1560 (1971).
14. G.N. Greaves, in The Physics of SiO_2 and its Interfaces, ed. S.T. Pantelides (Pergammon, New York, 1978), p. 268; Philos. Mag. B 37, 447 (1978); J. Non-Cryst. Solids 32, 295 (1979).
15. G. Lucovsky, Philos. Mag. B 41, 457 (1980).
16. N.F. Mott, Adv. Phys. 26, 363 (1977); J. Non-Cryst. Solids 40, 1 (1980).
17. J. H. Stathis and M.A. Kastner, Phys. Rev. B 29, 7079 (1984).
18. T.H. DiStefano and D.E. Eastman, Solid State Commun. 9, 2259 (1971).
19. R. Evrard and A.N. Trukhin, Phys. Rev. B 25, 4101 (1982).
20. Heraeus Amersil. Inc., Sayreville, New Jersey 08872.
21. M. Stapelbroek, D. L. Griscom, E. J. Friebele, and G.H. Sigel, J. Non-Cryst. Solids 32, 313 (1979).
22. E.J. Friebele, D.L. Griscom, M. Stapelbroek, and R.A. Weeks, Phys, Rev. Lett. 42, 1346 (1979).
23. J.W. Corbett and J.C. Bourgoin, in Point Defects in Solids, Vol. 2, ed. J.H. Crawford and L.M. Slifkin (Plenum Press, New York, 1975), p. 1.
24. D.L. Griscom, in The Physics of SiO_2 and its Interfaces, ed. S.T. Pantelides (Pergamon, New York, 1978), p. 232; J. Non-Cryst. Solids 40, 211 (1980).
25. D.L. Griscom, submitted to J. Non-Cryst. Solids (proceedings, N.J. Kreidl Honorary Symposium, Vienna, July 1984).
26. D.K. Biegelson and R.A. Street, Phys. Rev. Lett. 44, 803 (1980).
27. I.P. Kaminow, B. G. Bagley, and C.G. Olson, Appl. Phys. Lett. 32, 98 (1978).
28. E.W.J. Mitchell and E.G.S. Paige, Phil. Mag. 1, 1085 (1956).
29. A. Appleton, T. Chiranjivi, and M. Jafaripour-Ghazvini, in The Physics of SiO_2 and its Interfaces, ed. S.T. Pantelides (Pergamon, New York, 1978), p. 94.
30. E.P. O'Reilly and J. Robertson, Phys. Rev. B 27, 3780 (1983).
31. P.J. Gaczi, Philos. Mag. B 45, 241 (1982).
32. J.H. Stathis and M.A. Kastner, Philos. Mag. B 49, 357 (1984).
33. S. Lange and W.H. Turner, Appl Optics 12, 1733 (1973).
34. L.N. Skuja and A.R. Silin, Phys. Stat. Sol. (a) 56, K11 (1979).

PARAMAGNETIC STATES IN CHALCOGENIDE GLASSES INDUCED BY NEAR MID-GAP INFRARED RADIATION

S. G. Bishop, J. A. Freitas, Jr.* and U. Strom
Naval Research Laboratory, Washington, D.C. 20375 USA

ABSTRACT

Localized paramagnetic states have been optically induced in glassy As_2Se_3 and As_2S_3 by weakly absorbed light with photon energy (1.17 eV, 1.06 μm) much less than the band gap. Measured ESR line widths are comparable to those observed for centers induced by above gap light, but the estimated spin densities are at least three orders of magnitude lower than those achievable with above gap excitation. The optically induced spin densities scale with the magnitude of the below gap, one photon absorption coefficient at the 1.06 μm wavelength and there is no evidence for a two-photon excitation mechanism.

INTRODUCTION

Previous experimental studies[1,2] have demonstrated that metastable localized paramagnetic states, as evidenced by electron spin resonance (ESR), mid-gap optical absorption, and fatiguing photoluminescence (PL), can all be induced at low temperature by irradiation with band edge light (light with photon energy corresponding roughly to the Urbach tail of the absorption edge and for which the absorption coefficient of the glass is about 100 cm^{-1} (1.7 eV in As_2Se_3, 2.4 eV in As_2S_3)). In addition, subsequent prolonged irradiation of the sample with infrared light which corresponds to the induced mid-gap absorption band reduces the strength of or "bleaches" both the induced ESR and the induced midgap absorption and restores the fatigued PL efficiency. The observation of these optically induced metastable paramagnetic centers provided the most direct evidence then available for the existence of localized gap states in these amorphous semiconductors. The optically induced ESR centers were found to be unique to the amorphous phase and their density appeared to saturate at or below $10^{17} cm^{-3}$ in all of the glasses studied. Analysis of the optically induced ESR spectra led to the identification of a hole center which consists of an electron missing from a non-bonding chalcogen orbital and a center in As-containing glasses which is localized in an As p orbital.

In the low energy range below the Urbach tail, $\alpha < 1$ cm^{-1}, there is a broad absorption tail[3] extending deep into the gap which produces a weak low energy tail on the PL excitation (PLE) spectrum for these glasses.[4] In the present work it is shown that localized paramagnetic states can also be induced by 1.06 μm (1.17 eV) light which corresponds to an approximate absorption coefficient of

---

*Sachs-Freeman Associates, Bowie, MD; CNPq-Brazil.

0.1 cm^{-1} in glassy As_2Se_3 and 0.01 cm^{-1} in As_2S_3 glass. This inducing photon energy lies within the weak low energy absorption tail for both glasses and is near mid-gap for As_2S_3.

EXPERIMENTAL PROCEDURE

The ESR spectra were obtained at 10 K in a helium gas flow variable temperature Dewar system with a standard Varian E-9 X-band (9 GHz) bridge spectrometer. Data acquisition was facilitated by a signal averaging system which provided improved signal to noise and allowed the subtraction of background (cold-dark) resonances from the optically induced signals. Although the experimental procedure was in many respects identical to that described in Ref. 1 (in particular, ESR operating parameters were identical with the exception of gain), the optically induced spectra produced by the 1.06 μm light are nearly three orders of magnitude weaker than those induced by above gap light, and this fact necessitates some careful changes in procedure. The experiments were also complicated by the need to irradiate the samples at 77 K with a remotely located, 0.5 W, CW Nd:YAG laser, transport the samples under cold, dark conditions to the ESR spectrometer, cool to 10 K in the microwave cavity, and then obtain the ESR spectra. The cold-dark or background spectra were obtained after the optically induced spectra were recorded. This was accomplished by annealing the samples in a 50 C water bath for 15 min to remove all optically induced centers[5,6] and then reinserting in the cryostat/cavity and recooling to 10 K for ESR measurements. The samples of chalcogenide glass were sealed in evacuated quartz tubes; the As_2S_3 sample consisted of three 2 mm diameter cylinders stacked in the tube and the As_2Se_3 sample consisted of irregular fragments packed in the tube. The laser beam irradiated a cylindrical volume 2 mm in diameter and as long as the thickness of the samples which are quite transparent at 1.06 μm. Ten such volumes were subjected to successive exposures by translating the laser beam by one beam diameter after each exposure. Exposure times ranging from 1 to 120 sec and CW laser powers ranging from 10 to 100 mW were used. Because the absorption coefficients of the glasses at the 1.06 μm inducing wave length are at least three orders of magnitude less than the 100 cm^{-1} absorption coefficient for which above gap light induces centers most efficiently, exceptional care was required to exclude all above gap light from the cold samples in order to observe the weak signal induced by the YAG laser.

RESULTS AND DISCUSSION

The ESR spectra induced in glassy As_2Se_3 and As_2S_3 by a 15 sec exposure to 10 and 100 mW, respectively, of 1.06 μm light are shown in Fig. 1. Both the optically induced spectra and the much stronger background spectra are shown (note the disparate gain scales) to demonstrate clearly how weak the optically induced signals are. In

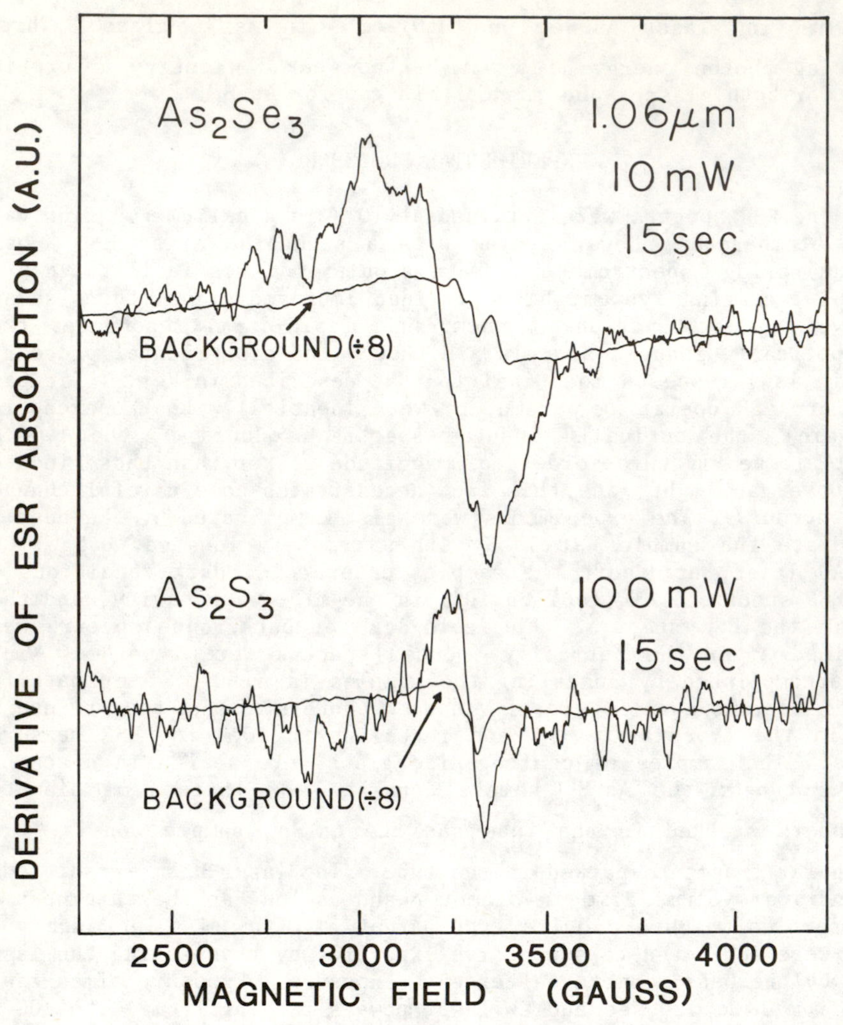

Fig. 1. The ESR spectra induced in glassy As_2Se_3 and As_2S_3 by a 15 sec exposure to 10 and 100 mW, respectively, of 1.06 μm light. The inducing was carried out at 77 K and the ESR spectra were obtained at 10 K. Also shown are the background (cold-dark spectra) obtained after a 15 min anneal at 50 C.

all cases the spectra represent the co-adding (signal averaging) of multiple passes through the resonance. Because of the low signal-to-noise ratios, it is not possible to place any significance on the detailed features of the observed spectra. For example, we have not attempted to do a computer simulation of the lineshapes and infer principal components of the g tensor. However, it seems apparent that the As_2Se_3 spectrum contains a major component with width of about 250 G which is in excellent agreement with the 250 G line width of the optically induced hole center associated with the chalcogen (Se) produced by above gap light.[1] In addition, it is possible to infer the existence of a narrower line (width ~40 G) near g=2. The relative intensities of these two lines is a function of the inducing conditions. The broader line (250 G) is accentuated by inducing at higher powers for shorter times and the narrower line (40 G) becomes increasingly dominant for increasing dosage of inducing light (e.g. longer exposures at lower powers). Similarly, the As_2S_3 spectrum of Fig. 1 contains a broad component of the order of 100 G wide which can be associated with the 75 G wide chalcogen (S) hole center reported for As_2S_3 glass in Ref. 1. Again, the presence of a sharper component (~40 G wide) can be inferred, and the relative strength of these two ESR lines also exhibits a dependence on inducing light power and dosage. (Because we are concerned primarily with the broader resonances in both glasses, the 40 G wide component has been numerically subtracted from the spectra shown in Fig. 1.)

The use of the calibration procedures described in Ref. 1 (including the same standard sample) established the observed spin densities associated with the broad (250 G) component of the spectra as approximately $10^{14} cm^{-3}$ for As_2Se_3 and $10^{13} cm^{-3}$ for As_2S_3. It should be emphasized that because of experimental uncertainties the estimated spin densities are at best accurate to within one order of magnitude. These spin densities are at least three orders of magnitude lower than those produced by above gap light in these materials as reported in Ref. 1.

In Fig. 2 the spin densities induced in As_2Se_3 glass by 15 sec exposures to 1.06 μm light are plotted for a range of laser powers. Within the error of the measurements, the optically induced spin density in As_2Se_3 appears to be independent of power, having reached a saturated value at the lowest power employed (for this exposure time).

In Fig. 3 the spin densities induced in As_2S_3 glass by 100 mW of 1.06 μm light are plotted for a range of exposure times. Although the data are rather limited, there is a high level of confidence for the critical data point at 120 sec because this "saturated" value of about $10^{13} cm^{-3}$ has been achieved many times for a variety of combinations of powers and exposure times. The data of Fig. 3 therefore demonstrate that the optically induced spin density for below gap light has a saturation or limiting value in As_2S_3 just as it does for As_2Se_3 (Fig. 2). Qualitatively speaking, it takes a greater dose of photons to reach saturation in As_2S_3 than in As_2Se_3, and for this reason it is possible to document the approach to saturation in As_2S_3 (Fig. 3).

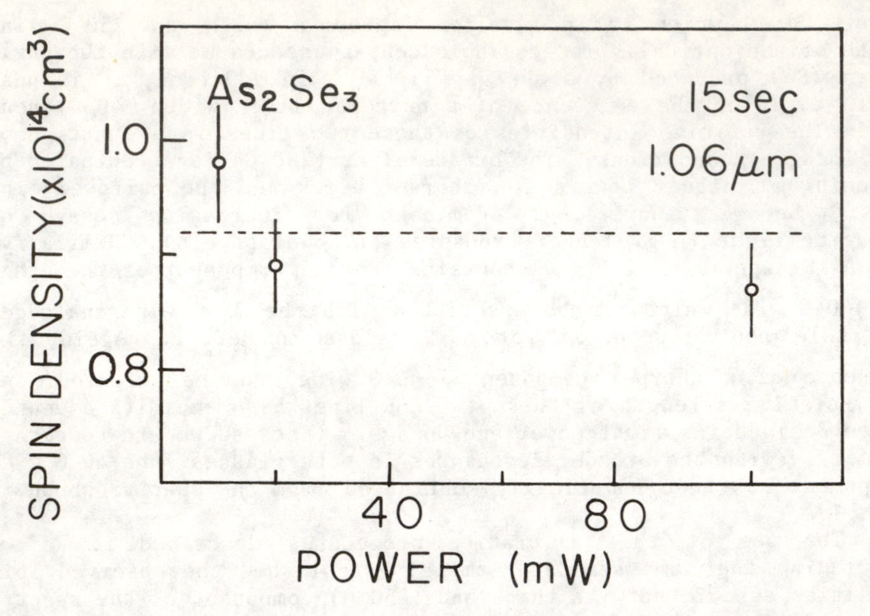

Fig. 2. The spin densities induced in As_2Se_3 glass at 77 K by 15 sec exposures to 1.06 μm light for a range of laser powers.

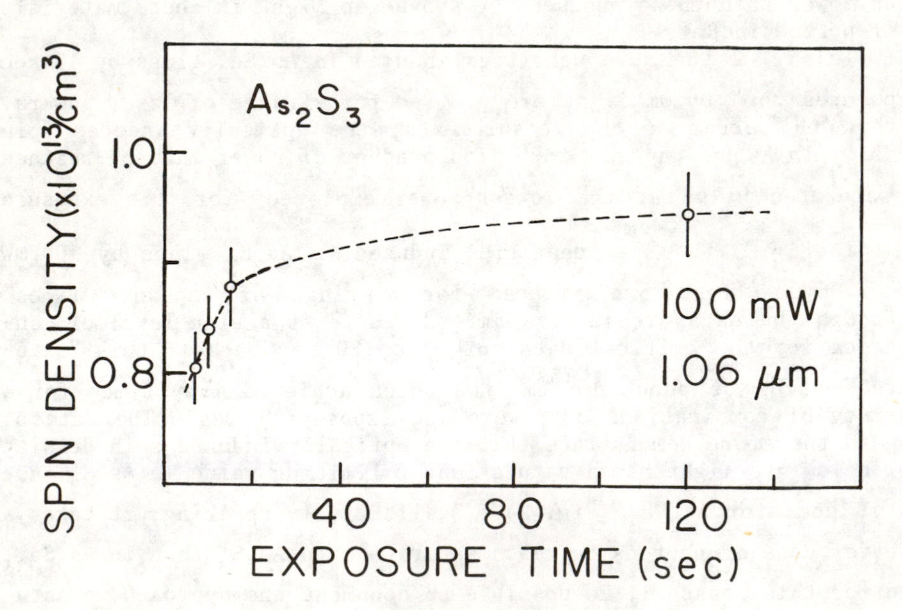

Fig. 3. The spin densities induced in As_2S_3 glass at 77 K by 100 mW of 1.06 μm light for a range of exposure times.

In order to simplify the discussion of these results, it is best to state at the outset that we have found no experimental evidence which would lead us to invoke a two-photon absorption mechanism to explain the inducing of paramagnetic states by below gap light. In both the CW results presented here and in preliminary studies of the use of higher power pulsed 1.06 μm light for the optical inducing process (which will not be discussed here) we have not found a nonlinear dependence of the inducing efficiency upon the power of the inducing light. In addition, as discussed below the saturation values of the optically induced spin densities in the two glasses scale with the apparent magnitude of the below gap, one photon absorption coefficient at the 1.06 μm wavelength.

Subsequent to the original studies of optically induced ESR in chalcogenide glasses reported by Bishop, Strom, and Taylor,[1] Biegelsen and Street[6] reported the observation of two distinct inducing mechanisms corresponding to fast and slow rates. Their interpretation of the fast rate coincides with that of Bishop et al.,[1] that photoexcited carriers are trapped at existing defects. Their slow inducing process involves photostructural changes in the glass which introduce additional defects and lead to saturated values of the induced spin density which are much higher than those achieved by the fast mechanism. Comparison of the relatively short times required to achieve the saturated spin densities observed in the present work with those reported in Ref. 6 indicates that the below gap excitation process also involves the fast rate, namely, changes in the occupancy of existing defects. Biegelsen and Street[6] also distinguished the two inducing mechanisms on the basis of different ESR lineshapes. However, the low signal-to-noise ratio for the optically induced ESR spectra presented in Fig. 1 precludes any such distinction in the present work.

Having suggested that the optically induced centers produced by the 1.06 μm light are closely related to those induced by above gap light, and that the inducing mechanism is a fast, one-photon process of the type discussed in Refs. 1 and 6, it is possible to propose an approximate relationship between the saturated values of the optically induced spin densities and the absorption coefficients of the glasses at the wavelengths of the inducing light. The optical absorption coefficients for glassy As_2Se_3 and As_2S_3 for the above gap inducing light reported in Ref. 1 were both about 100 cm^{-1}. In both cases saturated spin densities of approximately $10^{17} cm^{-3}$ were induced. This corresponds to one induced ESR center for every 1000 absorbed photons, on the basis of the inducing photon doses reported in Ref. 1. (Both glasses also exhibited an optically induced midgap optical absorption coefficient of approximately 10 cm^{-1}. Estimates of the density of optically induced centers inferred[1,7] from these optically induced absorption coefficients were in approximate agreement with the measured density of optically induced ESR centers.) The approximate absorption coefficients at 1.06 μm for glassy As_2Se_3 and As_2S_3 are .1 cm^{-1} and .01 cm^{-1}, respectively. If one assumes that the achievable saturated density of optically induced ESR centers simply scales with the absorption coefficient at the inducing wavelength, the 1.06 μm light would be expected to generate appxoimately $10^{14}cm^{-3}$ and $10^{13}cm^{-3}$ paramagnetic centers in As_2Se_3 and As_2S_3, respectively. This "prediction" is in reasonable agreement with the observed results.

This interpretation is complicated somewhat by the fact that the 1.06 μm inducing light can also bleach the ESR centers. This bleaching process has been well documented[1,6] in the chalcogenide glasses and it is possible to estimate its efficiency. In the work of Bishop, Strom, and Taylor the same optical source was used for the inducing and bleaching processes, indicating that the two mechanisms might be of comparable efficiency. A quantitative analysis of the bleaching data of Ref. 1 verifies that this is the case; approximately 1000 photons must be absorbed by the induced mid-gap absorption in order to bleach one paramagnetic center. Because of the comparable effficiencies for the inducing and bleaching processes, it can be expected that the inducing of ESR centers by below gap light can proceed only to that point where it is balanced or offset by the bleaching process which is introduced as the optically induced mid-gap absorption grows. For this reason the inducing process at 1.06 μm saturates at a very low concentration of optically induced spins as the optically induced absorption coefficient approaches the level of the "intrinsic" absorption tail for the glass. This explains the fact that the concentration of optically induced spins saturates at $10^{13}cm^{-3}$ in As_2S_3 and $10^{14}cm^{-3}$ in As_2Se_3; these values are apparently determined by the relative magnitudes of the weak band tail absorption coefficients in the two glasses at 1.06 μm.

The most important result of the present study is the observation that photons of such low energy, less than half the optical band gap in As_2S_3, can induce the localized paramagnetic states. This result is consistent with a previous report[8] that below gap light (1.92 eV) can produce PL fatigue in As_2S_3 glass. These results raise important questions about the nature of the low energy optical absorption tail in the chalcogenide glasses. Street et al.[9] have proposed a below gap absorption process which excites PL by raising an electron from a negatively charged intrinsic defect in the gap to the conduction band. They suggest that this mechanism alone gives rise to the low energy tail in the absorption edge of chalcogenide glasses which is also manifested in their PL excitation spectra.[2] However, the work of Tauc et al.[10] on Fe-doped glasses and the purification studies of Hilton et al.[11] have clearly demonstrated that impurities can contribute to the optical absorption tails in chalcogenide glasses. Unfortunately, the preliminary results presented here, which involve excitation at only one wavelength, provide little insight concerning the nature of the excitation mechanism. Our conclusions are limited to the facts that the ESR centers appear to be closely related to those induced by above gap light, and that the excitation progress apparently involves a one-photon absorption mechanism.

ACKNOWLEDGEMENTS

We wish to thank E.J. Friebele and R. Lidle for providing components of the instrumentaion and for valuable assistance with some aspects of the experiments.

REFERENCES

1. S.G. Bishop, U. Strom, and P.C. Taylor, Phys. Rev. B15, 2278 (1977).
2. R.A. Street, Adv. Phys. 25, 397 (1976).
3. J. Tauc, Amorphous and Liquid Semiconductors (Plenum Press, NY, 1974), p. 172.
4. R.A. Street, T.M. Searle, I.G. Austin, and R.A. Sussmann, J. Phys. C7, 1582 (1974).
5. P.C. Taylor, U. Strom, and S.G. Bishop, Phil. Mag. B37, 241 (1978).
6. D.K. Biegelsen and R.A. Street, Phys. Rev. Lett. 44, 803 (1980).
7. S.G. Bishop, U. Strom, and P.C. Taylor, Phys. Rev. Lett. 34, 1346 (1975).
8. S.G. Bishop, U. Strom, and P.C. Taylor, in Proc. 7th Int. Conf. on Amorphous and Liquid Semiconductors, edited by W.E. Spear (Univ. of Edinburgh, 1977), p. 595.
9. R.A. Street, T.M. Searle, and I.G. Austin, Phil. Mag. 32 431 (1975).
10. J. Tauc, F.J. Di Salvo, G.E. Peterson, and D.L. Wood, in Amorphous Magnetism, edited by H.O. Hooper and A.M. de Graaf (Plenum Press, NY, 1973), p. 119.
11. A.R. Hilton, D.J. Hayes, and M.D. Rechtin, Chalcogenide Glasses for High Energy Laser Applications, Report No. 08-74-44 (Texas Instruments).

TRANSIENT PHOTO-INDUCED ABSORPTION SPECTROSCOPY IN a-As_2Se_3 IN THE PRESENCE OF STRONG BIAS ILLUMINATION*

Don Monroe+ and M.A. Kastner
Department of Physics and Center for Materials Science and Engineering
Massachusetts Institute of Technology, Cambridge, MA 02139

ABSTRACT

Transient absorption spectroscopy provides a rich source of information about states in the gap of semiconductors. Unfortunately, the average power of the repetitively-pulsed excitation induces a background of excitations which may influence the transient measurements. The use of a strong, bias beam helps to clarify the role of this background, as we demonstrate for a-As_2Se_3. In particular, we report the first observations of transient photobleaching of the otherwise metastable absorption, as well as the observation of a sharp peak in the induced absorption spectrum.

INTRODUCTION

Photo-induced absorption (PA) has long been used as a probe of states in the gap of amorphous semiconductors[1,2,3], but various observations have been associated with markedly different physical processes by different authors. In this paper we will describe some preliminary results of an attempt to draw together some of the results for a-As_2Se_3, and to discuss similarities and differences. Our ability to distinguish different physical mechanisms is enhanced by the introduction of a bias beam, in addition to the pulsed excitation source and sub-bandgap probing light.

EXPERIMENTAL DETAILS

In transient photo-induced absorption (TPA) one monitors the changes in transmission of a sub-bandgap light source, as a function of time following pulsed excitation. The experiments described here were performed on glassy As_2Se_3, held in an optical-access cryostat.

The excitation source was a series of 10ns long pulses at a 10Hz repetition rate produced by a Quanta Ray YAG laser/dye laser/Raman shifter apparatus, which provides photons throughout the range $0.5eV < \hbar\omega_x < 5eV$. The laser beam was spread out and apertured to provide for uniform excitation density in a well-defined geometry.

The probe beam was derived from a 75W tungsten-halogen lamp which was predispersed using a prism monochromator to lower the average power. The transmitted light was collected and passed through a grating monochromator, which provided the ultimate resolution of the measurement (typically 20nm). The grating monochromator output was focussed onto a cooled InAs detector, operated in photocurrent mode, which provided satisfactory response down to $\hbar\omega_p = 0.4eV$. The tran-

*Supported by NSF Grant DMR81-15620, with facilities support from DMR81-19295
+IBM Predoctoral Fellow

0094-243X/84/1200094-08 $3.00

sient modulation in the transmission of the sample due to the excitation pulse was averaged using a Nicolet 4094 signal averager.

Previously, the induced absorption signal has been plotted as a fractional change in transmission, $-\Delta T/T$. This is always much less than unity, so normalizing $-\Delta T/T$ to the number of absorbed photons per unit area yields the average change in absorption cross-section at $\hbar\omega_p$ absorbed photon. This quantity provides a more useful indication of the magnitude than $-\Delta T/T$, which depends on the incident intensity. The absolute scales are uncertain to within factors of two, but the relative magnitudes of data taken under the same conditions, and thus the spectra and time dependences, are much more reliable (of order 10%).

PHYSICAL PROCESSES IN INDUCED ABSORPTION

The TPA experiment provides information about the changes in occupation of states in the gap due to the pulsed excitation. Fig. 1 shows how these changes contribute to changes in transmission, if the excitation pulse creates isolated electrons and holes. Either the electron or hole can be re-excited, or the filling (emptying) of a state may block excitation into (out of) that state. In general, the oscillator-strength sum rule insures that the total change in cross-section, integrated over $\hbar\omega_p$, must be zero: there is as much induced transmission as induced absorption. Induced transmission can only be important if the initial absorption at $\hbar\omega_p$ is large; for this reason, sub-bandgap probes usually exhibit induced absorption, and induced transmission is important only in the spectral region of band-to-band excitation, as seen in a-Si:H by Strait et al[4]. In the

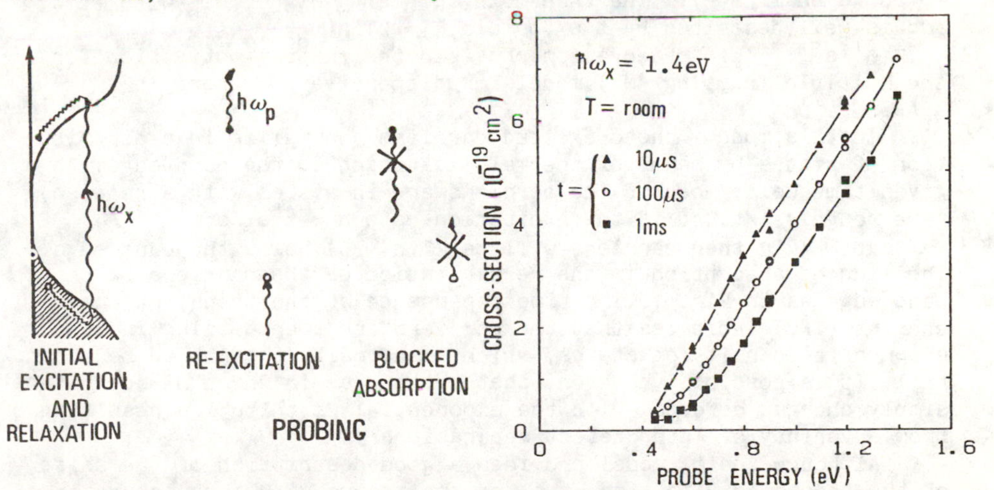

Figure 1. One-electron processes contributing to changes in transmission at $\hbar\omega_p$: re-excitation of electron and hole cause induced absorption, while blocked absorption causes induced transmission.

Figure 2. TPA spectra of thermalizing carriers at room temperature, with no recombination. The absorption spectrum shifts to higher energy as the carriers move into deeper gap states.

presence of steady-state or metastable sub-bandgap absorption, induced transmission can be important even for low $\hbar\omega_p$, as we show below. It should be noted that absorption due to excitons[5] cannot be represented on the same figure.

Several different categories of induced absorption have been described. We will not discuss photodarkening or other high $\hbar\omega_p$ PA processes. Of the midgap absorption processes, we mention the PA associated with the excited photoluminescence (PL) center[5] and the metastable PA below. First, however we will discuss the observations associated with thermalizing carriers.

BAND TAIL PA

The contribution of carriers in band-tail states to TPA was first reported by Orenstein and Kastner[3]; spectra were presented in references 6 and 7. Fig. 2 shows spectra taken at room temperature on a-As_2Se_3 for time delays differing by a factor of ten. This data is taken with $\hbar\omega_x$ = 1.4eV; as a result of the long penetration depth at this photon energy (~1cm), the density of excitations is very low and normalization to correct for recombination[7] is unnecessary. The similarity of the present results to those of Orenstein et al. demonstrates that (i) the normalization procedure was legitimate, (ii) recombination does not produce dramatic changes in the spectra, and (iii) the spectra are not very sensitive to $\hbar\omega_x$ at room temperature. This last observation does not hold at low T, as described below.

The spectra exhibit a linear dependence on $\hbar\omega_p$ above some threshold (the "foot" at the longest times near $\hbar\omega_p$ = 0.6eV is probably a indication of background effects, which become more dramatic at low T, as we shall see). The threshold shifts to higher $h\omega_p$ with time, and is well described by the formula $E = kT\ell n(\nu_o t)$, with ν_o ~$10^{12}s^{-1}$. This observation provided the initial motivation for the Multiple Trapping (MT) model for dispersive transport in a-As_2Se_3.[7]

In this model, photo-excited carriers thermalize in a distribution of states by means of thermal excitation to the band edge. At a given time delay most of the carriers are in states with a release rate roughly 1/t, so if the excitation rate has the form $\nu=\nu_o\exp(-E/kT)$ then carriers will be $kT\ell n(\nu_o t)$ below the band edge. The induced absorption is the re-excitation of the carriers to the band edge as in Fig. 1. The time dependence of the transient photocurrent (TPC) indicates that the carriers are thermalizing in an exponential density of states, which is normally thought to be a band tail. It is possible, however, that the states in which MT occurs are simply charged defects, with the exponential distribution resulting from a variety of inter-defect separations[7].

Although the MT model provides a good description of the shift of the threshold with time, the spectra change with temperature in a way which is not contained in the simple description. Probably there is more than one contribution to the spectrum, with the relative weight of the parts changing with temperature. This possibility will be discussed elsewhere.

TPA measurements have provided important insight into the nature of dispersive transport at low T: at 20K the spectra shift with time about half as fast as at room temperature[8]. Since the change in energy of the carriers due to thermal excitation ($kT\ell n(\nu_o t)$) should be negligible at these temperatures, this shift has been interpreted[9] in terms of a model in which carriers move deeper by hopping directly between localized states, a process which is also manifested in the TPC. In order to check this interpretation, we would like to be able to extract more detailed information from the TPA spectra. At low temperatures, unfortunately, the interpretation of the spectra is complicated by important new effects.

BACKGROUND EFFECTS AND BIAS BEAMS

These new effects result from the build-up of a steady-state background (SSB) of excitations which is significantly different from the thermodynamic ground state of the system at the temperature of the measurement. It should be emphasized that, although the experiments described below involve the deliberate manipulation of the SSB using a bias beam, the presence of a SSB is inherent in repetitively-pulsed transient measurements. An appreciation of the manifestations of such effects is therefore important to a large variety of measurements including TPC and transient PL.

We discussed in conjunction with Fig. 1 the possible electronic absorption processes which can contribute to the observed TPA signal. The description of these processes does not change when one includes SSB effects. What changes is the effective ground state in which these processes can take place. When SSB effects are included, one can no longer interpret the excited electrons and holes of Fig. 1 simply as excitations above or below the Fermi level; they must be interpreted as increases or reductions with respect to the SSB occupation of the states. This complicates the interpretation of the data. Furthermore, if the background is not in steady state and is changing with time, it may interfere with the reproducibility of the transient measurements.

To estimate the magnitude of SSB effects resulting from repetitive pulsing, let us examine the situation in which the average occupation of states has come into steady state with the average illumination. Once this steady state is reached, recombination insures that just before a pulse the number of carriers has returned to the value it had just before the preceding pulse. If N_p carriers are introduced per pulse, then N_p carriers will recombine by the time the next pulse comes, a time t_{rep} later. If the lifetime is longer than t_{rep} then these recombining carriers must come out of the SSB, which is decaying at a rate N_B/τ, where N_B is the number of carriers in the SSB and τ is their lifetime. Thus $N_B/N_p = \tau/t_{rep}$.

The SSB effects will be particularly important at low T (when the lifetimes are long) and when the repetition rate is high (as in many picosecond experiments). Of course, whether the large SSB actually affects the results of transient measurements depends on the system. This question can be rigorously answered in the negative only by varying the repetition rate. In a similar way, one can check for perturbing effects of the probe beam by changing its intensity. Unfortunately, these checks generally require a reduction in signal-to-noise ratio. The introduction of a third, bias beam allows exploration of SSB effects without the sacrifice of signal, and also provides interesting new information about the photo-excited state.

It is well known that illumination of chalcogenide glasses with bandgap light produces sub-bandgap absorption which persists for very long times at low T[1,2]. We will therefore refer to light with energies greater than about 1.6eV as "inducing" light, with a warning against taking this name too seriously. Correlated with the growth of the midgap absorption are the creation of an electron paramagnetic resonance (EPR) signal and the fatigue (reduced intensity) of the luminescence. These properties have traditionally been associated with the photo-induced occupation of existing defects, although photocreation of defects is also possible[10].

The very slow decay of the induced absorption at low T can be speeded up by raising the sample to higher temperatures, and also by illuminating the sample with light in the induced absorption band. Light with $\hbar\omega \lesssim 1.4$eV will therefore be referred to as a "bleaching".

In the present experiments, the bias light was provided by an an incandescent lamp. The "bleaching" beam was obtained with a Si filter which removes light with $\hbar\omega > 1.2$eV. The "inducing" beam was obtained using color filters to give a 30nm wide passband around 700nm (1.8eV). Both beams had a total power of order 100mW/cm^2.

It is important to note that the "inducing" and "bleaching" tendency of light with a particular photon energy is probably a question of degree, and depends on the density of centers already induced. For example, although exposing a virgin sample to "inducing" light will increase the absorption, the same light shining on a sample with a well-developed induced absorption may well have a bleaching effect.

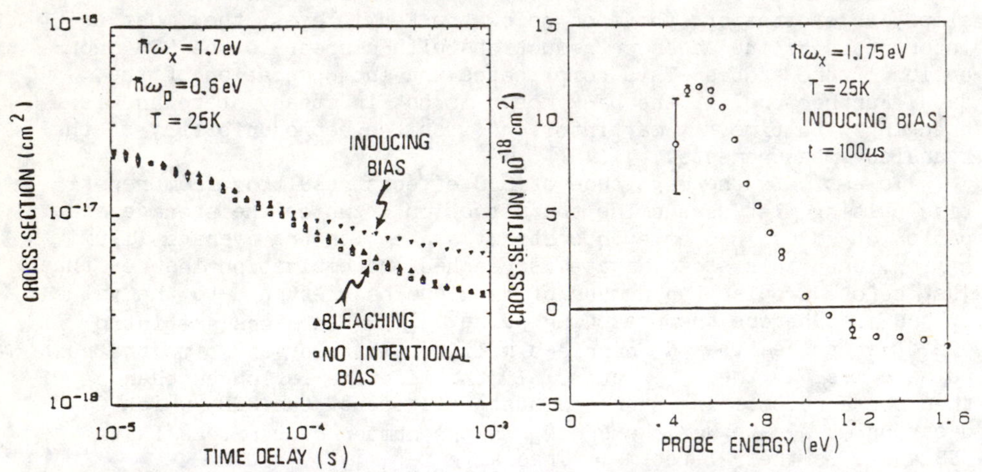

Fig. 3. PL-related PA (t<50μs) and band tail PA (t>100μs) Neither effect is suppressed by "saturation" of the induced absorption.

Fig. 4. Low-energy excitation at low temperatures in the presence of inducing bias produces induced transmission for $\hbar\omega_p > \hbar\omega_x$ and a sharp peak at low $\hbar\omega_p$.

The possibility of self-bleaching by the inducing light has important implications for interpretation of the observed saturation of photo-induced effects with only "inducing" light. The saturated value may not represent a true filling of all defects, but a balance between inducing and bleaching by the same light source, similar to the balance between obviously inducing and bleaching beams studied by Orenstein and Kastner[11]. This idea is supported by the following consideration: the saturated induced absorption corresponds to an absorption coefficient of roughly 20cm^{-1} in the roughly .01cm absorption depth of the inducing light. Since the absorption coefficient at $h\omega_x$ before irradiation is 100cm^{-1}, there are almost as many re-excitations per unit volume as primary excitations. Thus it is quite plausible that the beam is bleaching as well as inducing centers.

Further evidence is presented in Fig. 3, which shows the TPA signal as a function of time for three different bias light conditions. The excitation energy used is effective at exciting the PL-related PA, which represents the absorption from the excited state of the luminescence center (an exciton with strong coupling to phonons) as described by Robins and Kastner[5]. This PA decays with the ~50μs decay time of the luminescence, leaving behind the more slowly decaying PA. If the inducing bias were truly saturating the PA centers, we would expect to see a reduction in the TPA signal due to the unavailability of centers. No such effect is observed, in fact, the slowly-decaying PA even shows an increase, demonstrating the importance of re-excitation of the SSB. It is not surprising that the PL-related PA does not exhibit re-excitation, since tightly bound excitons must be created for PL to occur. This failure to observe saturation in the pulsed measurements, even when increased illumination does not increase the steady state absorption, suggests that the steady state is in fact a balance rather than a saturation, so the density of PA centers is at least an order of magnitude higher than the "saturated" value of around 10^{17}cm^{-3}. This conclusion must be verified by on more complete spectral studies, however.

METASTABILITY

We turn now to a discussion of the "metastable" induced absorption created by bandgap illumination at low T[1,2]. In this context, metastability means that the effects last longer than one is willing to wait - normally a time of order tens of minutes. It is important, however, to distinguish between truly metastable effects - those which do not decay at all except on long time scales - and apparently metastable effects - which exhibit a broad distribution of decay times which extend to very long times. In the former case, all of the relevant rates are slow, and no information can be gleaned from time dependence measurements. In the latter case, exemplified by power-law dependences, there is decay on all time scales. Transient measurements then provide information about the same distribution of rates which is responsible for the apparent metastability.

It is often difficult to tell to which class of metastability a given phenomenon belongs, but it appears that induced absorption may well be only apparently metastable. Support for this assertion comes from the fact that some decay of the absorption is observed on a time scale of minutes, after which the decay becomes painfully slow. On still shorter time scales, pulsed measurements exhibit power-law decays. Further evidence for a distribution of rates comes from annealing studies of the photoluminescence fatigue by Mollot et al.[12]. After inducing the fatigue at low T, they observed that isochronal anneals at higher T removed only a part of the fatigue within a reasonable amount of time. This observation strongly suggests that a distribution of activation energies is being sampled. This distribution should be manifested, at fixed T, as a distribution of decay times.

Thus we may be able to think of the transient measurements and the "metastable" measurements as accessing different regimes of a single fundamental process. Viewed in this light, it is not surprising that the induced absorption spectrum is quite similar on picosecond[13], microsecond[7], and several minute[2] time scales.

TPA IN THE PRESENCE OF BIAS

We discussed above the fact that the long-lived absorption could be bleached with light within the absorption band. Certainly such bleaching should be observable in transient experiments as well. The results of an experiment using an inducing bias beam and a low-energy ($\hbar\omega_x$=1.175eV) pulsed excitation are shown in Fig. 4. The fact that the change in cross section is negative for $\hbar\omega_p$>1.2eV shows that in fact transient photo-bleaching (TPB) has been observed. A study of the dependence of the response on $\hbar\omega_x$ should give important information about the bleaching process as well as the induced absorption itself. Interestingly, bleaching is not seen in the whole induced absorption band: at low probe energies the absorption is actually enhanced, showing a very sharp peak. The limitations of the InAs detector have so far prevented our resolving the low energy side of the peak.

The significance of this peak is not clear at this time. One possibility is that the peak is simply a sharp absorption feature corresponding to a highly localized internal excitation of the induced centers. This result would be quite exciting, although somewhat surprising, since the amount of energy absorbed by the center (1.175eV+0.6eV≈1.8eV) should be sufficient to ionize the center, resulting in a broad spectrum.

Another possibility is that the peak results from a competition between two processes with different thesholds: a TPA spectrum corresponding to absorption by the carriers which have been re-excited into shallower states, and a TPB spectrum resulting the absence of those carriers from deeper states. This hypothesis is supported by the observation that the threshold at which TPB takes over is very close to $\hbar\omega_x$. The complete elucidation of the processes underlying the TPB process will require more complete studies of the dependence of this behavior on excitation energy and pulse energy as well as time and temperature, but it is clear that a wealth of information is available from this sort of experiment.

It is interesting to contrast the present results with the observations of Pfost et al.[14], who studied the time dependence of the spectrally integrated TPA in a-Si:H in the presence of inducing light. Their measurements indicated an increased recombination rate of the band-tail PA as a result of optical population of recombination centers. Clearly several phenomena can contribute to biased TPA, illustrating the power as well as the risks of the technique.

We have not discussed in this paper the effects of bleaching bias on TPA, because the bleaching beam serves primarily to mitigate the effects of the SSB. In this respect, however, bleaching bias has been very successful, resulting in dramatic improvements in the reproducibility of otherwise standard TPA measurements at low temperatures, the results of which will be presented elsewhere.

SUMMARY

We have described the initial stages of an effort to connect the various observations of induced absorption in a-As_2Se_3. We have emphasized the important role that the average illumination can play by creating a steady state background, even in nominally transient measurements. By deliberately manipulating the background with bleaching light, we can reduce the importance of the SSB, and by using inducing bias, we can study in more detail the spectral and temporal behavior of the long-lived induced absorption band.

REFERENCES

1. J. Cernogora, F. Mollot, and C. Benoit a la Guillaume, Phys. Stat. Sol. (a) 15, 401 (1973).
2. S.G. Bishop, U. Strom, and P.C. Taylor. Phys. Rev. B15, 2278 (1975).
3. J. Orenstein and M.A. Kastner, Phys. Rev. Lett. 43, 161 (1981).
4. J. Strait, Z. Vardeny, and J. Tauc, Bull. Am. Phys. Soc. 27, 267 (1982).
5. L.H. Robins and M.A. Kastner, to be published, Phil. Mag. (1984).
6. J. Orenstein, M. Kastner, and D. Monroe, J. Non-cryst Solids 35&36, 951 (1980).
7. J. Orenstein, M.A. Kastner and V. Vaninov, Phil. Mag. B46, 23 (1982).
8. J. Orenstein, Ph.D. Thesis, MIT, 1980. unpublished.
9. D. Monroe, submitted to Phil. Mag.
10. D.K. Biegelson and R.A. Street, Phys. Rev. Lett. 44, 803 (1980)
11. M. Kastner, J. Non-cryst. Solids 35&36, 807 (1980).
12. F. Mollot, J. Cernogora, and C. Benoit a la Guillaume, Phys. Stat. Sol. (a) 21, 281 (1974).
13. R.L. Fork, C.V. Shank, A.M. Glass, A. Migus, M.A. Bosch, and J. Shah, J. Non-cryst. Solids 35&36, 963 (1980).
14. D. Pfost, Z. Vardeny, and J. Tauc, Phys. Rev. Lett 52, 376 (1984).

PICOSECOND OPTICAL GENERATION AND DETECTION OF PHONON WAVES IN a-As_2Te_3

C. Thomsen, J. Strait, Z. Vardeny*, H. J. Maris and J. Tauc
Department of Physics and Division of Engineering,
Brown University, Providence, Rhode Island 02912

J. J. Hauser
AT&T Bell Laboratories, Murray Hill, New Jersey 07974

ABSTRACT

Using the pump and probe technique we have observed oscillations of photoinduced transmission in thin films of a-As_2Te_3. The oscillations have periods of 70-240 ps, depending on sample thickness. We ascribe these to phonon propagation in the films. Thermal expansion and carrier deformation potentials are discussed as the origin of these waves. Their velocity in a-As_2Te_3 is measured to be $v = 1.6 \times 10^5$ cm/s. The effect on transmission of heating by the laser pulse is estimated and compared to experiment.

INTRODUCTION

In this paper we report on oscillations in photoinduced transmission of a chalcogenide glass. The results are interpreted in terms of an acoustic excitation propagating in the sample. Since optical methods are used for generation and detection of the phonon wave, the time resolution is enormously improved over conventional ultrasonic techniques. Limited only by the length of the light pulses ($\sim$ 1 ps), phonon propagation and attenuation can be studied in thin films. This is desirable when either the attenuation is very large, or the material can only be prepared as a thin film.

We will explain the detection technique and discuss two possible generation mechanisms for stress in our samples. The effect of heating by the laser pulse is included in the analysis.

EXPERIMENTAL

Light pulses with a photon energy of $\hbar\omega_x = 2$ eV with a duration of 1 ps were produced by a passively modelocked dye laser at a repetition rate of 0.5 MHz. Each pulse contained an energy of $\sim$ 1 nJ. The pump and probe method was used to measure the time dependence of transmission and reflection changes up to 1.8 ns following excitation.[1] Both pump and probe beam were incident onto the film from the substrate side and focussed to a $\sim$ 40 μm diameter spot.

The samples were amorphous films of As_2Te_3 sputtered onto sapphire substrates. The films with thicknesses 470, 900 and 1200 Å were prepared at substrate temperature θ_s = 300K, and film 1600 Å thick was prepared at θ_s = 77K. The absorption depth of

0094-243X/84/1200102-08 $3.00

the samples at 2 eV photon energy was ζ = 300 Å, the sample with θ_s = 77K was slightly less absorbing.

RESULTS

The changes in transmission as a function of time delay between pump and probe are plotted in Fig. 1 for samples of four different thicknesses at room temperature. The photoinduced response can be regarded as a superposition of two components:

(1) A steplike decrease in transmission ($-\Delta T/T \simeq 3 \times 10^{-3}$) at zero time delay followed by a monotonic relaxation and

(2) a damped oscillation with an amplitude of 3×10^{-5} to 3×10^{-4} and a period that depends on the thickness of the sample. The data in Fig. 2 show that only a sharp peak remains of response (1) at a sample temperature of 10K, while the oscillating part (2) remains the same within the experimental error.

The distinct dependencies of the two components on sample thickness and temperature will be shown to originate from two different processes in the material. The plot of oscillation period vs. sample thickness of Fig. 3 shows a linear relationship.

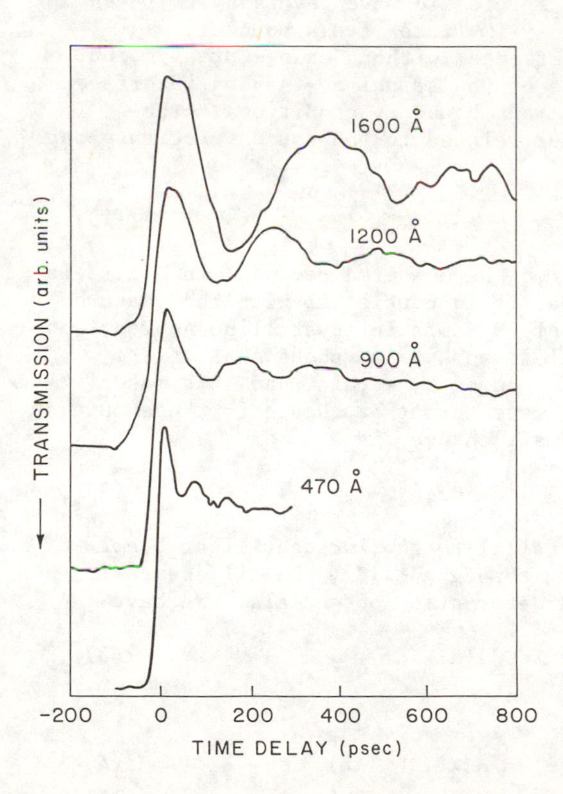

Fig. 1: Photo-induced transmission in a-As_2Te_3 at room temperature for films of various thicknesses.

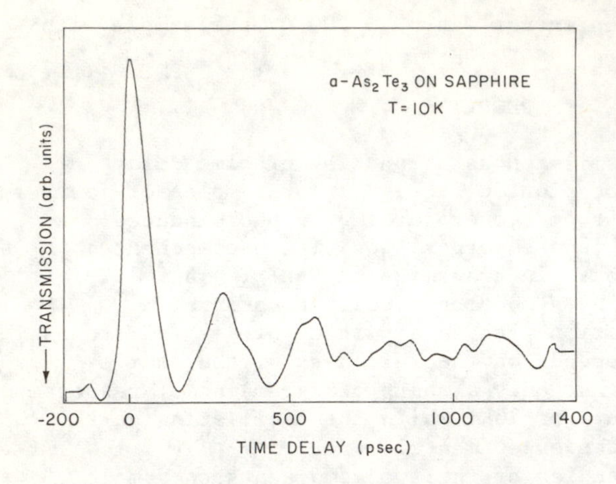

Fig. 2: Photoinduced transmission in a-As_2Te_3 at 10K of a film 1200 Å thick.

INTERPRETATION

We begin by discussing response (2). We suggest that this oscillatory component of the signal results from a stress wave propagating in the sample. An elastic wave reverses its phase only at the free side of the film (zero stress boundary condition), whereas it is reflected without change of sign, but with a decrease in amplitude by the sapphire - As_2Te_3 interface (approximately zero displacement boundary condition). The oscillation period τ_o is thus related to the sound velocity v and the film thickness by:

$$\tau_o = 4d/v \tag{1}$$

The velocity v determined from the measured periods[2] in Fig. 3 is $v = 1.6 \times 10^5$ cm/s. This result is consistent with the measured transverse sound velocity[3] of 10^5 cm/s in crystalline As_2Te_3, and the velocity of longitudinal sound[4] in amorphous As_2S_3. The absorption coefficient α is changed by strain, and this makes it possible to detect a stress wave through a change in the amount of transmitted light T. For small changes in α

$$\Delta T/T \simeq - \int_o^d \Delta\alpha dz, \tag{2}$$

where the z direction is parallel to the incident light beam.[5] α is modulated by the strain η through $dE_g/d\eta$, the difference in valence and conduction band deformation potentials. We have

$$\Delta\alpha = (d\alpha/dE_g)(dE_g/d\eta)\eta \ , \tag{3}$$

so that

$$\Delta T/T = -(d\alpha/dE_g)(dE_g/d\eta) < \eta > d, \tag{4}$$

where $<\eta>$ is the average strain in the sample. An oscillating

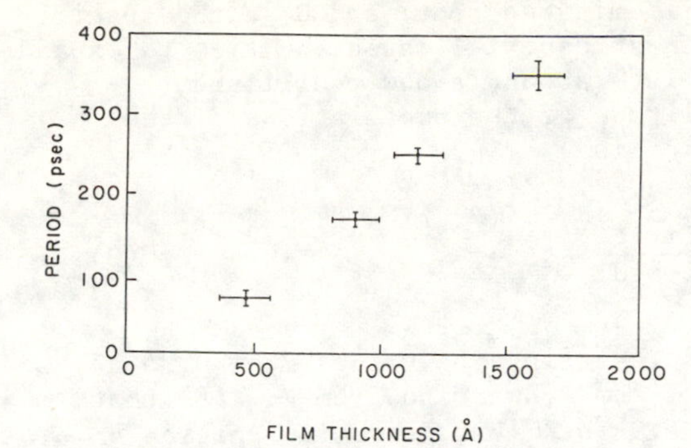

Fig. 3: Oscillation period vs. film thickness at room temperature.

change in transmission can thus result from an oscillating average strain.

Consider now the ways in which a stress wave can be generated through the laser pulse. We consider here two distinct mechanisms, both of which may play a significant role in the generation mechanism. The laser pulse creates a number of electron-hole pairs equal to the number N of absorbed photons. These will relax quickly (within a few picoseconds) to the band edges, losing the excess energy $E_x = (\hbar\omega - E_g)\,N$ by phonon emission. These emitted phonons produce a stress

$$\sigma_A = 3\beta B < \Delta\theta > \qquad (5)$$

where B is the bulk modulus, β is the linear thermal expansion coefficient and $<\Delta\theta>$ is the average temperature rise in the absorption region.

The second contribution σ_B arises from the electrons and holes, and is given by

$$\sigma_B = -\left(n_e \frac{\partial E_c}{\partial \eta} - n_h \frac{\partial E_v}{\partial \eta}\right) \qquad (6)$$

where n_e and n_h are the number of densities of electrons and holes, and E_c and E_v are the energies at the bottom of the conduction band and top of the valence band respectively. Since $n_e = n_h = N$ we have

$$\sigma_B = -N \frac{\partial E_g}{\partial \eta} \qquad (7)$$

We will first estimate the amplitude of the stress wave generated by the thermal expansion stress σ_A, and the magnitude of the resulting change in transmission $\Delta T/T$. The average temperature rise in the absorption region is

$$< \Delta\theta > = E_x/\zeta AC \qquad (8)$$

where A is the area of the illuminated spot, and C is the specific heat of the film. The sudden stress will cause the film to expand in the z-direction and oscillate around a new equilibrium thickness. The initial amplitude of the oscillations of the average strain in the film $<\eta>$ is

$$<\eta>_o = \frac{1+\nu}{1-\nu}\frac{\zeta}{d}\beta <\Delta\theta> \tag{9}$$

where ν is Poisson's ratio. Inserting the values $E_x = 0.35$ nJ, $A = 2 \times 10^{-5}$ cm^2, $C = 1.5$ JK^{-1} cm^{-3}, we find from Eq. (8) that $<\Delta\theta> = 4$K. Then using[6] $\beta = 1.9\ 10^5$K^{-1} and assuming Poisson's ratio of 0.3 we find $<\nu> = 1.3 \times 10^{-4}$ for 470 Å sample. Together with $d\alpha/dE_g = -3.5 \times 10^5$ cm^{-1} eV^{-1}, estimated from an absorption edge measurement, and a deformation potential of $dE_g/d\eta = 2.3$ eV[7] we find from Eq. (4) $\Delta T/T \simeq 3 \times 10^{-4}$. Both the absolute value and the initial direction of the oscillations are predicted correctly by this, so that thermal expansion is clearly a possible origin for the oscillations.

Consider now the stress σ_B due to the electrons and holes. We have calculated that this is numerically the same order of magnitude as σ_A. This term predicts the opposite sign of $\Delta T/T$ for the oscillations, which is apparently not in agreement with experiment. We have to be careful about this argument, however. To estimate the changes in absorption from $\Delta T/T$ we should correct for the effect of changes in reflectivity $\Delta R/R$. We have found that $\Delta R/R$ also contains an oscillatory component, which is of a magnitude comparable to $\Delta T/T$. Thus, it is conceivable that the phase of the absorption oscillations is not the same as is the phase of the oscillations of $\Delta T/T$. We are currently investigating this.

We now consider the origin of response (1). We propose that this comes from the change of the gap and the absorption coefficient due to the temperature rise in the film. One can divide $(\partial E_g/\partial\theta)_P$ (P = pressure) as follows:

$$\left(\frac{\partial E_g}{\partial\theta}\right)_P = \left(\frac{\partial E_g}{\partial\theta}\right)_V + 3\beta\left(\frac{\partial E_g}{\partial V}\right)_\theta \tag{10}$$

In chalcogenide glasses[7] the thermal expansion term (second term) is known to be small compared to the thermal term $(\partial E_g/\Delta\theta)_V$. If we keep just the first term $\Delta T/T$ is

$$\frac{\Delta T}{T} = -\left(\frac{d\alpha}{dE_g}\right)\left(\frac{\partial E_g}{\partial\theta}\right)_V \zeta <\Delta\theta> . \tag{11}$$

For $(\partial E_g/\partial\theta)_V$ we use the value[7] for As_2S_3, and obtain

$$\Delta T/T \simeq -6 \times 10^{-3}. \tag{12}$$

This is of the order of magnitude and sign of the experimental result (Fig. 1). $\langle\Delta\theta\rangle$, and hence $\Delta T/T$, should decay monotonically with time, as is observed experimentally.

This idea is consistent with the temperature dependence of responses (1) and (2). Consider first the oscillatory part of $\Delta T/T$ (response 2). The temperature dependent factors in Eq. (9) are β and $\langle\Delta\theta\rangle$. $\langle\Delta\theta\rangle$ changes with temperature because it is inversely proportional to the specific heat. However, the product β/C is temperature independent as long as Gruneisen's law holds. Thus, $\langle\eta\rangle$ and $(\Delta T/T)_2$ (related to response (2)) will be essentially temperature independent, as is found to be true experimentally (Fig. 2).

Response (1), however, is governed by the temperature behaviour of $(\partial E_g/\partial\theta)_V/C$ in Eq. (11). The specific heat is proportional to θ^3 for $\theta \ll \theta_D$, θ_D denoting the Debye temperature. An Einstein model has been fitted to the gap of several amorphous semiconductors[8], which would imply that $(\partial E_g/\partial\theta)_V$ decays faster than θ^3, such that $(\Delta T/T)$ in Eq. (11) decreases with temperature. It has been suggested[9] that optical and short wavelength acoustical phonons dominate $(\partial E_g/\partial\theta)_V$, and these high energy modes are frozen out at $\theta \ll \theta_D$. Thus the small relative strength of response (1) at 10K is consistent with its origin in the intrinsic change of α.

The remaining peak at 10K of zero time delay, which does not extrapolate back from the oscillation amplitude, we can at this point only speculate about. It is possible that we see either hot carrier thermalization or a decay of optical phonons into long wavelength acoustical phonons. Alternatively an unusually large thermal conductivity of the film at 10K could cause $\langle\Delta\theta\rangle$ of Eq. (9) to decay so rapidly.

Finally in Fig. 4 we show a computer simulation of a stress wave propagating through a 1000 Å film of As_2Te_3. Plotted in each figure is stress vs. distance in the film. The frames differ by 20 ps each. At t = 0 ps the exponential absorption of the light incident from the left determines the initial stress distribution. The wave then develops in time, and the amplitude is reduced by reflection losses at the sapphire-As_2Te_3 interface on the left. The experimentally measured quantity $(\Delta T/T)_2$ is proportional to the average strain shown in Eq. (4), which we plot in the bottom frame vs. time. It can be seen that response (2) is simulated rather well. Note that most of the damping of the wave occurs through reflection losses at the substrate interface, since the simulation assumes no attenuation in the film.

SUMMARY

We have presented a novel method of optically generating and detecting phonon waves. We discussed two mechanisms that can produce the observed strain and show that heat can make a significant contribution to photoinduced transmission measurements if the excitation energy is sufficiently larger than the bandgap.

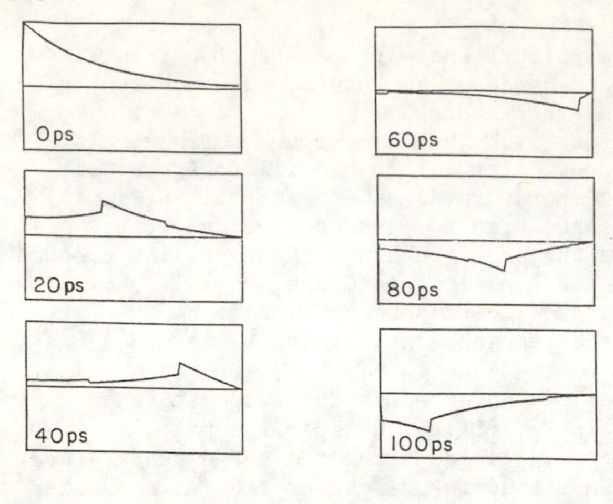

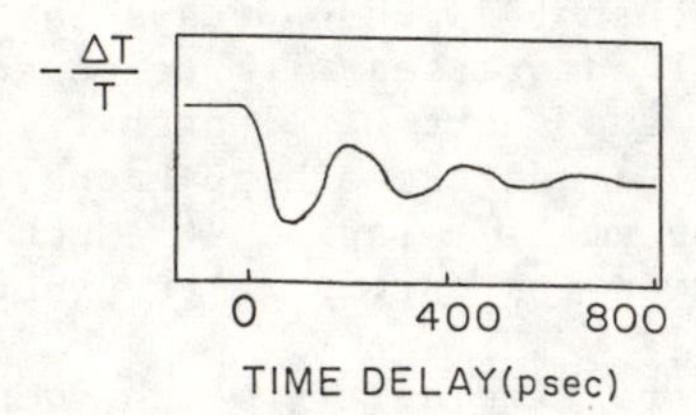

Fig. 4: Computer simulation of a propagating stress wave and of photo-induced transmission corresponding to response (2) (see text).

The phonon generation mechanism is a tool for studying high frequency (∿ 10 to 100 GHz) phonon velocity and attenuation in thin films and highly attenuating materials on a very short time scale, when a-As_2Te_3 is used as a transducer.[12]

ACKNOWLEDGEMENTS

We thank H. T. Grahn and T. R. Kirst for technical assistance. This work was supported by the National Science Foundation through the Materials Research Laboratory at Brown University.

* Present address: Solid State Institute, Technion, Haifa, Israel.

1. E. P. Ippen and C. V. Shank, Ultrashort Light Pulses, edited by S. L. Shapiro (Springer, New York, 1977), p. 102 ff.

2. We do at the present time not understand why the plot does not extrapolate precisely through the origin.

3. G. Cibuzar and C. Elbaum, private communication.

4. D. Gerlich, E. Litor and O. L. Anderson, Phys. Rev. B20, 2529 (1979).

5. This treatment does not yet include changes in reflectivity of the two surfaces.

6. We have used the value for $As_{0.45}Te_{0.55}$ given by J. Cornet and J. Schneider, <u>4th International Conference on the Physics of Non-Crystalline Solids</u>, edited by G. H. Frischat (Trans. Tech. Publications, Clausthal, 1976), p. 397.

7. J. M. Besson, J. Cernogora, and R. Zallen, Phys. Rev. <u>B22</u>, 3866 (1980).

8. G. O. Cody in <u>Hydrogenerated Amorphous Silicon</u>, ed. J. Pankove (Academic Press, 1984), Chapter 2, Vol. 21B, p. 11.

9. P. O. Persans, A. F. Ruppert, G. D. Cody and B. G. Brooks, to be published.

10. G. D. Cody, T. Tiedje, B. Abeles, B. Brooks and Y. Goldstein, Phys. Rev. Lett. <u>47</u>, 1480 (1981).

11. Z. Vardeny, J. Strait and J. Tauc, <u>Picosecond Phenomena III</u>. ed. S. L. Shapiro (Springer Verlag, New York, 1982), p. 372.

12. C. Thomsen, J. Strait, Z. Vardeny, H. J. Maris, J. Tauc and J. J. Hauser, Phys. Rev. Lett., in press.

ANALYSIS of $a\text{-}As_2Se_3$ TRANSIENT PHOTOCURRENTS IN THE LONG-TIME REGIME

Guy J. Adriaenssens and Herman Michiel
Laboratorium voor Vaste Stof-en Hoge Drukfysika
K.U.Leuven, Celestijnenlaan 200 D,B-3030 Leuven, Belgium

ABSTRACT

An earlier suggestion that the density of localized states should be resolvable from transient photocurrents in their long-time regime, has been tested for $a\text{-}As_2Se_3$. The field-independent time dependences that are observed indicate that charge release out of traps is being seen, as predicted by the model. However, no realistic result for the density of deep states emerges; possibly because the model depends too heavily on a questionable thermalisation approximation. The results further suggest that the mobility edge lies $\sim$ 0.035 eV deeper in the band in evaporated $a\text{-}As_2Se_3$ than it does in the bulk material.

INTRODUCTION

Over the last few years, transient photoconductivity techniques have increasingly been used to obtain information on the density of localized states in amorphous semiconductors[1]. One important reason for this has been the emergence of a simple-to-use thermalisation approximation[2,3] for the equilibration of excess charge carriers under multiple-trapping conditions. This method has been applied to $a\text{-}As_2Se_3$[4], as well as for a-Si:H[5]. A more rigourous treatment of the spectroscopic multiple-trapping problem[6] was also applied to $a\text{-}As_2Se_3$[7] and revealed deviations from the often postulated exponential density of states distribution. The method is computationally rather involved, but has since been greatly simplified[8]. All the above methods have in common that they make use of the transient photocurrents at times where loss of carriers due to either recombination or loss at the electrodes has not occurred. Michiel, Marshall and Adriaenssens[6] suggested that density of states information might also be obtained by analysing the transients at times which are longer than either recombination or transit time. In view of the controversy surrounding the true density of localized states distribution in $a\text{-}As_2Se_3$, we felt it worthwhile to explore those ideas further by applying them to a variety of experimental situations involving long-time $a\text{-}As_2Se_3$ photocurrent decays.

The basic idea is the following: At times long compared to what would be the transit time t_T for a given sample in a time-of-flight (TOF) experiment, the experimentally observed current will be essentially due to carriers which have been trapped up to that time. Making the usual assumption that carriers trapped since t=o at a depth E will to a good approximation be released at time

$$t(E) = \nu^{-1} \exp(E/kT), \qquad (1)$$

with ν the attempt-to-escape frequency, this means that current at $t >> t_T$ will be due to release out of traps of such depth that the probability for any carrier to be trapped there more than once will be negligible. Hence the photocurrent at these long times will be directly proportional to the waiting-time distribution function $\psi(t)$ of the multiple-trapping problem. Assuming further that the capture probability will have been energy-independent, the density of localized states may be written as[6]:

$$N(E) = (A/T)t.I(t), \qquad (2)$$

where A is a sample-dependent proportionality constant and the energy is given by

$$E = kT\ell n(\nu t). \qquad (3)$$

As will be explained in what follows, experiments do only partially bear out these ideas, and, most importantly, do not produce a believable N(E).

EXPERIMENTAL RESULTS

Both bulk and evaporated a-As_2Se_3 samples were measured in sandwich- as well as in gap-cell geometry, with applied field, temperature, and the illuminating spectrum as variable parameters. A pulsed Xenon lamp was used as basic light source. Fig.1 shows a few representative photocurrent decay curves after white light illumination, all at 301 K and under moderate applied fields. Full symbols refer to a 15 μm thick evaporated film measured in TOF geometry, the open circles indicate a bulk gap cell. Part of a true TOF, showing t_T, for the same film and under the same conditions is included for reference. Aside from some field-dependent features at shorter times, the three film traces clearly show a field-independent change of slope near 1000 s. A similar change is observed at a shorter time for the bulk sample. Multiplying these curves with time as required according to(2) to produce a N(E) vs. E diagram, the changes of slope translate into t.I(t) maxima as shown in fig.2. Even taking into account that the energy scale may slide to some extent by using a different value for ν, the result is difficult to accept as truly representing the a-As_2Se_3 density of localized gap states distribution.

The position of the maximum(t_M) proved to be independent of applied field as may be observed for the film data of fig.2, and as further illustrated for the same sample in fig.3.
The t_M value was also unaffected by whether a sandwich-type or gap-cell electrode configuration is used, by the use of sub-band-gap illumination or by whether the field was switched on just prior to the light pulse in TOF fashion or had been applied for days.

Temperature on the other hand does strongly influence the position of t_M as shown in fig.4. All available t_M information is

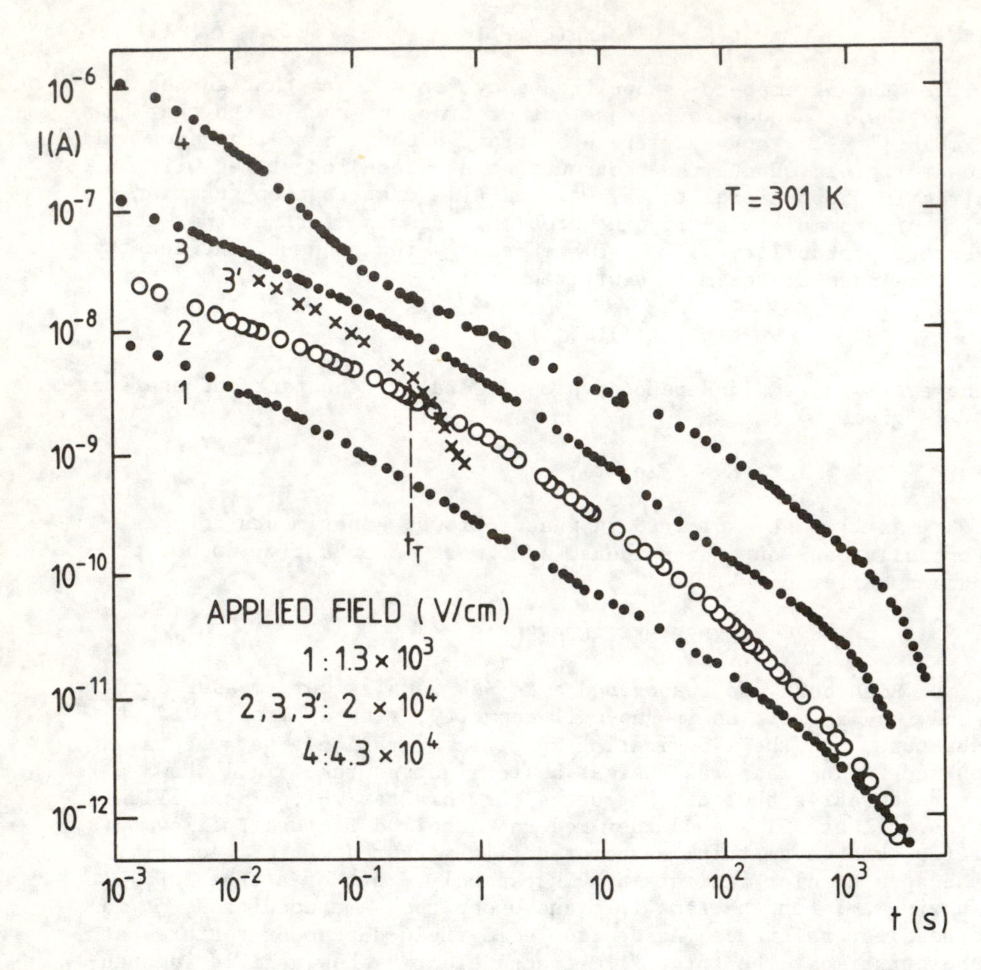

Fig.1. Photocurrent decay curves for a-As_2Se_3 evaporated film(•) and bulk (o) samples. The first and third curves from the top were shifted one half decade upward for clarity. Part of a TOF transient(x) has been added for reference.

displayed there. The different symbols in fig.4 represent the different samples which have been used; full symbols refer to evaporated films, open ones to bulk samples. It is obvious from the figure that there is a significant offset between the film and the bulk results; since both sandwich- and gap-cells were used for bulk as for evaporated materials, this offset is not a size effect.

Not displayed in any of the diagrams is information on the excess carrier recombination time, which in the model is assumed to be short compared to the release time out of deep traps. For the samples used in this study, we did measure, as a function of temperature, the times required for the photocurrent to decay to half value out of steady-state conditions. These times were never longer than a few seconds at the lowest temperatures, but for evaporated layers they do approach the measured t_M at the higher end of our temperature range(where the trapping center presumably turns into a recombination center). A valid test of the $t.I(t) \propto N(E)$ idea therefore needs to be made near room temperature.

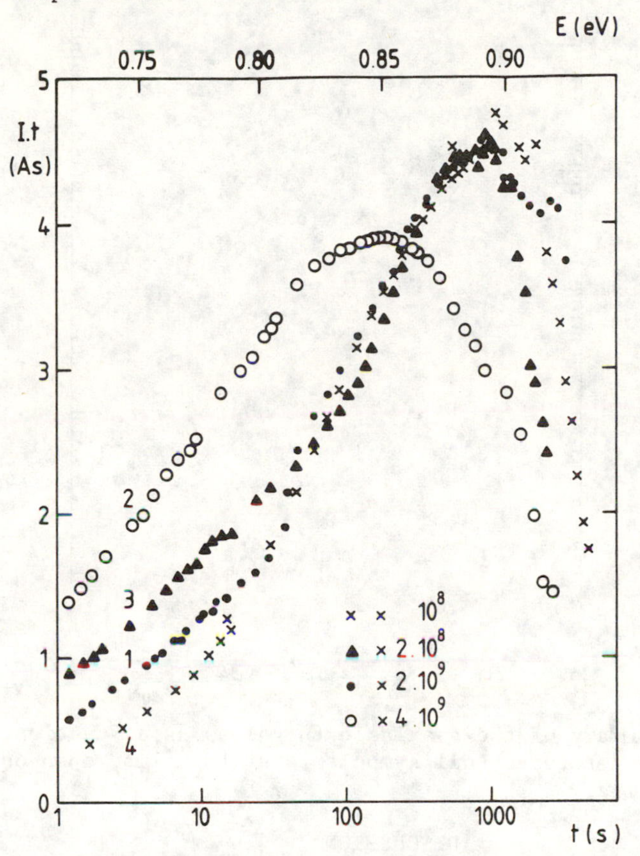

Fig.2. I(t).t products for the currents 1 through 4 of Fig.1. The energy scale was calculated according to equ.(3) with $\nu = 10^{12}s^{-1}$ and measures distance to the valence band

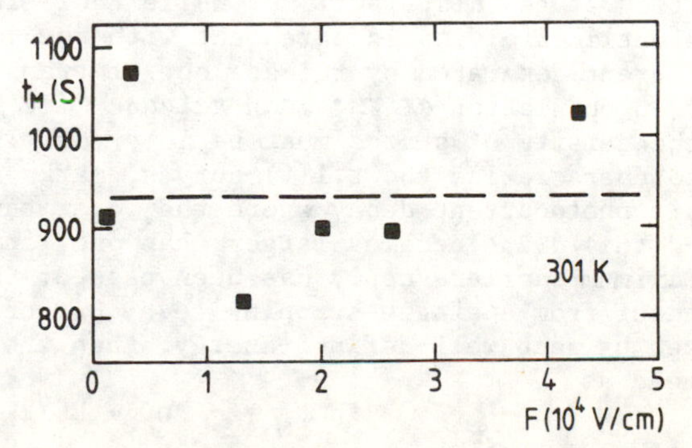

Fig.3 Time t_M at which maximum of t.I(t) occurs vs applied field.

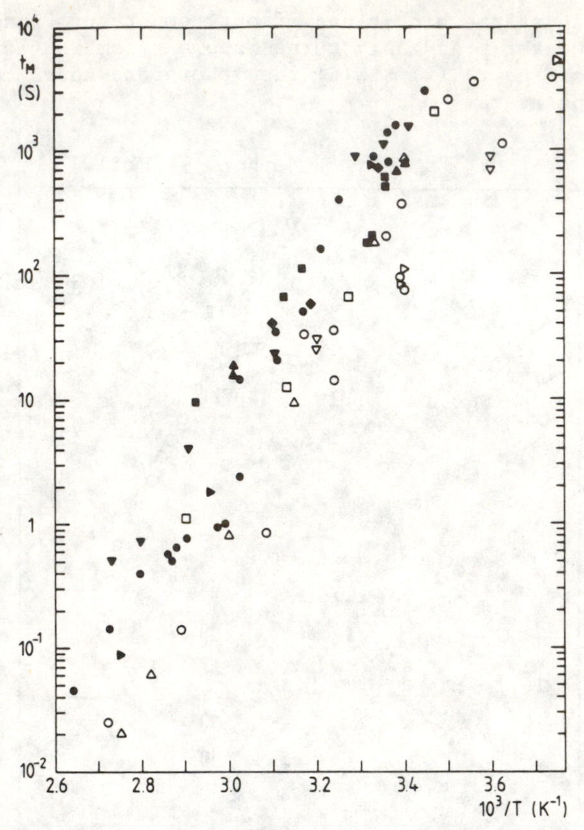

Fig.4. Summary of the available t_M values as a function of reciprocal temperature. Full symbols refer to films, open ones to bulk samples.

DISCUSSION

The measured t.I(t) curves show a too pronounced maximum near the a-As_2Se_3 gap center to be believable as actual density of localized states distributions. While their independence of the applied electric field is in agreement with the idea that we are seeing a current generated by release out of traps, the assumption that the trap population *at the mean release time for each trap*, would mirror the density of states must be in error. The t_M value, which we use to characterize the t.I(t) curves, marks the point on the post-transit photocurrent decay where the logarithmic time-rate-of-change $-d(\ell nI)/d(\ell nt)$ becomes larger than one, indicating that a point of maximum carrier supply has been passed. Should the carrier release occur from a single trapping level E_t, or a distribution characterized by such well-defined energy, then the data of fig.4 could be read as

$$\ell n\ t_M = -\ \ell n\nu + E_t/kT, \tag{4}$$

with E_t= 1.2 eV and ν = 10^{17} s^{-1}. Taken by themselves, these numbers are totally unrealistic. Introduction of activated trap and release parameters as outlined by Monroe and Kastner[9] would give

$$t_M^{-1} = \nu_o \exp(-\Delta E_b/kT)\exp(-E_t/kT) \qquad (5)$$

but would require a barrier to trapping of $\sim$ 0,4 eV to keep the characteristic trap depth E_t on the right side of the Fermi level in a-As_2Se_3. Such a barrier is not warranted by available evidence[9], which rather points at a fairly temperature-independent value of $\nu \simeq 10^{12}s^{-1}$.

Using this value of $10^{12}s^{-1}$ for all data, we find that, near 300 K, t_M corresponds to release out of traps located at an energy E_M some 0.85 to 0.90 eV from the band, as shown in fig.2, while at 370 K, the levels defining t_M are those with $E_M \simeq$ 0.78 eV. This change with temperature is considerably larger than the $\sim$ 0.02 eV that could be due to the temperature dependence of the band gap. Since the a-As_2Se_3 Fermi level energy of $\sim$ 0.9 eV defines a natural limit to any trapped charge thermalisation process, t_M data taken below room temperature should reveal that limit. The tendency of the results in fig.4 to level off at low temperatures seems to confirm this observation.

The change of E_M with temperature shows that on a time scale defined by the release rate(1), the trapped charge distribution moves down into the gap at a relatively slower rate at higher temperatures than it does at lower temperatures. The thermalisation approximation as outlined by Tiedje and Rose[2] can accomodate a small upward shift in the trapped-carrier center of gravity since the upper demarcation point $kT/(1-\alpha)$, with $\alpha = T/T_c$, moves up with temperature. But, given the T_c = 550 K obtained by Orenstein and Kastner[3] for a-As_2Se_3, which defines the downside demarcation point of kT_c, this cannot account for more than a 10^{-2} eV change between 300 K and 370 K, which is about one order of magnitude smaller than the observed E_M shift. These results consequently lend support to the Marshall and Main[10] analysis which revealed that the thermalisation approximation does not properly describe charge redistribution in the multiple trapping system.

A final comment should be reserved for the clear difference in measured t_M values originating from bulk material from those obtained with evaporated film samples, as illustrated in figs. 2 and 4. The difference amounts to a more or less constant factor 4 to 5 which, for $\nu = 10^{12}s^{-1}$, translates into an energy difference of 0.035 to 0.040 eV at room temperature. Since we are measuring the release of trapped charge to the transport band, it is the trap-to-band distance which is larger by this amount in films. We believe this difference reflects a widening of the mobility gap in evaporated a-As_2Se_3, where as a consequence of the increased disorder with respect to the bulk material, the electron states are being localized somewhat deeper into the bands.This difference may also be the reason why the dark current activation energy of a-As_2Se_3 is generally quoted as being $\sim$ 0,91 eV, while we have been consistently measuring 0.88 eV on bulk samples. The evaporated films we used for our experiments are not in any way special; they are standard, well-annealed specimen that reproduce well-known results when standard measurements are performed.

CONCLUSION

The underlying idea that in the long-time regime the transient photocurrent would be a replica of the trap dwell time distribution function, and hence lead to the density of localized states, could not be experimentally substantiated. From the invariance of the $t \gg t_T$ time-dependence with applied field we conclude that we are truly measuring release out of traps. But the temperature dependence of our results is seen as an indication that the assumption of a trap occupation directly proportional to the trap density does not hold. It should be feasible, however, to circumvent this assumption by solving the general multiple-trapping problem, not only for the excess free carrier density[8], but also for the occupation as a function of time, of traps at various depths. At which point it may be worthwhile to look at the long-time regime again.

ACKNOWLEDGMENTS

We thank J. Marshall for multiple discussions of multiple-trapping problems. This work has been supported by the Belgian Interuniversitair Instituut voor Kernwetenschappen.

REFERENCES

1. J.M. Marshall, Rep.Prog.Phys.46, 1235 (1983)
2. T. Tiedje & A. Rose, Solid State Commun 37, 49 (1980)
3. J. Orenstein & M.A. Kastner, Phys.Rev.Lett.46,1421(1981)
4. J. Orenstein, M.A. Kastner & V.Vaninov, Phil.Mag.B 46,23(1982)
5. T. Tiedje, B. Abeles & J.M. Cebulka, Solid State Commun 47,493 (1983)
6. H. Michiel, J.M. Marshall & G.J. Adriaenssens, Phil.Mag.B 48, 187(1983)
7. H. Michiel, G.J. Adriaenssens & J.M. Marshall, J.Phys.C 16, L 1005 (1983)
8. H. Michiel & G.J.Adriaenssens, to be published
9. D. Monroe & M.A. Kastner, Phil.Mag.B 47, 605 (1983)
10. J. M. Marshall & C. Main, Phil.Mag.B 47, 471 (1983).

METASTABLE PHOTOENHANCED THERMAL GENERATION IN a-SeTe ALLOYS

M. Abkowitz, G.M.T. Foley, J.M. Markovics, and A.C. Palumbo
Xerox Corporation, 800 Phillips Rd., Rochester N.Y. 14580

ABSTRACT

The mechanism of isothermal bulk carrier generation in the dark and the sensitizing effects of prior penetrating illumination and thermal history have been investigated in a composition series of a-Se:Te alloy films using xerographic techniques. In all cases, thermal generation is characterized by the simultaneous production of free holes and deeply trapped negative space charge. Results suggest that dark carrier generation is controlled by deep native (thermodynamic) defect centers whose population at a given temperature below T_g can be temporarily altered by photo and/or thermal excitation.

INTRODUCTION

Observations of light-induced changes in amorphous materials are well known. Examples of such phenomena are the photodarkening and photovolumetric effects in the chalcogenides[1] and the Staebler-Wronski effect in a-Si.[2] The phenomenon of light fatigue is a widely recognized problem in xerographic applications of amorphous chalcogenides. In this paper, the photoenhancement of bulk thermal generation rates in the a-Se:Te system is explored. Persistent conductivity, one manifestation of photoenhanced thermal generation, is normally associated with transient occupation of traps and the kinetics of trap emptying. There is increasing evidence, however, that illumination with bulk absorbed light can induce structural changes in the chalcogenides[3] and that accompanying changes in the integrated number of deep gap states are responsible for some of the striking observations in the electrical characteristics of samples so illuminated.[4] Tanaka suggests that large scale photoinduced structural changes are a feature of amorphicity.[1]

In these present studies, the mechanism of thermal generation is identified. Utilizing the depletion discharge technique,[5,6] a precise quantitative measure of the generation efficiency is possible as is a qualitative characterization of the gap state distribution with which carrier generation is associated. Experiments are described that lead to the conclusion that there are significant similarities between the photoenhanced and thermally enhanced bulk charge carrier generation phenomena in Se:Te alloy films.

BACKGROUND

The samples used in these studies were prepared by thermal evaporation of thermally blended Se:Te alloys at a rate of 2 μm/min. on to oxidized Al substrates held at 328K. They cover a range of compositions from pure Se to 15 at.% Te:Se.

In the experiments to be described, the sample is capacitively charged in the dark and the time resolved dark discharge is measured. That dark discharge of the surface potential can be driven by free carrier generation in the bulk or by surface-specific phenomena. Surface injection at the front and rear interfaces is largely eliminated by use of suitable blocking contacts, leaving carrier generation in the bulk as the primary decay mechanism. Bulk generation may, in principle, take two forms. In the first, generation and sweep out of equivalent numbers of mobile holes and electrons takes place. Under these conditions, the dark decay rate is described by the conventional capacitive discharge equation

$$\frac{dV}{dt} = -G_B \frac{L^2}{\varepsilon} \tag{1}$$

where G_B is the rate of generation of mobile electron-hole pairs, L is the sample thickness, and ε is the dielectric constant. A plot of the time dependence of dV/dt for such a process is basically featureless. The second form of bulk generation, depletion discharge, takes place via transport of a single mobile carrier. In the latter case, the process can involve either the deep trapping of electrons immediately after the thermal promotion of an electron and hole to their respective band (mobility) edges for example, or alternatively, the ionization of a deep center leaving behind a negatively charged core. Recent doping experiments in which the effect of monovalent additives, such as Cl, or bulk thermal generation in a-Se were studied,[6] indicate strongly that thermal generation is controlled by release from deep gap electronic states rather than by direct band-to-band transitions. The present observations on Se:Te alloys are consistent with the existence of emission centers and will be so interpreted.

In the depletion mode, if the uniform generation of carriers in the bulk is described by an algebraic time dependence, viz.

$$\rho = at^p \tag{2}$$

then the dark discharge is described by

$$\frac{dV}{dt} = -\frac{L^2}{2\varepsilon} apt^{p-1} \qquad t < t_d \tag{3}$$

and

$$\frac{dV}{dt} = -\frac{\varepsilon}{2a} (V_o/L)^2 pt^{-p-1} \qquad t > t_d \tag{4}$$

where V_o is the initial charge potential. The discharge curve on a log-log plot is characterized by a distinct change in slope at t_d, the depletion time. Physically, t_d is the time to generate an amount of charge in the bulk equivalent to the initial surface charge. As such, it reflects a convolution of the number of emission centers and their distribution in energy.

Figure 1 depicts typical dark discharge data for two different compositions (10.5 and 3.2 at.% Te:Se, respectively), which illustrate the predicted characteristics of depletion discharge behavior. Inflections in the log-log plots at the respective depletion times are readily identifiable.

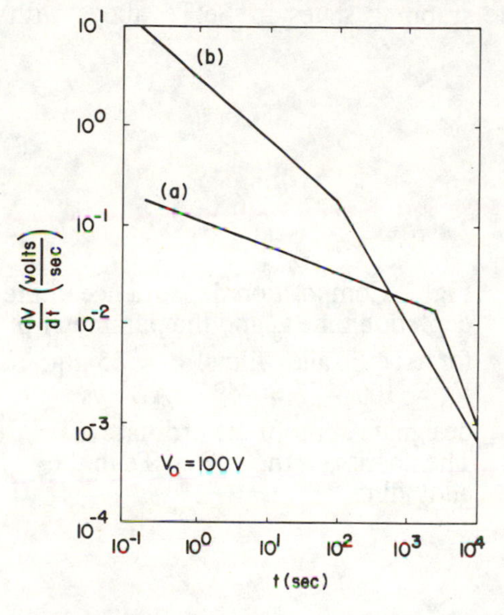

Fig. 1. Double log plot of dV/dt vs t immediately after charging to $V_o = 100$ V at 25°C for (a) 3.2 at.% Te:Se film and (b) 10.5 at.% Te:Se film, $d = 55\ \mu m$.

Analyses of the sign of the bulk space charge developed during dark decay can be made using the xerographic time-of-flight technique.[4,5] It can be readily demonstrated that thermal generation throughout the Se:As:Te system involves simultaneous production of mobile holes and deeply trapped electrons.

In addition to t_d, the parameter p provides a useful characterization of the dark discharge behavior and is readily obtainable from the slopes of the log dV/dt versus log t plots. Model calculations for different idealized distributions (in energy) of generation states indicate that, for a single discrete emission center, $p=1$. For a generalized exponential distribution, p is inversely proportional to the width of the distribution. Thus p is a measure of the energy distribution of the underlying emission centers.

RESULTS AND DISCUSSION

The data of Fig. 1 illustrate the applicability of the depletion discharge model and, simultaneously, serve to graphically illustrate the strong dependence of the dark discharge behavior on composition. This dependence is more explicitly shown in Fig. 2 in plots of t_d and p for the full range of alloy compositions studied. A significant increase in thermal generation rate is observed with increasing Te content, and this is reflected in a corresponding decrease in t_d. A value of $p=1$ is observed for pure selenium, representing emission from an essentially discrete generation center.[6] With addition of Te, p decreases linearly with increasing concentration of Te in the alloy, indicating a broadening of the distribution of emission centers consistent with increasing compositional disorder in the alloy film. This behavior parallels observations of prior studies,[7] which show progressive broadening of the manifolds of both shallow transport states and deep hole trapping states in Se:Te alloys with increasing Te content.

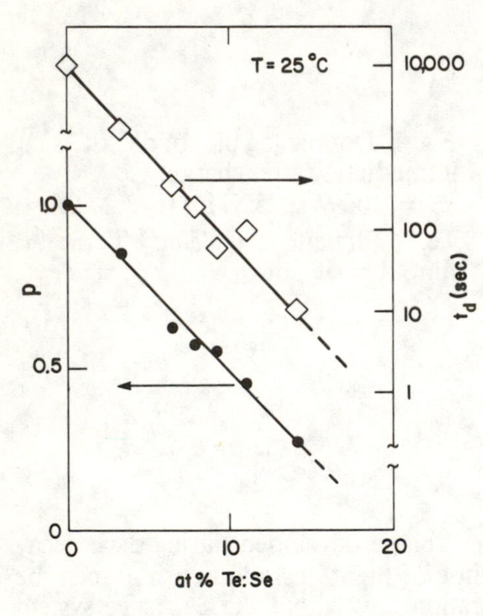

Fig. 2. Composition dependence of the depletion time t_d and the parameter p for a-Te:Se alloy films. d = 55 μm; V_o = 100 V, T = 25°C. Arrows designate appropriate ordinate scale. The abscissa is the at.% of Te in the alloy films.

Time resolved measurements of the response of the depletion time to step heating and cooling in these alloys provides evidence that the gap states that control thermal generation are thermodynamic defects. Figure 3 shows such data for a 7.8 at.% Te film. The observed time dependence of t_d during the step heating/cooling cycle illustrates some important characteristics:

(a)-(b) A rapid decrease is seen in t_d, reflecting the isostructural enhancement of the thermal generation rate for a fixed number and distribution of generation centers as a result of the increasing thermal energy of the carriers.

(b)-(c) Slow isothermal equilibration toward a stable asymptotic t_d is observed. This behavior is a clear-cut manifestation of the relaxation of a glassy structure toward a melt-like equilibrium, a well established characteristic of glassy systems annealing at temperatures $T_A(<T_g)$ not too far from their respective glass transition temperatures (T_g).[8] Measurements as a function of composition indicate that the rate of approach to equilibrium for a given T_A decreases with increasing T_g (higher Te content) as expected. Additional experiments verify that the relaxed value of t_d at any temperature T_A is independent of prior thermal history.

(c)-(d)/(d)-(e) On cooling to room temperature, the corresponding isostructural and isothermal changes for $T_A = 22°C$ are observed. The relaxation time for the isothermal phase at $T_A = 22°C$ is much longer than at $T_A = 35°C$, consistent with the expected relaxation behavior for a thermodynamic defect.

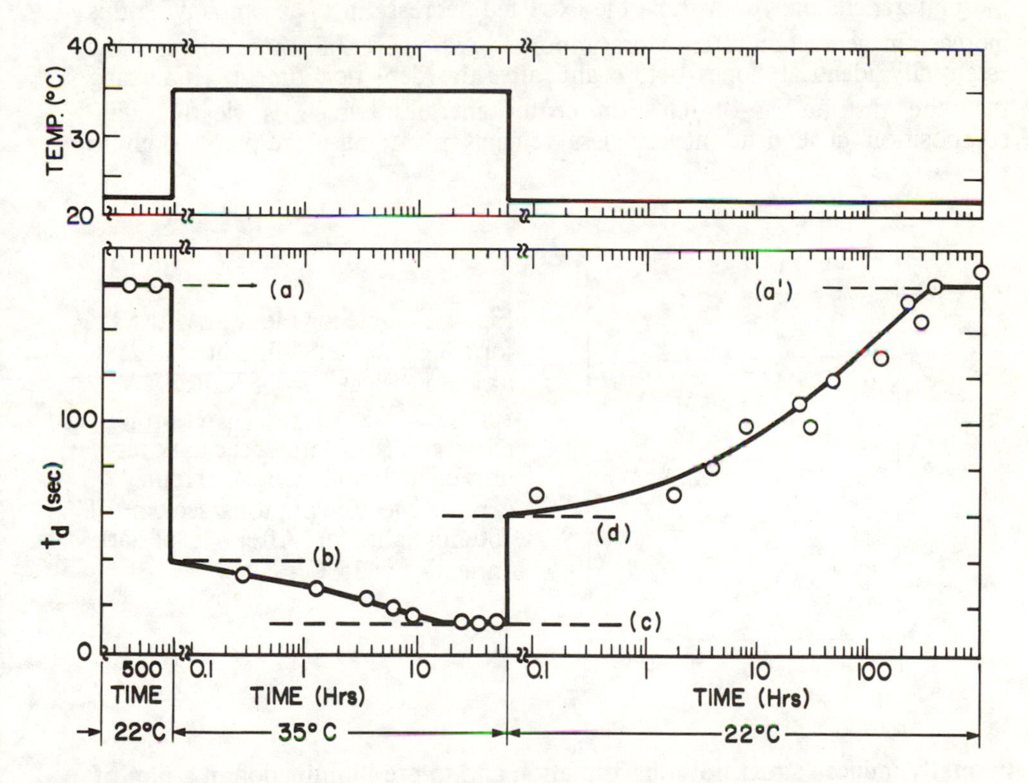

Fig. 3. Time-dependent variations in the depletion time t_d of a 7.8 at. % Te:Se film induced by step heating and cooling. The imposed thermal history is represented schematically in the top half of the figure. Response is divided into three time zones. In zone 1, specimen temperature is 22°C; in zone 2, it is stepped to 35°C. In zone 3, it is stepped back to 22°C. Shifts (a)-(b) and (c)-(d) occur under approx. isostructural conditions.

It can be shown that the observed asymmetry in the response of t_d to step heating at 22°C and step cooling from 35°C reflects the fact that the emission centers are distributed in energy.

Figure 4 compares depletion discharge data (V_o = 50 V) for a specimen of fixed composition (3.2 at.% Te) under different conditions. Curve (a) shows rested data for room temperature (25°C) dark decay. Curve (b) is room temperature data for the sample following careful pre-illumination with 750 nm radiation (~10^{14} photons/sec-cm^2 for 10 min) under isothermal conditions. (Note that the surface temperature of the sample was monitored to establish the absence of any heating effect.) Curve (c) is the dark discharge measured at 35°C for the same sample following a 4 hour anneal at 35°C. Curves (b) and (c) both show a striking increase in dark discharge rate, reflecting enhanced thermal generation. Both show the expected decrease in t_d accompanying this increase in generation rate. More importantly, however, all three curves show essentially identical slopes before and after the depletion time t_d, indicating that the distribution of emission center energies, which is clearly alloy-composition dependent, nevertheless remains invariant in response both to

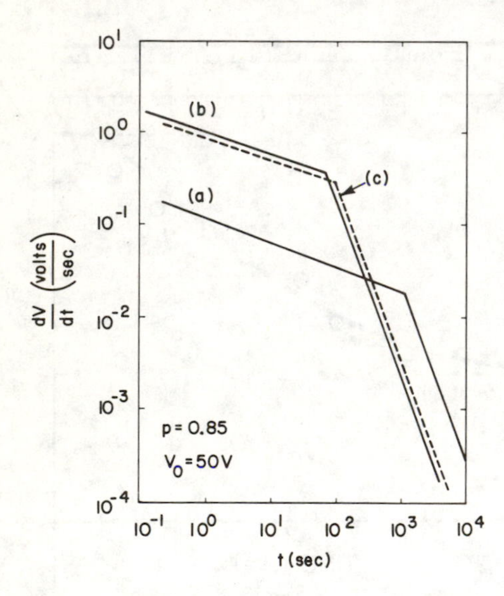

Fig. 4. Double log plot of dV/dt vs t for a 3.2 at.% Te:Se film at T = 25°C [(a) and (b)], at T = 35°C [(c)]. V_o = 50 V; d = 55 μm. (a). After prolonged dark rest. (b). After 2 sec dark rest immediately following a 30 min. exposure to 10^{14} photons/sec-cm^2 of 750 nm light. (c). After 4 hr of dark annealing at 35°C.

thermally induced structural transformation and to pre-illumination in a film of fixed composition. These observations are consistent with earlier xerographic studies in which deep trap distributions were measured directly in a-Se:As alloy films.[9] It was determined that thermal annealing of specimen films at various temperatures below T_g clearly effected changes in trap state densities, but not in their energy distributions.

Figure 5 examines the dark decay (depletion time t_d) isothermal relaxation-recovery behavior at room temperature for specimens of the fixed composition 2.5 at.% Te:Se, which have been respectively photo-(circles) and thermally (squares) sensitized. Similar data have been collected for films spanning the entire composition range 2.5 at.% Te:Se to 15 at.% Te:Se. It is observed that isothermal relaxation in the dark at room temperature becomes progressively slower for both thermally and photosensitized alloy specimen films as Te content is progressively increased. Furthermore, it is clear that when the respective t_d values immediately after sensitization do not differ greatly, then the time required for the isothermal recovery of a given film that has been either photo- or thermally sensitized is about the same. The most plausible interpretation of these data, given that the results of earlier xerographic studies in Se:As alloys identify deep trapping centers as native (thermodynamic) structural defects,[9] is to also identify hole thermal generation centers in Se:Te alloys as deep states arising from structural defects. The effect of thermal cycling (i.e., step heating and step cooling) or of sustained bulk photoexcitation is to temporarily alter the population of these native defects – a process clearly identified in an earlier study of halogenated a-Se.[6] The effect of dark isothermal annealing at temperature T_A is then to eventually restore the equilibrium native defect population with a rate of recovery that depends on T_g-T_A.

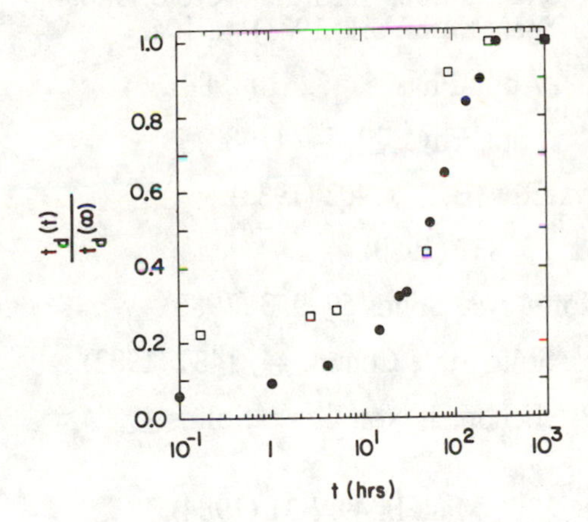

Fig. 5. Normalized relaxation behavior of depletion time $t_d(t)/t_d(\infty)$ as a 3.2 at.% Te:Se alloy film dark anneals at 24°C immediately after: (1) rapid quench from a well annealed state at 35°C (squares) (2) 10 min. illumination with ~10^{14} photons/sec-cm^2, 750 nm light at 25°C (circles). V_o = 50 V; d = 55 μm.

CONCLUSIONS

1. Dark decay in a series of a-Se:Te alloy films is clearly demonstrated to be a depletion process in which mobile holes are removed from the bulk, leaving behind a deeply trapped negative space charge.

2. Doping experiments with monovalent atoms, such as Cl, demonstrate that bulk thermal generation even in a-Se films is controlled by deep centers that are thermodynamic defects.[6] In the present study, this presumption

has been further verified by examining the time-dependent response of thermal generation, via the depletion time, to step changes in temperature.

3. It has been demonstrated that bulk thermal generation in the dark is enhanced by pre-illumination with penetrating light. The depletion characteristics of such sensitized specimens, and their relaxation recovery during dark annealing, bear a striking similarity to those in which bulk thermal generation is enhanced by a purely thermally induced structural change. In a series of earlier studies it was demonstrated that, for chalcogenide glasses of fixed composition, thermostructurally induced changes always involve a transient increase in population of native defect states whose position and distribution in the mobility gap is fixed.[9] Similar conclusions are suggested by the data of these present measurements. Specifically, long lived photoeffects, such as metastable enhanced deep trapping[4] and photoenhanced dark decay, involve a transient increase in the population of states that normally control these processes in the dark rested specimen.

REFERENCES

1. For an overview, see K. Tanaka, in Fundamental Physics of Amorphous Semiconductors, Proc. of the Kyoto Summer Institute, Kyoto, Japan, 1980, edited by F. Yonezawa (Springer, New York, 1981), p. 104.
2. D.L. Staebler and C.R. Wronski, J. Appl. Phys. 51, 3262 (1980).
3. H. Koseki and A. Odajima, Jap. J. Appl. Phys. 22, 542 (1983).
4. M. Abkowitz and R.C. Enck, Phys. Rev. B, 27, 7402 (1983).
5. A.R. Melnyk, J. Non-Cryst. Solids 35, 837 (1980).
6. M. Abkowitz and F. Jansen, J. Non-Cryst. Solids 59, 953 (1983).
7. M. Abkowitz and J.M. Markovics, Solid State Comm. 44, 1431 (1982).
8. J.P. Larmagnac, J. Grenet, and P. Michon, J. Non-Cryst. Solids 45, 157 (1981).
9. M. Abkowitz and J.M. Markovics, Phil. Mag. B. 49, L31 (1984).

IR-INDUCED TRANSIENTS OF PHOTOLUMINESCENCE AND PHOTOCONDUCTIVITY IN a-Si:H

R. Carius and W. Fuhs
Fachbereich Physik, University of Marburg, F.R. Germany*

ABSTRACT

We report on a study of transients of photoluminescence (PL) and photocurrent (PC) of a-Si:H-films which arise when at low temperature (10-110 K) the samples are excited by 1.92 eV-photons and are illuminated after a dark period t_D with IR-light ($h\nu < 0.7$ eV). We associate these transients with the photo-excitation of electrons and holes which are trapped in the band tails with long lifetimes. By variation of t_D and by comparison with the long time decay of the residual light induced ESR, the recombination of the metastable trap population is studied in differently doped films. The results indicate that below 80 K recombination proceeds by tunneling processes.

INTRODUCTION

The recombination of excess carriers in a-Si:H has intensively been studied in particular by photoluminescence (PL), photoconductivity (PC) and light-induced electron spin resonance (LESR). It is generally accepted that at low temperature carriers are trapped in band tail states from where they either recombine radiatively by tunneling to carriers in the other band tail or non-radiatively by tunneling to defect states. The relevant recombination centers have been identified as Si-dangling bonds, which depending on the doping level may be empty (D^+), singly occupied (D^0) or doubly occupied (D^-). When the temperature is raised thermal detrapping and diffusion to defects enhance the non-radiative recombination rate. Two conflicting models are discussed for radiative recombination. In the geminate model[1] it is assumed that the excited electrons and holes do not diffuse apart to great separations such that the recombination occurs between trapped geminate pairs. In the distant pair model[2] one assumes that the carriers are able to diffuse and are trapped at random in band tail states. In both models a broad distribution of intra-pair separations generates a broad distribution of lifetimes. It is an inherent property of both models[3,4] that at low temperature a metastable distribution of trapped electrons and holes builds up during illumination which have large intra-pair separations and correspondingly long lifetimes. These non-equilibrium distributions of trapped carriers persist after illumination and the long time decay of the residual LESR signal has been used to study their recombination[5]. These trapped carriers can be excited by IR-light. In a two-beam experiment this gives rise to the IR-quenching of the photocon-

* This work has been supported by the Bundesminister für Forschung und Technologie (BMFT).

ductivity above 100 K and at low temperature to quenching of the stationary photoluminescence and to enhancement of the photoconductivity[6]. In addition, when at low temperature the sample is kept in the dark after excitation with band gap light, the optical freeing of trapped carriers by IR-excitation causes transients of PC and PL[6]. We report here on a more detailed study of such transients in the temperature range 10-100 K.

EXPERIMENTS AND DISCUSSION

In these experiments a-Si:H-films were used which had been prepared by glow discharge decomposition in a capacitively coupled glow discharge system using the following parameters: 5 % SiH_4 diluted in He, rf-power 4W, frequency 13.56 MHz, total pressure 0.5 mbar, flow rate 80 sccm, substrate temperature 280°C. The deposition rate amounted to 2.5 Å/s. Doping was achieved from the gas phase by adding PH_3 or B_2H_6. The Fermi level positions E_C-E_F as determined from the dark conductivities were 0.87 eV for undoped films, 0.59 eV for films doped with 100 ppm PH_3 and 1.01 eV for p-type films at a doping level of 100 ppm B_2H_6.

The samples were excited with light of 1.92 eV photon energy. After switching off the illumination the decay of the LESR-signal was recorded by setting the magnetic field at the peak of the absorption derivative of the electron-dangling bond line. It is worth noting that the line shape of the LESR-spectrum did not change during the decay, i.e. the signal of the trapped majority carriers and that of

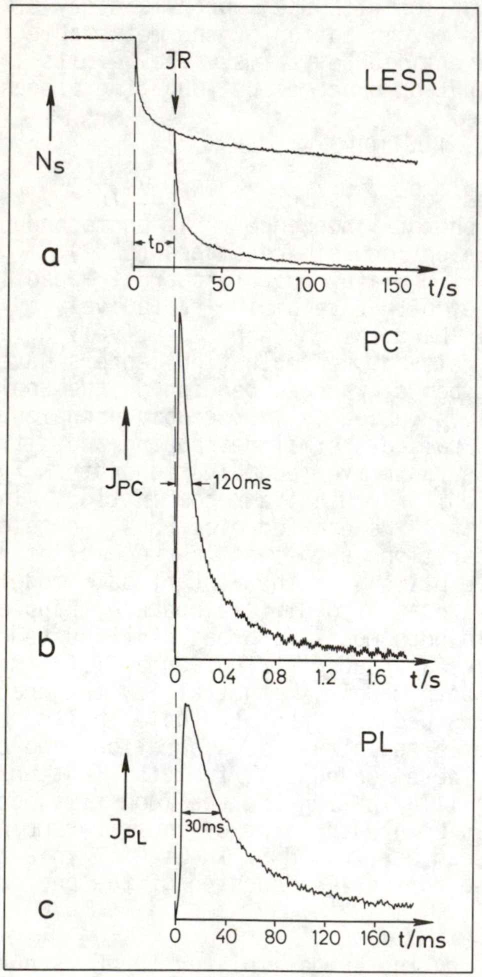

Fig. 1. Transients of LESR (a), PC (b) and PL (c), when after a dark period t_D the films are exposed to IR-light ($h\nu$ < 0.7 eV). The same light intensity is used for (b) and (c), whereas that for (a) was much lower. T = 15 K. Note different time scales.

the neutral dangling bonds decayed in the same manner. Fig. 1a displays such a transient recorded with an integration time of 0.1 s. It has the typical form with a rapid decay followed by an extremely slow long time decay, which indicates that a metastable trap population persists with a density of 10^{16} to $10^{17} cm^{-3}$. When after a dark period t_D the sample is exposed to IR-light ($h\nu < 0.7$ eV), which was obtained by passing the light of a tungsten lamp through a Ge-filter, the residual LESR is effectively quenched (Fig. 1a). Connected with this removal of the trapped carriers are transients of photoconduction (PC, Fig. 1b) and photoluminescence (PL, Fig. 1c). Such signals can be detected in undoped and doped a-Si:H-films of widely different doping levels and film quality. The detection system in these measurements had risetimes of below 30 ms for the PC and below 10 ms for the PL-transients. The total PL-intensity was detected by a S1-photomultiplier, hence the light emission is above 1.1 eV, i.e. in the intrinsic luminescence band.

The observation that IR-excitation quenches the LESR signal and thereby generates transients of PL and PC strongly indicates, that the IR-induced effects, reported here, originate from the same non-equilibrium carrier distribution. By IR-illumination trapped carriers are excited into states above the mobility edges where they contribute to conduction until they are immobilized in states near the mobility edges or recombine via defect states. At low temperatures ($T < 50$ K) thermal detrapping is negligible and the response time, τ_{resp}, of the photocurrent has been shown to be very short, $\tau_{resp} \ll 10^{-6}$s[7]. If, as shown in Fig. 2, the IR-excitation is removed, the PC-transient therefore decays almost instantly on the time scale of this experiment and when it is turned on again the transient recommences at the current level to which it has decayed during the interruption. From this behaviour it is evident that the large halfwidth of the signal does not arise from the transport and multiple trapping processes. In addition, if one calculates the number of carriers which are contained in such a current transient using the value of $\mu\tau \approx 10^{-11} cm^2/Vs$, which is typically found in the low temperature range[8], one obtains concentrations in excess of $10^{18} cm^{-3}$. This is much higher than what is indicated by the residual LESR. We believe therefore that the form of these signals has to be assigned to the decay of the generation rate of free carriers, which is proportional to the number of trapped electrons and holes. Retrapping and multiple excitation then are supposed to lead to the large halfwidths of these signals. The peak values of

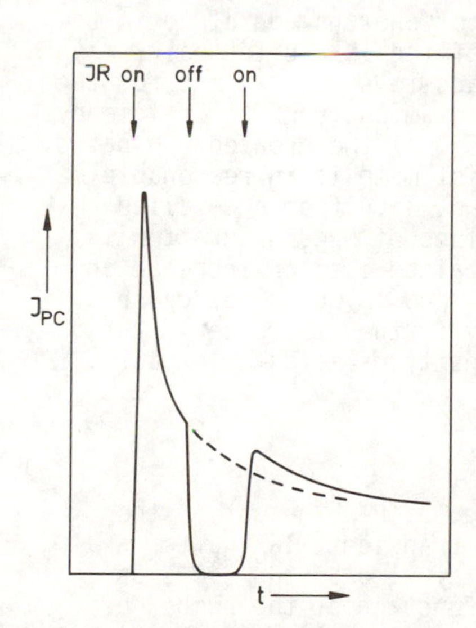

Fig. 2. PC-transient at 15K when the IR-excitation is interrupted.

the PL- and PC-transients, I_{PL} and I_{PC}, are determined by the IR-induced generation rates i.e. by the concentration of trapped carriers at $t = t_D$ and can be taken as a measure for the metastable trap population. With increasing intensity of the IR-light, I_{PL} and I_{PC} increase following a power law with an exponent near 1.

Obviously part of the excitation processes result in electron-hole pairs of shorter separation, when the carriers are retrapped, enabling radiative recombination. It is not possible to decide whether these carriers are still geminate. If they were, they would not contribute to photoconduction. On the other hand, the PC-transient can originate only from those electrons and/or holes which finally recombine radiatively or non-radiatively following distant pair kinetics. It seems thus possible, that although both transients are due to excitation from the same metastable distribution they represent only part of the excited carriers. For instance, the PL-transients may arise from those electrons and holes, which before excitation already had a small separation and which after being excited diffuse towards each other.

The building up of the metastable distribution can be examined by varying the number of absorbed photons, N_{abs}, by changing either the intensity G_{ex} of the exciting light (1.92 eV) or the exposure time t_{ex}. In Fig. 3 the peak height of the PC-transient after a delay time of t_D = 10 s is plotted as a function of N_{abs}. Similar behaviour is obtained for the PL-signal. After each exposure most of the electron-hole pairs have recombined after t_D = 10 s and the pairs with longer lifetimes are not yet saturated if N_{abs} is small. In this range it is possible to obtain from the PL-data an estimate for the production efficiency of pairs with long lifetime. This is achieved by comparing the number of photons, which are emitted during the excitation time t_{ex}, with the number of photons contained in the IR-induced PL-transient. Since not all of the IR-excited pairs may recombine radiatively, this procedure leads to a lower limit for the production efficiency. Assuming for the stationary PL an efficiency of 0.3 one finds that at low N_{abs} about 6 % of the created e-h-pairs have lifetimes in excess of 10 s. This estimate is in reasonable agreement with the conclusions of Street et al.[3] from a detailed comparison of PL and LESR. At higher values of N_{abs} saturation is attained i.e. an increase in G_{ex} and t_{ex} does not cause any further increase of the density of metastable pairs.

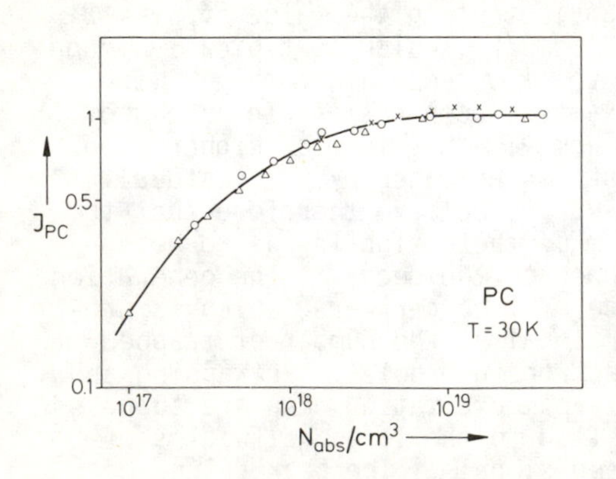

Fig. 3. Peak height of the PC-transient, I_{PC}, after a delay time of t_D = 10 s as a function of the number of absorbed photons (1.92 eV). Different powers and excitation times are used.

The number of absorbed photons where this occurs is roughly $10^{19}cm^{-3}$.

If recombination proceeds by tunneling, the lifetime of pairs with intra-pair separation r is given by[1]

$$\tau = \tau_o \exp(2\, r/r_o) \qquad (1)$$

In this expression r_o denotes the extension of the more extended state which in case of band tail electrons is supposed to be near 10 Å[5]. One usually assumes that the prefactor τ_o amounts to 10^{-12}s for non-radiative and to 10^{-8}s for radiative recombination. The excitation with 1.92 eV-photons leads to a random distribution of intra-pair separations and thus through relation (1) to a broad distribution of lifetimes. After t_D = 10 s all carriers with $\tau < 10$ s have recombined and according to relation (1) the remaining pairs have large separations $r > 125$ Å for non-radiative tunneling. This limits the concentration of trapped carriers to near $10^{17}cm^{-3}$, which is in accordance with the observed residual LESR-spin densities[5].

We are going to discuss now the decay of the metastable trap population which can be studied by variation of the delay time t_D. In these measurements the number of absorbed photons at each set of t_D was chosen such that the trap population had been saturated. Fig. 4a

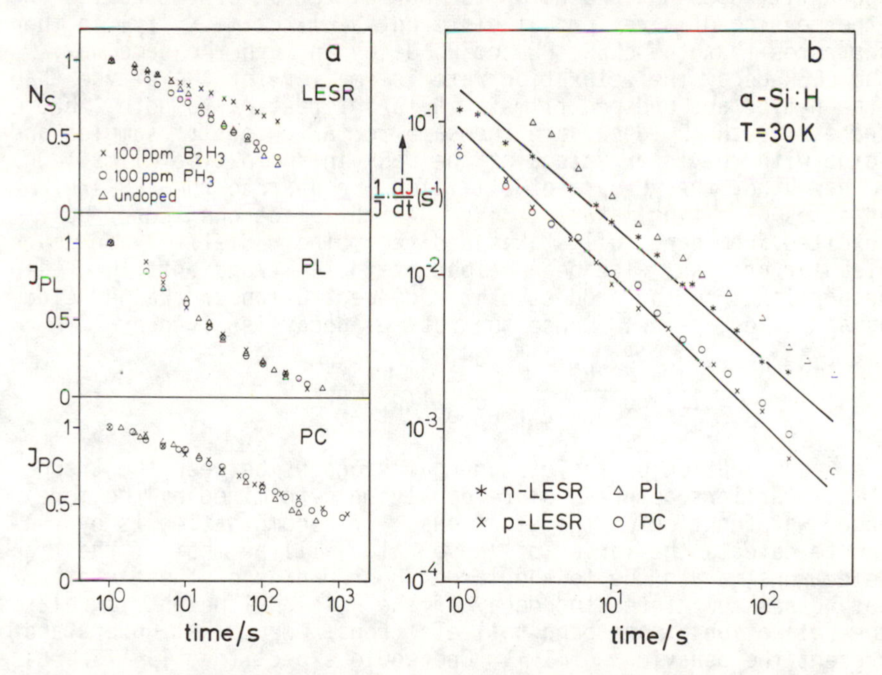

Fig. 4. (a) Decay of the LESR, PL- and PC-peak heights at 30 K normalized to their value after t_D = 1 s. (b) Plot of the relaxation rate 1/I dI/dt versus decay time. The full curves have been calculated from relation (4) using r_m = 200 Å, r_o = 10 Å for n-type and r_o = 8 Å for p-type films.

displays the decay of the residual LESR and the peak amplitudes of the PL- and PC-transients of differently doped samples. The LESR-decays are quite similar to those reported in the literature[5], the decay being considerably faster for n-type and undoped than for p-type films. In the doped films the LESR-spectrum consists of the signals of the trapped majority carriers and of the neutral dangling bonds. It is thus reasonable to assume that the recombination proceeds by tunneling of band tail carriers to the neutral dangling bond i.e. to the minority carriers which are trapped at the defect states. The differences between the decays of n- and p-type samples then indicate that the band tail holes are more strongly localized than the band tail electrons. It is surprising that in the decays of the PL- and PC-signal heights these differences do not reveal, n- and p-type films in these cases behave quite similarly.

In order to enable a comparison of the decays of these different quantities it is convenient to plot instead of the peak heights, I, the logarithmic derivative

$$\frac{1}{I} \cdot \frac{dI}{dt_d} = \tau_d^{-1} \tag{2}$$

This quantity does not depend on the normalization of the curves and has the meaning of a reciprocal differential lifetime. Fig. 4b then demonstrates that the LESR, PL and PC decay in much the same way. In the LESR-decay the relaxation rate is smaller for the p-type than for the n-type and undoped films. It is interesting to note, that the rates for the PC-decay are the same for all kinds of samples and coincide with the lower values of the LESR in the p-type films. On the other hand, the PL-data of all films lie near to the LESR-relaxation rates of n-type material. By IR-light electrons and/or holes are excited from band tail states and recombine radiatively in a bimolecular process. The peak height of the PL-transient, I_{PL}, then is proportional to the product of the concentration of trapped electrons and holes, $p_t \cdot n_t$. Hence the rate of decay is given by

$$\frac{1}{I_{PL}} \cdot \frac{dI_{PL}}{d\, t_D} = \frac{1}{n_t} \cdot \frac{dn_t}{dt_D} + \frac{1}{p_t} \cdot \frac{dp_t}{dt_D} \tag{3}$$

If the recombination of the distant pairs occurs between the trapped band tail carriers both rates are equal and determined by the more extended wavefunction of the electrons. If recombination is by tunneling to defects the larger of these rates will be that of the trapped electrons, again due to the larger Bohr radius of these states. In both cases therefore, the decay of the PL-signal height is related to the relaxation of the band tail electrons. We do not understand at present the behaviour of I_{PC}. One would expect that I_{PC} is proportional to $p_t + n_t$ and is determined by the majority carriers. As a consequence the decay of I_{PC} of the n-type sample would be dominated by the electrons and therefore should follow the n-type LESR-decay. The fact that this is not observed could mean that IR-light excites in all films predominantly holes and I_{PC} then should be considerably larger in p-type than in n-type films. This, however, is not found,

I_{PC} is of comparable magnitude in all kinds of samples.

The form of these decays is typical of a broad spectrum of time constants which originates in a natural way from a distribution of intra-pair separations according to relation (1). Assuming a simple rectangular distribution of intra-pair separations the relaxation rate can easily be calculated. If after a time t, all pairs with $r < r_t = r_o/2 \ln t/\tau_o$ have recombined the rate is given by

$$\frac{1}{N_p}\frac{dN_p}{dt} = \frac{r_o}{r_m}\frac{1}{t(1 - r_o/2r_m \ln t/\tau_o)} \tag{4}$$

where r_m is the maximum intra-pair distance in the distribution. N_p denotes the concentration of pairs which may be trapped band tail electrons and holes in undoped samples or trapped majority carriers and neutral dangling bonds in doped films. This simple model gives a surprisingly good description of the experimental results. The full curve through the n-type data in Fig. 4b has been calculated using $r_o/r_m = 1/20$ and $\tau_o = 10^{-12}$s. It coincides fairly well with the LESR-decay of the n-type films. The decay of the p-type films can be fitted using a value of r_o which is smaller by 20 %. If it is assumed that r_m is the same in both types of samples and amounts to r_m = 200 Å, the radius of the band tail electrons is 10 Å and that of the band tail holes 8 Å. Thus the localization is stronger in the valence band tail. These numbers agree satisfactorily with other estimates in the literature[5].

The long-time decay of the metastable trap population is essentially independent of temperature below T = 80 K. Fig. 5 shows LESR- and PC-decays of the p-type sample at various temperatures, n-type films behave quite similarly. Up to 80 K the curves coincide and only at 110 K the relaxation rate is considerably enhanced. This result supports the assumption that at low temperature recombination occurs by tunneling processes between localized states. Above about 80 K thermal detrapping of band tail carriers or hopping processes of carriers among the band tail states may enhance the recombination when the carriers diffuse towards defect states.

These results show that the residual light induced ESR and the IR-generated

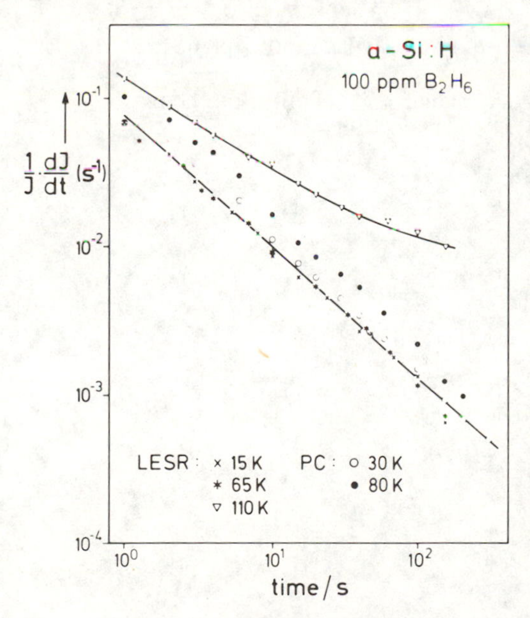

Fig. 5. Relaxation rates of LESR and PC of p-type films versus decay time at various temperatures.

transients of photoluminescence and photoconductivity originate from the same metastable trap population. It is found that the LESR, PC- and PL-transients decrease at the same rate with the delay time between the end of excitation and the onset of excitation by IR-light. A study of the IR- and PC-transients thus yields similar information as the residual LESR. In cases, where it is difficult to measure LESR, the simple method of IR-induced PC-transients therefore can give valuable information on the recombination of the trapped carriers[9]. The recombination of trapped carriers proceeds at low temperature by tunneling processes. The results indicate stronger localization for the states in the valence band tail than in the conduction band tail and the data are consistent with localization radii of 8 Å and 10 Å for band tail holes and electrons respectively.

REFERENCES

1. R.A. Street, Advances in Physics 30, 593 (1981)
2. D.J. Dunstan, F. Boulitrop, J. Phys. 42, C4-331 (1981)
3. R.A. Street, D.K. Biegelsen, Solid State Comm. 44, 501 (1982)
4. D.J. Dunstan, Solid State Comm. 49, 395 (1984)
5. D.K. Biegelsen, R.A. Street, W.B. Jackson, Physica 117B/118B, 899 (1983)
6. R. Carius, W. Fuhs, M. Hoheisel, Int. Top. Conf. on Transport and Defects in Amorphous Semiconductors (Bloomfield 1984, to be published in J. Non-Cryst. Solids)
7. M. Hoheisel, R. Carius, W. Fuhs, J. Non-Cryst. Solids 59/60, 457 (1983)
8. M. Hoheisel, R. Carius, W. Fuhs, J. Non-Cryst. Solids 63, 313 (1984)
9. M. Hundhausen, L. Ley, R. Carius, Proc. 17th Conf. on the Physics of Semiconductors, San Francisco 1984

INFRARED MODULATION OF PHOTOLUMINESCENCE IN GLOW DISCHARGE AMORPHOUS SILICON

C. Varmazis*, M. D. Hirsch** and P. E. Vanier
Metallurgy and Materials Science Division
Brookhaven National Laboratory, Upton, NY 11973

ABSTRACT

Dual beam photoluminescence (PL) from a-Si:H was measured at various temperatures by exciting the sample with a modulated infrared (IR) beam in addition to a steady beam of visible light. Varying the modulation frequency revealed that the IR caused a transient enhancement of the PL, followed by a slow quenching effect. The IR excitation spectrum showed quenching by photons with energies between 0.45 and 1.1 eV, similar to the IR quenching of photoconductivity. The fast process enhanced the emission peak equally at all energies, while the slow process quenched only the high energy side of the peak. These results can be explained by the re-excitation of holes from "safe traps" by the IR, followed by recombination at dangling bonds.

INTRODUCTION

When a film of a-Si:H is illuminated with a constant source of visible light, the occupancy of the gap states changes from the thermal equilibrium distribution to a quasi-equilibrium in which the rates of all transitions involving a given state are balanced. If a second beam of photons, with energy less than the optical gap, is directed onto the film the gap state occupancies are again changed. In such a two-beam experiment, the IR beam is absorbed more strongly than in the absence of the visible excitation because trapped photoexcited carriers can be re-excited into the bands by IR photons.[1] To a first approximation, it would seem that such re--excitation would cause an enhancement of photoconductivity, but would have little effect on photoluminescence. In fact, the enhancement of photoconductivity is only transient,[2] and the net effect in steady state for intrinsic films at low temperatures is a reduction in the photoconductivity by as much as a factor of 20.[3] In this work, both enhancement and quenching of photoluminescence by IR are observed. These results are similar to those recently reported by Carius et al,[4] but different from those of Bhat et al.[5]

*Present address: University of Crete, Iraklion, Greece.
**Present address: AT&T Bell Laboratories, Reading, PA.

EXPERIMENTAL DETAILS

Figure 1 shows a diagram of the apparatus for dual beam photoluminescence measurements. The samples were glow-discharge deposited on roughened sapphire substrates using pure silane at a pressure of 0.1 Torr and a substrate temperature of 225 C. A He-Ne laser was used for steady state excitation with a photon flux $\sim 10^{16}$ $cm^{-2} s^{-1}$, and a tungsten-halogen lamp was used for the chopped IR source ($\sim 10^{17}$ $cm^{-2} s^{-1}$). For measurement of the IR excitation spectrum, a source monochromator (not shown) was placed in the IR beam. Thick slices of crystalline Ge or Si were placed between the chopper blade and the sample to absorb both the visible part of the lamp spectrum and the near-IR emission from the chopper blade. Without adequate filtering, the former component of the beam would produce a spurious positive modulated signal while the latter would give a negative one. The amount of IR radiation reflected into the detection system was made negligible by having a hole in the sample support behind the illuminated area. Long pass filters were used to reject the scattered visible light. The luminescence was detected by a cooled S-1 photomultiplier and the modulated signal component was extracted by a lock-in amplifier.

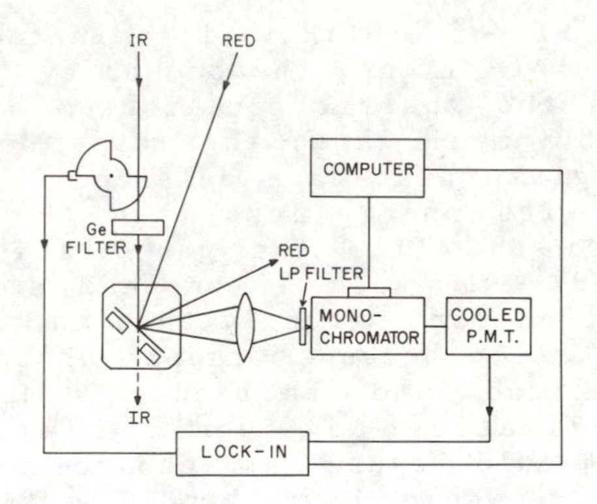

Fig. 1 Dual beam photoluminescence apparatus.

RESULTS AND DISCUSSION

The existence of both enhancement and quenching effects is illustrated by varying the chopping frequency, as shown in Fig. 2. For these measurements, the detection monochromator was set to detect 1.4 eV emission, and the temperature was held at 80 K. At low frequencies (DC-10Hz), the IR causes a net quenching of the PL signal, whereas at higher frequencies (200 Hz) the

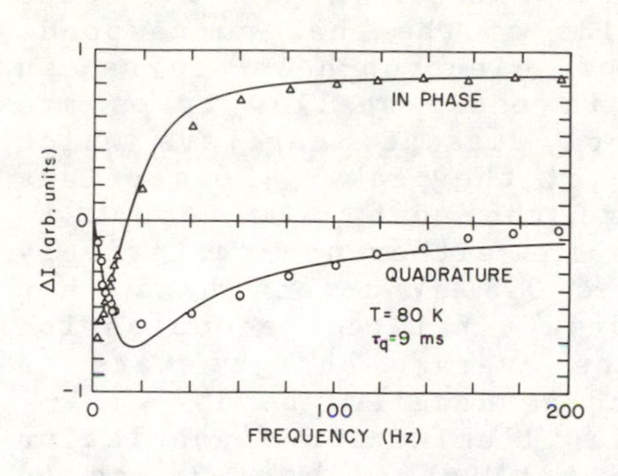

Fig. 2. Chopping frequency dependence of IR modulation of photoluminescence.

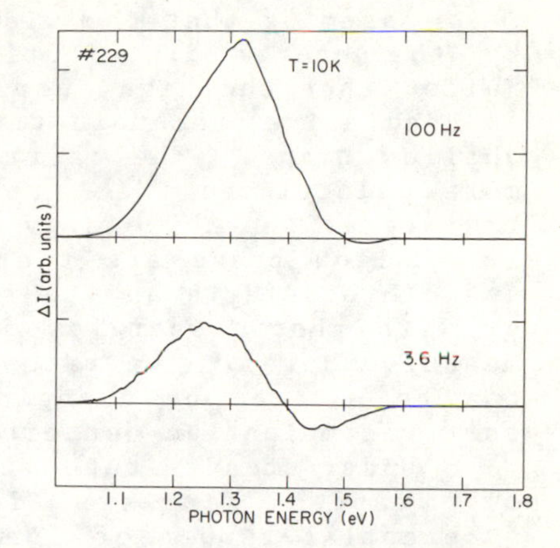

Fig. 3. IR-modulated PL emission spectrum of undoped film.

modulation is positive and fairly flat with further changes in frequency. The implication is that there is an immediate positive response to the IR, followed by a slow quenching effect with greater absolute magnitude than the positive effect.

The solid lines in Fig. 2 are calculated assuming that the quenching effect behaves like a forced harmonic oscillator with a single time constant τ_q=9 ms. The fit is not perfect, suggesting that in reality there is a distribution of time constants. At T=120 K, the best fit is obtained for τ_q=6 ms, using the same excitation conditions. If 1/ τ_q is taken as the rate of a characteristic transition, an activation energy of 8 meV is obtained. Such a weak temperature dependence indicates that the rate-limiting step in the quenching process is a tunneling transition rather than thermal emission into a band. Very similar behavior was reported by Persans for dual-beam photoconductivity, where the time constant was found to vary with temperature and with the intensity of visible light.[2]

Rather than detecting the PL emission at a fixed energy, it was found to be more revealing to measure the emission spectrum at various temperatures and chopping frequencies. For example, Fig. 3 shows the IR modulated part of the PL from an undoped film measured at T= 10 K. At 100 Hz, the spectrum consists almost entirely of the enhancement effect, which has the same lineshape as the single-beam PL peak. These spectra have not been normalized for the response of the photomultiplier, which falls off at low energies. At 3.6 Hz the slow quenching effect is evident, but it affects the high energy side of the peak more strongly than the low energy side.

Assuming that the width of the PL peak is determined by the energy distribution of the traps in the band tails, then the high energy side of the peak corresponds to radiative recombination of electron-hole pairs in shallow traps. The carriers in these shallow traps are more delocalized than the deeply trapped carriers which radiate at the low-energy side of the peak. The carriers in shallow traps are therefore the most likely ones to tunnel to defects and to recombine either non-radiatively or with photon energies in the 0.8 eV defect band (not measured here). The rate of defect-related recombination can be altered by changing the average charge state of fast recombination centers such as dangling bonds.

Under steady state visible illumination a population of trapped carriers is established containing a substantial number of long-lived pairs which are widely separated and also far from any defects. In a crystalline material these distant pairs can be distinguished spectrally from the close pairs with short lifetimes, because the Coulomb energy of the pair is the dominant source of line broadening (~0.02eV).[6] However in a-Si:H the PL from distant pairs is not clearly separable from that of the close pairs, because the Coulomb energy is small compared to the disorder-related broadening (~0.4eV).

Since the valence band tail is much broader than the conduction band tail in a-Si:H, it is suggested that the dominant effect of the IR is to liberate a large number of deeply trapped holes. Some of these re-excited holes are re-trapped closer to electrons than they were before the IR was turned on. This effectively shortens their radiative lifetime, increasing the intensity of PL detected at the higher chopping frequencies. On the other hand, many of the holes are retrapped closer to defects such as dangling bonds than they were before the IR re-excitation. For these holes the non-radiative recombination rate increases, and the PL signal at low chopping frequencies or in steady state is quenched.

The effect of moderate doping on the IR-modulated PL spectrum is shown in Fig. 4 (100 ppm PH_3) and Fig. 5 (100 ppm B_2H_6). It is well known that doped a-Si:H has a higher defect density than undoped material. Consequently the PL efficiency is lower and the signal-to-noise ratio poorer for these samples than for the intrinsic one. In addition, the higher defect densities result in a smaller enhancement effect and a more pronounced quenching effect. Thus at temperatures below 20 K the 3.6 Hz modulation is totally quenching. At 200 Hz the predominant effect in the n-type film is enhancement, but the p-type film still shows quenching for the high energy pairs.

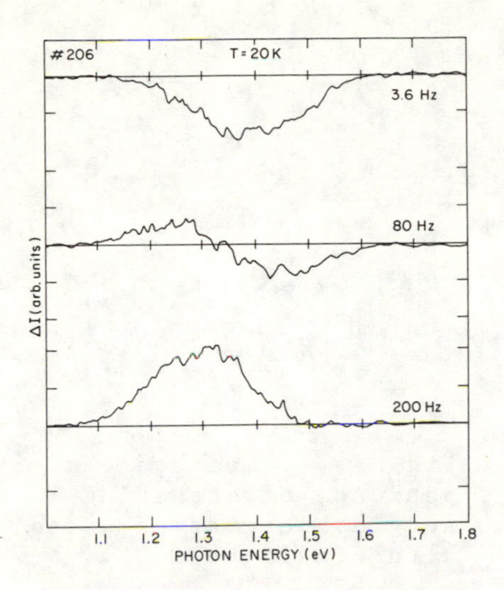

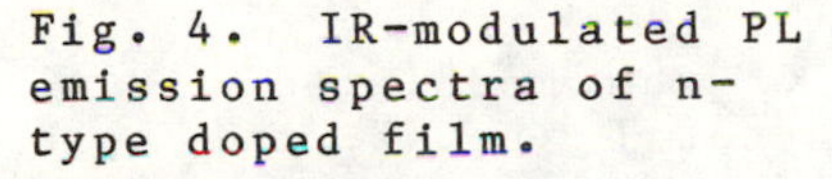
Fig. 4. IR-modulated PL emission spectra of n-type doped film.

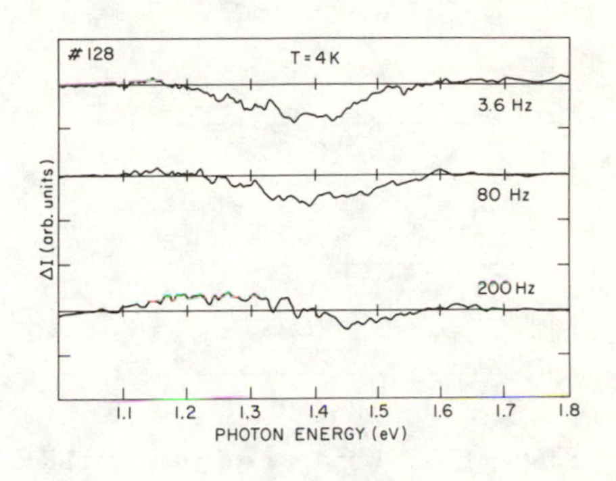

Fig. 5. IR-modulated PL emission spectra of p-type doped film.

The spectra shown above were all taken at very low temperatures (4-20 K) where the modulation effects are most clearly seen. At higher temperatures, the spectrum obtained at 3.6 Hz changes progressively as shown in Fig. 6. Both kinds of modulation are reduced with increasing temperature, but the enhancement effect disappears more rapidly, leaving a purely quenching effect at 120 K. In this sense, raising the temperature has a similar effect to doping -- it increases the non-radiative recombination rate.

In order to plot the temperature dependences of the enhancement and quenching effects separately, a deconvolution was performed in which it was assumed that the enhancement always had the same peak shape, independent of chopping frequency. The magnitude of the enhancement for each temperature was obtained from the spectrum measured at high chopping frequency. This spectrum was then subtracted from the low frequency spectrum to obtain that of the quenching effect by itself. This turned out to be a simple peak, slightly narrower than the single-beam peak and shifted to higher energy by about 80 meV. Figure 7 shows the magnitudes ΔI of the enhancement (■ 80 Hz and □ 100 Hz) and quenching (▲ 3.6 Hz) peaks for an undoped film as a function of temperature. Also plotted is the single beam PL(•), reduced by a factor of 50.

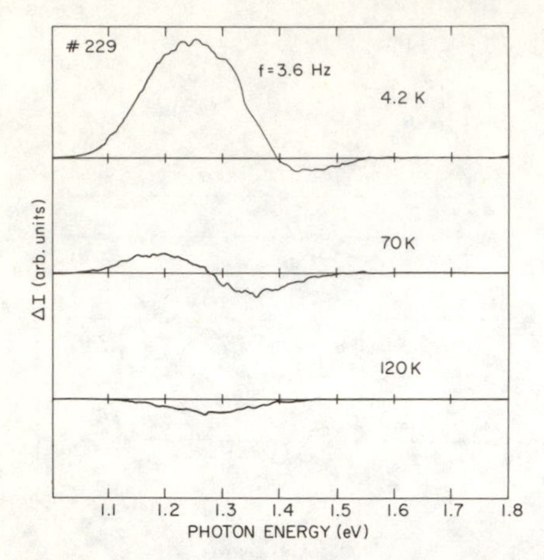

Fig. 6. Low frequency IR-modulated PL emission spectra of an undoped film at three temperatures.

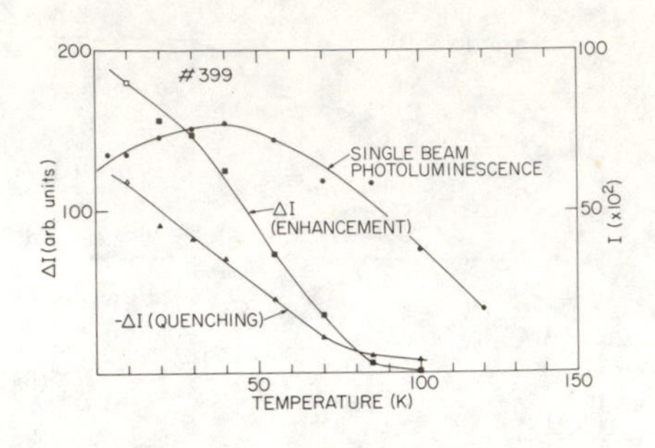

Fig. 7. Temperature dependence of quenching and enhancement compared with single beam PL.

At temperatures below about 30 K, the modulated signal for this sample is 2-3 % of the total signal, but both the enhancement and the quenching effects fall more rapidly with increasing temperature than does the single beam PL. This is another indication that the modulation effects involve the shallow traps in which holes are somewhat delocalized. Above 80 K the magnitude of the quenching becomes greater than that of the enhancement. At this point, thermal excitations assist the non-radiative recombination process by increasing carrier mobility, while the IR-induced enhancement becomes negligible.

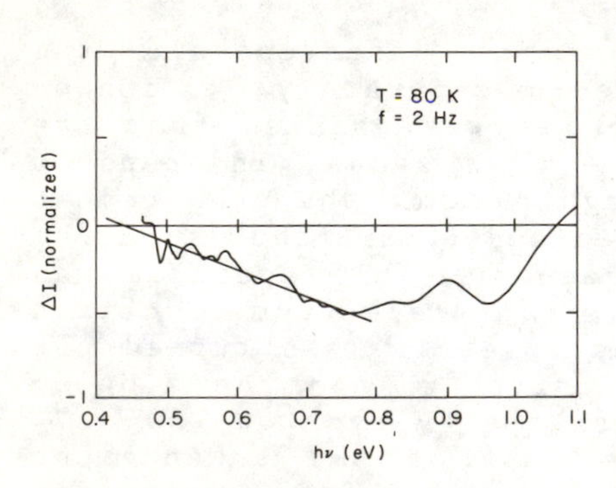

Fig. 8. Excitation spectrum for quenching of photoluminescence.

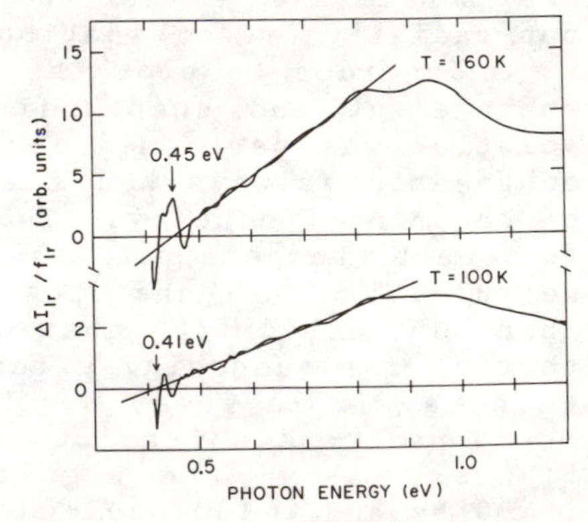

Fig. 9. Excitation spectrum for quenching of photoconductivity.

The IR excitation spectrum of the quenching effect shown in Fig. 8 was obtained by scanning a monochromator placed in the path of the IR beam, while keeping the detection monochromator fixed at 1.4 eV, and normalized with respect to the IR source spectrum. The film temperature was 80 K, and the chopping frequency was 2 Hz, so the quenching effect was dominant at this emission energy. This data indicates that a wide range of photon energies (0.5-1.1 eV) can cause excitations which lead to quenching of PL. The shape of the spectrum is remarkably similar to that previously measured for IR quenching of photoconductivity[3] which is plotted as a positive quantity in Fig. 9. One therefore concludes that the same excitations are involved in both types of dual beam experiment. The threshold energy of 0.44 eV is in close agreement with the photoconductivity threshold, which was found to shift slightly with temperature. It seems likely that this threshold represents the position of the quasi-Fermi level for trapped holes relative to the valence band.

These excitation spectra are quite different from that shown by Bhat et al,[5] who used high intensity lasers for both beams. Their range of photon energies for the "subgap" beam was 1.55 eV - 2.18 eV, which is outside the range of the IR beam used here. Their excitation spectrum indicated a sharp peak at 1.65 eV, which suggests that they are studying a different set of transitions than those discussed here.

A schematic of the transitions proposed to explain both the IR-modulated PL and the IR quenching of photoconductivity is given in Fig. 10. With intrinsic excitation G_{vc}, radiative transitions between shallow electron traps (t) and shallow hole traps (u) compete with recombination through defect states (r). Deeper safe hole traps (s) contain a large number of positive charges with very low rates of recombination. By charge neutrality, most of the r-states must be negative, and unable to capture electrons. IR excitation G_{vs} releases many holes which instantly enhance the t-u PL signal. On a longer time scale, tunneling transitions r-u neutralize many negative defects, and increase the

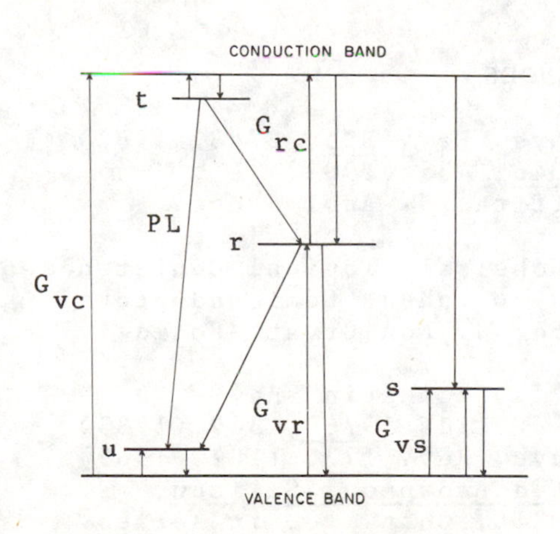

Fig. 10 Schematic of transitions in dual beam experiments on a-Si:H

non-radiative recombination (t-r) rate. Other excitations such as G_{vr} and G_{rc} are negligible.[7]

Dersch and Schweitzer[8] give evidence that the safe trap s is the dominant hole trap detectable by light induced electron spin resonance, but it is not clear whether these are disorder-shifted valence band tail states or specific defect structures such as divalent Si, or impurity states.

CONCLUSIONS

Photoexcited holes in a-Si:H at low temperatures accumulate in safe traps about 0.5 eV above the valence band edge, where they are strongly localized and have low rates of recombination. Charge neutrality is maintained by negative dangling bonds and trapped electrons. Infrared excitation tranfers the holes to shallower traps where they are less localized. The trapped electrons can then recombine directly with the shallow holes, giving an immediate enhancement of the luminescence. However, after a few milliseconds a new steady state is established (by tunneling transitions) in which there are more neutral dangling bonds. The rate of non-radiative recombination is then increased for those electron-hole pairs which are delocalized enough to interact with the dangling bonds.

ACKNOWLEDGEMENT

This research was performed under the auspices of the U.S. Department of Energy, Division of Materials Sciences, Office of Basic Energy Sciences under Contract No.DE-AC02-76CH00016.

REFERENCES

1. P. O'Connor and J. Tauc, Phys. Rev. B25, 2748 (1982)
2. P. D. Persans, Phil. Mag. B46, 435 (1982)
3. P. E. Vanier and R. W. Griffith, J. Appl. Phys. 52, 5135 (1982)
4. R. Carius, W. Fuhs and M. Hoheisel, Topical Conference on Transport and Defects in Amorphous Semiconductors, Bloomfield Hills, March 1984; J. Non-Cryst. Solids (in press)
5. P. K. Bhat, D. J. Dunstan, I. G. Austin and T. M. Searle, J. Non-Cryst. Solids 59/60, 349 (1983)
6. R. C. Enck and A. Honig, Phys. Rev. 177, 1182 (1969)
7. P. E. Vanier, in Hydrogenated Amorphous Silicon, J. I. Pankove, Editor, Vol 21B, Chap. 10, in series Semiconductors and Semimetals, R. K. Willardson and A. C. Beer, Editors, Academic Press, New York, 1984.
8. H. Dersch and L. Schweitzer, J. Non-Cryst. Solids 59/60, 337 (1983)

TIME-RESOLVED LUMINESCENCE, ODMR AND LIGHT-INDUCED EFFECTS IN a-Si:H FILMS PREPARED BY GLOW-DISCHARGE DECOMPOSITION OF DISILANE

M. Yoshida, K. Morigaki, I. Hirabayashi, H. Ohta and A. Amamou*
Institute for Solid State Physics, University of Tokyo,
Roppongi, Tokyo 106, Japan

S. Nitta
Department of Electrical Engineering, Gifu University,
Gifu 501-11, Japan

ABSTRACT

Time-resolved luminescence and time-resolved optically detected magnetic resonance measurements have been carried out for a-Si:H films prepared by glow-discharge decomposition of disilane. Those results are reported as well as light-induced effects on luminescence and ODMR.

INTRODUCTION

Hydrogenated amorphous silicon (a-Si:H) films prepared by glow discharge decomposition of disilane (Si_2H_6) exhibit interesting features compared to those prepared from monosilane (SiH_4)[1-3]; particularly a-Si:H films prepared at room temperature have large band gap energy, 2.4 eV, large hydrogen content, 30-40 at.%, low dangling bond density, 10^{17} cm^{-3} and strong luminescence extending towards 2.3 eV. These features are assosiated with the formation of polysilane, $(SiH_2)_n$. In this paper we present the experimental results on time-resolved (TR) luminescence and time-resolved optically detected magnetic resonance (TRODMR) as well as on steady state luminescence and steady state ODMR in these films. We also report light-induced effects on luminescence and ODMR. On the basis of the results obtained by these measurements, we discuss the recombination processes and the nature of recombination centres involved in a-Si:H films prepared from disilane, particularly for the high energy luminescence.

EXPERIMENTAL

TR luminescence measurements were carried out at 4.2 K, using a nitrogen-pumped pulsed dye laser of 1 ns width operating at 395 nm with peak power of 160 W and a boxcar-integrator. The sampling gate width was arranged for each time delay in such a way as to obtain enough resolution for luminescence measurements. TR luminescence spectra were taken, using a single prism monochromator, an S-1 type photomultiplier and a cooled germanium detector. Steady state luminescence spectra and spectral dependence of ODMR signals were taken at 2 K under excitation by unfocused argon ion laser light at 488 nm,

*Present address: Departement de Physique, Université de Tunis, Tunis, Tunisia

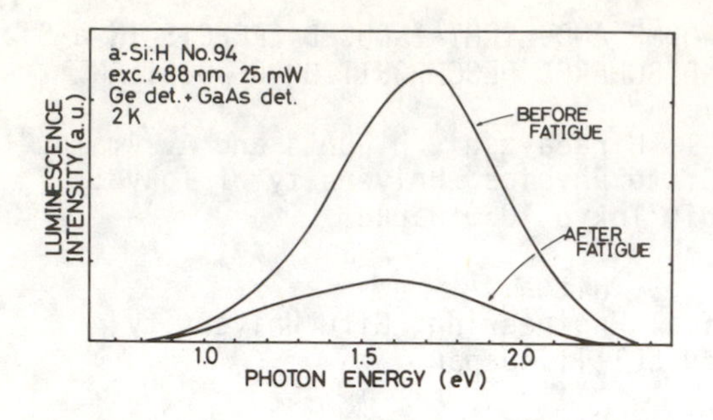

Fig. 1. Luminescence spectra taken before and after prolonged illumination for sample No.94.

using a single prism monochromator, a cooled Ge detector and a GaAs photomultiplier. TRODMR measurements were carried out, using pulsed light formed by an acousto-optic modulator from argon ion laser light at 488 nm and a two channel boxcar integrator. The microwave source for the ODMR measurements was a combination of klystron (9 V 54 or 9 V 58) and travelling wave tube (11 W 71) operating at 9.6 GHz. The microwave power generally used was about 500 mW for the steady state ODMR measurements and about 5 W at peak for the TRODMR measurements. The microwave power was chopped at 1 kHz for the former measurements.

Samples used were prepared by glow discharge decomposition of disilane, using an inductively coupled system. a-Si:H was deposited onto fused silica substrates, whose surface was roughened to avoid interference effects, with thickness of 3.5 μm. The substrate temperatures were 36°C, 16°C and 250°C for samples Nos.93, 94 and 99, respectively. The spin densities of dangling bonds were 1.8×10^{17} cm^{-3}, 2.5×10^{17} cm^{-3}, and 5.6×10^{16} cm^{-3}, for samples Nos.93, 94, and 99, respectively.

EXPERIMENTAL RESULTS

A. Luminescence

The steady state luminescence spectrum extends over the wide range from 0.8 eV to 2.4 eV for samples Nos.93 and 94, as shown in Fig. 1, where the spectrum was taken under excitation by unfocused argon ion laser light of 25 mW at 488 nm before and after prolonged illumination, as will be mentioned later. The time-resolved luminescence spectra were measured at delay times, t_d, taken after the laser pulse was switched off, ranging from 0 to 10 ms. The peak energy of the luminescence spectrum quickly shifts from 1.93 eV at $t_d = 0$ to 1.65 eV at $t_d = 500$ ns, then increases to 1.85 eV at $t_d = 5$ μs and gradually decreases to 1.6 eV at $t_d = 2$ ms. The luminescence intensity decay curves measured at 1.60 eV and 2.02 eV are shown in Fig. 2. For both energies, the decay curves for long delay times longer than 1 ms are fitted by exponential curves.

B. Optically Detected Magnetic Resonance

The steady state ODMR spectrum consists of three lines, i.e., two enhancing lines and one quenching line. One of two enhancing

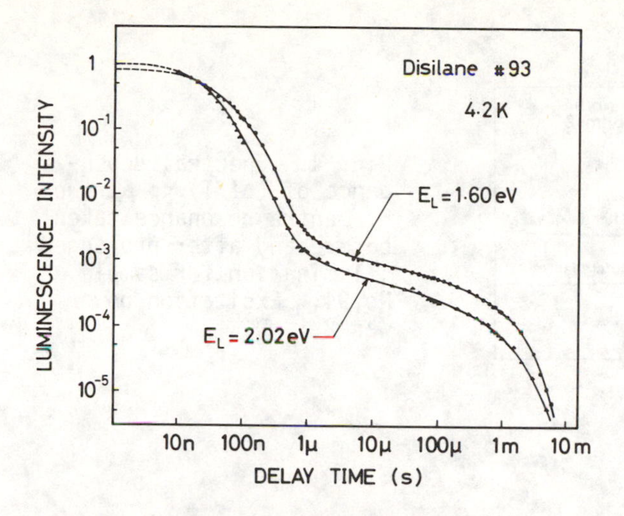

Fig. 2. Luminescence decay curves observed at 1.60 eV and 2.02 eV for sample No.93.

lines with g = 2.002 is identified as being due to the A centres (trapped hole centres) commonly observed for a-Si:H samples[4,5] prepared from monosilane, but the other with g = 2.00 is very broad ($\Delta H_{1/2}$ = 370 G), being a new ODMR signal that has not yet been observed for a-Si:H samples prepared from monosilane. This line we call the B line dominated over the A line as enhancing lines. A quenching line with g = 2.004 and $\Delta H_{1/2}$ = 17 G is identified as being due to dangling bonds (D_2 centres). Figure 3 shows a typical example of ODMR spectra observed, monitoring total emitted light under excitation by unfocused argon ion laser light of 40 mW at 488 nm, where a very broad enhancing line (B) and a narrow quenching line (D_2) are seen. Spectral dependences of changes in the luminescence intensity at resonance, $(\Delta I/I)_{ESR}$, for the D_2 and B lines were measured, as shown in Figs. 4 and 5. For the D_2 line, its spectral dependence was measured before and after prolonged illumination, as will be described later. The A line was observed only in the low energy part of the luminescence spectrum. This is consistent with the identification of this line by the A centres that show such spectral dependence

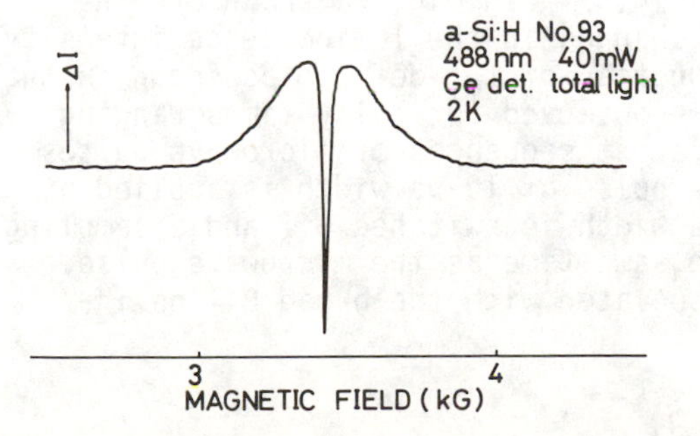

Fig. 3. ODMR spectrum observed, monitoring total emitted light for sample No.93.

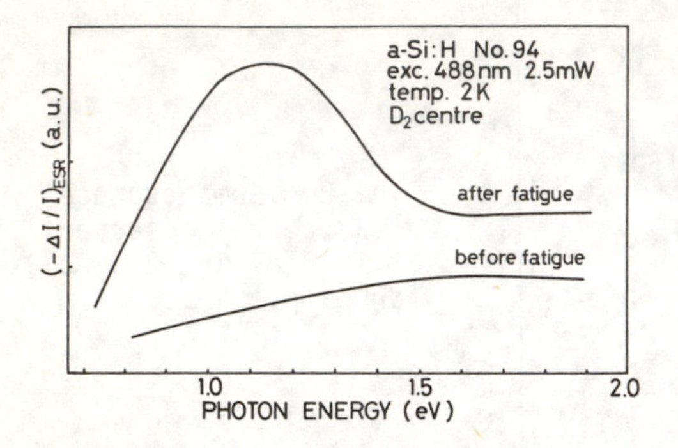

Fig. 4. Spectral dependence of $(\Delta I/I)_{ESR}$ at the D_2 centre resonance taken before and after prolonged illumination for sample No.94. Excitation power was 2.5 mW.

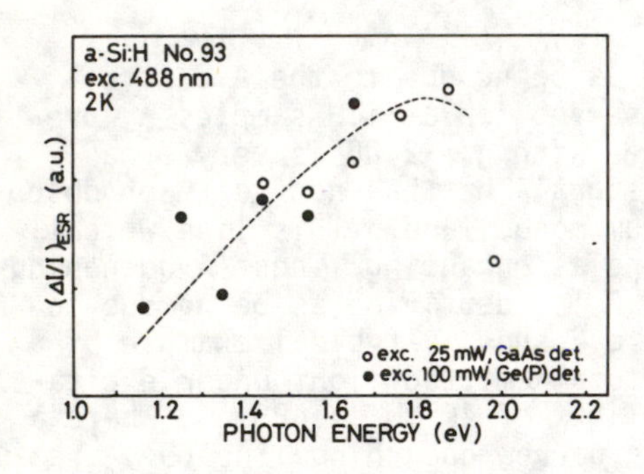

Fig. 5. Spectral dependence of $(\Delta I/I)_{ESR}$ at the B centre resonance for sample No.93. Excitation power is indicated for each detector.

for a-Si:H samples prepared from monosilane.[5]

In the following, we present the results of TRODMR measurements carried out for samples Nos.93 and 99, monitoring total emitted light under pulsed excitation, as described in section Experimental. Figure 6 shows the TRODMR spectra for various delay times whose definition is shown in the inset of Fig. 7. In Fig. 7, shown are the absolute magnitudes of relative changes in the luminescence intensity at resonance, $(\Delta I/I)_{ESR}$, as a function of t_d, defined as shown in the inset of Fig. 7. The A lines are observed for delay times ranging from 10 μs to 200 μs, using different sequences of microwave pulses and gate pulses, where microwave pulse of 10 μs width is applied at t_d after a laser pulse of 200 μs width is switched off and a sampling gate is open during 10 μs at the same time as the microwave pulse. As seen in Fig. 7, $(\Delta I/I)_{ESR}$ associated with the broad B line increases with t_d.

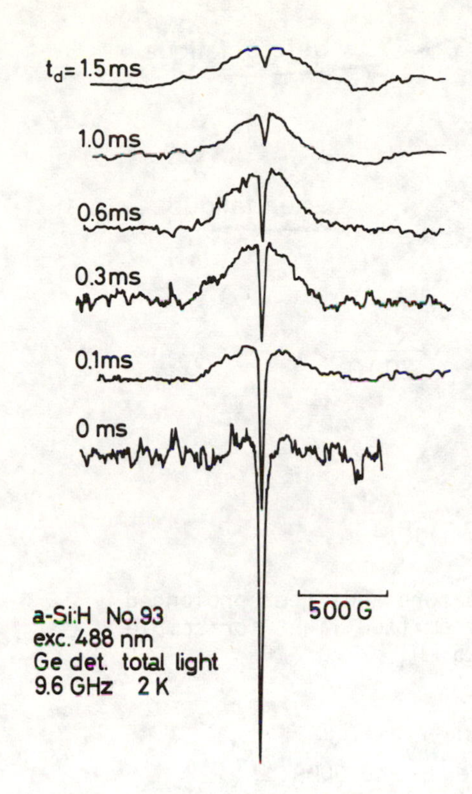

Fig. 6. TRODMR spectra observed at various delay times for sample No.93. Peak excitation power was 240 mW.

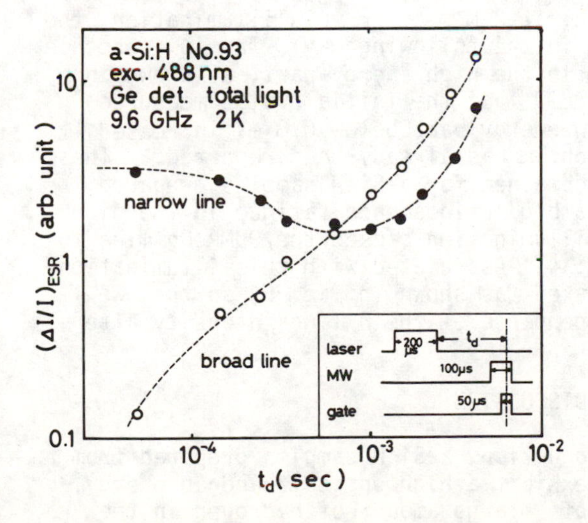

Fig. 7. $(\Delta I/I)_{ESR}$ vs. delay time, t_d, for sample No.93. Pulse sequence is indicated in the inset.

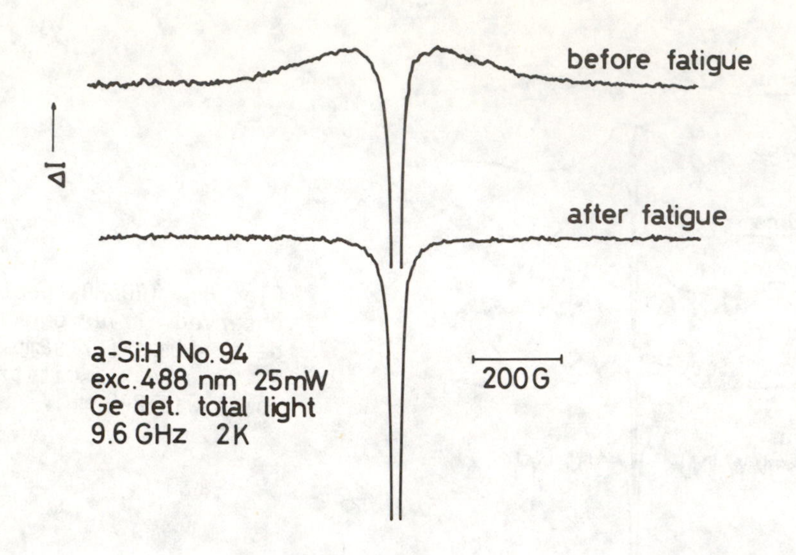

Fig. 8. ODMR spectrum taken before and after prolonged illumination, monitoring total emitted light for sample No.94. Excitation power was 25 mW.

C. Light-Induced Effects on Luminescence and ODMR

We have observed light-induced effects on luminescence and ODMR at 2 K.[6] Luminescence spectra were measured for sample No.94 before and after prolonged illumination by unfocused argon ion laser light of 360 mW at 488 nm for 30 min. at 2 K. After this illumination, the luminescence spectrum changed in the following way: The luminescence intensity decreased very much in the high energy part, as shown in Fig. 1. This decrease reached 90 % of the initial value measured before illumination.[6] The low energy part below 0.8 eV increased compared to before illumination, as has already been reported.[6] These results are similar to those obtained for a-Si:H samples prepared from monosilane[7,8] except for a big luminescence fatigue in the high energy part. Such prolonged illumination caused the ODMR D_2 line to be increased, as shown in Fig. 4. Associated with this illumination, the B line completely disappeared, as shown in Fig. 8, so that we could not measure spectral dependence of the B line intensity after prolonged exposure to light.

DISCUSSION

As was mentioned in Introduction, a-Si:H samples prepared from disilane at room temperature exhibit a high optical bandgap energy. This is due to incorporation of a large amount of hydrogen in the form of polysilane, $(SiH_2)_n$.[1-3] The features of the results obtained from the luminescence and ODMR experiments are also associated with such incorporation of hydrogen into the samples. The high energy part of the luminescence spectrum exhibits a big luminescence fatigue and a new broad ODMR line (B line). It is quite reasonable to consider that the high energy part of the luminescence spectrum arises from the region containing a large amount of polysilane. This can also explain a big luminescence fatigue observed in the high energy part of the luminescence spectrum, because previous investigations[9,10] reveal that a large amount of hydrogen prevails in the luminescence fatigue as a result of flexible amorphous network containing onefold coordinated hydrogen atoms. The nature of the B centres is not clear only from the ODMR measurements, but a significant broadening of the B line is also consistent with the above model for the high energy luminescence, because the B line is mainly observed in the high energy luminescence, as indicated from its spectral dependence shown in Fig. 5 and thus the B centre wavefunction possibly extends over hydrogen atoms, so that hyperfine interaction with hydrogen nuclei may be expected to cause a line-broadening.

As was pointed out above, the B centre is correlated with the high energy luminescence that is one of the characteristics of a-Si:H samples prepared from dislane. The B centre ODMR signal appears as an enhancing line, so that this centre obviously acts as a radiative centre responsible for the high energy luminescence. The TRODMR of the B centre suggests that recombination of electrons with holes at the B centres is of distant pair type, in which we assume that the B centre is a trapped hole centre like the A centre generally observed in undoped a-Si:H samples.[4,5] A detailed explanation on the results of TR luminescence measurements will be given elseshere.

Light-induced effects on ODMR have also been discussed in ref. (6).

CONCLUSION

The luminescence and ODMR measurements reported in this paper allow us to suggest that the high energy part of the luminescence spectrum observed for a-Si:H samples prepared from disilane is mainly associated with the region containing a large amount of hydrogen in the form of polysilane. It is also concluded that the high energy part of the luminescence spectrum arises from recombination of electrons trapped in the tail- and gap-states with holes mainly trapped at the B centre whose ODMR signal has been observed for the first time in the present experiment. We have observed a big luminescence fatigue in the high energy part of the luminescence spectrum which is due to creation of dangling bonds by exposure to light. This is consistent with our previous suggestion[9,10] that defect creation by exposure to light is correlated with a flexible amorphous network containing onefold coordinated hydrogen atoms.

REFERENCES

1. B. A. Scott, R.M. Plecenik and E. E. Simonyi, Appl. Phys. Lett. 39, 73 (1981).
2. D. J. Wolford, J. A. Reimer and B. A. Scott, Appl. Phys. Lett. 42, 369 (1983).
3. S. Furukawa and N. Matsumoto, Solid State Commun. 48, 539 (1983).
4. For a review, K. Morigaki, in Hydrogenated Amorphous Silicon, ed. J. I. Pankove, Semiconductors and Semimetals Vol.21 part C, Academic Press, New York (in press).
5. K. Morigaki, Y. Sano and I. Hirabayashi, Solid State Commun. 39, 947 (1981).
6. M. Yoshida, K. Morigaki and S. Nitta, Solid State Commun. 51, 1 (1984).
7. K. Morigaki, I. Hirabayashi, M. Nakayama, S. Nitta and K. Shimakawa, Solid State Commun. 33, 851 (1980).
8. J. I. Pankove and J. E. Berkeyheiser, Appl. Phys. Lett. 37, 705 (1980).
9. Y. Sano. K. Morigaki and I. Hirabayashi, Solid State Commun. 43, 439 (1982).
10. I. Hirabayashi, K. Morigaki and M. Yoshida, Solar Energy Mat. 8, 153 (1982).

THE ROLE OF DANGLING BONDS IN RADIATIVE AND NONRADIATIVE PROCESSES IN a-Si:H

B. A. Wilson and A. M. Sergent
AT&T Bell Laboratories, 1D-465, Murray Hill, N.J. 07974

J. P. Harbison
Bell Communications Research, Inc., Murray Hill, N.J. 07974

ABSTRACT

New time-resolved photoluminescence (PL) measurements in annealed a-Si:H films offer new insight into the low energy luminescence band at 0.7 eV and the role of the dangling bond in radiative and nonradiative recombination. While the absolute intensity of this band rises and then falls with spin density, N_S, when compared to the main PL band the *relative* intensity grows linearly with N_S over 3 decades. Since this ratio holds both at early times as well as for cw data, it is clear that the quenching mechanisms for both bands are similar. We propose that the 0.7 eV band arises from radiative recombination between an electron trapped in a conduction band tail state and a hole trapped at a dangling bond. A quantitative model based on this picture accounts for the optical data presented here and previously reported optically detected magnetic resonance results.

INTRODUCTION

Annealing of good a-Si:H films in the temperature range 300 to 600°C drives off hydrogen from the material and creates dangling bond (DB) defects.[1,2,3] Associated with this degradation of sample quality is the quenching of the main luminescence band. An additional band at 0.7 eV appears in the annealed samples,[4] suggesting an origin of radiative recombination at DB defects. Recent optically detected magnetic resonance (ODMR) measurements of this band, however, record only a quenching signal at the DB resonance,[5] apparently precluding this hypothesis. In this paper we report time-resolved luminescence data on this band which confirm a dual role - both radiative and nonradiative - for the DB. We propose that the 0.7 eV band arises from radiative recombination between an electron in a conduction band tail state and a hole trapped at a dangling bond, and present a simple quantitative model that reconciles the optical and ODMR results.

EXPERIMENTAL TECHNIQUES

The samples used for these measurements were deposited simultaneously on roughened fused silica substrates. The deposition parameters shown in Fig. 1 are known to produce optimal materials which degrade monotonically with annealing temperature above the original substrate temperature of 250°C.[6] All but one of the films were subsequently annealed for 20 minutes in vacuum at one of the following temperatures: 300, 400, 500, 525, 550, 570 and 600°C. The amount of hydrogen evolved and the spin density created during each anneal

0094-243X/84/1200149-08 $3.00

are shown in Fig. 1. The hydrogen evolved was calculated from the pressure rise in the vacuum chamber during annealing. The spin signal was measured at room temperature with a Varian E-4 spectrometer. For optical measurements the samples were placed in a temperature-controlled flowing-He gas cryostat.

The cw luminescence spectra were excited with the 5145 Å line of an argon ion laser and detected with Ge and InAs photodiode detectors by lock-in techniques. The spectra have been corrected for system

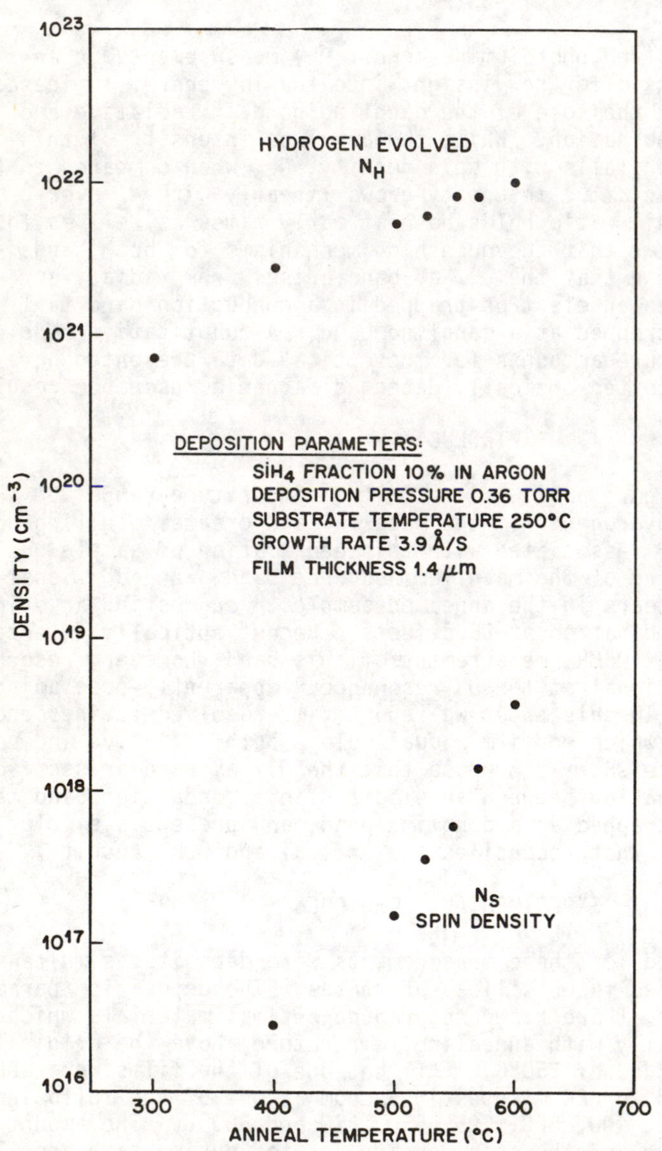

Fig. 1. Cumulative densities of evolved hydrogen and spin formation with annealing.

response. We used 10 ns pulses from a doubled Nd:YAG laser at 5320 Å to excite the samples for luminescence decay measurements. Spectrally integrated data were obtained by focussing the PL directly on the Ge photodiode. The two PL bands were isolated with interference filters. The Ge photodiode covers the main PL band with approximately flat spectral response, but somewhat favors the high energy side of the 0.7 eV band. For both cw and pulsed experiments the laser power was kept as low as possible to minimize sample degradation due to light exposure.

EXPERIMENTAL RESULTS

The evolution of the cw PL spectra with anneal temperature is shown in Fig. 2. In the as-grown sample we observe only the broad main band, peaking at $\sim$ 1.35 eV. The peak shifts to lower energy and the intensity falls monotonically with increasing anneal temperature. The band at 0.7 eV, which becomes apparent only above an anneal temperature of 500°C, first grows in absolute magnitude and subsequently

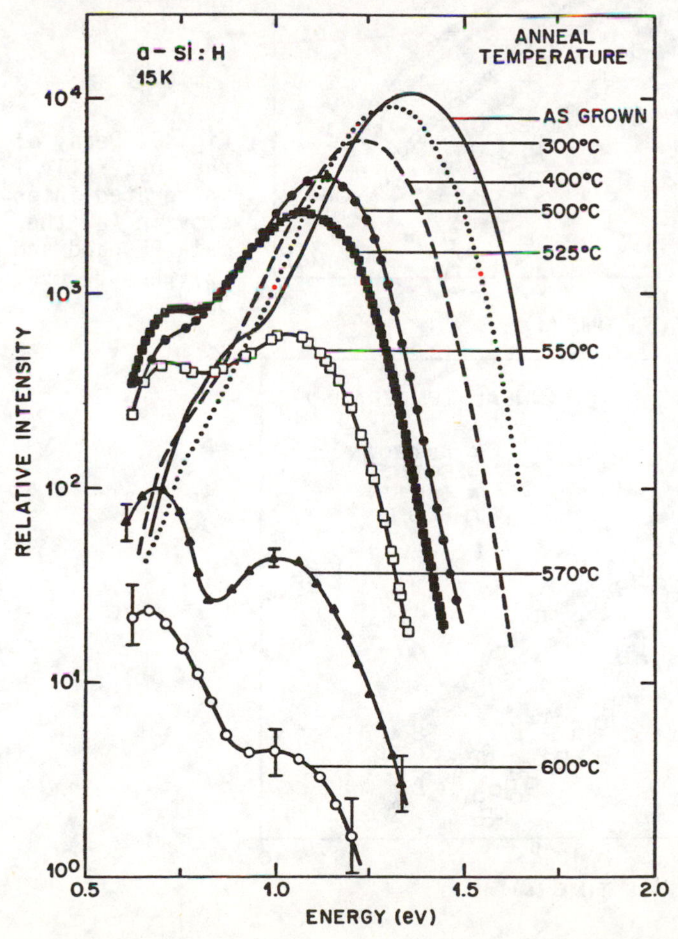

Fig. 2. cw photoluminescence spectra as a function of anneal temperature. Error bars represent frequency dependent scatter in the data.

is quenched at higher temperatures. This quenching is less rapid than for the main band. The peak position does not shift noticeably.

The spectrally integrated decay curves of both bands are displayed in Fig. 3. The behavior of the main band is monotonic, and thus simpler to follow. It is clear from the progression of decay curves that there are two quenching mechanisms operant with very different time scales; one faster than the resolution of this experiment and one at much later times $\gtrsim 1$ μs. Subnanosecond experiments[7] have

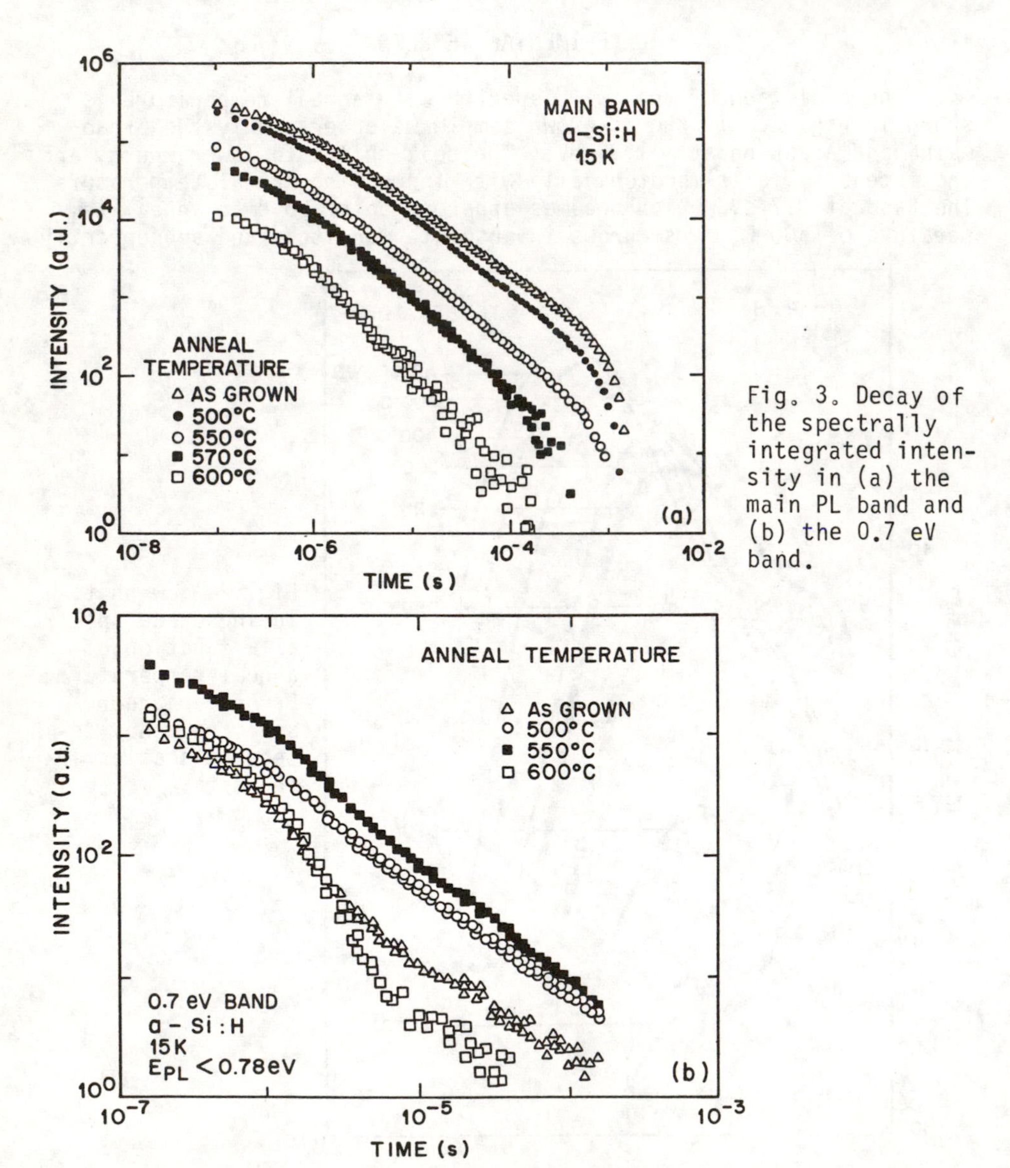

Fig. 3. Decay of the spectrally integrated intensity in (a) the main PL band and (b) the 0.7 eV band.

revealed that the fast process occurs on a time scale < 100 ps, and is most likely associated with the initial trapping of carriers out of the photoexcited state into deep defects. The initial intensity thus monitors the outcome of the competition for trapping at radiative versus nonradiative sites. The slower process which dominates the loss in cw quantum efficiency at high DB density, N_s, has been ascribed to tunnelling[8] or diffusing of carriers from radiative band tail states to nonradiative DB sites.

Although the evolution of the 0.7 eV band is more complex, it is clear that there are changes occurring on the same fast and slow time scales. In this case, however, the early-time process first raises the entire decay curve, and subsequently quenches it at higher anneal temperatures. The slow process appears to affect this band similarly to the main band, monotonically reducing the intensity at long times.

In order to isolate the effects of the fast mechanism, we have examined the progression of the spectrally integrated intensity at 100 ns, well before the slow process becomes significant. The 100 ns intensities of the two bands are shown individually in parts a and b of Fig. 4. It is the plot of the *ratio* of the intensities, however, which simplifies the description of these results. The ratio of the 0.7 eV band to the main band shown in Fig. 4c exhibits a linear dependence on N_s over three decades. This relationship holds both for the 100 ns spectrally integrated data as well as for the peak intensities of the cw spectra. The cw data provide an absolute scale for the ratio since the spectra are normalized for system response. The values of the ratio at 100 ns are relative, and have been normalized to the same magnitude. Thus the fast process results in a branching ratio of the two radiative populations that is proportional to N_s, and the slow quenching process depletes both populations similarly.

DISCUSSION

The model we propose to describe these results involves only band tail (BT) states and DB states. DB states can be unoccupied, T_3^+, singly occupied, T_3^0, or doubly occupied T_3^-. In undoped materials, such as the films used here, most DB sites are singly occupied in the ground state. Photoexcitation creates mobile electron-hole pairs that are rapidly trapped in BT and DB states. We propose that while the main PL band results from recombination between electrons and holes both trapped in BT states, as usually assumed, the 0.7 eV band requires that the electron be trapped in a BT state and the hole at a DB, i.e. a T_3^+. The inverse process, recombination between a hole trapped in a BT state and a T_3^-, may also be radiative, but below the energy range of our detection system. If both carriers are trapped at DB sites, recombination is most likely nonradiative since the energy gap is only a few tenths of an eV.[9]

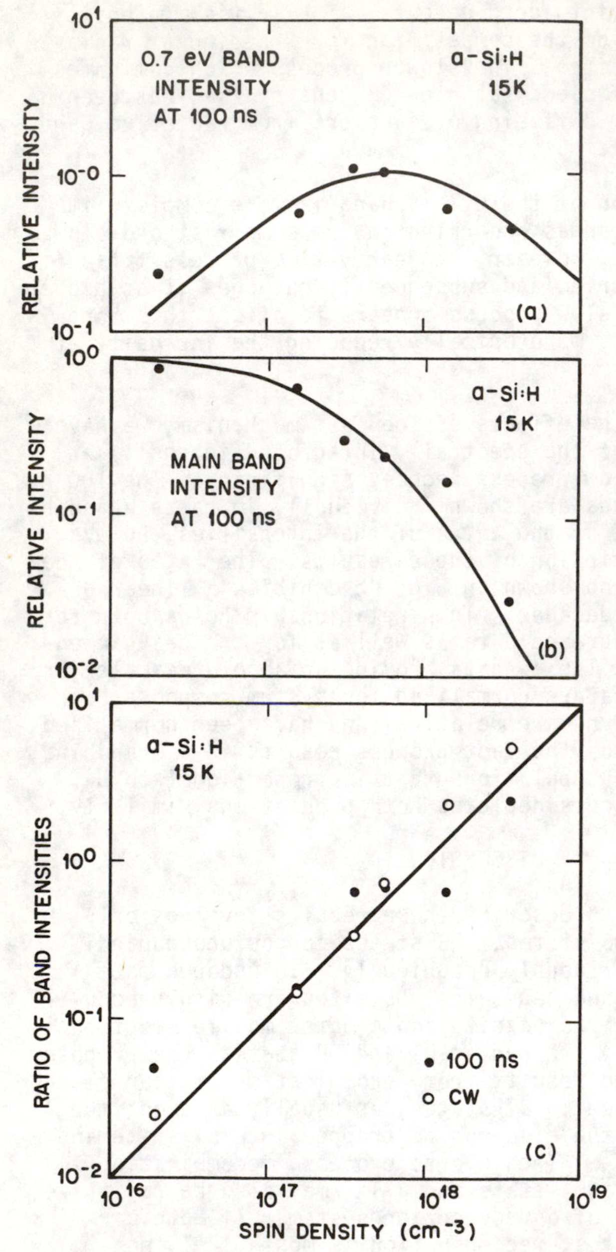

Fig. 4. Spectrally integrated PL intensities versus spin density at 100 ns delay for (a) the 0.7 eV band, (b) the main band. The ratio of the two band intensities is plotted in (c) with closed circles representing the ratio at 100 ns delay, open circles the cw ratio. Solid lines represent fits to the branching ratios of equation 1.

Assuming that the BT densities, N_{BT}, do not change drastically with annealing, we can write down the branching ratio for the two radiative populations:

$$f_{mainband} = \frac{1}{(1+\alpha^{e}N_{s})(1+\alpha^{h}N_{s})} \; ; \qquad f_{0.7eV} = \frac{\alpha^{h}N_{s}}{(1+\alpha^{e}N_{s})(1+\alpha^{h}N_{s})} \qquad (1)$$

where we define

$$\alpha^{e} \equiv \frac{\nu_{s}^{e}}{\nu_{BT}^{e}N_{BT}^{e}} \; ; \qquad \alpha^{h} \equiv \frac{\nu_{s}^{h}}{\nu_{BT}^{h}N_{BT}^{h}}$$

and ν_s and ν_{BT} refer to the competing capture rates into DB and BT states, and the superscripts e and h to the values for electrons and holes. We note immediately that the ratio of the populations is correctly proportional to N_s. In addition, the model predicts the observed rise and fall of the 0.7 eV band intensity with N_s. Quantitatively, with only 2 parameters, the model is able to self-consistently fit the decay curves of both bands as well as the absolute ratio, as demonstrated by the solid lines in Fig. 4. The best fits provide values for the trapping parameters: $\alpha^{e} \sim 3\times10^{-18}cm^{3}$, $\alpha^{h} \sim 1\times10^{-18}cm^{3}$.

The similarity observed in the effects of the slow quenching process, and the ODMR results are also consistent with this model. Both bands require a photoexcited electron trapped in a BT state, and are quenched when this electron tunnels to a neighboring T_3^0. Since this process is spin-dependent, while the recombination with a spinless T_3^+ is not, one would expect to observe only a quenching signal at the DB resonance in ODMR measurements of the 0.7 eV band. Comparison with photoconductivity data[10,11] finds consistency in the relative sizes of α^e and α^h but not in their absolute magnitudes. Since the time of flight measurements were too slow to isolate initial trapping mechanisms, however, the discrepancy is difficult to interpret.

The cw spectra of Fig. 2 lend further support to this picture. Photoemission data[12] have shown that most of the band gap narrowing with hydrogen evolution occurs at the valence band edge, leaving the spacing between the conduction band and DB states relatively constant. Thus BT to BT transitions would be expected to reflect the full narrowing of the gap, while the 0.7 eV transition between an electron in the conduction band tail and a T_3^+ should shift little with annealing, as observed. The narrowness of the 0.7 eV band in comparison with the main band leads one to speculate that, as suggested by Phillips,[13] the luminescing DB states may lie predominantly on internal surfaces rather than in the strained bulk. Low energy bands in the 0.8-0.9 eV region commonly observed in doped a-Si:H are not generally narrower than the main PL band, and may have a different origin than the 0.7 eV band studied here. Recent luminescence studies[14] of these bands report other differences in behavior, tending to reinforce this suggestion.

SUMMARY

We have presented time-resolved luminescence measurements in annealed a-Si:H films which strongly support a dual radiative and nonradiative role for dangling bond defects. A hole trapped at a DB may recombine radiatively with an electron in a BT state giving rise to the observed 0.7 eV band. An electron trapped at a DB, however, quenches both the 0.7 eV band as well as the main band. A simple branching ratio expression based on this picture quantitatively accounts for the observed dependence of the band intensities on spin density. Both bands are quenched if the BT electron tunnels to a DB prior to radiative recombination, explaining the similarity in the long-time quenching of the bands, and reconciling the ODMR data on the 0.7 eV band which revealed only a quenching signal at the DB resonance. Thus this simple model provides a consistent picture of all the optical and ODMR results.

REFERENCES

1. D. K. Biegelsen, R. A. Street, C. C. Tsai and J. C. Knights, Phys. Rev. B 20, 4839 (1979).
2. H. Fritzsche, M. Tanielian, C. C. Tsai and P. J. Gaczi, J. Appl. Phys. 50, 3366 (1979).
3. K. Zellama, P. Germain, S. Squelard, B. Bowden, J. Fontenille and R. Danielon, Phys. Rev. B. 23, 6648 (1981).
4. U. Voget-Grote, W. Kümmerle, R. Fischer and J. Stuke, Phil. Mag. G 41, 127 (1980).
5. M. Yoshida and K. Morigaki, Proceedings of the 10th International Conference on Amorphous and Liquid Semiconductors, eds. Kazunobu Tanaka and Tatsuo Shimizu, p. 357 (1983).
6. B. A. Wilson, A. M. Sergent, K. W. Wecht, A. J. Williams, T. P. Kerwin, C. M. Taylor and J. P. Harbison, to be published in Phys. Rev. B.
7. B. A. Wilson, P. Hu, T. M. Jedju and J. P. Harbison, Phys. Rev. B 28, 5901 (1983).
8. R. A. Street, Advances in Physics 30, 593 (1981).
9. J. D. Cohen, J. P. Harbison and K. W. Wecht, Phys. Rev. Lett. 48, 109 (1982).
10. R. A. Street, Appl. Phys. Lett. 41, 1060 (1982).
11. T. Tiedje, J. M. Cebulka, D. L. Morel and B. Abeles, Phys. Rev. Lett. 46, 1425 (1981).
12. B. von Roedern, L. Ley and M. Cardona, Phys. Rev. Lett. 39, 1576 (1977).
13. J. C. Phillips, Phys. Stat. Sol. B 101, 437 (1980).
14. R. A. Street and D. K. Biegelsen, Bulletin of the APS 29, 507 (1984).

TIME RESOLVED RADIATIVE AND NONRADIATIVE RECOMBINATION IN a-Si:H

U. Strom, J.C. Culbertson, P.B. Klein and S.A. Wolf
U.S. Naval Research Laboratory, Washington, D.C. 20375*

ABSTRACT

Time resolved phonon and photoluminescence measurements are reported for a-Si:H films which are illuminated with pulsed near bandgap light. It is found that heat and light emission exhibit strikingly different time and excitation intensity dependences. The results demonstrate that the radiative and nonradiative channels are essentially uncoupled for times greater than 10^{-7} sec and that the radiative quantum efficiency is close to unity. The results further show that for injected electron-hole densities $<10^{18}cm^{-3}$ the a-Si:H is an efficient thermalizing medium in which the optically generated heat pulse travels ballistically for a distance of ~1 μm.

INTRODUCTION

Photoluminescence (PL) studies have been an important tool for describing recombination processes in a-Si:H. In a recent review Street[1] has summarized the role of various luminescence variables in defining plausible recombination mechanisms. Generally, nonradiative processes are an important aspect of any proposed model. One of the experimental observations which provides indirect information about nonradiative processes is the time dependent shift of the PL peak position 0.1 eV toward lower energies. This has been interpreted[2,3] in terms of thermalization of carriers by tunneling between localized states just below the mobility edge. Another important observation is the sublinear dependence of the PL intensity with increasing excitation intensity for $>10^{18}cm^{-3}$ injected electron hole pairs.[4,5] This has been interpreted as evidence for a competing nonradiative (Auger) process which becomes more likely as the radiative recombination changes from mono to bimolecular at high excitation intensities.[5,6] There are also some conflicting experimental results[2,7] pertaining to the effect of decreasing radiative efficiency on the PL decay. In all of these cases the availability of direct information about the nonradiative decay could be beneficial. One primary motivation of the present work is to measure, essentially simultaneously, both luminescence and heat emission in order to examine the validity of proposed models for the nonradiative decay in a-Si:H.

EXPERIMENTAL ARRANGEMENT

The experimental arrangement is shown in Fig. 1(a). The 8 ns long pulse from a pulsed dye laser was focused to a ~200 μm diameter

*Supported by DoE Contract No. DE-AI02-80CS83116 administered by SERI.

0094-243X/84/1200157-06 $3.00 American Institute of Physics

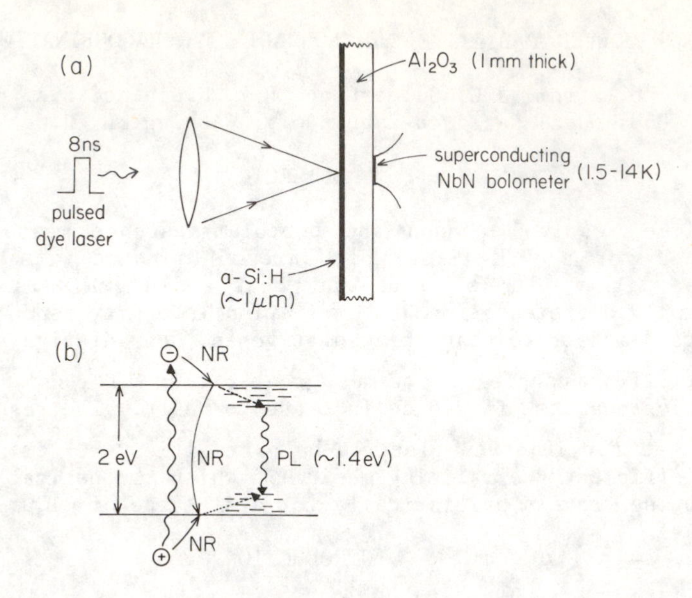

Fig. 1. (a) Experimental arrangement. (b) Schematic of excitation and recombination. NR = nonradiative; PL = photoluminescence.

spot on an a-Si:H film. The a-Si:H was deposited by plasma decomposition of silane onto 1 mm thick sapphire substrates. The films had thicknesses of 1 and 4 μm and were comparable to "NRL" samples studied by nuclear magnetic resonance.[8] PL was measured with a cooled S-1 photomultiplier and phonons were detected with a superconducting NbN bolometer.[9] The response times of both heat and light detectors were considerably less than the laser pulse width. The CW PL measured at 2 K exhibited a peak near 1.4 eV.

Figure 1(b) shows a schematic of the relevant excitation and decay processes. The bandgap of a-Si:H has been variously estimated to range from 1.7 to 2.1 eV. The value of 2 eV corresponds to the energy at which light penetrates ~1 μm. Typically, a somewhat higher energy ~2.4 eV has been used for excitation. This light has a penetration depth of ~0.2 μm in a-Si:H. The generated electron-hole (e-h) pairs are expected to thermalize quickly (on a ps time scale) to the band or mobility edges and then recombine radiatively or nonradiatively. The observed PL energy is ~0.6 eV less than the bandgap. About half of this difference is accounted for by the width of the localized state distributions, and the other half by lattice distortions of the electrons and holes (Stokes shift).[1] As shown in Fig. 1(b) the branching between radiative and nonradiative channels occurs at the band edge. It is conceivable that further nonradiative recombination could occur at longer times, competing directly with the PL. As will be shown, the time-resolved phonon measurements preclude this possibility, at least for times less than ~5 μs.

RESULTS AND DISCUSSION

Typical time resolved phonon and luminescence signals for low excitation intensity (2.9 nJ) are shown in Fig. 2. The PL signal exhibits "fast" and "slow" components typical for a-Si:H. We have measured the PL decay into the 10^{-3} s range and find agreement with previous measurements.[1] The phonon signal is delayed with respect to the PL signal by the sound propagation time through the 1 mm thick sapphire substrate. The ballistic heat flow is shared by longitudinal acoustic (LA) and transverse acoustic (TA) phonons. The transport times of the smaller and larger peaks agree well with the sound velocities of LA and TA phonons in sapphire, respectively. The relative magnitudes of LA and TA phonon signals are determined by phonon focusing for this propagation direction in sapphire. There is also a time-delay of the heat signal due to the a-Si:H film, but for 1 μm propagation distance this delay corresponds to less than 1 ns. The halfwidth of the TA phonon peak is ~30 ns which is somewhat larger than the 8 ns wide laser pulse. The added width is most likely due to a spread in the phonon pathlength because of a finite bolometer size (~1x1 mm^2). In fact, the phonon pulse widths are considerably reduced compared to what is expected from geometric phonon optics, primarily because of strong focusing effects.[10]

There are two important aspects of the data in Fig. 2 which are relevant to thermalization and recombination models in a-Si:H. One, is that the observed phonon signal is essentially ballistic through the remainder of the a-Si:H film (~0.8 μm). The phonon mean free path in glasses decreases rapidly with increasing phonon energy.[11] If a-Si:H is comparable to a-SiO_2 in this respect then the optically

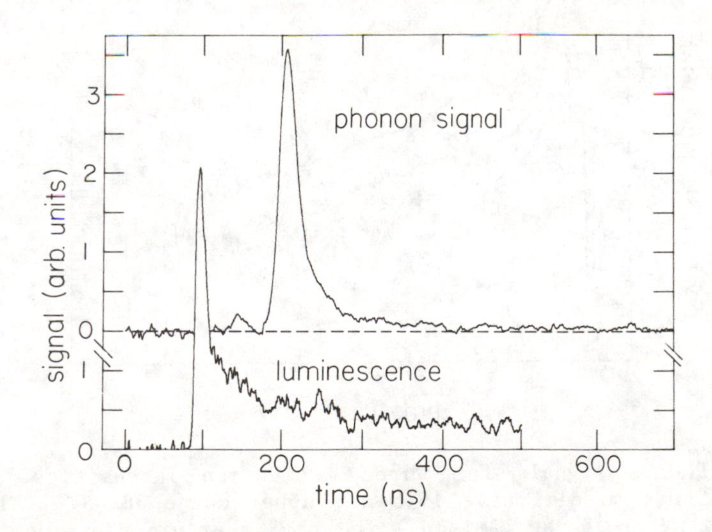

Fig. 2. Luminescence and phonon signals measured under identical conditions. Temperature 2.1 K; Excitation 5100Å, 2.9 nJ/pulse; PL energy 1.43 eV; a-Si:H thickness 1 μm.

generated acoustic phonons in a-Si:H must down convert to frequencies below ~200 GHz within a time period of less than 10 ns. This is in contrast to photoexcited phonon studies in a-As_2S_3 where the phonon signal is dominated by a broad diffuse-like peak.[12] It is likely that the different microscopic structure of the two amorphous materials plays an essential role in determining differing phonon transport properties.

The other relevant aspect of the data in Fig. 2 is that the magnitude of the PL observed at longer times is not matched by a comparable late heat signal. In fact, the small tail of the heat signal is most likely diffusive in nature, as evidenced by the fact that the phonon signal at these times is greatly increased for a thicker 4 μm a-Si:H film. This observation implies that the competing nonradiative recombination occurs primarily for times shorter than 10^{-7} sec. This conclusion is in agreement with suggestions by Collins et al.[7] on the basis of time resolved PL measurements for a-Si:H samples with widely different radiative efficiencies.

We also find, in agreement with others,[4,5] that the PL intensity increases slower than linearly with excitation intensity for $\gtrsim 10^{18}$ cm^{-3} absorbed photons. Figure 3 shows phonon signals for excitation levels above this threshold. The excitation levels correspond respectively to 3×10^{18}, 6×10^{18} and 1.3×10^{19} cm^{-3} photons/pulse. It can be observed that as a function of excitation intensity the

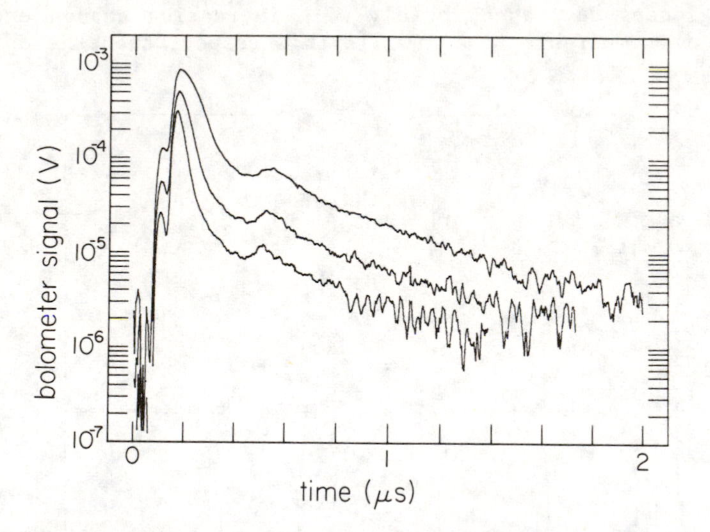

Fig. 3. Phonon signals for three excitation intensities. Lower curve 10.5 nJ; middle curve 19.5 nJ; upper curve 38 nJ. Bolometer signals $\lesssim$1 μV are noise limited. The peak near 0.6 μs is due to the reflection of the 0.2 μs phonon pulse at the sapphire/bolometer interface and its subsequent reflection at the a-Si:H/sapphire interface.

heat signal is shifted from short times (~200 ns) to longer times and that the relative amount of the signal increased more rapidly than linearly. It is also important to note that the shape of the heat decay signal at long times is not significantly dependent upon excitation intensity. The decay is approximately exponential in time, with a phonon signal reduction by one order of magnitude for each ~1 μs interval. Thus, the time integral of these phonon signals up to a sufficiently long time interval (~2 μs) is a meaningful measure of the total nonradiative emission. The normalized phonon signals (integrated phonon signal divided by excitation pulse energy) have been plotted in Fig. 4 together with results for other pulse energies. The phonon data (triangles) corresponding to the lowest and highest pulse energies are the most uncertain because of limited signal to noise and too short an integration time interval, respectively. The integrated PL intensities are shown in the lower part of the figure. We have determined that the PL decay shape does not depend on pulse energy. Instead the entire PL response is reduced by an energy dependent factor. This observation is in agreement with other studies.[7,13] From Fig. 4 note that the reduced PL intensity is matched by an increased phonon signal. This observation can in principle lead to a direct determination of the internal quantum efficiency. If for a given pulse energy we define as $p>1$ the enhancement factor of the phonon signal (enhanced over what would have been expected if the phonon signal increased linearly with pulse energy) and $r<1$ the reduction factor of the PL then the internal quantum efficiency η (defined at low excitation energies) can be shown to be expressed as

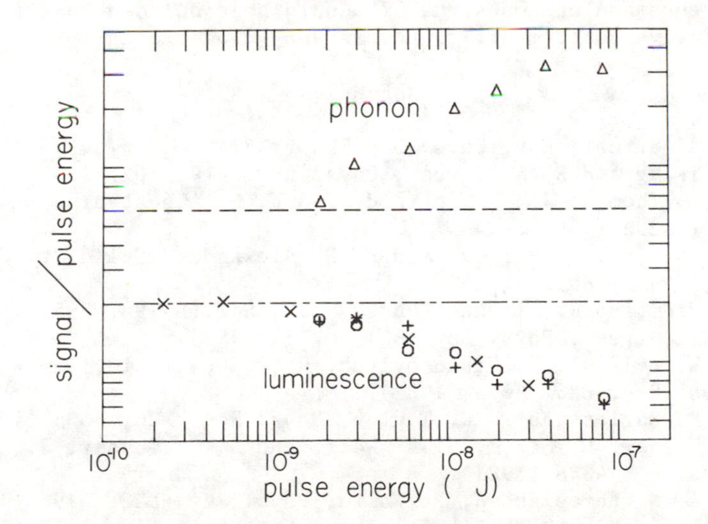

Fig. 4. Normalized phonon and luminescence integrated intensities (up to ~2 μs) versus excitation pulse energy (2.43 eV). Excitation energy of ~3 nJ corresponds to ~1×10^{18} photons/cm^3.

$$\eta \simeq \frac{p-1}{p-r}\frac{E_{exc}}{E_{gap}} \tag{1}$$

This is an approximate expression since there is generally a small negative correction to E_{gap} of order Δ/E_{gap}, where Δ represents possible phonon emission processes due to thermalization in band tails and relased lattice distortions. In order to define p and r the limiting phonon and PL signals for low pulse energies must be established. This has been accomplished for the PL signal as shown by the dashed-dotted line in Fig. 4, but has not been convincingly done for the phonon signal. Nevertheless, our results place meaningful bounds on η. Using E_{exc} = 2.4 eV, E_{gap} = 2.0 eV and r = 0.5 (at J = 1.5×10^{-8}J) we find for $\eta \equiv 1$ from eq. (1) that p = 3.5 (at J = 1.5×10^{-8}J) which leads to the dashed line in the upper part of the figure. Given our experimental uncertainties, we conclude that the internal quantum efficiency of our a-Si:H films is ~0.8-1.0.

In summary, simultaneous measurements of radiative and nonradiative emission from optically excited a-Si:H have yielded direct information about the role of nonradiative recombination processes. These measurements demonstrate conclusively that the radiative and nonradiative channels are essentially uncoupled for times $>10^{-7}$s. It was further shown that the thermally generated phonons, for sufficiently small photon densities, down convert to very long wavelength phonons which propagate ballistically over distances ~1 μm. Finally, on the basis of the intensity dependence of the luminescence and phonon signal it could be concluded that the internal radiative quantum efficiency is very close to unity.

REFERENCES

1. R. A. Street, Adv. Phys. 30, 593 (1981).
2. C. Tsang and R. A. Street, Phys. Rev. B19, 3027 (1979).
3. D. G. Thomas, J. J. Hopfield, and W. M. Augustyniak, Phys. Rev. 140, A202 (1965).
4. J. Shah, B. G. Bagley, and F. B. Alexander, Solid St. Commun. 36, 199 (1980).
5. W. Rehm and R. Fischer, Phys. Stat. Sol.(b) 94, 595 (1979).
6. R. A. Street, Phys. Rev. B21, 5775 (1981).
7. R. W. Collins, P. Viktorovitch, R. L. Weisfield, and W. Paul, Phys. Rev. B26, 6643 (1982).
8. W. E. Carlos and P. C. Taylor, Phys. Rev. B26, 3605 (1982).
9. K. Weiser, U. Strom, S. A. Wolf, and D. U. Gubser, J. Appl. Phys. 52, 4888 (1981).
10. G. A. Northrop and J. P. Wolfe, Phys. Rev. B22, 6196 (1980).
11. W. Dietsche and H. Kinder, Phys. Rev. Lett. 43, 1413 (1979).
12. U. Strom, P. B. Klein, K. Weiser, and S. A. Wolf, J. de Physique C6, 30 (1981).
13. B. A. Wilson, P. Hu, J. P. Harbison, and T. M. Jedju, Phys. Rev. Lett. 50, 1490 (1983).

PICOSECOND PHOTOLUMINESCENCE AS A PROBE OF BAND-TAIL THERMALIZATION IN a-Si:H

T.E. Orlowski and B.A. Weinstein
Xerox Webster Research Center, Rochester, NY 14644

H. Scher
Sohio Research Center, Cleveland, OH 44128

ABSTRACT

Recent picosecond time-resolved photoluminescence measurements have obtained the initial decay dynamics in a-Si:H over the range (10 psec - 4 nsec) in the high energy (>1.5 eV) part of the spectrum over a temperature range (T) from 20K to 180K at an excitation energy of 2.33 eV. The key features of the data are an initial unresolved decay (system response 10 psec) which evolves into an algebraic decay, $t^{-\alpha}$, for 50 psec $< t <$ 3 nsec, with $\alpha(T) = \alpha_o + \beta T$. The onset of the power law occurs in a time range which is less than the fastest radiative lifetime. We interpret the results with a comprehensive model of nonradiative relaxation through electron band-tail thermalization. Correlation between the thermalization process and radiative decay as well as T-dependent local diffusion is discussed and shown to be in excellent agreement with the data.

INTRODUCTION

Present understanding of carrier transport and recombination mechanisms in amorphous semiconductors has benefitted substantially from time-resolved photoluminescence (PL) measurements. Early work[1] revealed a broad distribution of recombination rates in a-Si:H and chalcogenide glasses generally thought to arise from radiative and nonradiative tunneling processes with pair-separation dependent rates. Time-resolved spectra in these systems show a characteristic shift of the PL band to lower energy with time which has been attributed to recombination at randomly separated charged (donor-acceptor type) sites in the chalcogenides[2] and carrier thermalization within band-tail states[3] in a-Si:H. Much more detailed information can be learned as the time resolution in PL experiments improves. Recent picosecond PL experiments[4] have obtained the fastest radiative recombination rate in As_2S_3 glass and the temperature dependence of the competing nonradiative processes. In those experiments band-tail excitation was used and the results were interpreted within the framework of a localized exciton model. In this paper we report the results of recent picosecond PL measurements[5] in a-Si:H. Here the excitation energy (2.33 eV) is well above the optical gap (1.8 eV) resulting in the creation of more highly mobile carriers. Because of the time (10 psec - 4 nsec) and spectral (2.1 - 1.5 eV) regimes associated with the streak camera detection scheme, our measurements select the closest pairs emitting at the highest PL energies. We find rapid PL decay in this time/spectral regime with power law behavior. A comprehensive model, which correlates thermalization with radiative decay and T-dependent local diffusion, is developed and shown to be in excellent agreement with the data. Recently subnanosecond (.25 - 10 nsec) PL measurements[6] at much lower PL energies were reported in a-Si:H. We compare those findings with the present results.

EXPERIMENTAL

The experimental apparatus used to obtain our results has been described earlier[7]. Single pulses (7 psec) from a passively mode-locked Nd^{3+}:glass laser were frequency doubled in KD*P to 2.33 eV and weakly focused on the sample housed in a variable temperature helium refrigerator. Photoluminescence was collected and time-resolved using a streak camera and optical multichannel analyzer. The maximum excitation density was $\sim 10^{17}$ photons/cm^3. Using a weak reference pulse for timing synchronization of the streak

0094-243X/84/1200163-07 $3.00

camera with the excitation pulse, we were able to average data from up to 20 individual laser shots. Modification of the streak camera was necessary to protect the photocathode from sample PL emitted at times comparable to the retrace time (60 μsec) of the streak voltage circuit. This was accomplished with a high voltage gating pulse which gated the photocathode voltage on ~1 μsec before the arrival of the excitation pulse at the sample and off 2 μsec later. This procedure protects the streak camera from saturation and reduces background considerably. Calibration of the streak camera streak rate (timebase) was accomplished using an etalon. Care was taken to keep PL intensities well within the dynamic range of the streak camera. Sharp-cut blocking filters at 2.1 eV (total attenuation $\sim 10^{15}$) were used to prevent detection of scattered light from the excitation pulse. Because of weak PL signals, no other filters were used. We observe the entire PL band within the limits of the streak camera photocathode response which, from 2.1 eV to 1.7 eV, varies by <50%. At lower energies the response drops quickly (down 10X at 1.5 eV) so our measurements select high energy PL predominantly in the range (2.1-1.5 eV). Samples used in this study (1 μm thick) were prepared by plasma decomposition of undiluted SiH_4 at low rf power (1 W) deposited on roughened fused silica. The low temperature time-integrated PL quantum efficiency of these samples is high (0.25-0.50, 10K). Extreme care was taken in sample cleaning prior to mounting in the sample chamber to prevent surface contamination emission.

RESULTS

Figure 1 displays the PL intensity I(t) measured at 20K and averaged over 20 laser shots for fast (500 psec) and slow (4 nsec) streak camera total sweep times. The pulse preceding t=0 in the plots is a timing reference used for signal averaging purposes and for characterizing the detection system response at each sweep speed. In all of our measurements the PL buildup time is less than the system response time which places an upper limit of ~10 psec on the time for photoexcited carriers to relax into states emitting within our spectral observation window (2.1 - 1.5 ev). Since the exponential optical absorption edge (Urbach edge) in a-Si:H extends to ~1.9 eV at 20K[3,8], and the fastest radiative lifetimes are ~8 nsec[6], these experiments probe the closest pairs emitting at the highest PL energies as they thermalize within the manifold of band-tail states. This thermalization process, which manifests itself as a PL spectral shift, represents relaxation of carriers to lower energy states and not recombination. Recent experiments[6] have shown that at 15K the PL spectrum at 500 psec following excitation is quite broad with significant intensity at energies ≥1.6 eV. Using a high energy cutoff filter at 1.8 eV we examined the PL buildup and decay characteristics and, within experimental accuracy, found no variation in these times, although the peak PL intensity decreased about a factor of three with the reduced spectral bandwidth. This indicates that the PL spectrum extends to energies ≥1.8 eV

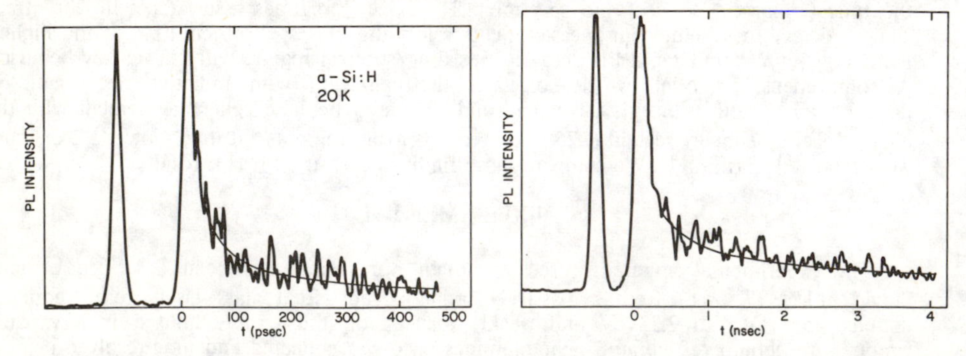

Fig. 1. Measured buildup and decay of PL (2.1 - 1.5 eV) in a-Si:H at 20K. Also shown are least squares fits assuming a power law decay.

for t < 100 psec.

The PL decay data at 20K is best fit by a power law, $t^{-\alpha}$, for 50 psec < t < 3 nsec. At earlier times the decay is unresolved from the system response. From the least squares fits in Fig. 1 we find for the fast sweep $\alpha = 0.59 \pm 0.03$ and for the slow sweep $\alpha = 0.61 \pm 0.03$. Thus, within experimental uncertainties the decay is characterized from 50 psec to 3 nsec by $\alpha = 0.60$. Attempts at fitting the data to an exponential decay gave poorer fits in all cases.

Measurements of the temperature dependence of the initial PL decay were performed over the range 20 to 180K. We find that the power law exponent, α, is temperature dependent taking the form $\alpha(T) = \alpha_o + \beta T$. Shown in Fig. 2 is a plot of α vs. T where we find from the intercept a value of $\alpha_o = 0.55 \pm .03$ and from the slope a value of $\beta = .0022 \pm .0003\ K^{-1}$. No change in the initial (peak) PL intensity is found up to 100K where it begins to decrease gradually, down a factor of two at 140K. At 180K the peak PL intensity is down a factor of four from that at 20K. Although the power law decay behavior seen in these experiments differs from the exponential decay reported in Ref. 6 (for the spectral regime below 1.6 eV), the temperature dependence of the initial intensity does agree. As emphasized earlier, our measurements use PL as a probe of thermalization processes which are much faster than the fastest radiative recombination rates. In the next section we develop a model which accounts for the observed power law behavior of the initial PL dynamics in a-Si:H in terms of electron band-tail thermalization.

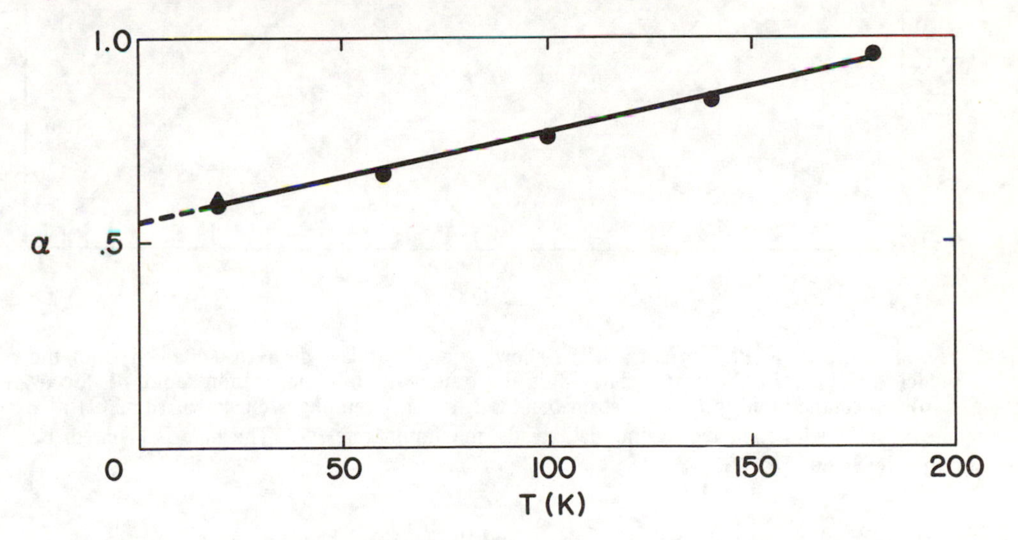

Fig. 2. Temperature dependence of the power law exponent, α, for the initial PL decay in a-Si:H.

MODEL

We assume, along with a number of authors[1], that the higher energy luminescence is due to a band-tail electron recombining with a band-tail hole. A typical configuration for this interaction is shown in the inset of Fig. 3. The arrows indicate the possible electron transfer steps (among sites, O) and the hole is taken to be immobile (at site ⊗). As discussed above, these experiments select the closest pairs emitting at the highest PL energies. The pair can be geminal , a localized exciton, or randomly separated as a result of relaxation of an electron with as much as 0.4 eV energy above the conduction band (CB) mobility edge.

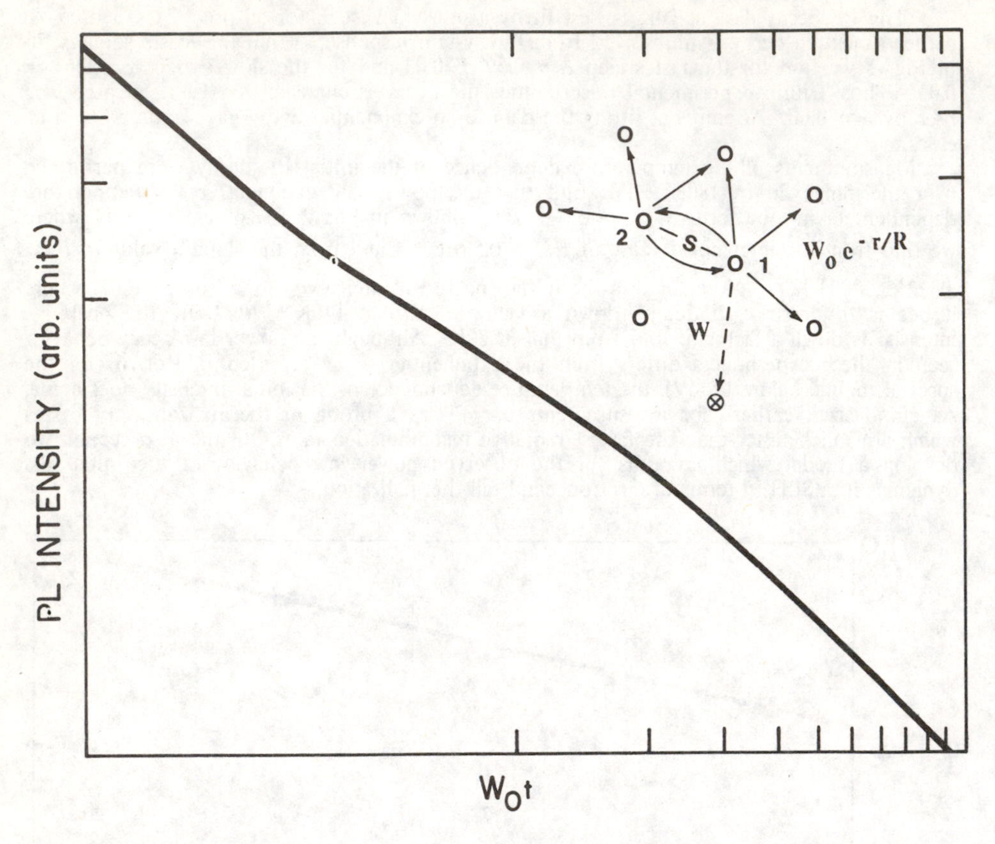

Fig. 3. Theoretical PL intensity, I(t) , showing a power law decay as obtained from the model described in the text. The inset illustrates the mechanism for rapid PL decay involving competition between electron band-tail hopping (among a cluster of sites, O) at a rate $W_0e^{-r/R}$ with radiative recombination at a maximum rate ,W. The hole is taken to be immobile at site ⊗.

The single (fastest) radiative recombination rate is designated as W. Competing with W are the electron transfers away from the hole offset by the back transfers. One can readily assume that the back transfer occurs preferentially from a small number of nearby sites. This local diffusion involving interaction among a cluster of sites imbedded in a fixed random background of additional sites is solved exactly. The electron can hop back and forth an arbitrary number of times to the sites within the cluster and from each of these, hop once to the fixed random background of sites. The time to return to the cluster after leaving is on the order of a large sampling time of all the other sites. The configuration average (c.a.) over these latter site positions/energies can be carried out exactly, retaining all the correlations between the cluster geometry and the background. Referring again to Fig. 3, we are interested in determining the (averaged) probability $\langle P(1,t)\rangle$ for the electron to be found at site 1 at time t if it started there at t=0. The luminescence intensity is then $I(t) = W\langle P(1,t)\rangle$. We specialize the cluster to two sites {1,2} separated by s and obtain[9]

$$W\langle P(1,t)\rangle = W\exp\{-t(W_{21} + W)\}\Phi_1(t) +$$

$$W(W_{12}W_{21})^{1/2} \, {}_0\!\int^t d\tau \, \exp\{-(t-\tau)W_{12} - \tau(W_{21}+W)\}\tau^{1/2}(t-\tau)^{-1/2} \, I_1(\sigma)\Phi_2(t-\tau,\tau) \, , \qquad (1)$$

$$\Phi_1(t) = \exp\{ -\int d^3r \, d\varepsilon \, p(r, \varepsilon) [1-\exp(-t \, W(r-r_1))] \}, \qquad (2)$$

$$\Phi_2(t-\tau,\tau) = \exp\{ -\int d^3r \, d\varepsilon \, p(r,\varepsilon) [1-\exp(-(t-\tau) \, W(r-r_2) - \tau W(r-r_1))] \}. \qquad (3)$$

Here, $p(r, \varepsilon)$ is the probability of finding a site at r with energy difference ε and $W(r)$ is the transition rate to this site (the ε-dependence is implicit). $I_1(\sigma)$ is the modified Bessel function of order unity and $\sigma \equiv \{4W_{12}W_{21}(t-\tau)\tau\}^{1/2}$, where W_{12}, W_{21} are the transition rates between $\{1,2\}$. The function Φ_2 is a generalization (to two sites) of the more familiar[10] Φ_1, the c.a. probability to remain on a single site in a random distribution of sites.

DISCUSSION

In order to calculate I(t) from Eqns. (1)-(3) we must specify W(r) and $p(r, \varepsilon)$. To initially simplify the calculation of Φ_1 and Φ_2 we consider the transition rate to be non-zero and a function of r only in an energy width $\hbar\omega$, an effective phonon energy, for $\varepsilon < 0$ (hopping down) and κT for $\varepsilon > 0$ (hopping up). The basic parameters can now be defined:

$$W(r) = W_o \exp(-r/R) \qquad (4)$$

$$p(r, \varepsilon) = g(\varepsilon_1-\varepsilon) \equiv g \exp\{-(\varepsilon_1 -\varepsilon)/\kappa T_o\} \qquad (5)$$

$$n = {}_{-\hbar\omega}\!\int^{\kappa T} d\varepsilon \; p(r, \varepsilon) = \kappa T_o g(\varepsilon_1) \{ \exp(T/T_o) - \exp(-\hbar\omega/\kappa T_o) \} . \qquad (6)$$

Here ε_1 is the energy, measured from the CB mobility edge, of the electron at 1 and κT_o is the width of the exponential density of band-tail states (~ 30 meV in a-Si:H). The estimates for the parameters are: $R \simeq 7Å$ for $\varepsilon_1 \leq 0.1$ eV, $W_o \simeq 10^{13}$ sec^{-1} (order of magnitude estimate obtained from donor state calculations in x-tal Si, cf. Refs. 10-12), and $g \sim 4 \times 10^{20}$ $eV^{-1}cm^{-3}$.

Now, using the large argument asymptotic form for $I_1(\sigma) \sim \exp(\sigma)\,(2\pi\sigma)^{-1/2}$ and the Laplace method one can evaluate the expression in Eq. (1) analytically as a function of $\eta \equiv 4\pi nR^3$ and W_ot and compare with the PL I(t) in Fig. 1. A representative theoretical I(t) is shown in Fig. 3. Note the power law decay $I(t) \sim t^{-\alpha}$ for nearly two decades of W_ot with a departure toward increasing α at large W_ot. An excellent fit of the theory to the PL I(t) (Fig. 1) over the entire experimental range (50 psec $\leq t \leq$ 4 nsec), is obtained using $\eta=5 \times 10^{-3}$ and $W_o = 3 \times 10^{13}$ sec^{-1}. This comparison is shown in the log-log plot of I(t) vs. t in Fig. 4. The departure of the measured I(t) from power law behavior, with $\alpha = 0.6$, for $t > 3$ nsec is in agreement with the same feature in the theoretical curve.

The theoretical values for α show a linear dependence on η over the range of α values shown in Fig. 2. Using Eq. 6 one can obtain the relative change in η as a function of T for various values of $\hbar\omega$, and therefore $\alpha(T)$. The linear relation (solid line) shown in Fig. 2 corresponds to $\hbar\omega = 20$ meV. Thus 20 meV is an effective measure of the energy width for relaxation involving single hops in the band-tail. Since there is a peak in the acoustic phonon density of states in a-Si at 20 meV[13], it is tempting to ascribe the relaxation we probe to interaction with acoustic phonons.

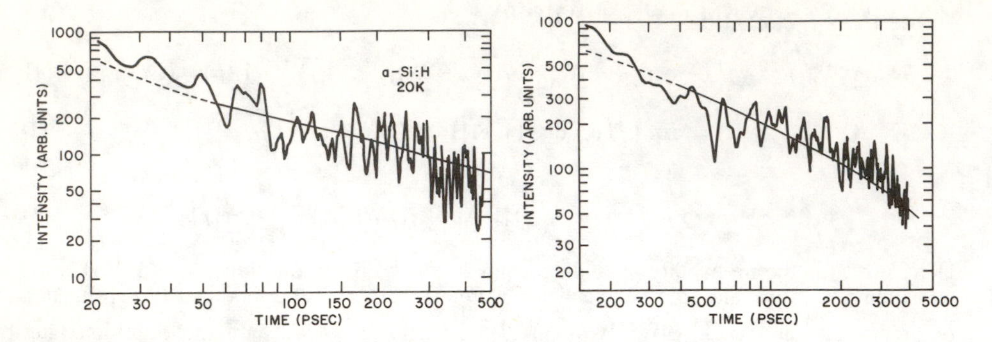

Fig. 4. Log-log plots of the PL decay data from Fig. 1 with theoretical fits obtained using Eq. 1 to calculate I(t).

One can now determine the electron position in the band tail by again using Eq. 6 with the value of $\hbar\omega$ obtained from the $\alpha(T)$ data and the above value of η at T= 20K. We obtain $\varepsilon_1 = 0.05$ eV. The nearest neighbor hopping time at ε_1 (with width $\hbar\omega$) is 60 psec. This time is consistent with the notion that the electron relaxes, by hopping down, to a level where the hopping time is on the order of the observation time of the experiment (~ 100 psec). At ε_1 the release time to the CB edge at T= 20K is 10 sec! Only at T= 180K is the release time comparable to the observation time. Therefore, multiple trapping is unlikely to play any role in the nonradiative transitions (via local diffusion) in the temperature range of Fig. 2. With the model presented here , one can obtain a linear $\alpha(T)$ with a nonzero T=0 value simply by considering relaxation via electron hopping!

If the initial fast relaxation of the electrons in the band-tail produces an occupation proportional to $g(\varepsilon)$ then it can be presumed that the distribution is peaked at ε_1 (with a width $T_o + T$) where ε_1 is relatively independent of T ($\Delta\varepsilon_1 < \kappa T_o$) for T < 180K. An important additional measurement would be the time resolved PL build-up at lower energy.

SUMMARY

Picosecond time-resolved PL measurements have obtained the initial decay dynamics in a-Si:H in the high energy (> 1.5 eV) portion of the spectrum from T= 20 to 180K. The data is best fit by a power law, $t^{-\alpha}$, over nearly two decades in time (50 psec < t < 3 nsec) at 20K with $\alpha(T) = \alpha_o + \beta T$; at earlier times the decay is unresolved from the system response (10 psec). The onset of the power law occurs in a time range much shorter than the fastest radiative lifetime. We assume that the higher energy PL is due to a band-tail electron recombining with a band-tail hole. Since our samples have a high PL quantum efficiency we identify the observed PL decay behavior with nonradiative relaxation processes which bring electrons to lower energy states within the band-tail manifold and assume that the hole is immobile. We develop a comprehensive model of nonradiative relaxation which correlates the thermalization process with radiative decay and T-dependent local diffusion. The model incorporates both the site energy and position in determining the electron transfer efficiency. By considering hopping down to states within an effective phonon energy, $\hbar\omega$, and hopping up to states within κT, the model can explain both the power law nature of the PL decay and the linear temperature dependence of the power law exponent as well as its T=0 intercept. Applying this new model to all of our PL, data we conclude the following about the initial PL decay dynamics in a-Si:H: 1) within 50 psec, electrons have moved down the band-tail 0.05eV from the conduction band edge; 2) the maximum hopping rate, W_o, is 3×10^{13} sec^{-1}, and 3) the effective energy width for relaxation involving single hops in the band-tail is ~ 20 meV which is close to the peak in the acoustic phonon density of states for a-Si.

REFERENCES

1. R.A. Street, Adv. Phys. 25, 397 (1976); ibid., 30, 593 (1981), and references therein.
2. G.S. Higashi and M. Kastner, Phys. Rev.B 24, 2295 (1981).
3. C. Tsang and R.A. Street, Phys. Rev.B 19, 3027 (1979).
4. B.A. Weinstein, T.E. Orlowski, W.H. Knox, T.M. Nordlund and G. Mourou, Phys. Rev.B 26, 4777 (1982).
5. T.E. Orlowski, Bull. Am. Phys. Soc. 28, 239 (1983).
6. B.A. Wilson, P. Hu, J.P. Harbison and T.M. Jedju, Phys. Rev. Lett. 50, 1490 (1983); Phys. Rev. B28, 5901 (1983).
7. T.E. Orlowski and H. Scher, Phys. Rev.B 27, 7691 (1983).
8. G.D. Cody, T. Tiedje, B. Abeles, B. Brooks and Y. Goldstein, Phys. Rev. Lett. 47, 1480 (1981).
9. H. Scher and T.E. Orlowski, to be published.
10. H. Scher and M. Lax, Phys. Rev. B 7, 4502 (1973).
11. A. Miller and E. Abrahams, Phys. Rev. 120, 745 (1960).
12. M. Pollak and T.H. Geballe, Phys. Rev. 122, 1742 (1961).
13. M.H. Brodsky and M. Cardona, J. Non-Crystalline Solids 31, 81(1978).

A COMPARISON OF PHOTOINDUCED ABSORPTION AND PHOTOLUMINESCENCE MEASUREMENTS IN a-Si:H

R.W. Collins and W.J. Biter
The Standand Oil Co., 4440 Warrensville Ctr. Rd.,
Cleveland, Ohio 44128

ABSTRACT

Measurements of the steady state excitation intensity, temperature and spectral dependence of the photoluminescence and photoinduced absorption are reported on the same glow discharge deposited hydrogenated amorphous silicon samples. Similarities in the results of the two measurements suggest possible alternative models for the details of the photoinduced absorption process.

INTRODUCTION

Both photoluminescence[1] (PL) and to a lesser extent photoinduced absorption[2] (PA) have been used as a tool to study the electronic structure as well as thermalization, relaxation and recombination processes of photoexcited carriers in hydrogenated amorphous silicon (a-Si:H) prepared by different techniques. Steady state PA and PL are both observed under the same conditions of sample temperature and excitation (pump) photon energy and intensity. As a result, it is informative to compare the results of the two measurements on the same a-Si:H samples.

EXPERIMENTAL DETAILS

The experimental apparatus used for the PL and PA closely resembles that described elsewhere.[3] Both PA and PL were measured under identical conditions by opening a shutter to expose the sample to focused light from a 1000W tungsten source in order to measure (PA-PL) and closing it to obtain PL. To ensure that the contribution to ΔT, the change in transmission, reflected changes in k, the extinction coefficient, rather than changes in n, the index of refraction, the interference fringe criteria of O'Connor and Tauc were employed.[3]

The samples used in this study were deposited in a capacitively coupled rf glow discharge system. Films of different defect density could be obtained by varying deposition conditions such as rf power and substrate platform bias. The relative defect density was assessed on the basis of the PL intensity, using the well established relationship for a-Si:H.[1]

RESULTS

Figures 1 and 2 show the excitation (pump) intensity dependence of the PL and PA efficiencies (PL intensity or ΔT divided by excitation intensity) for one of the samples of lowest defect

0094-243X/84/1200170-08 $3.00

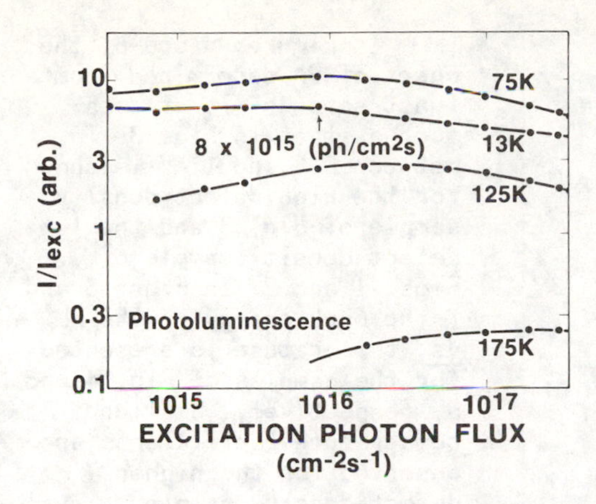

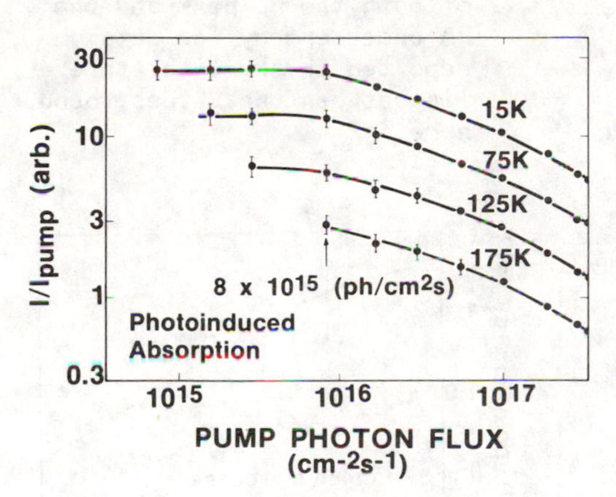

Figs. 1 (upper) and 2. PL and PA efficiency vs. excitation flux for low defect density a-Si:H (at 1.38 and 1.29eV, respectively).

density. The PL data are similar to previous results.[4,5] The superlinearity at low excitation intensity for temperatures > 75K is not observed in samples of high defect density and is attributed to an increasing probability of non-geminate pair formation as the excitation intensity is increased.[4] The sublinearity at high excitation intensity has been discussed in terms of saturation of band tail states[6] or Auger recombination.[4] The transition to this high intensity regime is clearly seen in the 13K data where the non-geminate component is negligible.

The PA data in Fig. 2 show a transition to sublinearity in the pump intensity dependence of the PA intensity, and the shape of this curve is independent of sample temperature. The data in the initial sublinear regime and at the highest intensity correspond to slopes of 0.6 and 0.3 if plotted as PA intensity vs. pump intensity. Similar slopes have been reported by Olivier *et al.*[7] whereas O'Connor and Tauc[3] have reported a value of 0.5 throughout this range.

In Figure 3, the spectral dependence of the PA is displayed for a sample with a high defect density (a factor of 20 lower in PL intensity than the sample of Fig. 1), plotted as $(\Delta\alpha E)^2$ vs. E to compare with data reported elsewhere.[2] Linearity on such a plot indicates consistency with a model of the PA as a transition between carriers trapped in a delta function density of states and a parabolic band, with the E intercept equal to the energy onset. At all temperatures consistency with the model is obtained for the accessible energy range of 0.6 to 1.0eV. At higher energies and predominantly at low temperature, significant deviations from linearity are observed.

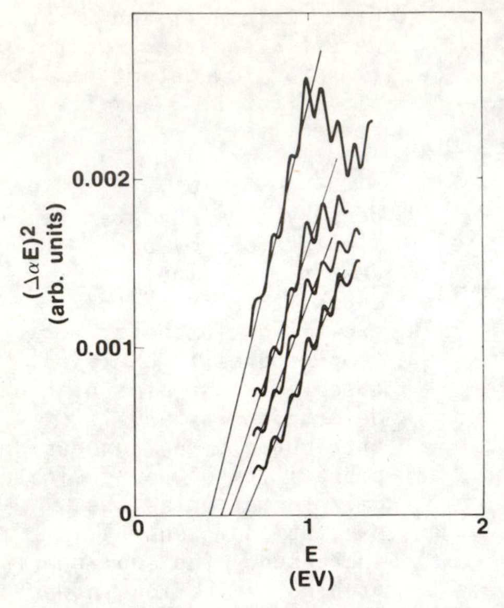

Fig. 3. PA spectral dependence for high defect density a-Si:H (l. to r. T=15,100,160,250K).

In Figures 4 and 5, the onset of PA determined from a least squares fit of data such as that in Fig. 3 between 0.6 and 1.eV is shown for the high defect density sample of Fig. 3 and the low defect density sample of Figs. 1 and 2. In Figs. 6 and 7 the peak energy of the PL vs. temperature is presented for the samples of Fig. 4 and 5, respectively. Smaller temperature coefficients are observed for the higher defect density sample.

The temperature shifts of both the PL peak and the PA onset energy for data reported in the literature lie between the values found here.[1,2]

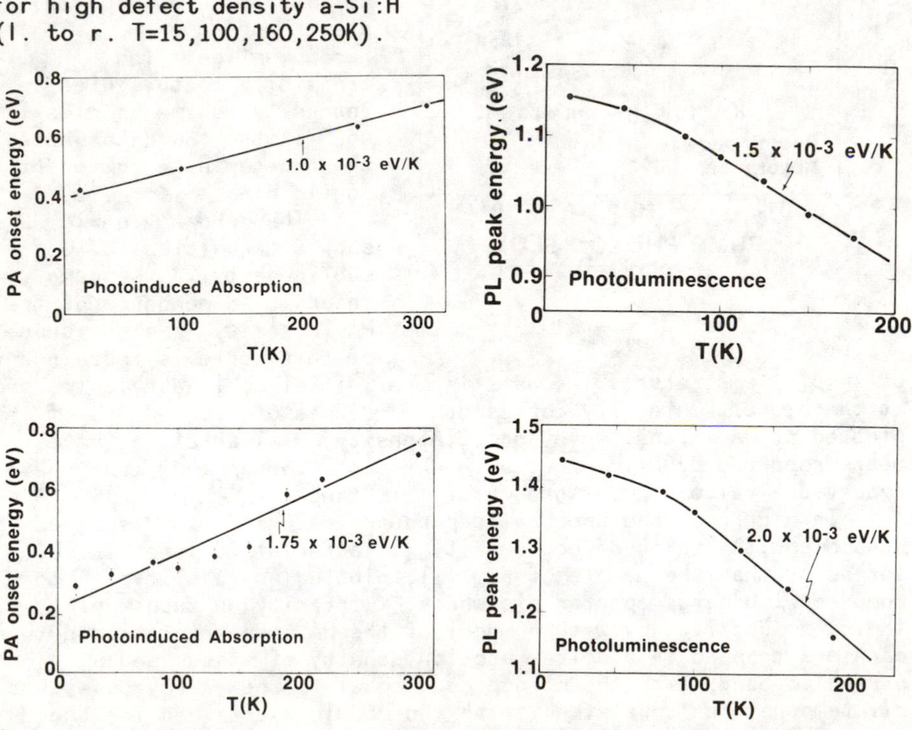

Figs. 4,5 (left) 6, and 7 (right). Onset energy of PA (left) and peak energy of PL vs. temperature for high defect density sample of Fig. 3 (top) and low defect density sample of Fig. 1 (bottom).

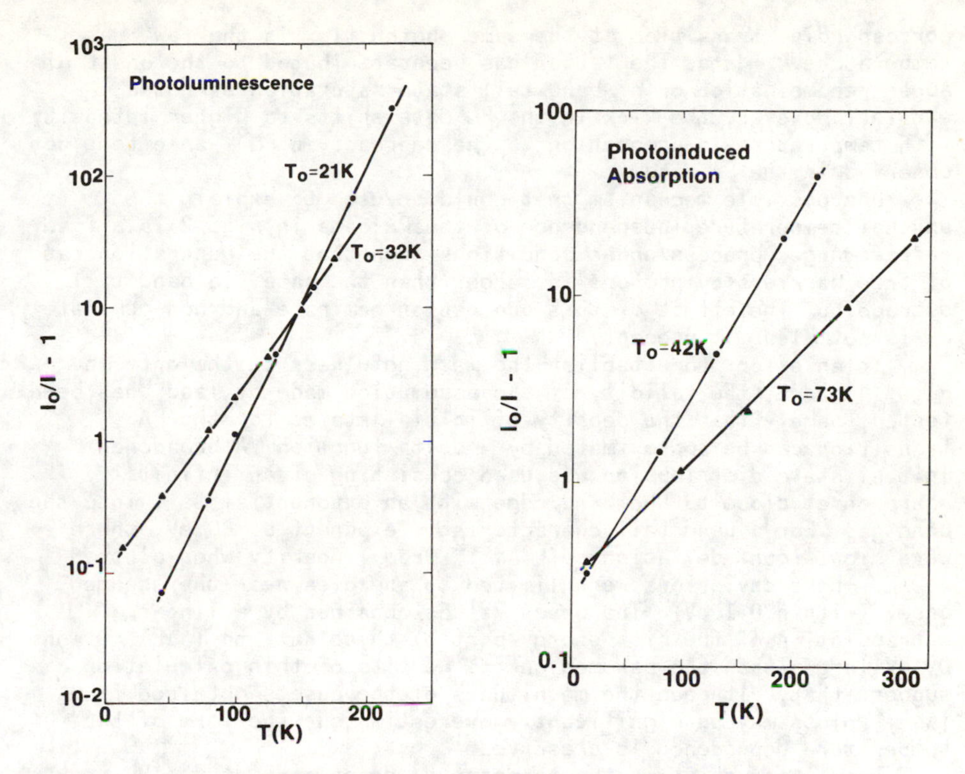

Figs. 8 and 9. $I_o/I-1$ for peak PL (left) and PA (at 0.8eV) vs. temperature for high and low defect density samples of Figs. 3 and 1.

The temperature dependence of PL intensity for these two samples obeyed the exponential form shown in Fig. 8 as described in detail elsewhere.[8] The higher defect density sample exhibited a larger T_o value. For the temperature dependence of the PA, a simple exponential form for ΔT vs. temperature over a wide temperature range is not observed as has been reported elsewhere.[3] However, as shown in Fig. 9, if $(I_o/I-1)$ is plotted vs. temperature for a properly chosen I_o, a strict exponential law is found just as for the PL data. Again the T_o value for the higher defect density sample is larger in approximately the same ratio as for the PL T_o values. The PA T_o values, however, are about twice as large.

DISCUSSION

The data in Fig. 2 indicate that at low pump intensities, the PA process appears to obey monomolecular statistics. The bimolecular recombination model of Tauc[2] does not describe the higher intensity data since the slope of the decrease in PA efficiency with pump intensity is temperature independent, itself a surprising result. It is tempting to associate the kink in the PA data in Fig. 2 with the corresponding transition at the same photon flux in the low temperature PL data. The latter has been attributed to the onset of Auger recombination or of band tail state saturation as noted earlier. However, the kink in the PL data shifts to higher intensity with temperature in proportion to the decrease in PL[6]--an effect not observed in the PA data.

One possible mechanism that could be used to explain the unusual temperature independence of the PA data in Fig. 2 is a free carrier Auger process under conditions such that the generation rate of free carriers by photons is larger than the rate via band tail detrapping. The effect of this process on geminate and non-geminate PL is not clear at present.

In an effort to establish the meaningfulness of the data in Figs. 4 and 5, the validity of the assumption made by Tauc[2] has been tested, namely that the density of initial states for the PA transition can be approximated by a delta function. A broadened initial state distribution was used consisting of an infinitely sharp onset close to the band edge with an exponential tail into the band gap. For a band tail characteristic exponent of 80meV, there were significant deviations of $(\Delta\alpha E)^2$ from linearity when plotted vs. E. These deviations were limited to energies near the assumed onset (within 0.15eV). The onset values obtained by a linear extrapolation of the high energy part of these data to $(\Delta\alpha E)^2=0$ were 0.1eV larger than the assumed onset. Results of this calculation suggest that, although the magnitudes of the onsets obtained from $(\Delta\alpha E)^2$ plots may be significantly overestimated, the form of the temperature dependence is preserved.

The linear shift in the temperature dependence of the PL peak energy has been described in terms of a balance between the radiative probability, P_R and the probability of thermal activation of the carrier, $p_{oE}\exp(-\epsilon/kT)$, (presumably the electron) out of the band tail.[9] However, sample-to-sample variations are observed in the PL peak temperature coefficient. Since for an intrinsic transition, P_R should not depend on the defect density, it may be concluded that a competitive T-independent non-radiative process is affecting the balance. If it is assumed that the process is due to the tunnelling of band tail electrons to gap states, a more general equation can be written:

$$E_{PL}(T) = E_{PL}(0) - kT\ln\{P_{oE}/(P_R+\gamma_{TE}N_{RCE})\} \qquad (1)$$

where $\gamma_{TE}N_{RCE}$ is the probability of band-tail tunnelling. Correcting the data of Fig. 7 for an assumed temperature shift of the band gap of 4×10^{-4}eV/K and using $P_{oE}=10^{13}s^{-1}$, it is found that $P_R=8\times10^4s^{-1}$ and for the high defect density sample of Fig. 6, $\gamma_{TE}N_{RCE}= 3\times10^7s^{-1}$. This model, however, assumes that this process occurs at one band edge only.

We suggest that the band tail tunnelling observed in the glow discharge sample reported here is a result of extensive band tails and a high defect density. The PL spectrum for the sample of Fig. 6 is significantly wider than that of Fig. 7.

Since monomolecular recombination has been proposed to explain the results in Fig. 2, a similar model may be used to explain the shift in the onset of the PA. Again both PL and non-radiative tunnelling may deplete the hole population available for PA (using the currently accepted model for the PA transition as due to holes[2]). At high temperatures (T>125K), enough electrons are detrapped to make the radiative process negligible. Thus one can write an equation similar to Eq. 1 with $P_R=0$ and "E" replaced by "H". Using this equation for the two samples of Figs. 4 and 5 with $P_{oH}=P_{oE}$ and using an additive band gap correction, we find that $\gamma_{TH}N_{RCH}=9\times10^5 s^{-1}$ and $1\times10^2 s^{-1}$.

The different values obtained for $\gamma_{TE}N_{RCE}$ from PL and $\gamma_{TH}N_{RCH}$ from PA for the high defect density sample support the assumption that the two measurements are dominated by processes occurring at opposite band tails. They also support the assumption that in considering mechanisms involved in the statistics of PL, the holes can be neglected.

The temperature dependence of the PL intensity shown in Fig. 8 is consistent with a distribution of activation energies for non-radiative recombination.[10] When band tail tunnelling of electrons to non-radiative recombination centers can be neglected, the width of this distribution is proportional to $T_o \ln(P_{oE}/P_R)$. Otherwise, for the high defect density sample of Fig. 8 for which the peak energy data suggested non-radiative tunnelling, it can shown that the width is now proportional to $T_o \ln\{P_{oE}/(P_R+\gamma_{TE}N_{RCE})\}$. Measurements of PL(T) at low excitation intensity (where recombination is predominantly geminate) indicate that the tunnelling term in the high defect density sample does not completely cancel the increase in T_o. As a result, it can be concluded that the higher defect density sample exhibits a broader distribution of activation energies as its larger spectral width would indicate.

These results for glow discharge a-Si:H are qualitatively different than the results for sputtered a-Si:H of different defect density. For the sputtered samples, it was concluded that band tail non-radiative tunnelling was not a dominant mechanism.[11] Consistent with this conclusion, it was observed for the sputtered samples that the temperature coefficient of the PL peak energy was relatively independent of the defect density.

The PA data of Fig. 9 can be interpreted similarly to the PL data of Fig. 8. The fact that the PA obeys the $I_o/I-1$ relationship is consistent with a distribution of activation energies of the same functional form as for the PL. This fact is also consistent with a temperature independent band-tail tunnelling contribution which competes with the recombination that occurs via thermal activation. It can be shown[12], for T>125K, using monomolecular rate equations for the trapped (N_{TH}) and free (N_{FH}) hole density, that

$$N_{TH}= \gamma_v N_v G\{P_{IH}\gamma_{CH}N_{RCH}+ \gamma_{TH}N_{RCH}(\gamma_v N_v+\gamma_{CH}N_{RCH})\}^{-1} \quad (2)$$

where G is the hole generation rate by photons, $\gamma_v N_v$ is the probability of hole trapping in the band tail, $\gamma_{CH}N_{RCH}$ is the probability of direct recombination through deep centers, P_{IH} is the detrapping probability (the only strongly temperature dependent term), and $\gamma_{TH}N_{RCH}$ is the probability of band tail-tunnelling, presumably through the same centers as direct capture. Using this equation, the empirical form of the PA, and allowing P_{IH} a range of activation energies, the distribution of activation energies is similar to that of PL except with a width $\Delta\epsilon$ given by:

$$\Delta\epsilon \;\alpha\; T_o \ln \left\{ \frac{P_{oH}\gamma_{CH}}{\gamma_{TH}(\gamma_v N_v + \gamma_{CH}N_{RCH})} \right\} \tag{3}$$

Thus, depending on the relative rate coefficients, a larger T_o may imply larger recombination center density or it may imply broader band tails. For reasonably high quality samples, it is expected that $\gamma_v N_v >> \gamma_{CH}N_{RCH}$ and, for this case, T_o provides a relative measure of the band tail extent.

A comment on the absolute magnitude of the PA and PL signals is in order. Only about a factor of about 5 is exhibited in the sample-to-sample variations in the low temperature PA magnitude among a larger set of these samples spanning a wide range of defect density whereas the PL is known to be a sharply decreasing function of defect density over orders of magnitude. We will briefly consider the reason for this effect.

The PA yield is proportional to the trapped carrier density. If PL did not deplete the hole population available for PA then, at low temperature, Eq. 2 would be valid with $P_{IH}=0$. For samples of sufficiently high defect density such that $\gamma_v N_v < \gamma_{CH}N_{RCH}$, Eq. 2 would predict that the PA yield should drop roughly as the square of the defect density. This is not observed. An analysis of the coupled equations for free and trapped electrons and holes including the possibility of radiative recombination of trapped carriers predicts that in the limit of low temperature:[12]

$$N_{TH} = \frac{\gamma_{CH}N_{RCH}}{\gamma_{CE}N_{RCE}P_R} G \tag{4}$$

It might be expected that N_{RCH} and N_{RCE} roughly scale with one another, resulting in a sample independent PA yield to first order. However, it is not clear how radiative recombination would effect the intensity of PA for 50<T<125K where the trapped electron density is falling rapidly.

Note that in these analyses it is assumed that once a carrier tunnels into or is captured by a recombination center it cannot be released by a probe photon. This may not be a valid assumption in all a-Si:H samples and inclusion of this effect would inevitably complicate our interpretation. The rather well defined, reasonably sharp distribution of activation energies tends to support the assumption that the PA is due primarily to carriers in a sharply falling density of states distribution. This is not true for doped a-Si:H, material which gives significantly different PA results than those presented above.

CONCLUSION

In conclusion, measurements of the excitation, temperature, and spectral dependence of both photoluminescence and photoinduced absorption on the same glow discharge a-Si:H samples are used to provide detailed information on the steady state statistics of carrier recombination and the density of states distribution. It is found that the excitation intensity dependence of the PA is consistent with monomolecular recombination statistics. A temperature independent sublinearity at high pump photon flux is seen in PA but not clearly in PL. Both the shift in the onset of the PA and the PL peak energy with temperature can be described by very similar models which invoke the quasi-equilibrium between band tail detrapping and tunnelling to deep traps or radiative recombination. The temperature dependence of the PA and PL magnitudes imply distributions of activation energies of band tail carriers which are similar in shape. The T_o value obtained from the temperature dependence of PA is not as straightforwardly indicative of the band tail slope as is the case for PL. The relatively sample independent magnitude of PA in a-Si:H of different defect densities can be explained by including radiative recombination in the monomolecular statistics.

REFERENCES

1. R.A. Street, Adv. Phys. 30, 593 (1981).
2. J. Tauc, Festkorperprobleme: Advances in Solid State Physics, edited by P. Grosse (Vieweg, Braunschweig, 1982), Vol. XXII, p. 85.
3. P. O'Connor and J. Tauc, Phys. Rev. B 25, 2748 (1982).
4. R.A. Street, Phys. Rev. B 23, 861 (1981).
5. R.W. Collins and W. Paul, Phys. Rev. B 25, 5263 (1982).
6. R.W. Collins and W. Paul, J. Non-Cryst. Solids 58 & 60, 369 (1983).
7. M. Olivier, J.C. Peuzin, and A. Chenevas-Paule, J. Non-Cryst. Solids 35 & 36, 693 (1980).
8. R.W. Collins, M.A. Paesler, and W. Paul, Solid State Commun. 34, 833 (1980).
9. R.A. Street, J. Phys. (Paris) 42, C4-575 (1981).
10. R.W. Collins and W. Paul, Phys. Rev. B 25, 5257 (1982).
11. R.W. Collins, P. Viktorovitch, R.L. Weisfield, and W. Paul, Phys. Rev. B 26, 6643 (1982).
12. R.W. Collins, unpublished.

METASTABLE CARRIERS IN a-Si:H A STUDY BY IR STIMULATED PHOTOLUMINESCENCE AND PHOTOCONDUCTIVITY

F. Boulitrop
Laboratoire Central de Recherches,
THOMSON-CSF, F91401 Orsay

ABSTRACT

IR reexcitation of carriers which remain metastable in a-Si:H at low temperature after an optical excitation, induces a transient pulse on the Photoluminescence (PL) and the Photoconductivity (PC). From the PL pulse a metastable carrier density of $\sim 10^{17}$ cm^{-3} is deduced while the PC pulse shows that carriers are retrapped and reexcited ~ 100 times before recombining. It is also shown that at least some of the metastable carriers are in long lifetime radiative states and that they have a logarithmic time decay. This and comparison with ESR data strongly suggests that the metastable carriers are trapped in tail states and that some tail states are radiative states for the 1.2-1.4 eV luminescence band in a-Si:H.

INTRODUCTION

ESR[1, 2] and Thermostimulated Currents[3] (TSC), have shown that some 10^{16}-10^{17} cm^{-3} non-equilibrium carriers remain in the tail states of a-Si:H at low temperature, seconds and even hours after the end of the excitation. Kinetics studies of the band edge Photoluminescence (PL) have also measured some 10^{17} cm^{-3} such metastable carriers[4]. Photoinduced absorption (PA)[5] and the IR quenching of the metastable ESR signal due to holes[2] have shown that trapped holes can be optically reexcited to the valence band. It has been reported that this IR reexcitation induces transient pulses of the Photocurrent (PC) and of the PL intensity[6]. However integration of the PC transient pulse and comparison with the PC value under a given CW visible light excitation yields 10^{19} cm^{-3} for the metastable carrier density[6]. This large value presents two problems. The first is experimental, since this value is almost two orders of magnitude larger than the one given by ESR, TSC or PL. The second is theoretical, since the tunnelling model of recombination[7,8] predicts $\sim 10^{17}$ cm^{-3} metastable carriers in the tail states. Although at low temperature the contribution of each photoexcited carrier to the photocurrent under CW excitation arises from its drift before trapping by tail states and subsequent recombination, metastable carriers reexcited by IR may be retrapped and optically reexcited several times before recombination. If this is not taken into account the metastable carrier density is overestimated using PC. Since PL observes carriers only when they recombine, measurements of the metastable carrier density from the PL transient pulse should be more reliable. Such measurements and comparison of the PC and PL transient pulses are the subject of this paper.

EXPERIMENTAL TECHNIQUES AND RESULTS

The samples studied are n^+in^+ sandwich structures prepared by glow discharge on ITO coated glass substrates with an Al top contact. The undoped layer is ∿ 2 μm thick and contains ∿ 10^{16} cm^{-3} dangling bonds. The samples were held at low temperature in an Oxford helium flow cryostat (5 K-500 K). Excitation was provided by an He-Ne laser and reexcitation by a 150 W tungsten lamp with long pass filters. PC measurements were made with a load resistor and a fast voltmeter while PL was detected by a Si avalanche detector. The energy spectrum of the IR induced PL was measured by a spectrometer and a cooled Ge detector. Light pulses and data taking were provided by a microcomputer.

The effect of a visible light and an IR light pulse sequence on the luminescence intensity and on the photocurrent through the n^+in^+ structure with 3 V bias is shown on Fig.1. PL and PC intensities are normalised to their values under CW visible excitation. The integral with time of the transient pulses are also shown. Note that the integral of the pulse is 70 times larger for PC than for PL.

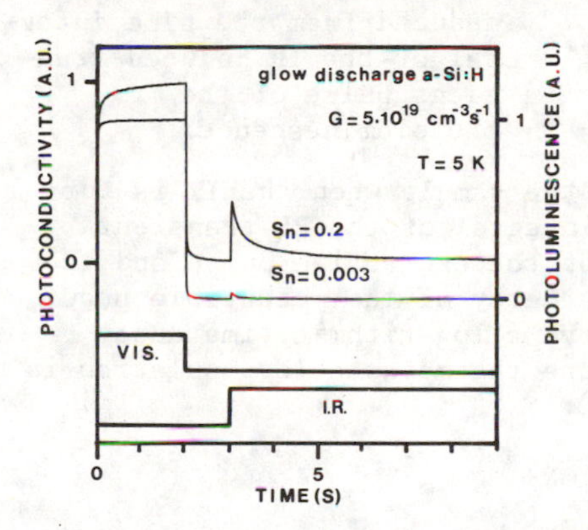

Fig.1 : Effect of an excitation-reexcitation sequence on the photoconductivity and on the photoluminescence. S_n is the time integral of the pulses normalised to the signal intensity under CW excitation.

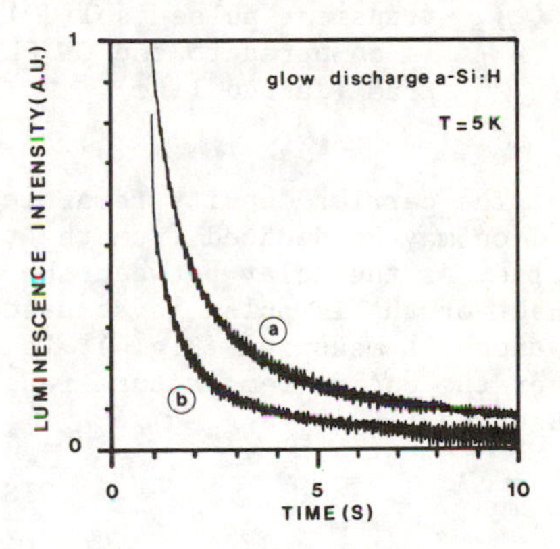

Fig.2 : Decay of the PL at long times, without (a) and with (b) an IR (hν < 0.8 eV) reexcitation during the first second of the decay.

In Fig.2, we compare the photoluminescence decay at long times (1-10 s) with and without IR reexcitation during the first part of the decay (0-1 s). Note that IR strongly decreases the PL intensity at long times.

The energy spectrum of the luminescence emitted by the IR induced transient pulse is compared to the CW PL spectrum. Both spectra are similar in shape and width, although we note that the transient PL spectrum is shifted 150 meV toward lower energies.

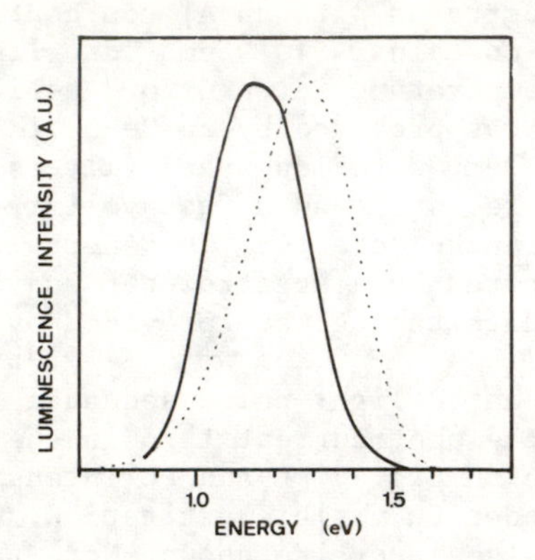

Fig.3 : The energy spectrum of the PL transient pulse (solid line) is compared to the CW PL spectrum (dashed line).

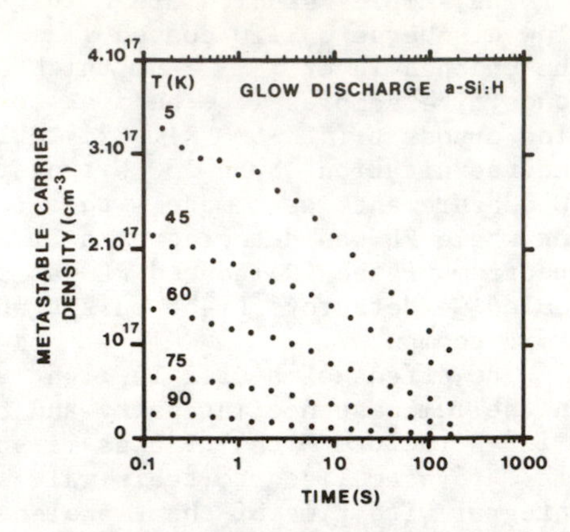

Fig.4 : Time decay of the metastable carrier density as deduced from the time integral of the IR induced transient pulse of the photoluminescence.

The carrier density remaining in the sample when the IR is turned on may be deduced from the time integral of the PL transient pulse. As the delay between the end of the excitation pulse and the onset of the IR pulse is scanned, the decay of the metastable populations is measured (Fig.4). We observe a logarithmic time decay over the entire temperature range where the metastable population is observed.

DISCUSSION

As suggested by the IR quenching of the metastable ESR signal due to holes[2] we assume that at the end of the IR induced PL transient pulse, all carriers have recombined. Since the probability for a recombining carrier pair to emit a photon is the PL quantum efficiency η, the density ρ_{met} of metastable carriers is related to the total number N_{ph} of photons emitted during the PL transient pulse by :

$$\rho_{met} = \frac{N_{ph}}{\eta V} = \frac{SG\eta V}{\eta I_{CW} V} = S_n G$$

V is the excited volume of the sample.
G is the generation rate.
$\frac{G\eta V}{I_{CW}}$ is a calibration coefficient computed from the CW PL intensity I_{CW} for a given G.
$S_n = \frac{S}{I_{CW}}$ is the time integral of the normalized transient PL pulse.

Note that ρ_{met} is independent of the PL quantum efficiency. The PL data of Fig.1 yield a carrier density of $\rho_{met} = 1.5\times10^{17}$ cm^{-3} at 5 K and 1 s after the end of the excitation.

Applying the same calculation to the PC data of Fig.1 gives $\rho_{met} = 10^{19}$ cm^{-3} which is consistent with the value reported by Hoheisel et al.[6] but which is almost 100 times larger than what is obtained from the PL data. However, applying this calculation to the PC transient pulse requires that IR reexcited carriers are reexcited only once, then recombine. The discrepancy between PC and PL clearly shows that this is not the case and that carriers are retrapped and reexcited an average of $\rho_{met}(PC)/\rho_{met}(PL) \sim 100$ times before recombining.

The similarity of the CW PL spectrum and the energy spectrum of the IR transient PL suggests that the IR reexcited carriers which recombine radiatively recombine by band edge luminescence. The logarithmic time decay of the carrier density at all temperatures is similar to what is observed by ESR[2] and is consistent with the experimentally measured t^{-1} decay of the PL intensity at long times[9], the theoretical results of the Distant Pair model of recombination at low temperature[8], and with the Multiple Trapping model of transport[10, 11] at higher temperature (T > 50 K). This and the quenching of the metastable ESR signal by IR[2] strongly suggests that the metastable population is responsible for the metastable ESR signal. The identity between the components at g = 2.004 and g = 2.012 of the metastable ESR signal and the ESR spectra due to tail states[12] shows that the metastable carriers are trapped in the tail states. The large value of the photoinduced absorption threshold (0.4 eV)[5] and the complete IR quenching of the metastable ESR signal due to holes[2] shows that the IR induced transient pulses of the PL and of the PC are due to reexcitation of the metastable holes to the valence band.

The results of Fig.2 show that at least some of the metastable carriers are in radiative states although they have very long lifetimes. The logarithmic time decay of the metastable population implies that the number of carriers which recombine per unit of time, decays such as t^{-1}. Now the t^{-1} decay of the PL intensity at long times[9] confirms that the metastable carriers are in radiative states although they are in a long lifetime configuration (large pair separation). We may conclude that some tail states are radiative states for the 1.2-1.4 eV luminescence band in a-Si:H. This result has been previously suggested by the similarity between the temperature dependences of the LESR signal and the mean decay time of the PL[13] and by the explanation of all the features of the band edge PL by a simple model of radiative tail states[14], but has never been proved before.

CONCLUSION

We have measured $\sim 10^{17}$ cm^{-3} metastable carriers in low defect density a-Si:H at 5 K, using the IR induced transient pulse of the photoluminescence ; and from the transient pulse of the photoconductivity we have shown that these carriers were reexcited an average of 100 times by the IR before recombining. Comparison with ESR strongly suggests, that the metastable carriers are trapped in tail states, that holes are reexcited to the valence band by the IR, and that some tail states are radiative states for the band edge luminescence in a-Si:H.

ACKNOWLEDGMENTS

I wish to thank Pr. I. Solomon and Mrs N. Proust for providing samples, Mr. E. Criton for skilled technical assistance, Dr. D.J. Dunstan, Dr. J. Harrang and Dr. J. Magarino for helpful discussions.

REFERENCES

1. R. A. Street and D. K. Biegelsen, Solid State Commun. 44, 501 (1982).
2. F. Boulitrop, J. Dijon, D. J. Dunstan and A. Hervé, International Conference on the Physics of Semiconductors, San Francisco (1984).
3. J. Dijon, Solid State Commun. 48, 79 (1983).
4. D. J. Dunstan, S. P. Depinna and B. C. Cavenett, J. Phys. C 15, L425 (1982).
5. S. Ray, Z. Vardeny and J. Tauc, Journal de Physique 42, C4-555 (1981).
6. M. Hoheisel, R. Carius and W. Fuhs, J. Non-Cryst. Solids, 59-60, 457 (1983)
7. C. Tsang and R. A. Street, Phys. Rev. B 19, 3027 (1979).
8. D. J. Dunstan, Philos. Mag. B 46, 579 (1982).
9. F. Boulitrop (unpublished).
10. J. Orenstein, M. A. Kastner and V. Vaninov, Philos. Mag. B 46, 23 (1982).
11. D. R. Wake and N.M. Amer, Phys. Rev. B 27, 2598 (1983).
12. M. Stutzmann and J. Stuke, Solid State Commun. 47, 635 (1983)
13. R. A. Street and D. K. Biegelsen, J. Non-Cryst. Solids 35-36, 651 (1980).
14. F. Boulitrop and D. J. Dunstan, Phys. Rev. B 28, 5923 (1983).

TRIPLET PHOTOLUMINESCENCE IN CRYSTALLINE CHALCOGENIDE SEMICONDUCTORS

L. H. Robins and M. A. Kastner
Department of Physics and Center for
Materials Science and Engineering
Massachusetts Institute of Technology, Cambridge, Mass. 02139

ABSTRACT

The similarity of the photoluminescence (PL) spectra and time decays in several amorphous and crystalline chalcogenide semiconductors suggests that the excited state of the PL center is similar in all these materials. Evidence that the PL arises from a triplet excited state is reviewed. PL time decays in c-As_2Se_3, c-Se, and c-$GeSe_2$ are compared. The temperature and magnetic field dependence of the PL time decay in c-As_2Se_3 at temperatures less than 4 K is reported. It is suggested that these low-temperature effects arise from changes in the spin-lattice relaxation rates between different triplet levels.

It is well-known that similar photoluminescence (PL) spectra are found in the crystalline and amorphous forms of several chalcogenide semiconductors, including As_2Se_3, As_2Se_3, Se, and $GeSe_2$. The PL in these materials has a peak energy close to half the bandgap and a roughly Gaussian lineshape with a width of 0.3-0.4 eV[1-3]. The principal PL decay times are on the order of 10^{-4} to 10^{-3}s in each of these materials[2,4,5]. The similarity of the spectra and decay times suggests that the excited state of the PL center is similar in all these materials. It has been suggested that the slowly decaying PL arises from triplet exciton recombination[6,7]. In this paper, we will first review the main evidence for triplet recombination. Then some recent experiments which provide additional information about the spin substates of the triplet excited state in c-As_2Se_3 will be discussed.

The strongest evidence for triplet PL has been found in c-As_2Se_3.[6] In this system the PL decay time decreases by a factor of 20 and the initial intensity increases by the same amount between 6 and 50 K, so that the PL quantum efficiency does not change. Because the quantum efficiency is constant, the change in the decay time must arise from an increase in the radiative rate rather than in the rate of a competing non-radiative process. The increase in the radiative rate follows a thermally activated form with an activation energy of $4x10^{-3}$ eV (Figure 1). This behavior can be explained by the following model. The temperature-independent rate seen at low temperature is the radiative decay rate of a triplet excited state. This rate is slow because optical transitions from a pure triplet to the ground state are strictly forbidden, and the actual triplet excited state is weakly coupled to singlet states by spin-orbit interaction. As the temperature is increased, thermally assisted transitions from the triplet state to the lowest excited singlet, followed by singlet-to-singlet recombination, becomes the main decay channel. The upward triplet-to-singlet transition is the rate-limiting step for the

0094-243X/84/1200183-06 $3.00

latter process. Therefore the observed decay rate remains much smaller than a typical singlet radiative rate even at high temperature.

The observation of temperature-dependent linear polarization of the PL provides additional evidence for the triplet model.[6] This model predicts that the triplet PL will be polarized perpendicular to the singlet PL arising from the same anisotropic spatial orbital, for the following reason. Radiative decay from the triplet occurs because of the spin-orbit coupling to singlet excited states. Therefore the polarization of the triplet PL is determined by the selection rules for optical transitions from the singlet states to which the triplet state is coupled. Since the spin-orbit interaction couples only spatially orthogonal states, it follows that the triplet PL from a particular orbital state will be polarized perpendicular to the singlet PL from the same orbital. It was observed[6] that the polarization axis of the PL at low temperature, at which only triplet recombination occurs, is perpendicular to the polarization axis at high temperature, at which singlet recombination is more probable.

The PL decay time in c-$GeSe_2$ was found to have a similar temperature dependence, decreasing from 300μs at 4 K to 25μs at 80K.[2] In this system the quantum efficiency was found to increase by a factor of four from 4 to 40 K, and to remain constant from 40 to 80 K. Thus, it appears that the radiative decay rate increases with temperature in this material also. However, the increase in the rate is not as great as in c-As_2Se_3, and does not seem to be described by a single activation energy. Recently midgap PL has also been observed in trigonal c-Se.[3] A report of the time decay of this newly found PL is in preparation.[5] In contrast to the other two crystalline chal-

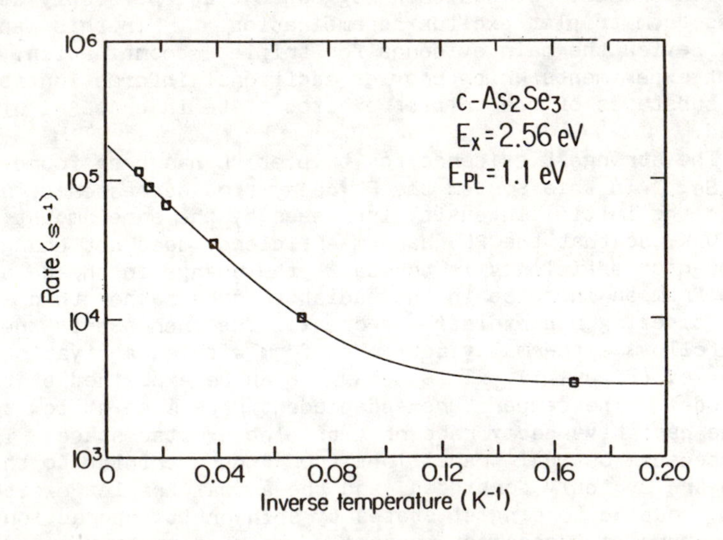

Figure 1. Temperature dependence of radiative rate in c-As_2Se_3. Solid line is function $\nu_r=3300+1.8\times10^5 e^{-\Delta E/kT}$ where $\Delta E=4\times10^{-3}$ eV.

cogenides, the low-temperature decay time of 700μs does not decrease with temperature up to 100 K. At higher temperatures the decay time decreases because of an activated non-radiative process, but no increase in the radiative rate can be observed. It may be that the crossing rate from the lowest excited triplet to singlet state is very small in c-Se.

The time decay of the PL in c-As_2Se_3 changes at temperatures below 4 K so that it can no longer be fitted by a single time constant (Figure 2). However, the data can be fitted reasonably well by a sum of two exponentials in this range, as shown in Figure 2. The two rates found at the lowest temperature (T=2 K) are $\nu_+ = 5100\ s^{-1}$ and $\nu_- = 1600\ s^{-1}$. The single rate of 3300 s^{-1} found at slightly higher temperatures (T=4 to 6K) is approximately equal to the average of ν_+ and ν_-. It also appears that the quantum efficiencies of the faster and slower decaying components of the PL are equal at T=2 K. However, it should be noted that the PL decay deviates slightly from an exponential form at short times even for T>4 K, because of non-radiative processes which affect a fraction of the PL centers.[6] This deviation may cause a systematic error in the values calculated for ν_+ and for the relative quantum efficiencies of the two processes.

Using the triplet model, the change in the PL decay can be explained by the temperature dependence of the spin-lattice relaxation (slr) rates. The radiative decay rates of the three triplet levels will in general differ because of the difference in spin-orbit coupling. If the slr rates among these levels are faster than the radiative rates, then one average rate will be observed. If the spin-lattice relaxation is slow, then the radiative rate of each

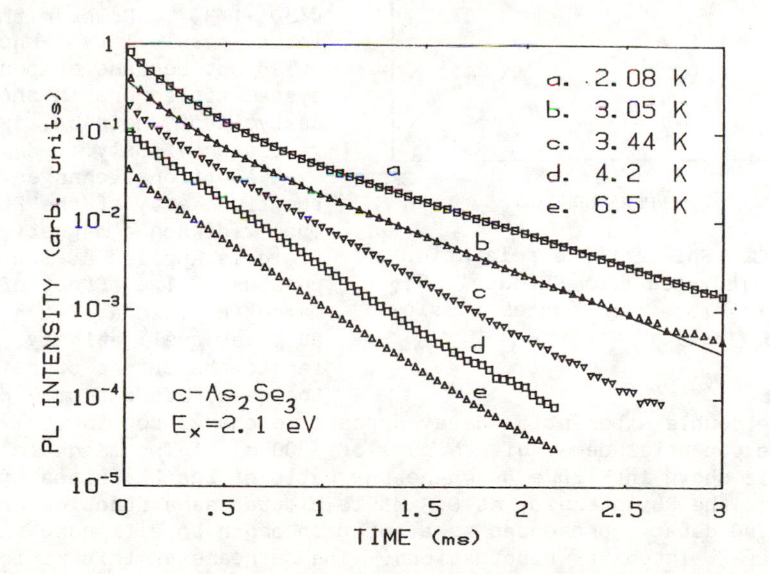

Figure 2. Temperature dependence of PL time decay from 2 to 6.5 K. Solid lines are sums of two exponentials.

level can be observed separately. The slr rates can be obtained from the data of Figure 2 by assuming that at the lowest temperature the radiative rates are much larger than the slr rates. The slr rates calculated using a simple rate equation model are plotted in Figure 3.

The single rate found at T>4 K is close to the average of the two low-temperature rates, the quantum efficiencies of the two low-temperature processes appear to be equal, and the calculated upward and downward slr rates are approximately equal. One possible explanation of these observations is that only two of the three triplet levels contribute to the observed PL. It may be that the radiative rate of the third level is much smaller than the other two rates. This would be the case for a σ-orbital in an axially symmetric potential; for this symmetry the spin-orbit interaction does not couple the $S_z=0$ triplet level to any singlet level. Another possibility is that the fast low-temperature rate arises from two triplet levels which have approximately equal radiative rates, and the slow rate arises from the third level. Then one would expect that the ratio of the slr rates would be two to one and the average decay rate at T>4 K would be equal to $2/3\nu_+ + 1/3\nu_-$. However the latter possibility cannot be ruled out because of possible systematic errors in the analysis of the data, as suggested previously.

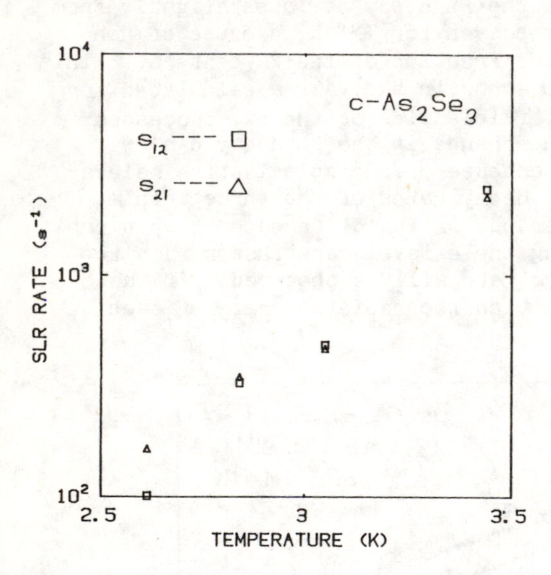

Figure 3. Spin-lattice relaxation rates calculated from PL data. S_{12} is the slr rate from faster to slower decaying level.

Additional changes in the time decay of the PL are observed when a magnetic field is applied at low temperature. The effect of the magnetic field at T=2 K is, at least qualitatively, similar to the effect of raising the temperature from 2 K to 4 K: the double exponential decay appears to change continuously to a single exponential decay with a rate of 3300 s^{-1}. The magnetic field effect is shown in Figure 4, where the ratio of the PL intensity at 2.0 ms to the PL intensity at 0.3 ms is plotted as a function of field. The data is presented as a ratio in order to eliminate the effect of PL intensity fluctuations. The decrease of this ratio with increasing field is caused mainly by the decrease in PL intensity at longer time. The nature of the field dependence provides additional evidence for the triplet exciton model. If the spins of the electron and hole were not strongly coupled then one would expect the applied field to partially align the spins and therefore make the PL decay

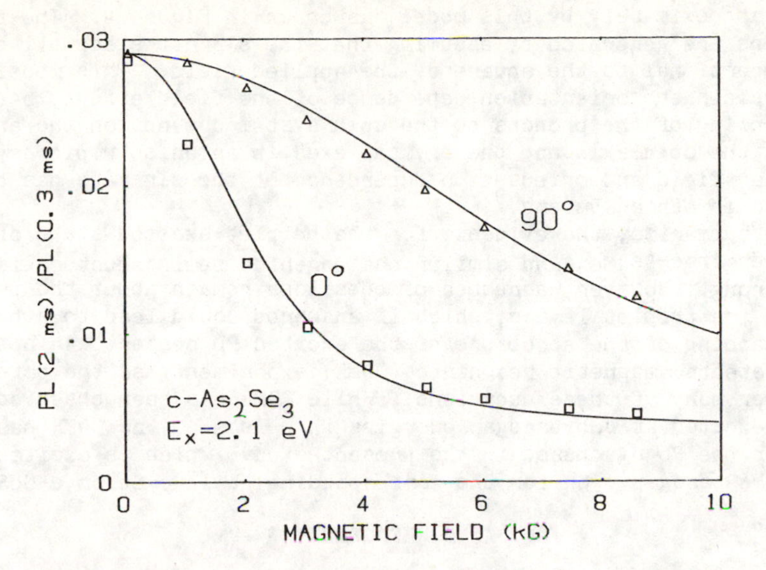

Figure 4. Effect of magnetic field on PL time decay at T=2 K for two field directions relative to crystal. The solid lines represent quadratic field dependence of slr rate, as described in text.

slower, rather than quenching the slow component.

The magnetic field effect also depends on the direction of the field relative to the crystal axes (Figure 4). It appears that the field effect scales with orientation; a given field directed along the axis where the effect is strongest causes the same change in the form of the PL time decay as a field three times as large directed along a perpendicular axis.

The similarity of the field dependence and temperature dependence of the PL suggests that the field effect also arises from an increase in the slr rates. Other mechanisms, such as a mixing of the radiative rates of the zero-field levels by the applied field distinct from spin-lattice relaxation, cannot be ruled out at present. However the field effect can be analyzed in a simple way in terms of a particular spin-lattice relaxation mechanism.[8] The slr rate arising from the absorption (emission) of one phonon is proportional to the product of the following quantities: the density of phonon states at the energy equal to the energy splitting of the levels, the energy per phonon, and n (1+n), where n is the phonon occupation number. The product of the latter two quantities is just the thermal energy, provided that this is large compared to the energy per phonon. At low phonon energies the density of states varies quadratically with energy. The energy difference between the levels is directly proportional to B as long as B is large compared to the zero-field splitting of the triplet levels. Therefore the slr rate arising from one-phonon transitions is proportional to B^2 if the conditions mentioned above are satisfied. The magnetic field dependence can be fitted approximately by this model, as shown in Figure 4. The fitted functions are generated by assuming that the slr rates are all equal and proportional to the square of the applied field. This model may also explain the orientation dependence of the field effect, because the coupling of the phonons to the spin system depends on the angle between the spin axis and the crystal axes in an anisotropic crystal. A similar field and orientation dependence of the slr rate has been observed in other systems.[9]

To summarize, the evidence for the triplet excited state of the PL center in c-As_2Se_3 and similar chalcogenide semiconductors is now very strong. However, a number of questions remain about the properties of the triplet levels, which if answered could lead to a better understanding of the structure of the excited PL center. An optically detected magnetic resonance (ODMR) experiment has the potential to answer many of these questions. While ODMR has been observed for the PL excited at sub-bandgap energies in c-As_2Se_3,[7] no ODMR has been seen for the PL discussed in the present study, which is excited above the bandgap[7], or for the corresponding PL in c-Se or c-$GeSe_2$.

ACKNOWLEDGEMENT

This work is supported by NSF Grant DMR81-19295 with facilities support from NSF Grant DMR81-15620.

REFERENCES

1. R. A. Street, Adv. in Phys. 25, 397 (1976).
2. V. A. Vassilyev, M. Koos and I. K. Somogyi, Phil. Mag. B39, 333 (1979).
3. G. Weiser and H. Lundt, Sol. St. Comm. 48, 827 (1983).
4. G. S. Higashi and M. A. Kastner, Phil. Mag. B47, 83 (1982).
5. C. Y. Chen, L. H. Robins and M. A. Kastner, to be published.
6. L. H. Robins and M. A. Kastner, Phil. Mag. B (1984), to be published.
7. S. P. Depinna and B. C. Cavenett, Phil. Mag. B46, 71 (1982).
8. G. E. Pake, Paramagnetic Resonance (W. A. Benjamin, N.Y., 1962), pp. 120-134.
9. K. F. Renk, H. Sixl and H. Wolfrum, Semicond. and Insul. 4, 265 (1978).

LUMINESCENCE AND PHOTO-EFFECTS IN OBLIQUELY DEPOSITED a-Se-Ge FILMS

P.K. Bhat, T.M. Searle and I.G. Austin
Department of Physics, The University, Sheffield S3, UK.

K.L. Chopra
Department of Physics, Indian Institute of Technology,
Delhi, India.

ABSTRACT

Photoluminescence studies are reported on normal deposited and obliquely deposited amorphous $Se_{75}Ge_{25}$ films. These studies include spectra, efficiency versus temperature and lifetime distributions. The results are discussed in the context of structural differences.

INTRODUCTION

Many amorphous chalcogenide thin films show large changes in properties such as change in volume (39%), refractive index (~10%) and shift in optical absorption edge (10%) etc on exposure to light and Helium ions [1,2]. There is evidence from electron microscopy[1] and low angle neutron diffraction experiments[3] that these chalcogenide films contain a large number of voids and dangling bonds. There is also evidence from X-ray diffraction[4] and infrared absorption measurements[5] for photo-induced bonding changes after irradiation with gap light. The connection between the changes in the physical properties of the chalcogenide films and the defects both intrinsic and extrinsic is not well understood. Several models[6,7,8] have been proposed to explain the microscopic origin of the photo induced changes in amorphous chalcogenides in terms of defects localized states in the gap of amorphous chalcogenides.

Photoluminescence (PL) is a powerful tool for studying these localized states. In the present paper we present a systematic study of amorphous $Se_{75}Ge_{25}$ films prepared both at normal (ND) and 80° oblique angle of deposition (OD). The reason for choosing this composition is that 80° deposited films of a-$Se_{75}Ge_{25}$ show maximum photo effects[1]. Luminescence studies include spectra, efficiency versus temperature and life time distributions. We show that investigation of the dynamics of PL processes provide new insight into the nature of defects in ND and OD a-$Se_{75}Ge_{25}$ films.

EFFICIENCY AND SPECTRAL MEASUREMENTS

We have examined a range of films grown by thermal evaporation of bulk alloy of $Se_{75}Ge_{25}$ in vacuum ≈ 10^{-6} torr. Preparation details can be found elsewhere[1]. Films were deposited at a rate of about 20 Å/sec at room temperature on roughened glass substrates. The photo- luminescence spectra were excited with the Krypton laser line at 2.18eV and the samples were mounted in an exchange gas cryostat. Most of the PL measurements by other groups have been made using similar photon-energies (2.41-1.9eV), but with powers

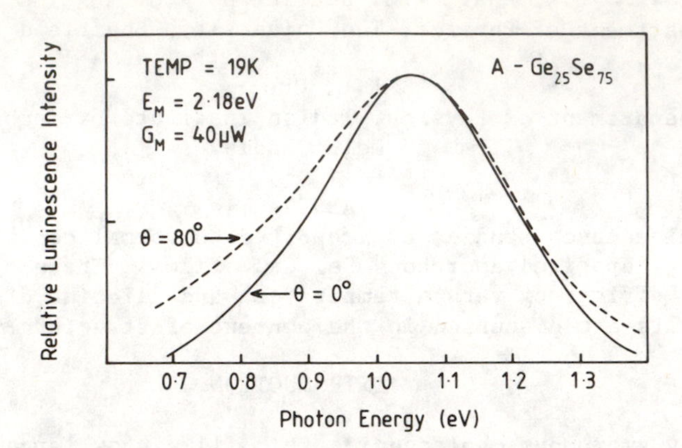

Figure 1 PL spectra of OD and ND films. E_m and G_m are excitation energy and intensity respectively.

between 1 and 5 mW. It is well known that those intensities produce strong fatigue effects and secondly induce a shift in the emission band towards higher energies with increasing intensity. The intensity of the unfocussed laser beam used in our experiments was <50μw and the PL spectrum was taken in <60 secs to minimise fatigue. The laser beam was chopped at ~ 40Hz and luminescence was detected with a monochromator and cooled Ge detector.

Figure 1 shows typical low temperature luminescence spectra of a-$Se_{75}Ge_{25}$ films. The spectrum for ND film is in general agreement with previous reports, showing a broad peak at 1.05eV, about half the band gap energy. The PL spectrum for OD films also peaks at about 1.05eV but is broadened at lower energies. The PL efficiency, η of ND films is close to unity, falling by 25-35% in OD films. η seems independent of annealing (at Tg ~ 240°C) and exposure to light (Hg lamp of about 55mwcm^{-2} intensity) both in normal and 80° deposited films, within experimental error.

The temperature dependence of PL emission in both normal and 80° deposited a-$Se_{75}Ge_{25}$ films is shown in Figure 2. An interesting aspect of the temperature dependence is the increase in peak energy E_L with increasing temperature in normal deposited a-$Se_{75}Ge_{25}$ films. In contrast, E_L in OD films show relatively slow or no dependence on temperature up to 180°K. It was difficult to measure PL at higher temperatures because of the low powers used and the fast decrease in PL intensity with increasing temperature. Figure 3 shows the relative PL efficiency measured at the peak maximum normalised to the same value at low temperature. Both OD and ND films show similar behaviour. The temperature dependence of η cannot be described by a single activated process, but is closer to the exponential temperature dependence seen in a number of glasses[9]. Thus $\eta \approx \exp - T/T_o$, with $T_o \approx 50K$. Figure 3 also shows the

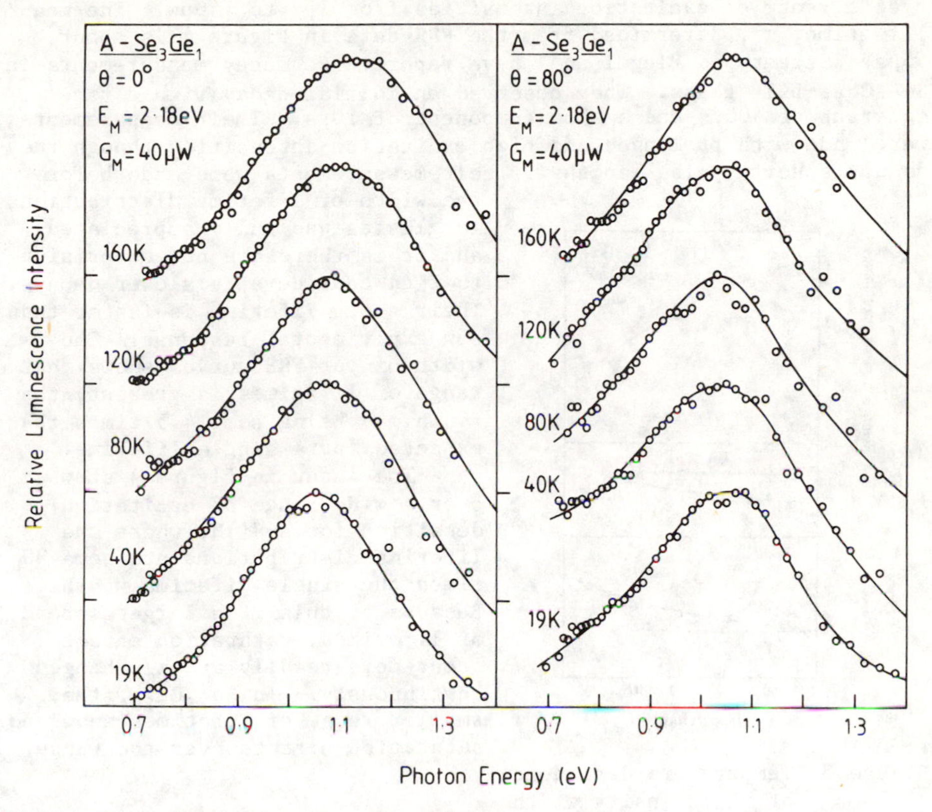

Figure 2 Temperature dependence of PL spectra.

temperature dependence of the peak width (f.w.h.m.) and position. In normal deposited films, both width and E_L increase continuously with rising temperature. However, there is little change in f.w.h.m. and E_L with rising temperature in 80° deposited films.

LIFETIME MEASUREMENTS

We have used frequency response spectroscopy (FRS) to determine the PL life time distribution. The FRS technique uses lock in detection in quadrature, with a sinusoidal modulation of the PL excitation source, to observe those components with life time close to the inverse modulation frequency. The details of FRS technique can be found elsewhere[10]. The main feature is that the quadrature response gives the lifetime distribution under the same conditions as in the efficiency measurements. The method has limited time resolution, a single lifetime giving a peak with a f.w.h.m of about one decade. All our FRS are of the total PL without the monochromator.

Figure 4 shows a low temperature FRS spectrum for normal deposited a-$Se_{75}Ge_{25}$ film. The FRS spectrum does not show any shift over a range of excitation intensities from 1μw to 140μw. The mean life time, τ_m, determined from the FRS data in Figure 4 is about 55μs. Street and Biegelsen[11] have reported PL decay measurements in $Se_{75}Ge_{25}$ bulk glass. They observed an initial decay with a time constant of 400ns and a slow component of 10μs. Their measurements were made with prolonged and high excitation intensities though they do not specify pulse lengths. Their measurements were made before the width of lifetime distributions in glasses was fully appreciated, and it is therefore not surprising to find considerably slower decays. Their short lifetime is faster than our experimental response. The width of our FRS curves shows that a range of lifetimes is present, the f.w.h.m. being some 4.5 times that expected for a single lifetime.

The inset in figure 4 shows τ_m over a wide range of excitation densities for a-Si:H, where the lifetime distributions are some 30 times the single lifetime width. Because of this much larger spread of lifetimes, saturation effects occur more readily and τ_m changes continuously. In $Se_{75}Ge_{25}$, the smaller range of lifetimes shows no saturation effects over the range

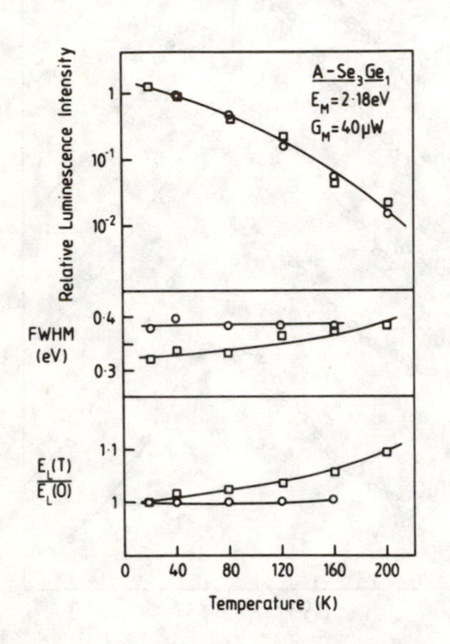

Figure 3 Temperature dependence of PL intensity width and peak psoition. OD circles, ND squares.

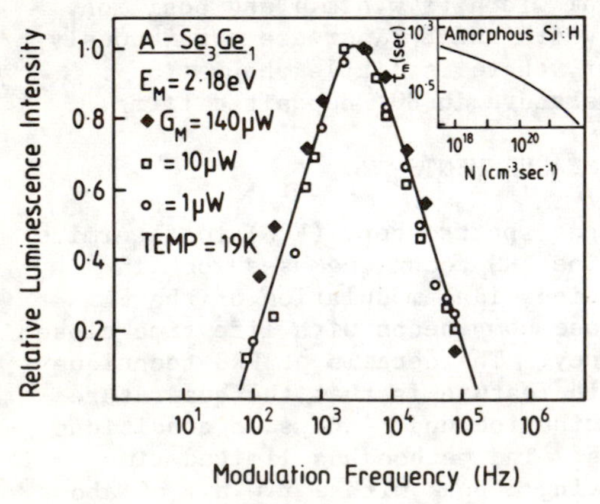

Figure 4 Dependence of lifetimes on excitation density.

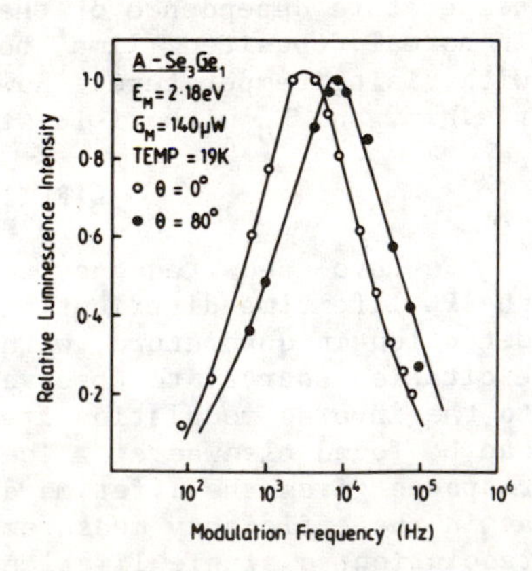

Figure 5 Comparison of lifetimes in ND and OD films.

of excitation intensity used here. Presumably such effects do occur at higher intensities, as suggested by the blue shift of the spectrum, but measurements are complicated by fatigue effects.

Figure 5 shows FRS spectra for normal deposited and 80° deposited a-$Se_{75}Ge_{25}$ films. The FRS spectrum of OD films looks very similar to ND films but shifted to shorter times. As τ_m does not shift with the excitation density, the above shift could not be explained by the differences of excitation densities because of the change in the absorption coefficient of ND and OD films.

DISCUSSION

By analogy with the arsenic chalcogenides it is usually assumed that the PL band width in Ge_xSe_{1-x} alloys is determined by the electron phonon interaction. As discussed above, time resolved measurements show that there is a range of lifetimes present, and the high efficiency at low temperatures indicates these are radiative. A range of radiative lifetimes usually indicates distant pair recombination. However the much narrower and more symmetric distribution in $Se_{75}Ge_{25}$ compared with a-Si:H suggests a much more limited range of pair separations. The resulting centres may thus be more like isolated defects than randomly distributed pairs. In As_2Se_3 there is no correlation between pair spacing and recombination energy[12], and so the width is determined by the electron phonon interaction alone. In crystalline donor-acceptor pair systems, in contrast, the pair lifetime and energy are correlated by the Coulombic term, with closer and hence faster pairs having higher energies. In these systems high excitation densities shift the PL peak to higher energies, as the slower pairs are saturated. The observation of such a shift in the Ge_xSe_{1-x} system may suggest that here too the electron and hole centers are charged when empty, and that pairing effects may contribute to the band width.

The main difference in the spectra of OD and ND films is the broadening of the former at low energies. This could be due either to a change in electron phonon coupling, to a change in the spatial distribution of pairs, or to the introduction of a new radiative center. Changes in the electron phonon coupling parameter S would produce a symmetric broadening, contrary to what is observed. Also, since the f.w.h.m. varies as $\sqrt{S}$, and separation (E_s) of the peak and the zero phonon line is proportional to S, we expect an increase in S to shift the peak to the red. Estimating E_s to be about 0.3eV, and with an 11% change in width, we would predict a red shift of some 66meV between the OD and ND films, whereas the experimental value is 0±20 meV. If changes of pair separation were responsible, a broadening of the lifetime distribution would also be expected. Though this distribution does shift, it does not appear to broaden, nor to become asymmetric. The spectral changes require more low energy pairs, which probably have longer lifetimes.

The remaining possibility is that the OD films contain a new centre. To explain the observed spectral changes, we would require the additional PL band at about 0.8eV. We note that this is also the energy of the PL band in pure Se[8,13], and we therefore suggest that OD films contain a large proportion of selenium rich regions compared to normally deposited films. This is also supported by electron microscopic measurements.

The quantum efficiency of PL from selenium is known to be low[13], and since the values of η shown in figure 3 were measured at 1.05eV the extra PL component in OD films would not be expected to make a significant contribution to the temperature dependence.

The temperature dependence of the PL peak to higher energies is surprising, as the gap decreases with temperature. It may be related to the blue shift with high excitation densities, since the latter will increase as the absorption coefficient increases at fixed excitation energy. Figure 3 shows that in OD films E_L is almost constant, and we suggest this is because the blue shift of the main band is roughly balanced by the increasing 0.8eV component.

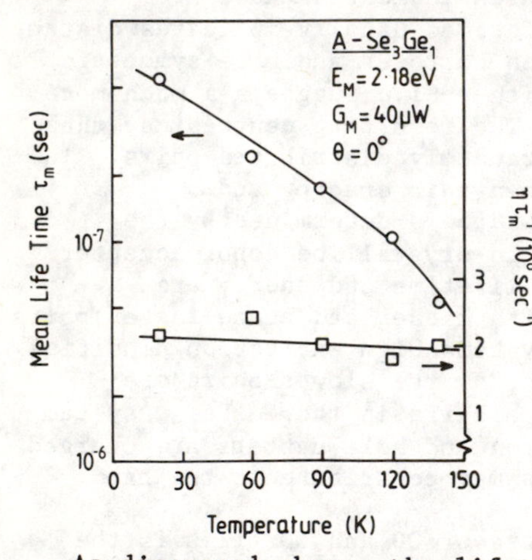

Figure 6 The mean lifetime τ_m and the product $\eta\tau_m^{-1}$ as a function of temperature in normal deposited films. η is normalised to unity at OK.

As discussed above, the lifetime data suggest that the main PL comes from pairs with a range of separations small compared with a-Si:H. One advantage with FRS is that η and τ_m are measured under the same excitation conditions, and this can be used to understand the temperature quenching. If P_r and P_{nr} are radiative and non radiative rates in competition from the same excited state then the efficiency η can be written as $\eta = \dfrac{P_r}{P_r + P_{nr}} \approx P_r\tau_m$. In this case $\eta\tau_m^{-1}$ gives a mean radiative rate $\overline{P}_r$. Such rates are usually temperature independent. Figure 6 shows the temperature dependence of τ_m and $\eta\tau_m^{-1}$ in normal deposited a-$Se_{75}Ge_{25}$ films. $\eta\tau_m^{-1}$ is essentially independent of temperature as expected in this model. Estimating $\eta = 0.5$ gives the mean radiative lifetime $1/\overline{P}_r \approx 1\mu s$. P_{nr} is most probably the rate of thermal release of one of the carriers from its excited state, or of tunnelling out of it. It has the characteristic exp T/T_o dependence mentioned above, a form usually associated with thermal release from levels distributed in energy by more than a few kT.

However if we use the same analysis to compare $\eta\tau_m^{-1}$ in ND and OD films, we find that at low temperatures this product changes from $1.9 \times 10^4 s^{-1}$ to $3.1 \times 10^4 s^{-1}$. The change in efficiency between OD and ND films comes therefore from a combination of two processes, one which competes directly with the radiative process, lowering the lifetime, and one which competes for carrier capture into the radiative states. A shunt process of this latter kind lowers the efficiency without changing the lifetime. Such processes may be due to non radiative centers produced by the OD process, or possibly by the radiative recombination in the Se rich regions. PL in Se has a low efficiency and also shows a PL band at ~ 0.6 eV, which we would not detect[13].

Finally we have looked at the PL life time in fatigued films of 80° deposited a-$Se_{75}Ge_{25}$. The fatigued samples were exposed for ~ 15 mins with intensities ~ 50 $mwcm^{-2}$ at 568nm. There is no significant change in τ_m despite a fall in η by a factor of 20-50. This shows very clearly that exposure to light creates non-radiative centres which compete for carrier capture into the radiative excited state. It is likely that the centres responsible for this non-radiative shunt process are the same as those produced in OD films.

SUMMARY

In conclusion, steady state luminescence and FRS data in a-$Se_{75}Ge_{25}$ films suggest that the luminecence involves two defect states. We speculate that one is probably a Se chain and the other a Ge dangling bond. The films deposited at 80° angle of incidence have a defect structure different from that of deposited normally and may be rich in Se. This may suggest that the origin of the photostructural changes in amorphous chalcogenide films may have more connection with the intrinsic defects present in the films than the photo created defects.

REFERENCES

1. Bhanwar Singh, S.Rajagopalan, P.K.Bhat, D.K.Pandya and K.L.Chopra J.Non.Cryst.Solids, 35 & 36, 1053 (1980).
2. K.L.Chopra, K.Solomon Harshavardhan, S.Rajagopalan and L.K.Malhotra. Appl.Phys.Letts, 40,428 (1982); T.Venkatesan and B.J.Wilkens. Appl.Phys.Letts., 41,839,(1982); K.Solomon Harshavardhan, S.Rajagopalan, L.K.Malhotra and K.L.Chopra. J.Appl.Phys.54,1048 (1983).
3. T. Rayment and S.R. Elliott. Phys.Rev. 28,1175 (1983).
4. K.Tanaka. Appl. Phys.Letts. 26,243 (1975).
5. Yasushi Utsugi and Yoshihiko Mizushima. J.Appl.Phys.49,3470, (1980).
6. R.A.Street and N.F.Mott. Phys.Rev.Letts. 35,1293 (1975).
7. M.Kastner and S.J. Hudgens. Phil.Mag. B,37,665 (1978).
8. F.Mollot, J.Cernogora and C.Benoit à la Guillaume. Phil.Mag.B 42, 643, (1980).
9. R.A.Street. Advances in Phys. 30, 593, (1981).

10. P.K.Bhat, I.G.Austin and T.M.Searle. J.Non.Cryst.Solids. 59 & 60, 381, (1983).
11. R.A.Street and D.K.Biegelsen. J.Non.Cryst.Solids. 31,339, (1979).
12. S.P.Depinna and B.C.Cavenett. Phil.Mag. B46,71, (1982).
13. S. G. Bishop and P. C. Taylor, J. Non. Cryst. Solids, 35 and 36, 909 (1980).

RECOMBINATION KINETICS IN AMORPHOUS SEMICONDUCTORS

Marvin Silver
University of North Carolina, Chapel Hill, NC 27514

David Adler
Massachusetts Institute of Technology, Cambridge, MA 02139

ABSTRACT

We calculate the time dependence of the band-tail luminescence in amorphous semiconductors, and show that there should be an initial rise preceding the typical power-law decay. Similar calculations indicate that the defect luminescence should initially grow only above a critical temperature. Recent experiments in a-Si:H show no initial increase in the band-tail luminescence intensity even at very short times. We suggest that these results are further evidence for a high value of the band-mobility of electrons in a-Si:H.

INTRODUCTION

In general, the transient response of an amorphous semiconductor to a pulse excitation is dispersive in nature, due to a broad distribution of event times.[1-4] The most investigated process involves the multiple trapping of excess free carriers in an exponential density of localized states.[5,6] In this case, the photoresponse e.g. in a time-of-flight experiment decays as $t^{-(1-\alpha)}$ before a time τ_t and $t^{-(1+\alpha)}$ afterwards, where $\alpha = T/T_o$ and T_o characterizes the exponential distribution:

$$g(E) = g_o \exp (E/kT_o). \tag{1}$$

The time τ_t can be identified as the transit time, which is not $L^2/\mu_o V$, where L is the film thickness, V the applied voltage, and μ_o the carrier mobility, as expected in non-dispersive materials; instead,

$$\tau_t = \frac{1}{\nu} \left(\frac{\nu}{1-\alpha} \frac{L^2}{\mu_o V}\right)^{1/\alpha} \tag{2}$$

where ν is the attempt-to-escape frequency of the traps. This increase in transit time arises from the numerous trapping events which retard the motion of the carriers as they drift through the material.

We should expect that the effects of multiple trapping strongly influence an array of other experiments in addition to photoconductivity, and indeed they have been observed e.g. in transient field-effect observations.[7] Similarly, pulsed photoluminescence measurements should be characterized by an anomalous time dependence, because the various luminescence centers must be populated before light emission can take place. In this paper, we calculate the build-up and decay of the photoluminescence in an amorphous semiconductor which exhibits dispersive transport due to multiple trapping. We apply the results to analyze the electronic structure of hydrogenated amorphous silicon (a-Si:H).

CONVENTIONAL MODEL FOR HYDROGENATED AMORPHOUS SILICON

It is conventionally believed[8,9] that high-quality films of a-Si:H have a density of localized states as shown in Fig. 1. The major characteristics are a conduction band tail with $T_o \approx 300K$, a valence band tail with $T_o \approx 500K$, and two narrow bands arising from dangling-band ($T_3°$) defects with a positive value of effective correlation energy, $U_{eff} \approx 0.4$ eV. (We have omitted the possible flattening out of the localized-state distributions near the mobility edges[10,11] for simplicity.) It is assumed that photoexcitation with photon energies in excess of the optical gap is followed by the rapid trapping of electrons in the conduction band tail and holes in the valence band tail. Initially most of these trapped carries are rereleased and retrapped continually. However, close pairs have a large probability of recombining. The $T_3°$ states near midgap also act as recombination centers.

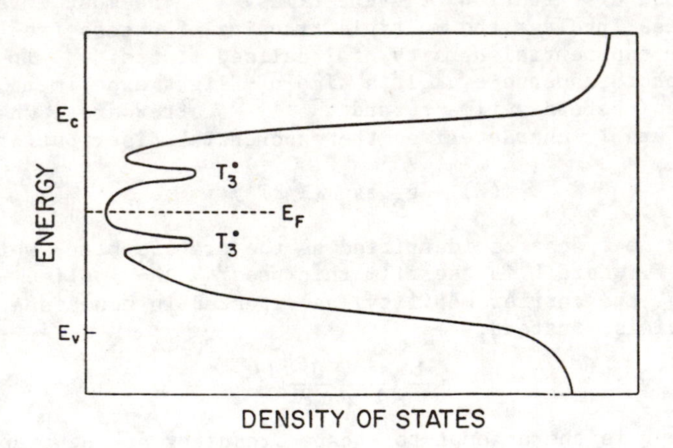

Fig. 1. Conventional model for the density of localized states in undoped a-Si:H films.

TRAPPING AND RECOMBINATION PROCESSES

We refer to free electrons and holes as e^- and h^+, respectively. Localized band tail states are denoted as $T_4°$ when neutral. Then the most important trapping processes can be represented as:

$$T_4^o + e^- \rightarrow T_4^- \tag{3}$$

$$T_4^o + h^+ \rightarrow T_4^+ \tag{4}$$

$$T_3^o + e^- \rightarrow T_3^- \tag{5}$$

$$T_3^o + h^+ \rightarrow T_3^+ . \tag{6}$$

Thus, carrier trapping creates positively and negatively charged localized states. The inverse of processes (3) and (4), carrier release from traps in the band tails, yield the observed dispersive transport. The major recombination processes are:

$$T_4^+ + e^- \rightarrow T_4^o \tag{7}$$

$$T_4^- + h^+ \rightarrow T_4^o \tag{8}$$

$$T_3^+ + e^- \rightarrow T_3^o \tag{9}$$

$$T_3^- + h^+ \rightarrow T_3^o \tag{10}$$

$$T_4^+ + T_4^- \rightarrow 2\, T_4^o . \tag{11}$$

The last of these is presumed to yield the observed "band tail" luminescence, near 1.3 eV, while (9) and (10) are supposed to account for the "defect" luminescence, near 0.9 eV.[12]

BAND-TAIL LUMINESCENCE

We consider the case in which the concentration of photo-excited carriers is relatively low, 10^{15} cm^{-3}, so that the average distance between an electron and a hole is ~ 1000 Å. (We initially assume no correlation exists between the photogenerated electrons and holes; the case of geminate recombination, in which such a correlation exists, will be briefly discussed later.) Since the density of populated band-tail states is much smaller than the density of available states, it is more likely that oppositely charged carriers diffuse towards each other between trapping events than they move closer by long-range tunneling; consequently, we neglect tunneling processes. We assume that rapid luminescence occurs whenever the pair separation becomes 10 Å or less.

If the free carrier mobility is small, $\mu_o \sim 10$ $cm^2/V\text{-}s$, as is commonly believed,[9] the diffusion length of the excited carriers is much less than the initial pair separation, and essentially all of the initial (i.e. $t < 10^{-11}$s) luminescence would arise from geminate recombination. At later times, the bulk of the luminescence must come from close pairs. Let n_p be the concentration of such pairs and τ be their lifetime. Then, we can write:

$$\frac{dn_p}{dt} = b_n n_f p + b_p p_f n - \frac{n_p}{\tau} \quad (12)$$

where n_f and p_f are the free electrons and holes; p and n are the carriers trapped in band tail states, and b_n and b_p are the electron and hole recombination coefficients, respectively. Since diffusion of the oppositely charged carriers towards each other is the rate-limiting step, we ascribe band-tail luminescence to processes (7) and/or (8) rather than (11). For times short compared to the deep trapping and recombination times, dispersive

$$n_f = \alpha_n(1 - \alpha_n)\ (\nu t)^{\alpha_n - 1}\ n \quad (13)$$

$$p_f = \alpha_p(1 - \alpha_p)\ (\nu t)^{\alpha_p - 1}\ p \quad (14)$$

For times also short compared to τ, the luminescence intensity L must be proportional to n_p/τ. Thus

$$L \propto [(1 - \alpha_n)\ (\nu t)^{\alpha_n} b_n + (1 - \alpha_p)\ (\nu t)^{\alpha_p} b_p]\ n_o^2/\nu\tau. \quad (15)$$

It is clear from (15) that the luminescence intensity should initially <u>increase</u> with time.

For times long compared with τ, we must take the decay of the concentration of carriers trapped in tail states into account. Orenstein and Kastner[5] have shown that n_f and p_f decay as t^{-1} and that n_t and p_t decay as $t^{-\alpha_n}$ and $t^{-\alpha_p}$, respectively, providing $n \simeq p$. This can be seen in a simple manner by considering the approximate relations:

$$\frac{dn}{dt} \simeq - b_n n_f n_t = b_n n^2\ (\nu t)^{\alpha_n - 1} \quad (16)$$

$$\frac{dp}{dt} \simeq p_p p_f p_t = b_p p^2\ (\nu t)^{\alpha_p - 1} \quad (17)$$

Where the factors α(1-α) has been incorporated into the rate constants. Thus, the trapped and free carrier concentrations vary

$$n \propto (\nu t)^{-\alpha_n} \qquad p \propto (\nu t)^{-\alpha_p} \quad (18)$$

$$n_f \propto t^{-1} \qquad p_f \propto t^{-1}. \quad (19)$$

Substituting Eqs. (18) and (19) into (12), we find that the luminescence intensity in this regime decays as:

$$L \propto b_n n_f p + b_p p_f n \propto b_n (\nu t)^{-(\alpha_p + 1)} + b_p (\nu t)^{-(\alpha_n + 1)} \quad (20)$$

This decay is similar to that of the free carriers in the monmolecular regime, which is also proportional to $t^{-(1 + \alpha)}$.[13]

We must also consider the effects of trapping by the deep

defect states. In the absence of recombination, at short times:

$$n_f \propto n_o(\nu t)^{\alpha_n - 1} [1/\alpha_n(1 - \alpha_n) + (\nu t)^{\alpha_n}(N_d/N_c)]^{-1} \quad (21)$$

where n_o is the initial concentration of electrons.

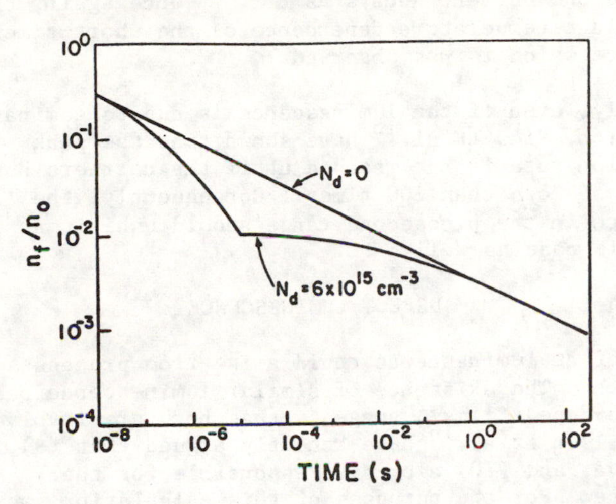

Fig. 2. Time dependence of the free carrier population with the addition of a sharp distribution of deep states. We use $\alpha = 0.70$, $N_d = 6x10^{15}$, $t_d = 10^{-5}s$, and $E_d = 0.4$ eV.

At long times:

$$n_f \propto n_o(\nu t)^{\alpha_n - 1} [1/\alpha_n(1 - \alpha_n) + (N_d/N_c)(t_d/t)(\nu t)^{\alpha_n}]^{-1} \quad (22)$$

where $t_d = (1/\nu) \exp(E_d/kT)$ and E_d is the energy separation between the defect and the conduction band mobility edge. Clearly, the presence of deep traps greatly complicates the time dependence of the luminescence. The time dependence of n_f for $\alpha_n = 0.70$, $E_d = 0.4$ eV, $N_d/N_c = 3 \times 10^{-4}$, and $\tau_d = 10^{-5}$ is sketched in Fig. 2. It is clear that even small concentrations of deep defects greatly affect the variation of the luminescence intensity at intermediate and long times. At short times, n_f and p_f are the same as in the absence of the deep centers. However, at intermediate times, i.e. $t \sim \nu^{-1} [(N_c/N_d)/\alpha(1-\alpha)]^{-1/\alpha}$, n_f and p_f decay as t^{-1}, even though there is no significant loss of carriers. At times long compared to $\nu^{-1} \exp(E_d/kT)$, n_f is relatively constant (although trap saturation is not occurring). Finally, at very long times, n_f decays as $t^{-(1-\alpha_n)}$, the expected power-law decay of the band-tail luminescence. Note that we have assumed even at these long times that not much loss of carriers via recombination has occurred. In the case of low excitation and for the low temperatures at which luminescence is ordinarily observed, α is very small and the bimolecules recombination time is extremely long.

At very low temperatures, the dispersion is primarily due to hopping transport within the localized states rather than to multiple trapping. In this case, at short times, the photogenerated carriers move by hopping to states of equal or lower energy, since few phonons are available for increasing their energy, and the current decays as t^{-1}.[14] Once again, this should yield a temperature dependence of the short-time luminescence which is not observed.

Finally, even if the luminescence is due to geminate recombination, Ries et al.[15] have shown that the peak recombination rate is not reached until the carriers have hopped an average of more than 100 times. Consequently, the luminescence in the picosecond range should exhibit an initial rise in this case as well.

DEFECT LUMINESCENCE

The defect luminescence could arise from processes (9) or (10) or both. The existence of similar luminescence peaks in P-doped and B-doped films[12] suggests that both are involved. However, Wilson et al.[16] have recently argued that (9) is nonradiative, and (10) alone is responsible for the luminescence. For the purposes of this calculation, we assume that both processes contribute; however, even if one is nonradiative, it will not affect the derived time dependence.

Let n_d and p_d be the concentrations of T_3^- and T_3^+ centers, let b_n' and b_p' be their respective recombination coefficients, and let n_p' be the sum of the concentrations of free electrons within 10 Å of a T_3^+ center and that of free holes within 10 Å of a T_3^- center. Then, at early times, when $p = p_o$ and $n = n_o$:

$$\frac{dp_d}{dt} = b_p' N_d p_o (\nu t)^{\alpha_p - 1} \tag{23}$$

$$\frac{dn_d}{dt} = b_n' N_d n_o (\nu t)^{\alpha_n - 1} . \tag{24}$$

Equations (23) and (24) yield:

$$p_d = b_p' N_d p_o \nu^{-1} (\nu t)^{\alpha_p} \tag{25}$$

$$n_d = b_n' N_d n_o \nu^{-1} (\nu t)^{\alpha_n} . \tag{26}$$

Noting that $p_o = n_o$, we can write the luminescence intensity as:

$$L \propto \frac{dn_p'}{dt} = n_f p_d b_n' + p_f n_d b_p'$$

$$= \frac{n_o^2 N_d}{\nu} (b_n b_p' + b_p b_n') (\nu t)^{\alpha_n + \alpha_p - 1} . \tag{27}$$

Equation (27) indicates that the defect luminescence initially decays at low temperatures, when $\alpha_n + \alpha_p < 1$. However, above a critical temperature, given by $\alpha_n + \alpha_p = 1$, the defect luminescence should grow in intensity at short times. For a-Si:H, this temperature is approximately 190K. At long times the luminescence will decay as carrier depletion due to recombination becomes significant. We have not attempted to calculate this regime but one would expect that the luminescence would decay as $t^{-(1+\alpha)}$ similar to band tail luminescence. A sketch of the predicted time dependence of the defect luminescence for several different values of the temperature is given in Fig. 3.

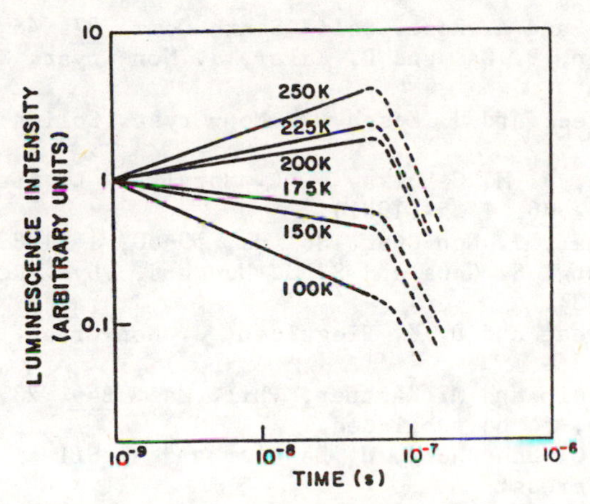

Fig. 3. Time dependence of the defect luminescence at several temperatures.

DISCUSSION

Wilson et al.[16] have performed careful time resolved measurements of the band-tail luminescence in a-Si:H over a wide range of times, and find no initial rise in intensity. In view of the calculations carried out in the present work, this is difficult to understand within the conventional model. However, there is now a growing mass of evidence[17,18] that the band mobility of electrons in a-Si:H is high. If so, the trapping time necessary to form the appropriate charge-transfer states is very short, less than 10^{-12}s for excitation concentrations greater than 10^{15} cm^{-3}. In this case, the initial rise in the band-tail luminescence predicted here occurs at times too short to observe, and only the decay should be seen, in agreement with the experimental results. We conclude that the time-resolved luminescence measurements provide still another piece of evidence in favor of a high band mobility for electrons in a-Si:H.

ACKNOWLEDGEMENTS

This research was supported in part by the U. S. National Science Foundation under Grants DMR 79-20023 and DMR 80-19295.

REFERENCES

1. H. Scher and E. W. Montroll, Phys. Rev. B12, 2455 (1975).
2. M. Silver and L. Cohen, Phys. Rev. B15, 3276 (1977).
3. M. Silver, G. Schonherr and H. Bassler, Phys. Rev. Lett. 48, 352 (1982).
4. G. Pfister and H. Scher, Adv. Phys. 27, 747 (1978).
5. J. Orenstein and M. Kastner, Phys. Rev. Lett. 43, 161 (1981).
6. T. Tiedje and A. Rose, Solid State Comm. 37, 48 (1981).
7. B. A. Khan, P. Bai and D. Adler, J. Non-Cryst. Solids, in press.
8. R. A. Street and J. Zesch, J. Non-Cryst. Solids 59-60, 449 (1983).
9. T. Tiedje, J. M. Cebulka, D. L. Morel, B. L. Abeles, Phys. Rev. Lett. 46, 1425 (1981).
10. W. E. Spear, J. Non-Cyst. Solids, 59-60, 1 (1983).
11. C. Y. Huang, S. Guha and S. J. Hudgens, Phys. Rev. B27, 7460 (1983).
12. R. A. Street and D. K. Biegelsen, J. Non-Cryst. Solids, in press.
13. J. Orenstein and M. Kastner, Phil. Mag. B46, 23, 1983.
14. D. Monroe, to be published.
15. B. Ries, G. Schonherr, H. Bassler and M. Silver, Phil. Mag., in press.
16. B. A. Wilson, A. M. Sergent, K. W. Wecht, A. J. Williams and T. P. Kerwin, to be published.
17. M. Silver, E. Snow, M. Aiga, V. Connella, R. Ross, Z. Yaniv, M. Shaw and D. Adler, J. Non. Cryst. Solids 59-60, 445 (1983).
18. M. Silver, E. Snow and D. Adler, Solid State Comm., to be published.

DEPENDENCE OF THE METASTABLE OPTICALLY-INDUCED ESR IN a-Si:H ON TEMPERATURE AND POWER

Charles Lee, W.D. Ohlsen, and P.C. Taylor
University of Utah, Salt Lake City, UT 84112

H.S. Ullal and G.P. Ceasar
ARCO Solar, Inc., P.O. Box 4400 Woodland Hills, CA 91365

ABSTRACT

The kinetics of a metastable, optically-induced increase in the ESR in a-Si:H have been examined between 77 and ~ 500 K. Isothermal annealing (decay) and inducing (growth) curves indicate that the decay is bimolecular and thermally-activated with an activation energy of ~ 1 eV, and the growth is relatively temperature independent. At high incident light intensities the ESR increases monotonically roughly as a power law in time ($I \alpha t^{\beta}$, $\beta < 1$). At lower intensities β decreases with intensity.

INTRODUCTION

Metastable light-induced changes in both the transport and optical properties of hydrogenated amorphous silicon (a-Si:H) are well known. The best-known metastable change in the transport properties is a decrease in the photoconductivity first observed by Staebler and Wronski.[1]

Dersch et al.[2] were the first to observe a metastable, optically-induced increase in the electron spin resonance (ESR) signal at 300 K in a Si:H. The lineshape of this signal was, within experimental accuracy, the same as that observed before irradiation. This ESR signal is generally attributed to a "dangling bond" on a silicon atom.[3,4] A metastable increase in the ESR at 300 K can also be produced by x-irradiation.[5] These optically-induced increases in the ESR can be annealed by cycling to elevated temperatures.[2,6]

At low temperatures there is an additional component to the metastable optically-induced ESR.[7,8,9] This component can be induced with small light intensities ($\lesssim$ 50 mW cm^{-2}), but it decays above ~ 80 K. At greater intensities ESR centers can be induced at 77 K which decay only above ~ 300 K.[10,11] There are clearly several different contributions to the metastable, optically-induced ESR in a-Si:H. Which contribution one emphasizes depends upon such parameters as the temperature of irradiation, the power density and spectral distribution of the exciting light.

In this paper we examine the dependence of the metastable, optically-induced ESR on the incident power density of the exciting light and on the temperature. Both isothermal annealing and inducing experiments have been performed. In the first kind of experiment the isothermal decay of the increase in the ESR is monitored at temperatures between 300 and 500 K, and in the second kind of experiment the increase in the ESR is monitored as a function of irradiation time over the same temperature range.

0094-243X/84/1200205-08 $3.00

EXPERIMENTAL DETAILS

Films of a-Si:H between 1 μm and 3 μm thick were deposited on 250°C (523 K) substrates using standard glow-discharge techniques. Samples were deposited on aluminum foil and quartz substrates in the same run. The samples on aluminum foil were removed from the substrates using dilute hydrochloric acid. The resulting "flakes" were washed in distilled water, thoroughly dried, and sealed in evacuated quartz tubes. The high-temperature experiments were performed mainly on the powdered samples, but detailed experiments at 300 K have been performed on films on quartz substrates. The ESR spin densities n_s of both kinds of samples before irradiation were $\leq 5 \times 10^{15}$ spins/cm^3.

Irradiation was performed with either a tungsten source at power densities over the approximate range from 0.1-1 W/cm^2 or with a Kr^+ (at 6471 Å) or an Ar^+(5145 Å) ion laser. The quartz tubes containing the powdered samples were rotated in a horizontal position during irradiation in order to illuminate all the flakes evenly. Elevated temperatures were obtained with a nitrogen gas-flow system. Samples were typically annealed at 480 K for 20 minutes upon completion of each experiment.

The ESR experiments were performed using a Varian model V4500 spectrometer with Varian model V4012 12" electromagnet. Signal enhancement was accomplished with a Nicolet model 1070 signal averager.

RESULTS AND DISCUSSION

Figure 1 shows the isothermal time decay of the optically-induced portion of the ESR in a powdered sample of a-Si:H supplied by Brookhaven National Laboratory (BNL). All data were taken after 90 minutes of irradiation at a nominal 300 K with ~ 1 W cm^{-2} from a tungsten source. (Under irradiation the temperatures of the powdered samples are greater than 300 K because of poor thermal contact with the quartz tube. The film samples are not heated appreciably). The decay curves are not well described by simple exponentials, but they are well fitted by bimolecular statistics (second-order kinetics). In a bimolecular process the density of optically-induced ESR spins is given by

$$(n_s/n_o)^{-1} = 1 + n_o \gamma t \qquad (1)$$

where n_o is the saturated density of metastable ESR spins after long-time irradiation and where the bimolecular decay rate γ is well fitted by the form

$$\gamma = \gamma_o \exp(-E_a/kT) \quad . \qquad (2)$$

In this expression γ_o is the infinite temperature intercept and E_a is the activation energy.

From Eq. (1) it is apparent that the reciprocal of the ESR intensity increases linearly with time. The data of Fig. 1 approximate this dependence very well (except possibly at short times, $t \leq 5$ min). From the temperature dependence of the slopes of the lines in Fig. 1 one can estimate an activation energy for the decay rate. These data

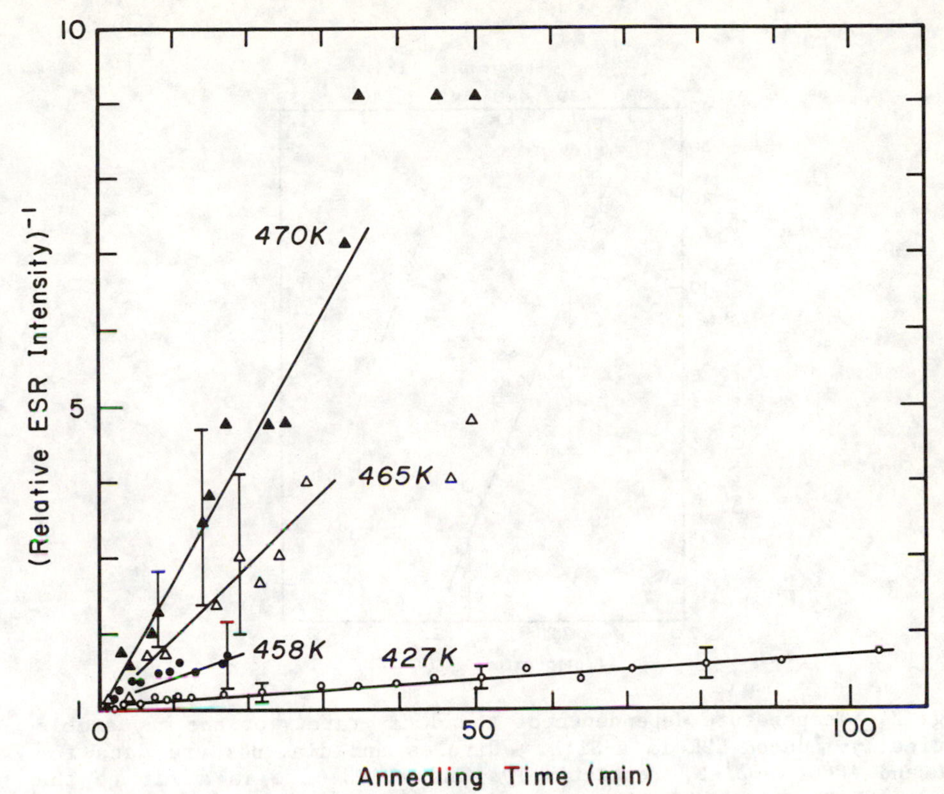

Fig. 1. Inverse of the optical-induced ESR intensity in a-Si:H as a function of annealing time at various temperatures.

are shown in Fig. 2 for the BNL sample (circles) and a sample made by ARCO Solar, Inc. (ARCO; diamonds). The activation energies obtained from the slopes of the lines in Fig. 2 are ~ 1 eV.

A second type of experiment which further tests the kinetics of the metastable optically-induced ESR is the isothermal growth under constant illumination. The ESR intensities measured at 77 K as functions of time after irradiation at 297 and 405 K are plotted in Fig. 3 for the same sample as indicated by the diamonds in Fig. 2. Similar results are obtained for irradiation at 77 K. The trend which is the most apparent from these data is the existence of an apparent long-time asymptote in the ESR intensity and the decrease of this asymptotic value with increasing irradiation temperature. Qualitiatively, this trend is the result of a competition between the inducing rate and the decay rate at the elevated temperatures.

It is difficult to tell the difference between saturation and very slow growth at long times, but two tests were performed to check for saturation. First, the powdered samples were irradiated for 18 hours and checked against the induced ESR responses after ~ 2 hours. These two ESR intensities were found to be the same within experimental error (± 10%). Second, film samples on quartz substrates were irradiated and observed at 300 K for times up to 3,000 minutes and

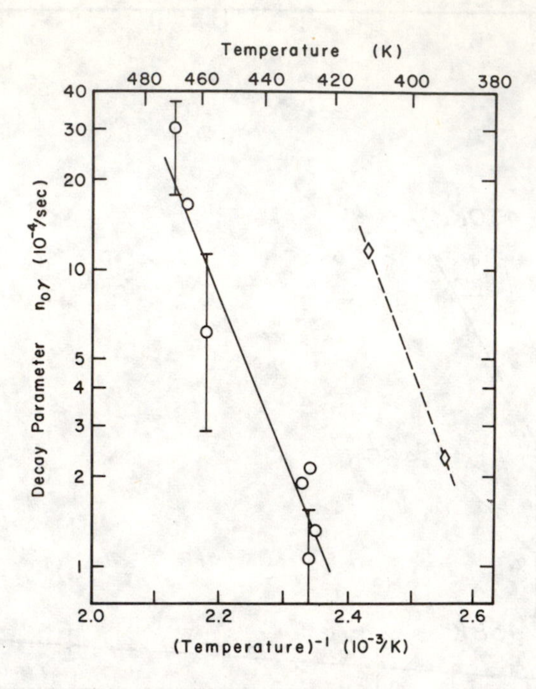

Fig. 2. Temperature dependence of the decay rate γ of the metastable optically-induced ESR in a-Si:H. Circles and diamonds are data for BNL and ARCO samples, respectively. The solid line is a fit to the data with E_a = 1.1 eV.

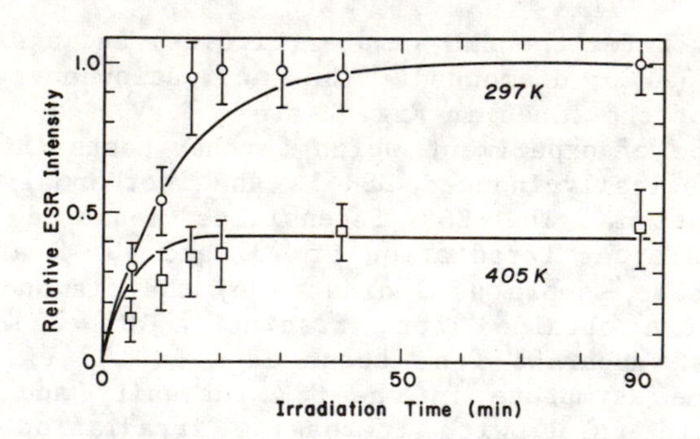

Fig. 3. Metastable, optically-induced ESR intensity as a function of irradiation time at two representative temperatures (powdered sample). Irradiation intensity was ~ 1 W cm^{-2}. Temperatures are approximate due to sample heating. The saturated spin density at 300 K is approximately 10^{16} cm^{-3} for the 3 μm sample.

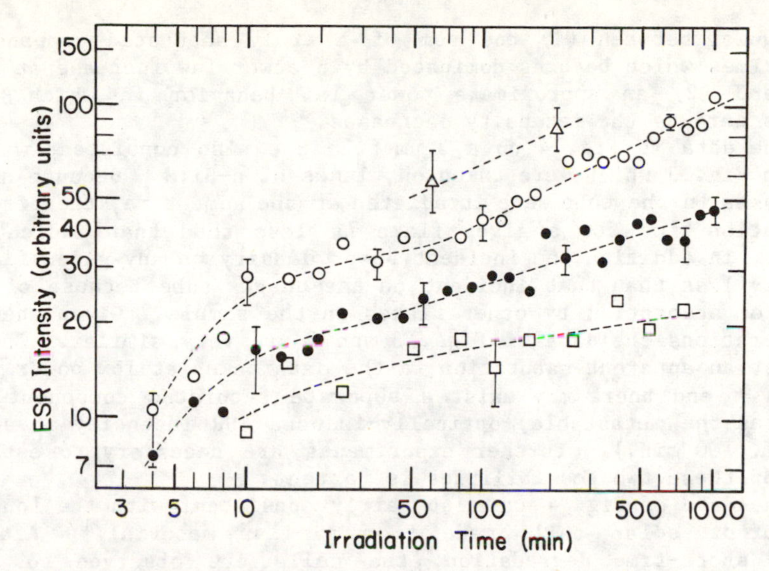

Fig. 4. Metastable, optically-induced ESR intensity as a function of irradiation time at 300 K (3 μm film). Open circles, closed circles and open squares represent data taken with 1.3, 0.4 and 0.1 W cm^{-2} from a tungsten source, respectively. Open triangles are data taken at 6471 Å and 700 mW cm^{-2}. The spin density at 500 min for 0.1 W cm^{-2} irradiation density (open squares) is approximately 10^{16} cm^{-3}.

intensities of the tungsten light between ~ 0.1 and ~ 1.0 W cm^{-2}. Results are shown on a log-log plot in Fig. 4 for a sample which is 3 μm thick.

It is clear from Fig. 4 that at long times and high powers there is a slow growth in time, but the growth becomes less prominent as the incident power density decreases. Also shown in Fig. 4 are data taken at 6471 Å with a Kr^+ laser at ~ 700 mW cm^{-2} (open triangles). The dashed line through the data is drawn for the power law $I \propto t^{\beta}$ with $\beta \simeq 1/3$. This result is very similar to that previously observed by Stutzmann et al.[12] from experiments on films irradiated with 6471 Å light. The major difference is that in the previous data[12] the sum of dark and optically-induced spin densities was plotted, but in our results the dark spin densities are at least 5 times smaller than the lowest optically-induced signals plotted in Fig. 4. Measurements were also performed at 5145 Å from an Ar^+ laser at similar power levels and the power law behavior with $\beta \simeq 1/3$ was also observed. At this wavelength the optically-induced ESR intensities are down by approximately a factor of ten, presumably at least in part due to the decreased penetration depth of the light at this wavelength.

In most respects the data on the ARCO sample in Fig. 4 are in good agreement with the previous results of Stutzmann et al.[12] at the higher power levels, but they depart significantly at lower power levels. The accuracy of the measurements is not sufficient to distinguish between (1) the sum of a truly saturated component at short times which becomes dominated by a power law increase at longer times and (2) an approximate power law behavior in which β tends towards zero as the intensity decreases.

The data in Fig. 4 on a 3 μm film are also consistent with the data in Fig. 3 which were taken on flakes of a-Si:H. Because not all the flakes in the tube were irradiated at the same time, the effective irradiation time for a given flake is less than that indicated in Fig. 3. In addition the incident power density on any given flake is probably less than that incident on the quartz tube because of scattering or absorption by other flakes in the sample. Given these two considerations the data of Figs. 3 and 4 are very similar. There is at least an apparent saturation in the data taken at low power levels in Fig. 4 and there may exist a separate saturated component which dominates the metastable, optically-induced ESR intensity at shorter times ($\lesssim$ 100 min.). Further experiments are necessary to establish which of these two possibilities is correct.

The data of Fig. 4 are also fairly consistent with the long term behavior of solar cells made from similar material.[17] After an initial short-time degradation, the cells are observed to retain their efficiencies for times approaching at least two years.

Currently there are two classes of models which have been invoked to explain metastable changes in the electronic and optical properties of a-Si:H. One type of model suggests that there is an increase in the density of silicon dangling bonds when "weak bonds" are broken optically.[2,13] A second class of model assumes that the metastable changes result from an optical rearrangement of electrons and holes in existing diamagnetic (negative U) defects.[14]

The bimolecular nature of the isothermal decay of the optically-induced ESR, which was measured at relatively low incident light intensities, is easier to explain on the basis of the rearrangement of charge in existing, negative U defects than it is on the basis of creating new defects.[15] The bond-breaking hypothesis requires the "diffusion" of one of the formerly-bonded electrons to a distant trapping site through the motion of hydrogen atoms.[2,12] The major difficulty with the hydrogen diffusion model is that it requires thermally-activated generation and annealing processes whereas the observed ESR results only exhibit significant thermal activation in the annealing process. It may also be difficult to explain the low-intensity data of Fig. 4 with the bond-breaking approach.

One possible explanation which is consistent with the results observed to date is that there are two processes involved in the optically-induced ESR just as has been observed in the chalcogenide glasses.[16] At low power densities one is rearranging charge in a fixed number of existing (negative U) defects while at higher power densities the dominant process is the creation of new defects, presumably by bond breaking. Further results are clearly necessary before this hypothesis can be either confirmed or refuted.

There is evidence from many other experiments that more than one center may be involved in the optically-induced degradation of the electronic properties of a-Si:H.[18-20] In particular, Han and Fritzsche[19] have found that at least two states are necessary to explain the generation and annealing behavior of optically-induced changes in the photoconductivity of a-Si:H. One type of state affects carrier lifetime and the other produces an increase in the below gap absorption.

By irradiating samples at different temperatures, Guha et al.[20] have found that the annealing behavior of photoconductivity in a-Si:H depends on the temperature at which the samples were initially exposed to light. These authors have found that the room temperature photoconductivity decreases much less after irradiation at 160 K than it does after irradiation at 300 K. In addition, the changes produced at 300 K are much more stable than those produced at 160 K. This behavior has a parallel with the fatigue of the 1.2 eV PL peak and the growth of the 0.9 eV peak.[21] The magnitudes of the photoconductivity and PL effects appear to peak for irradiation at about 300 K and decrease at higher and lower temperatures. In this sense these particular effects may well be better understood in terms of a model which invokes the hydrogen-assisted diffusion of a broken bond. At temperatures below 300 K the hydrogen diffusion is inhibited and at temperatures above 300 K the defects anneal. At the very least one can say that the correlation which exists between the isothermal annealing of the optically-induced 0.9 eV PL (or fatigue of the 1.2 eV PL)[21] and the ESR[6] does not imply that these two effects are directly related. Because the low temperature behavior of these two effects is very different, more than one localized electronic state must be invoked in any complete explanation of both the PL and ESR results.

Similar conclusions can be drawn from comparisons with other experiments, such as below gap optica l absorption,[22,23] optically detected magnetic resonance (ODMR),[24] and time resolved photoconductivity.[25] Consistent explanations for these experiments and the PL and ESR results appear to require optically-induced changes in more than one type of below-gap electronic state in a-Si:H.

ACKNOWLEDGEMENTS

Some of the authors (CL, WDO, PCT) gratefully acknowledge ARCO Solar, Inc. for partially supporting this research. This research is partially supported by the National Science Foundation under Grant DMR-83-04471. J. Hautala is acknowledged for kindly helping with some of the ESR experiments.

REFERENCES

1. D. L. Staebler and C. R. Wronski, Appl. Phys. Lett. 31, 292 (1977).
2. H. Dersch, J. Stuke and J. Beichler, Phys. Status Solidi B105, 265 (1981); B107, 307 (1981); Appl. Phys. Lett. 38, 456 (1981).
3. D. K. Biegelsen, Proc. Electron Resonance Soc. Symp. 3, 85 (1981).
4. For a recent review, see P. C. Taylor in Semiconductors and Semimetals, Vol. 21C, Willardson and Beer, eds. Academic Press, NY (1984), in press.
5. W. M. Pontuschka, W. E. Carlos, and P. C. Taylor, Phys. Rev. B25, 4362 (1982).

6. P. C. Taylor and W. D. Ohlsen, Solar Cells 9, 113 (1983).
7. R. A. Street, D. K. Biegelsen and J. C. Knights, Phys. Rev. B24, 969 (1981).
8. K. Morigaki, T. Sano and I. Hirabayashi, J. Phys. Soc. Jpn. 51, 147 (1982).
9. Y. Sano, K. Morigaki and I. Hirabayashi, Solid State Commun. 43, 439 (1982).
10. I. Hirabayashi, K. Morigaki and M. Yoshida, Solar Energy Mat. 8, 153 (1982).
11. I. Hirabayashi, K. Morigaki and S. Nitta, Jpn. J. Appl. Phys. 19, L357 (1980).
12. M. Stutzmann, W. B. Jackson and C. C. Tsai, unpublished.
13. R. A. Street, Appl. Phys. Lett. 42, 507 (1983); J. I. Pankove, Solar Energy Mat. 8, 141 (1982).
14. D. Adler, Kinam C4, 225 (1982); D. Adler and R. C. Frye, AIP Conf. Proc. 73, 146 (1981); D. Adler and F. R. Shapiro, Physica 117B & 118B, 932 (1983).
15. C. Lee, W. D. Ohlsen, P. C. Taylor, H. S. Ullal and G. P. Ceasar, Phys. Rev. B (1984), unpublished.
16. D. K. Biegelsen and R. A. Street, Phys. Rev. Lett. 44, 803 (1980).
17. H.S. Ullal, D.L. Morel, D.R . Willett, D. Kanani, P.C. Taylor and C. Lee, Proc. 17th IEEE Photovoltaic Specialists Conf. (1984), in press.
18. M.H. Tanielian, N.B. Goodman and H. Fritzsche, J. de Phys. 42, C4-375 (1981).
19. D. Han and H. Fritzsche, J. Non-Cryst. Solids 59 & 60, 397 (1983).
20. S. Guha, C.-Y. Huang, S.J. Hudgens and J.S. Payson, J. Non-Cryst. Solids (1984), in press (Proc. Int. Top. Conf. on Transport and Defects in Amorph. Semicond. Detroit, 22-24 Mar., 1984).
21. J.I. Pankove and J.E. Berkeyheiser, Appl. Phys. Lett. 37, 705 (1980).
22. W.B. Jackson and N.M. Amer, J. de Phys. 42, C4-293 (1981).
23. S. Yamasaki, N. Hata, T. Yoshida, H. Oheda, A. Matsuda, H. Okushi and K. Tanaka, J. de Phys. 42, C4-297 (1981).
24. K. Morigaki, T. Sano and I. Hirabayashi, J. Phys. Soc. Jpn. 51, 147 (1982).
25. R.A. Street, Appl. Phys. Lett. 42, 507 (1983).

THE KINETICS OF FORMATION AND ANNEALING OF LIGHT INDUCED DEFECTS IN HYDROGENATED AMORPHOUS SILICON

M. Stutzmann, W.B. Jackson, C.C. Tsai
Xerox Palo Alto Research Center, Palo Alto CA 94304

ABSTRACT

Light induced creation and thermal annealing of metastable dangling bonds in undoped hydrogenated amorphous silicon have been investigated by electron spin resonance and photoconductivity measurements. The kinetics of the formation process can be described consistently by a self-limiting mechanism based on nonradiative tail-to-tail recombination of photoexcited carriers as the defect inducing step. The kinetic behaviour of the annealing process is consistent with an exponential decay of the metastable dangling bonds with a wide distribution of thermally activated decay constants.

INTRODUCTION

Following its discovery in 1977,[1] the existence of reversible light induced changes in the transport properties of hydrogenated amorphous silicon (a – Si:H) ("Staebler – Wronski effect") has attracted considerable attention, mostly because of the negative consequences of this phenomenon for the stability of amorphous silicon based photosensitive devices. Moreover, reversible changes have been observed in many other electronic, optical, and magnetic properties of a – Si:H, thus emphasizing the importance of the light induced changes for our understanding of this material.

In the case of undoped a–Si:H it has been shown conclusively that the recombination of excited carriers is responsible for the light induced changes.[2] Many of these changes can be explained at least qualitatively by a reversible increase of defect states in the mobility gap, in particular by an increase of neutral silicon dangling bonds (D^o) as detected by electron spin resonance (ESR).[3,4] The observation of a characteristic activation energy of about 1.5 eV for the annealing of the metastable changes has led to the conclusion that hydrogen is involved in the microscopic process.[4,5] In this paper we report on a systematic investigation of the influence of illumination time, light intensity, and sample temperature on metastable defect creation and annealing in high purity a – Si:H and present a model that can quantitatively explain our experimental data. The role of sample parameters (thickness, impurity concentration, surface properties) is subject of a second study.[6]

EXPERIMENT

For our investigation of the Staebler – Wronski effect in undoped a – Si:H we have combined the ability of ESR to identify specific microscopic defects with the sensitivity of photoconductivity (PC) to detect changes in the density of gap states. All measurements were performed on 3μm thick undoped a – Si:H samples deposited on Corning 7059 substrates (0.5 x 1 cm^2). For the PC measurements, Cr bottom electrodes (coplanar, 0.5 mm gap) were evaporated onto the substrates prior to sample deposition. The amorphous silicon was prepared in a UHV glow discharge system,[7] using pure silane, low rf power, and a substrate temperature of 230 °C. All samples used for the experiments reported here were produced in the same deposition.

ESR measurements were performed in the X-band, determining the Si dangling bond resonance at the g-value 2.0055. The sample was illuminated or annealed inside the microwave cavity without being removed during an experiment. The sample temperature was controlled with a continuous flow cryostat using dry nitrogen. Since the lineshape and the saturation behaviour of stable and metastable dangling bonds are identical, the changes of the ESR signal amplitude with time could be measured continuously (and conveniently) using a double field-modulation technique.

Only monochromatic light was used for sample illumination, either defocused laser light (1.9 eV) or monochromatized light with variable photon energy from an arc lamp. Sample heating due to the illumination was determined to be negligible ($\Delta T < 15$ K for intensities below 0.5 Wcm^{-2}).

RESULTS AND DISCUSSION

A) Kinetics of defect creation

In this section we will discuss the dependence of the PC and of the dangling bond density determined by ESR in the context of a new kinetic model for the metastable defect creation. Rather than presenting the experimental data first, we would like to begin with the development of this model and compare it to the experiment as we proceed. Our model is outlined schematically in Fig. 1.

Illumination with photons of energy $h\nu$ generates excited electrons and holes with a rate g. These carriers thermalize rapidly into the tail states of the conduction and valence band which act as shallow traps. The trapped carriers then contribute to the PC by a mechanism of multiple thermal excitation to extended states and retrapping, until they finally recombine. Recombination can occur via recombination centers or by direct tail-to-tail transitions. In a-Si:H at or above room temperature it is known that the dangling bond states act as the dominant recombination centers and that direct recombination is negligible in comparison.[8]

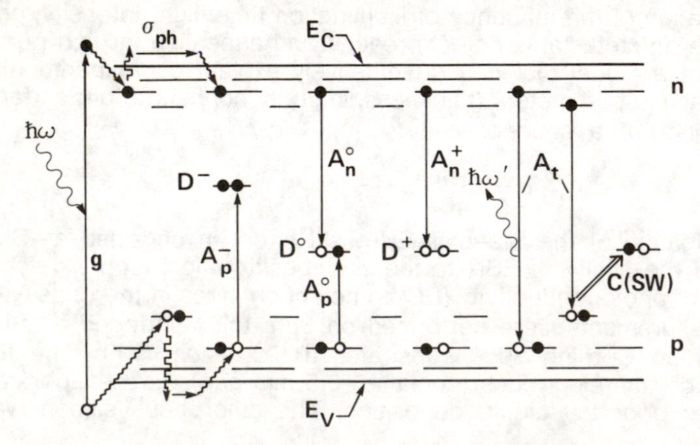

Fig.1: Schematic model for the recombination. See text for details.

Depending on the different charge states of the dangling bonds the following capture processes are possible: $h + D^- \rightarrow D^o$, $h + D^o \rightarrow D^+$, $e + D^o \rightarrow D^-$, and $e + D^+ \rightarrow D^o$. The respective transition probabilities, $A_{n,p}^{+,o,-}$, indicated in Fig. 1 have recently been estimated from time-of-flight measurements and are of the order 10^{-9} to 10^{-8} cm^3s^{-1}.[9] However, apart from these transitions there will exist a small fraction of direct recombination events which will occur with a transition probability A_t and are either luminescent or nonradiative. We propose that the small fraction of *nonradiative direct tail–to–tail transitions* is responsible for the creation of new metastable dangling bonds with an efficiency c(sw) as shown in Fig. 1. A physical justification for this assumption will be given at the end of this paper.

It is now straightforward to write down the rate equations determining the occupancy of the different levels in our model, leading to the following expressions for the total densities, n and p, of excited electrons and holes under steady state illumination:

$$n = g/(A_n^+ \cdot N^+ + A_n^o \cdot N^o + A_t \cdot p) \qquad (1a)$$

$$p = g/(A_p^o \cdot N^o + A_p^- \cdot N^- + A_t \cdot n) \qquad (1b)$$

Here N^+,N^o, and N^- are the densities of the different dangling bond charge states. It should be mentioned that a steady state approach is reasonable for a model of the Staebler–Wronski effect, because the time necessary for n, p, and $N^{+,o,-}$ to reach a steady state (~ 10^{-6} s at room temperature) is much shorter than the time scale of the observed reversible changes (>1s).

At this point it is necessary to look more closely at the relative occupation of the three different dangling bond levels under illumination. The densities N^+, N^o, and N^- are determined by the following equations:

$$(N^o)^2/(N^+N^-) = A_n^+ \cdot A_p^-/(A_n^o \cdot A_p^o) \sim 20 \qquad (2a)$$

$$N^+ + N^o + N^- = N \qquad (2b)$$

$$n + N^- = p + N^+ \qquad (2c)$$

Eq. (2a) follows from the rate equations for the dangling bond levels, (2b) is due to the fact that the densities of the different charge states have to add up to the total density N of dangling bond states, and (2c) is the charge neutrality condition. Using this set of equations together with Eqs. (1a) and (1b) and our knowledge of the relevant transition probabilities, it is possible to show that for light intensities < 1 Wcm^{-2} and dangling bond densities $\geq 10^{16}$ cm^{-3} the following holds.

– The total densities n and p of excited carriers is always much smaller than N. Consequently the efficiency of the dangling bonds to act as recombination centers is independent of n and p (no saturation occurs).

– Regardless of the light intensity used for illumination (within the limits given above) the fractions N^+/N and N^-/N of positively and negatively charged dangling bonds are constant and of the order 10% to 20%. This last point is corroborated by the experimental observation that the ESR signal in undoped a–Si:H, which is due to the neutral states N^o, remains unchanged upon illumination of the sample.

Because the fractions of charged states, $N^{+,-}/N$, are constant during the prolonged illumination, it is possible to define time– and intensity independent effective transition probabilities A_n and A_p for electron and hole capture by the dangling bonds as:

$$A_n = (A_n{}^o \cdot N^o + A_n{}^+ \cdot N^+)/N \ , \ A_p = (A_p{}^o \cdot N^o + A_p{}^- \cdot N^-)/N$$

Using the values for $A_{n,p}{}^{+,o,-}$ derived from ref. 9, one obtains for A_n and A_p the numerical values $A_n \sim 3 \times 10^{-9}$ cm^3s^{-1} , $A_p \sim 7 \times 10^{-10}$ cm^3s^{-1}. With these effective transition probabilities, Eq. (1a) and (1b) can be rewritten in the simpler form:

$$n = g/(A_nN + A_tp) \sim g/(A_nN) \quad (3a)$$

$$p = g/(A_pN + A_tn) \sim g/(A_pN) \quad (3b)$$

The approximations in (3a) and (3b) are possible because according to the discussion above n, p $\ll$ N, and the transition probability A_t for tail–to–tail transitions will be much smaller than A_n or A_p.

The validity of Eqs. (3) can be tested experimentally by measuring the PC , σ_{ph}, as a function of light intensity, I, and dangling bond density, N. Since the PC at a constant temperature is proportional to the densities of the excited carriers and the ESR spin density $N_S = N^o \propto N$, one expects a dependence of the form $\sigma_{ph} \propto I/N_S$. Since the dangling bond density increases during the illumination by more than one order of magnitude, the dependence of σ_{ph} on N_S can be tested by measuring these two quantities simultaneously while the sample is illuminated inside the microwave cavity. This is shown in Fig. 2. A good agreement with Eqs. (3) is observed, both for the dependence of σ_{ph} on N_S as

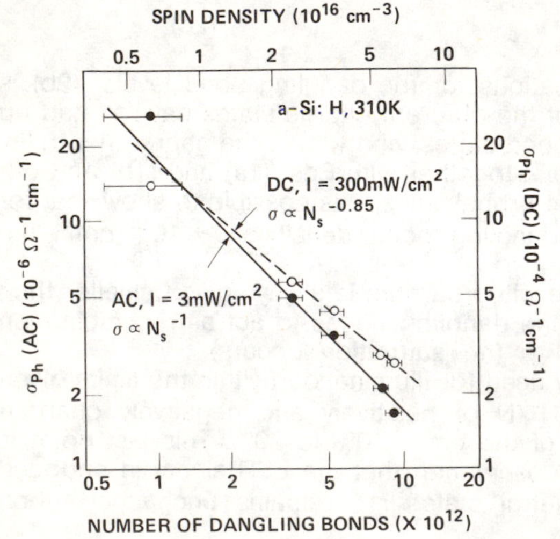

Fig.2: Photoconductivity as a function of ESR spin density for two different intensities. AC and DC refer to measurements with chopped light and constant illumination, respectively.

well as on the intensity. A more detailed study of the intensity dependence confirmed this nearly linear behaviour of the PC ($I^{0.92}$ in the annealed state and $I^{1.04}$ after prolonged illumination).

We can now derive the kinetic behaviour of the defect creation during the illumination. According to the model (Fig. 1), the rate at which new dangling bonds are created is given by the tail–to–tail recombination rate $A_t \cdot n \cdot p$ and the efficiency constant c(sw). Substituting for n and p from Eqs. (3) we obtain:

$$dN/dt_i = c(sw)A_t np = c(sw)\,A_t g^2/(A_n A_p N^2) \qquad (4)$$

Here t_i is the illumination time. This differential equation for the dangling bond density N can be integrated to:

$$N^3(t_i) - N^3(0) = g^2 \cdot t_i \cdot 3c(sw)A_t/(A_n A_p) \qquad (5)$$

Eq. (5) completely characterizes the dependence of the defect density on the kinetic parameters illumination intensity and time. The following comments can be made.

– According to Eq. (4), the creation of metastable dangling bonds is self–limiting in the sense that the creation rate decreases with the square of the existing defect density. This follows from the fact that in our model the defect inducing direct transitions compete with the transitions via the dangling bonds. It also explains why a–Ge:H and highly defective samples of a–Si:H do not show a Staebler–Wronski effect.

– For $N(t_i) > 2N(0)$, i.e. after sufficiently long illumination, the term $N^3(0)$ in Eq. (5) can be neglected. One obtains the long time limit behaviour $N(t_i) \propto g^{2/3} \cdot t_i^{1/3}$, showing the sublinearity of the reversible changes in exposure, $I \cdot t$. [5,10]

– The time and intensity dependence of the ESR spin density, N_S, and of the PC can be derived from Eq. (5) using $N_S = \delta \cdot N$ ($\delta \sim 0.7$), $\sigma_{ph} = en\mu$, and Eq. (3a) (μ is a suitably chosen trap dominated electron drift mobility).

In Fig. 3 and 4 the kinetic behaviour of the spin density and PC as predicted by the theory (solid lines) is compared to our experimental results (points). The agreement is very good, especially if one keeps in mind that the only adjustable parameter in our model is the product $c(sw) \cdot A_t$. All other quantities appearing in Eq. (5) are known. A value of $c(sw)A_t \sim 1.5 \times 10^{-15}$ cm^3s^{-1} was determined by fitting the calculations to the ESR data for an intensity of 400 $mWcm^{-2}$ in Fig. 3. With the same value we were not only able to fit the spin density curves for other intensities, but also the PC data in Fig. 4, in this case allowing for small variations in the effective light intensity. There is some discrepancy between the theory and the ESR data for the highest intensity used. This could indicate that either sample heating due to the illumination becomes important or that for this high intensity some assumptions in our model are no longer valid. We would like to mention also that the good consistency between the ESR and PC data in the context of our model seems to indicate that no electronically active defects other than dangling bonds are created during prolonged illumination near the center of the gap.

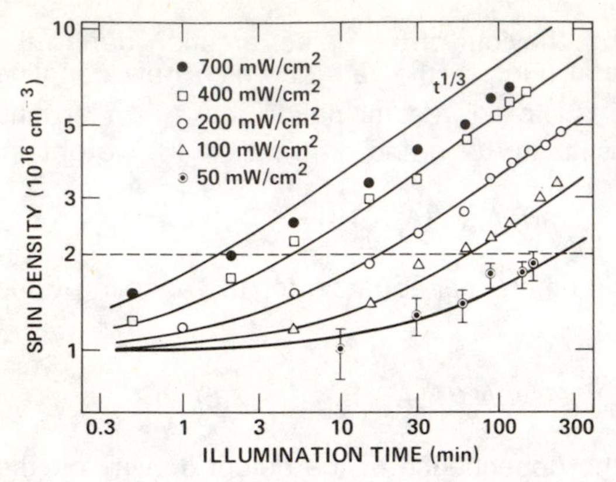

Fig.3: Experimental (points) and calculated (solid lines) dependence of the ESR spin density on illumination time for different light intensities.

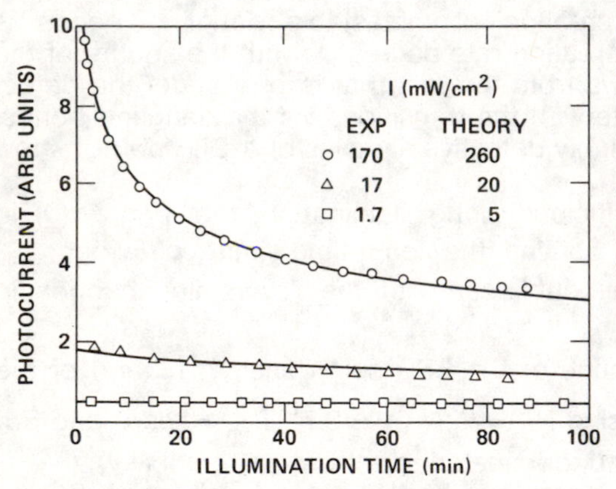

Fig.4: Comparison of the experimental (points) and theoretical (solid lines) decrease of the photoconductivity as a function of illumination time. The fits were obtained using the same parameters as for Fig.3, but allowing small variations of the experimental vs. theoretical intensity.

B) Kinetics of the annealing process

It follows from the picture developed so far that the most direct way of studying the annealing kinetics of the metastable defects is to monitor the decay of the ESR dangling bond signal as a function of time for different annealing temperatures T_A. In contrast to section A, however, the quantity of interest is not the total density of dangling bonds, N, but the density N_{ind} of metastable or induced defects, $N_{ind} = N - N(0)$, since the stable defects with density N(0) should not be affected by the annealing process.

In the simplest case, the decay of N_{ind} can be described by a monomolecular form:

$$dN_{ind}/dt = -\gamma N_{ind} \Rightarrow N_{ind}(t) = N_{ind}(0)\exp(-\gamma t) \qquad (6)$$

Another form proposed[11] is the bimolecular decay:

$$dN_{ind}/dt = -\gamma N_{ind}^2 \Rightarrow [1/N_{ind}(t)] - [1/N_{ind}(0)] = \gamma t \qquad (7)$$

In both cases the observed temperature dependence can be explained by a thermally activated decay constant γ:

$$\gamma = \gamma_o \exp(-E_a/kT_A) \qquad (8)$$

where E_a is the energy barrier for the thermal decay of a metastable state.

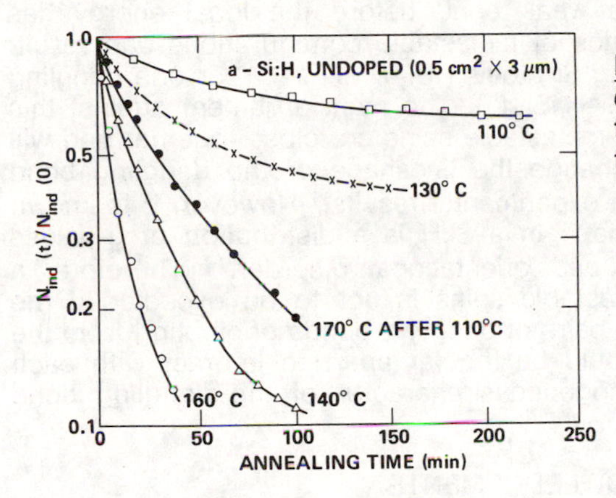

Fig.5: Decay of the induced spin density as a function of annealing time for different annealing temperatures.

The experimental data are shown in Fig. 5 assuming a monomolecular type of decay. The curves clearly deviate from the expected straight lines. The same is true if we try to plot our data according to Eq. (7). We will therefore change our approach by allowing for a distribution, $P(\gamma)$, of decay constants at a fixed temperature. Especially in an amorphous material such a distribution can be caused by a distribution, $P'(E_a)$, of activation energies for the thermal decay of the metastable defects. This assumption is consistent with the following experimental result. Annealing of a sample at 170°C after an initial anneal at 110°C leads to a decay which is closer to a single exponential and has a smaller decay constant than a similar anneal without previous low temperature treatment (cf. Fig. 5). The width of the distribution $P'(E_a)$ can be estimated by determining the activation energies of the slopes in Fig. 5 in the initial stage and after most of the metastable dangling bonds have been annealed out. We obtain the values 0.8 eV and 1.4 eV for the lower and upper end of $P'(E_a)$, respectively. This range is in agreement with the different activation energies for the annealing process reported in the literature.

C) Microscopic model

The results presented in sections A and B allow the following conclusions for a microscopic mechanism of the Staebler–Wronski effect in undoped a–Si:H. 1) The defect inducing step is the nonradiative tail–to–tail recombination of carriers. 2) Primarily dangling bonds are created by this process. 3) The energy barrier for thermal anneal of the metastable configuration is distributed between approximately 0.8 and 1.4 eV. A simple atomic configuration consistent with these results is Si---Si–H, i.e. a weak Si–Si bond with a hydrogen atom at one of the back bonds. When a hole is trapped in such a weak bond it becomes even weaker. Moreover, its energetic position in the tail of the valence band will make it a possible final state for a tail–to–tail transition. If such a transition is nonradiative, an energy of about 1 eV will become available and transfered to local phonons involving the weak bond and, especially, the Si–H back bond, because Si–H can absorb a larger amount of energy per phonon. As a consequence, the weak Si bond can break. This new configuration can become stabilized by switching of the hydrogen atom from the back bond position into the weak bond before the local energy has dissipated. In this picture two kinds of metastable configurations can result: two dangling bonds on neighboring Si atoms (Si–↑ H–Si–↑), or one dangling bond and a three center bond (Si–H–Si–↑). The main argument against this kind of process has been that the metastable spins are close together and will interact. This was expected to change the lineshape of the dangling bond resonance, in contradiction to the experimental results.[4] However, it is known that the origin of the ESR lineshape in a–Si:H is a distribution of g–values caused by potential fluctuations and orientational disorder.[12] Therefore, a change in lineshape for the metastable spins is not to be expected if the interaction is restricted to isolated pairs of dangling bonds originating from the same weak bond. These pairs could be too far apart to interact with each other, thus preserving the inhomogeneous character of the dangling bond resonance.

ACKNOWLEDGEMENTS

This work was supported by SERI under contract number XB–3–03112–1.

REFERENCES

1. D.L. Staebler, C.R. Wronski, Appl. Phys. Lett. 31, 292 (1977)
2. D.L. Staebler, R.S. Crandall, R. Williams, Appl. Phys. Lett. 39, 733 (1981)
3. I. Hirabayashi, K. Morigaki, S. Nitta, Jap. J. Appl. Phys. 19, L357 (1980)
4. H. Dersch, J. Stuke, J. Beichler, Appl. Phys. Lett. 38, 456 (1981)
5. D.L. Staebler, C.R. Wronski, J. Appl. Phys. 51, 3262 (1980)
6. C.C. Tsai, M. Stutzmann, W.B. Jackson, this conference
7. C.C. Tsai, J.C. Knights, R.A. Lujan, B. Wacker, B.L. Stafford, M.J. Thompson, J. Non–Cryst. Sol. 59&60, 731 (1983)
8. e.g.: J. Stuke, Proc. Conf. MRS Europe, Strasbourg (1984)
9. R.A. Street, Phil. Mag. B49, L15 (1984)
10. H. Dersch, PhD Thesis, Marburg (1983)
11. H.S. Ullal, D.L. Morel, D.R. Willett, D. Kanani, P.C. Taylor, C. Lee, 17th IEEE Photovoltaic Specialists Conf., Kissimmee (1984)
12. M. Stutzmann, J. Stuke, Sol. State Comm. 47, 635 (1983)

LIGHT-INDUCED CREATION OF DEFECTS AND RELATED PHENOMENA IN SILICON-BASED AMORPHOUS SEMICONDUCTORS

A. Morimoto, H. Yokomichi, T. Atoji,
M. Kumeda, I. Watanabe and T. Shimizu
Dept. of Electronics, Kanazawa University, Kanazawa 920, Japan

ABSTRACT

Light-induced creation of defects and their annealing process were studied by ESR, photoconductivity and photoluminescence measurements for a-$Si_{1-x}N_x$:H and a-$Si_{1-x}C_x$:H films besides a-Si:H films with various spin densities and H contents. The results show that the degradation of the photoconductivity and the photoluminescence is mainly attributed to creation of dangling bonds due to bond breaking.

INTRODUCTION

A number of studies on the Staebler-Wronski (S-W) effect in hydrogenated amorphous Si (a-Si:H) have been carried out. Various models have been proposed for the degradation due to a strong illumination , i.e., the decrease in the photoconductivity , the fatigue of the photoluminescence, and the increase in the ESR spin density. The origin of the S-W effect, however, is still controversial. The S-W effect for a-$Si_{1-x}N_x$:H or a-$Si_{1-x}C_x$:H films is scarcely investigated until now. In this paper, we present the results of our investigation on the changes of the ESR spin density, the photoluminescence and the photoconductivity due to white light illumination at room temperature for a-$Si_{1-x}N_x$:H and a-$Si_{1-x}C_x$:H films. The relation among these changes is investigated.

EXPERIMENTAL

a-Si:H films with various H contents and a-$Si_{1-x}N_x$:H films were prepared both by glow discharge decomposition (GD) and reactive sputtering (SP). a-$Si_{1-x}C_x$:H and a-$Si_{1-x}Ge_x$:H films were prepared only by GD. ESR and photoconductivity measurements were carried out both at liquid N_2 temperature (LNT) and room temperature (RT). Photoluminescence measurement was carried out only at LNT within a few minutes. He-Ne laser light (632.8 nm) and monochromatic light derived from a Xe lamp through interference filters with 0.1 mW/cm^2 were used for excitation of photocurrent. The He-Ne laser light and Ar^+ laser light (488 and 514.5 nm) with a few hundred mW/cm^2 were used for excitation of the photoluminescence. Strong illumination was carried out for more than 1 h with more than 120 mW/cm^2 by using white light from the Xe lamp through an IR cut filter.

RESULTS

I. a-$Si_{1-x}N_x$:H films

0094-243X/84/1200221-08 $3.00

In order to clarify the nature of defects in a-$Si_{1-x}N_x$:H films, we carried out the light-induced ESR at LNT. Figure 1 shows how the spin density N_s changes with white light illumination at LNT for SP a-$Si_{0.68}N_{0.32}$:H film[1]. I_0 in the figure is about 0.1 mW/cm^2. The g-value and the linewidth do not change with illumination. Illumination at LNT shows a remarkable increase in the light-induced ESR signal intensity. After cessation of illumination, a considerable part of the increased N_s remains at LNT, but disappears at RT. In the case of strong illumination, a part of N_s increased at LNT remains even at RT. Similar results are obtained for a-$Si_{1-x}N_x$:H films with various x.

All the present results can be explained as follows. A large number of Si dangling bonds with negative electron correlation energy U (D_n) exist in a-$Si_{1-x}N_x$:H films ($x \neq 0$) besides Si dangling bonds with positive U (D_p), and the sign of U is changed into a positive one due to illumination at LNT for some of D_n. The light-induced ESR signal arises both from D_p^0 centers created by the change of the sign of U and from D_n^0 centers created by trapping photo-excited electrons and holes. D_n^0 centers return to the ESR inactive $D_n^+ + D_n^-$ centers rapidly at LNT after cessation of illumination, whereas D_p^0 centers created by the former mechanism remain at LNT and return to $D_n^+ + D_n^-$ centers with raising the temperature to RT. In a-Si:H films, such phenomena do not appear, because no defects with negative U exist.

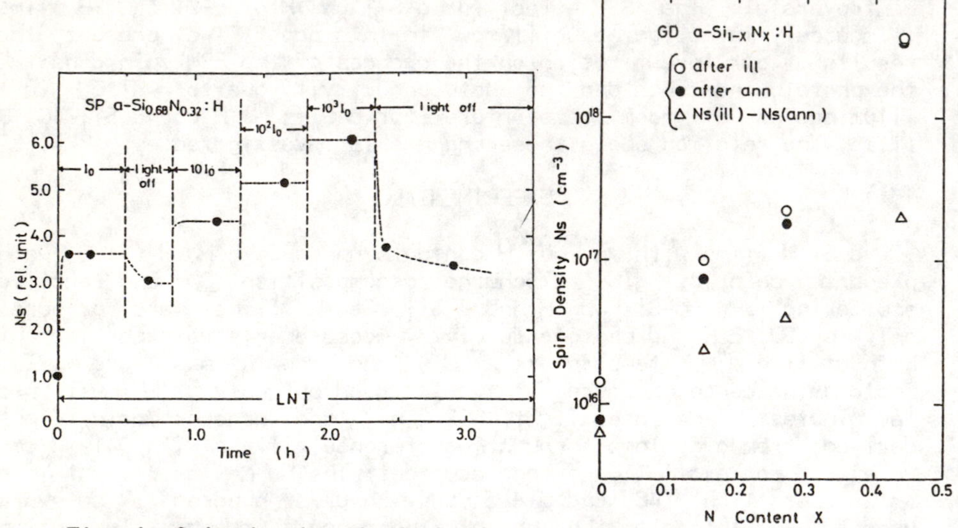

Fig. 1 Spin density N_s derived from light-induced ESR signal at LNT with various illumination intensities and N_s derived from ESR at LNT after cessation of illumination for SP a-$Si_{0.68}N_{0.32}$:H film.

Fig. 2 Spin density N_s for annealed and strongly illuminated GD a-$Si_{1-x}N_x$:H films and the difference between them as a function of x.

In a-Si:H films, it is known that a prolonged and strong illumination creates ESR centers originating from dangling bonds[2-4]. In a-$Si_{1-x}N_x$:H films, the photo-created ESR centers remain at RT as described above. Figure 2 shows N_s in the dark for a-$Si_{1-x}N_x$:H films annealed at 200 ° C and strongly illuminated at RT as a function of N content x. The density of photo-created ESR centers ΔN_s stable at RT increases with an increase in x, whereas $\Delta N_s/N_s$ decreases with x because of a large increase in N_s before illumination with x as shown in Fig. 2.

The relation between the increase in N_s in the dark and the decrease in $\eta\mu\tau$ due to strong illumination is shown in Fig. 3. Data for various a-$Si_{1-x}N_x$:H films without illumination are also shown together. Changes due to the illumination are indicated by the arrows. The relation between $\eta\mu\tau$ and N_s for unilluminated films is represented by a formula $\eta\mu\tau \propto N_s^{-\gamma}$ with $\gamma \fallingdotseq 2$. Changes due to strong illumination are also represented by a formula $\eta\mu\tau(ill)/\eta\mu\tau(ann) = \{N_s(ill)/N_s(ann)\}^{-\gamma}$ with $\gamma \fallingdotseq 2$. Since the reciprocal lifetime $1/\tau$ is expected to be proportional to N_s, $\gamma \fallingdotseq 2$ suggests that the decrease in the mobility μ with an increase in N_s. In a-$Si_{1-x}N_x$:H films without strong illumination, the decrease in $\eta\mu\tau$ with an increase in x can be attributed to the increase in the density of dangling bonds[5] and in the density of tail states. The similarity of γ value between those with and without strong illumination suggests that the decrease in $\eta\mu\tau$ with strong illumination can be attributed to the increase in the density of dangling bonds accompanied by the increase in the density of tail states.

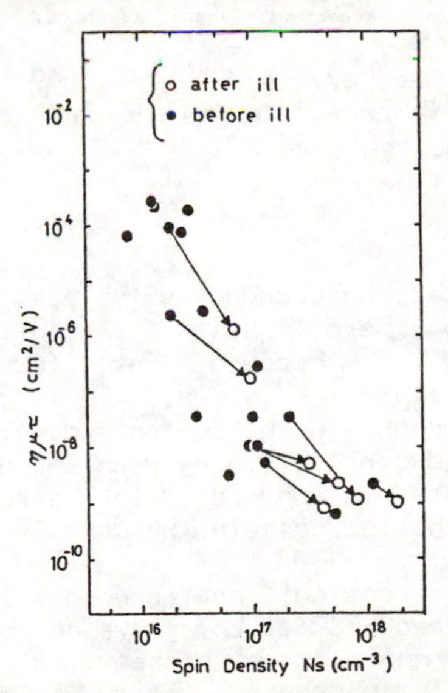

Fig. 3 Relation between spin density N_s and $\eta\mu\tau$ for a-$Si_{1-x}N_x$:H films before illumination and after strong illumination at RT.

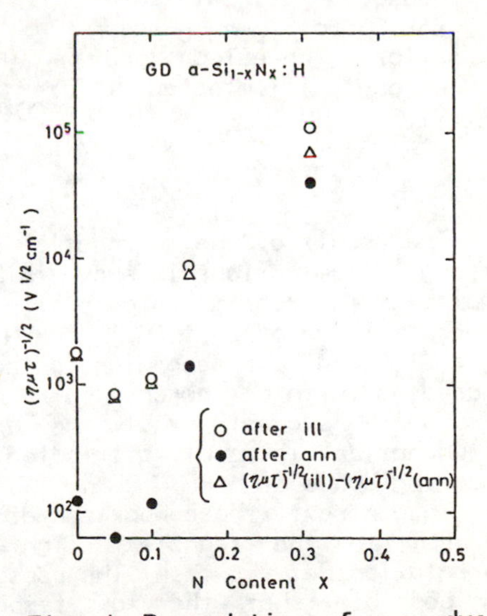

Fig. 4 Degradation of $\eta\mu\tau$ due to strong illumination for GD a-$Si_{1-x}N_x$:H films. $(\eta\mu\tau)^{-1/2}$ is expected to be approximately proportional to N_s from Fig. 3.

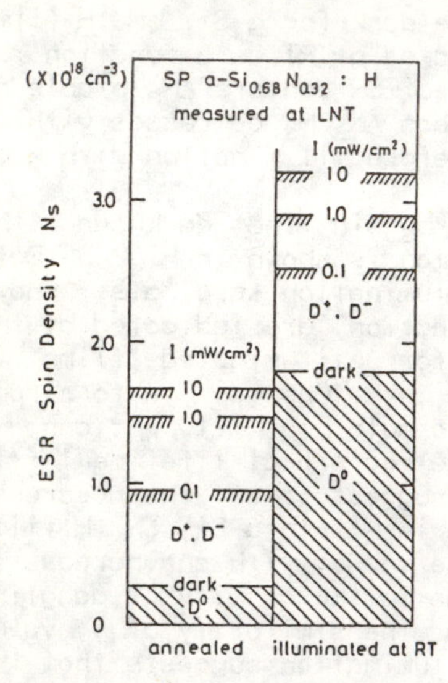

Fig. 5 Spin density derived from ESR and light-induced ESR measured at LNT both for SP a-$Si_{0.68}N_{0.32}$:H films annealed and strongly illuminated at RT.

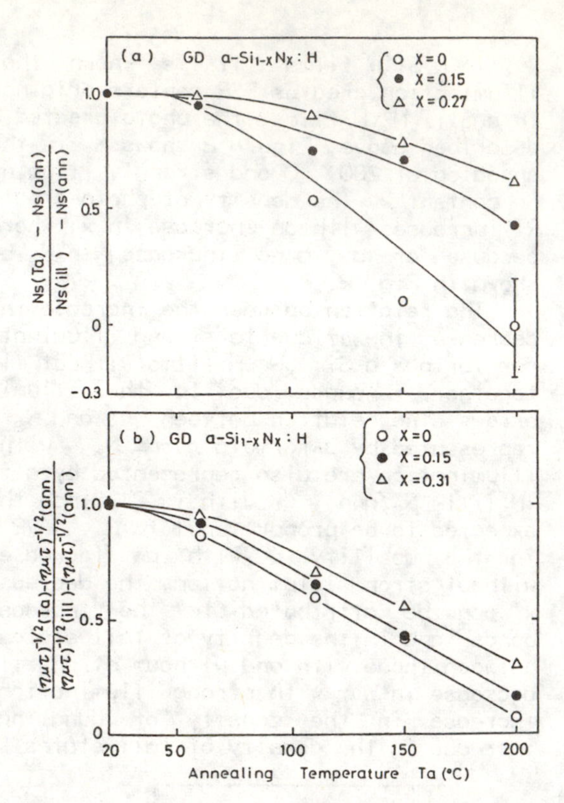

Fig. 6 Recovery of N_s (a) and $\eta\mu\tau$ (b) due to annealing for 10 min for GD a-$Si_{1-x}N_x$:H films.

The result of the degradation of the photoconductivity due to strong illumination is shown in Fig. 4. Here, $(\eta\mu\tau)^{-1/2}$ is plotted against x in order to compare with Fig. 2, because $(\eta\mu\tau)^{-1/2}$ is approximately proportional to N_s. $\Delta(\eta\mu\tau)^{-1/2} = (\eta\mu\tau)^{-1/2}(ill) - (\eta\mu\tau)^{-1/2}(ann)$ corresponding to the increase in the defect density increases with an increase in x. A good correlation between ΔN_s and $\Delta(\eta\mu\tau)^{-1/2}$ suggests that the decrease in $\eta\mu\tau$ due to strong illumination is mainly attributed to the increase in the density of dangling bonds.

In contrast, the comparison of the transient photoconductivity at LNT to the transient light-induced ESR at LNT shows no good correlation between them. The reason is not clear at present, but it is possible that the low temperature photoconductivity is not largely affected by the density of recombination centers[6].

In order to clarify what kind of dangling bonds are created by strong illumination, we carried out light-induced ESR both for the films annealed and strongly illuminated at RT. The result is shown in Fig. 5. For SP a-$Si_{0.68}N_{0.32}$:H film annealed at 200 ° C for 1 h, the density of D_p^0 is 2.7 × 10^{17} cm^{-3} and that of D_n^+ and D_n^- is 1.4 × 10^{18} cm^{-3}. Strong illumination causes a remarkable increase only in D_p density, though it does not cause any increase in D_n density.

An experiment of the photoluminescence fatigue was also carried out. The relative decrement of the photoluminescence intensity due to strong illumination, $\Delta I_{PL}/I_{PL} = \{I_{PL}(ann)-I_{PL}(ill)\}/I_{PL}(ann)$, increases with x. It might appear that the increase in $\Delta I_{PL}/I_{PL}$ with an increase in x is inconsistent with the decrease in $\Delta N_s/N_s$ with an increase in x. However, we can show that the changes of $\Delta I_{PL}/I_{PL}$ and $\Delta N_s/N_s$ with x are qualitatively consistent on the basis of the relation between I_{PL} and N_s, $I_{PL} = I \exp(-4\pi R_c^3 N_s/3)$, proposed by Street et al.[7], because N_s increases largely with x. Here, R_c is the critical distance between defects and photo-excited carriers at which radiative and non-radiative recombination rates are equal. The transition from $D^+ + D^-$ to $2D^0$ is expected to occur during the photoluminescence measurement. It is, however, not clear at present how such a transition affects I_{PL}.

It is important to investigate a recovery process as well as the degradation process of $\eta\mu\tau$ and N_s in order to clarify the origin of the degradation. Figure 6 shows the recovery processes of N_s (a) and

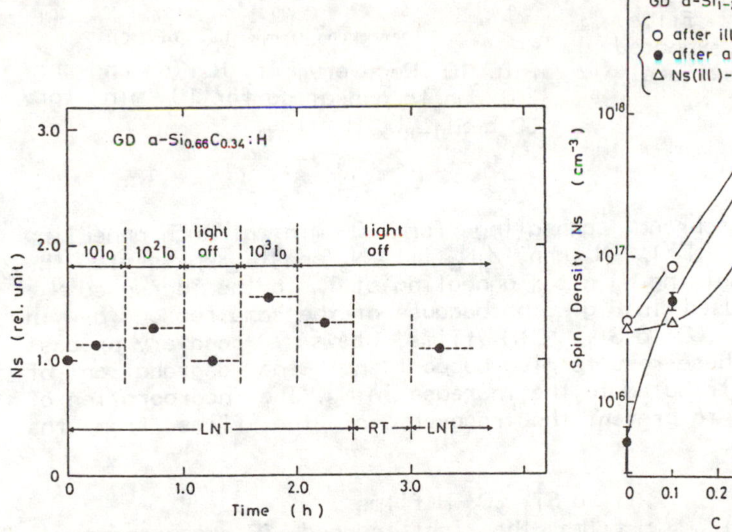

Fig. 7 Spin density N_s derived from light-induced ESR signal at LNT with various illumination intensities and N_s derived from ESR at LNT after cessation of illumination for SP a-$Si_{0.66}C_{0.34}$:H film.

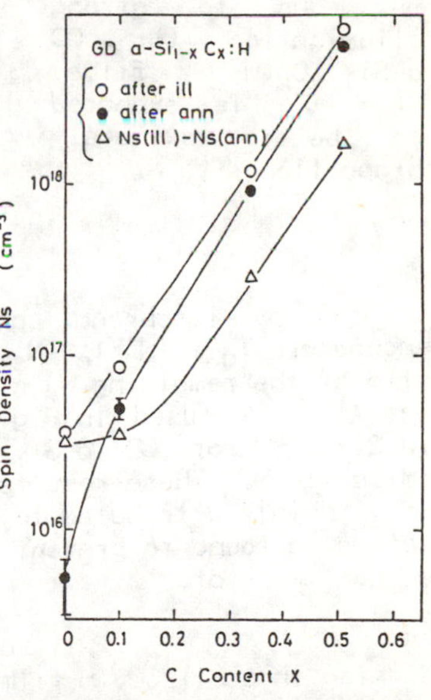

Fig. 8 Spin density N_s for annealed and strongly illuminated GD a-$Si_{1-x}C_x$:H films and the difference between them as a function of x.

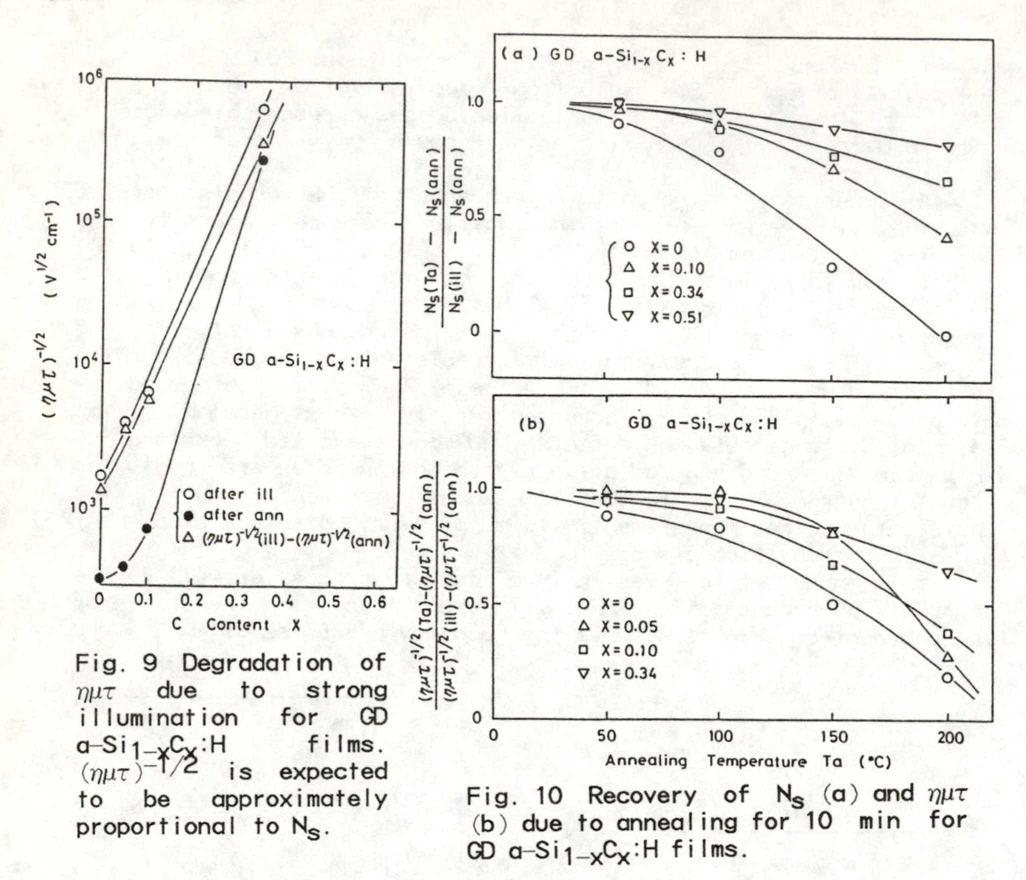

Fig. 9 Degradation of $\eta\mu\tau$ due to strong illumination for GD a-$Si_{1-x}C_x$:H films. $(\eta\mu\tau)^{-1/2}$ is expected to be approximately proportional to N_s.

Fig. 10 Recovery of N_s (a) and $\eta\mu\tau$ (b) due to annealing for 10 min for GD a-$Si_{1-x}C_x$:H films.

$\eta\mu\tau$ (b) by isochronal annealing for 10 min at each annealing temperature T_a. $\{N_s(T_a)-N_s(ann)\}/\{N_s(ill)-N_s(ann)\}$ represents the ratio of the remaining N_s after annealing at T_a to the increased N_s. $(\eta\mu\tau)^{-1/2}$ is used in Fig. 6(b) because of the same reason as in Fig.3. $\eta\mu\tau$ for GD a-$Si_{1-x}N_x$:H films shows a recovery process similar to N_s. These results also support that the degradation of $\eta\mu\tau$ is mainly attributed to the increase in N_s. The incorporation of N atoms is found to prevent the recovery of the films from the degradation state.

II. a-$Si_{1-x}C_x$:H films

Also in a-$Si_{1-x}C_x$:H films, the light-induced ESR measurement was carried out at LNT. Figure 7 shows how the spin density changes with a white light illumination for GD a-$Si_{0.66}C_{0.34}$:H film. I_0 in the figure is about 0.1 mW/cm^2. Though the g-value and the linewidth do not largely change with illumination, a slight asymmetry appears with strong illumination. This shows that the light-induced ESR signal possibly contains the band tail ESR signals as well as the ESR signal of D_n^0 and D_p^0 centers. Therefore the density of total

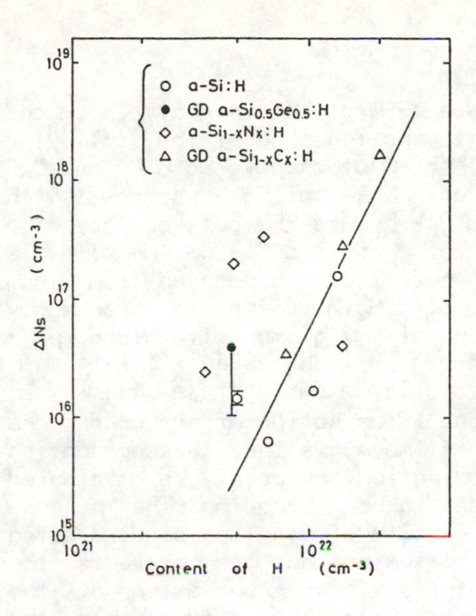

Fig. 11 Increment of spin density ΔN_S as a function of H content N_H for various films.

dangling bonds, D_p and D_n, may be less than the N_s values shown in this figure.

Figure 8 shows N_S in the dark for annealed and strongly illuminated a-$Si_{1-x}C_x$:H films as a function of x. ΔN_S increases with x, whereas $\Delta N_S/N_S$ decreases with x, as in the case of a-$Si_{1-x}N_x$:H films.

The result of the degradation of the photoconductivity due to strong illumination is shown in Fig. 9. Here, $(\eta\mu\tau)^{-1/2}$ is plotted as in the case of a-$Si_{1-x}N_x$:H films. $\Delta(\eta\mu\tau)^{-1/2} = (\eta\mu\tau)^{-1/2}(\mathrm{ill}) - (\eta\mu\tau)^{-1/2}(\mathrm{ann})$ corresponding to the increment of the defect density increases with an increase in x. The result of the photoluminescence measurement also shows a close relation between I_{PL} and N_S as in the case of a-$Si_{1-x}N_x$:H films. The good correlations between ΔN_S and $\Delta(\eta\mu\tau)^{-1/2}$ and between ΔN_S and I_{PL} suggest that the degradation due to strong illumination is mainly attributed to the increase in the density of dangling bonds.

The results of the experiment on the recovery process from the degradation state are shown in Figs. 10(a) and 10(b). The results are similar to those shown in Fig. 6 for a-$Si_{1-x}N_x$:H films.

III. EFFECT OF HYDROGEN

In order to reveal what causes the creation of dangling bonds, we investigated a relation between ΔN_S due to strong illumination at RT and the H content of the films N_H. The result is shown in Fig. 11. ΔN_S and N_H for a-Si:H films, a-$Si_{1-x}Ge_x$:H and a-$Si_{1-x}C_x$:H films have a good correlation, indicating that H atoms in the films play an important role in the degradation due to strong illumination. In the case of a-$Si_{1-x}N_x$:H films, N atoms , in addition to H atoms, play an important role in the degradation. This is probably attributed to the low coordination number of N atoms.

DISCUSSION AND CONCLUSION

The results of the light-induced ESR at LNT in $a\text{-}Si_{1-x}N_x$:H and $a\text{-}Si_{1-x}C_x$:H films without strong illumination suggest the following idea. The binary alloy films with N or C atoms have both D_p and D_n. Without illumination, only $D_p{}^0$ are ESR active, whereas with illumination at LNT, both $D_p{}^0$ and $D_n{}^0$ become ESR active. The ratio of the density of D_n to the density of D_p for $a\text{-}Si_{1-x}N_x$:H film is much larger than that for $a\text{-}Si_{1-x}C_x$:H film as can be seen from Figs. 1 and 7. The presence of D_n in $a\text{-}Si_{1-x}N_x$:H films is presumably attributed to 3-fold coordination of N atoms with lone pair electrons. An origin of the presence of D_n in $a\text{-}Si_{1-x}C_x$:H films is not clear, though it might be due to a large N_H in these films.

The investigation of the strong illumination effect on $\eta\mu\tau$ and I_{PL} for $a\text{-}Si_{1-x}N_x$:H and $a\text{-}Si_{1-x}C_x$:H films shows that the degradation of $\eta\mu\tau$ and I_{PL} due to strong illumination can mainly be attributed to the creation of dangling bonds due to bond breaking. The increase of N or C content x increases the density of dangling bonds created by bond breaking due to strong illumination and prevents the recovery from the degradation state. This suppression of the recovery has a good correspondence with the result that the incorporation of N or C atoms makes the film heat-resistent[8,9].

In conclusion, ESR with various illumination intesities at LNT in the present work can reasonably be explained by both the change of $D^+ + D^-$ centers into $2D^0$ centers and bond breaking. The degradation of the photoconductivity at RT and the photoluminescence can mainly be attributed to the creation of dangling bonds due to bond breaking in contrast with the Adler's proposal[10] that the S-W effect is brought about by the change of D^+ and D^- into $2D^0$.

REFERENCES

1. M. Kumeda, H. Yokomichi and T. Shimizu, Jpn. J. Appl. Phys., 23, No. 7, Part 2 (1984) (in press).
2. H. Dersch, J. Stuke and J. Beichler, Appl. Phys. Lett. 38, 456 (1981).
3. I. Hirabayashi, K. Morigaki and S. Nitta, Jpn. J. Appl. Phys., 19, L357 (1980).
4. P. C. Taylor and W. D. Ohlsen, Solar Cells, 9, 113 (1983).
5. T. Shimizu, S. Oozora, A. Morimoto, M. Kumeda and N. Ishii, Solar Energy Materials, 8, 311 (1982).
6. M. Hoheisel, R. Carius and W. Fuhs, J. Non-Cryst. Solids, 63, 313 (1984).
7. R. A. Street, J. C. Knight and D. K. Biegelsen, Phys. Rev., B18, 1880 (1978).
8. A. Morimoto, T. Miura, M. Kumeda and T. Shimizu, J. Appl. Phys., 53, 7299 (1982).
9. A. Morimoto, T. Kataoka, M. Kumeda and T. Shimizu, Phil. Mag. B (in press).
10. D. Adler, Solar Cells, 9, 133 (1983).

SEARCH FOR REVERSIBLE LIGHT-INDUCED CHANGES IN THE Si-H ABSORPTION BANDS OF a-Si:H

H. Fritzsche, J. Kakalios* and D. Bernstein
Department of Physics and James Franck Institute
The University of Chicago, Chicago IL 60637

ABSTRACT

We tried to verify the reversible light-induced changes in the strengths of the Si-H vibrational absorption bands in hydrogenated amorphous silicon (a-Si:H) that were reported by Zhang et al. and by Carlson et al. We did not find any such changes in a-Si:H samples prepared under several different conditions. Our detection sensitivity was more than an order of magnitude higher than that of the previous authors.

INTRODUCTION

One of the most challenging problems in the field of amorphous semiconductors is finding the origin of the reversible light-induced changes in the properties of hydrogenated amorphous silicon (a-Si:H). This effect, named after its discoverers Staebler and Wronski[1] is presently believed to be associated with local changes that affect or produce about 1-3X10^{17} cm^{-3} dangling bond gap states. As far as the overall structure is concerned these changes are quite subtle since they affect only 1 out of 10^5 silicon atoms. However, since the density of gap states in this material is very low, even 10^{17} cm^{-3} gap states have a strong effect on the electronic properties and the recombination kinetics of photo-excited carriers.

It came as a great surprise when reversible light-induced changes were reported recently[2] that involve 1 part in 10 instead of 1 part in 10^5 silicon atoms. Reversible light-induced increases in absorption coefficient of about 35% were reported[2] to occur in three infrared absorption modes associated with the Si-H bond at wavenumbers near $\bar{\nu}$=2000 cm^{-1}, 850 cm^{-1} and 600 cm^{-1}. The samples for which these effects were reported were prepared by sputtering in the presence of hydrogen. This report by Zhang et al.[2] was preceded by an earlier work of Carlson et al.[3,4] who observed reversible changes in absorbance of the Si-H vibrational modes of a few percent in glow discharge deposited a-Si:H.

It is important to follow up these leads because a relatively large change in bonding is not necessarily inconsistent with a relatively small change in localized gap states since saturated covalent bonds do not yield states in the gap. If these larger light-induced changes in bonding indeed occur whenever the Staebler-Wronski effect is observed then a relation of the two effects is indicated. The photo-induced appearance of 10^{17} cm^{-3} Staebler-Wronski defects is then

* Bell Labs. Ph.D. Scholar

0094-243X/84/1200229-05 $3.00

only a small by-product of a much larger bond rearrangement.

This paper reports a detailed search for a reversible photo-induced change in the Si-H infrared absorption bands. Samples were prepared by plasma deposition under normal as well as under various extreme conditions. We also studied a sputtered a-Si:H sample from the Harvard group. Our detection sensitivity ranged between 0.1 and 1 percent in absorbance. This is a factor 20-200 more than required to observe the effects reported by Zhang et al.[2]. In none of our samples did we detect a reversible change in the Si-H vibrational absorption bands.

EXPERIMENTAL DETAILS

We used a Perkin-Elmer 180 double beam spectrophotometer. A bare crystalline Si substrate was placed in the reference beam and an identical Si substrate carrying the a-Si:H film in the sample beam. Throughout the experiment the positions of the sample and of the referance remained unchanged. The sample was annealed at 460°K for 1-2 hours in situ in a flow of dry N_2 and in darkness. Without changing its position it was exposed to about 75 mW/cm^2 heat-filtered white light for one hour. During infrared measurements the crystalline Si substrate was facing the light source of the spectrometer in order to protect the a-Si:H film from light having $h\nu > 1.1$ eV. The infrared spectrum of each sample was measured after mounting, after a first anneal, after light-soaking, and after a second anneal. A measurement included a full scan of the wavenumber range $\bar{\nu}$=4000 cm^{-1} to 350 cm^{-1} and five slow and 5x expanded scans of the Si-H absorption peaks near $\bar{\nu}$=2000 cm^{-1} and 630 cm^{-1}. This sequence of measurements takes about 10 hours per sample. For this reason great care was taken to learn about long term drifts of the spectrometer and about ways to correct for them. One potential source of error was found to be a change in the relative phase of the choppers in the spectrometer over a period of hours. This could largely be corrected by re-phasing the choppers before scans. Baseline drifts probably arise from a change in the local surface brightness of the glow-bar light source. This causes a different change in the intensity profile of the sample and the reference beam. By placing the 0.8 cm diameter openings for the sample and the reference Si at a maximum of the intensity profile, the baseline drift was reduced. Despite these precautions we sometimes observed a change in transmittance of about 1 percent after annealing and of about 0.2 percent after light-soaking. The different magnitudes of these two numbers is probably due to the fact that annealing with heating and cooling requires about 4x as much time as light exposure. Taking into account the signs of the observed changes we conclude that there is no reversible light-induced change in absorbance A larger than the value quoted for each sample in Table I.

SAMPLES AND RESULTS

All samples were prepared by r.f. plasma deposition except for one that was obtained from the Harvard group and prepared by sputtering in a partial pressure of hydrogen in argon. Table I lists some

Table I Sample characteristics

		T_s °K	dilutant	Rate Å/s	d μm	C_H %	H-bond	ΔA/A %
1	306A	525	--------	1.5	1.35	11	SiH	<0.1
2	U015A	525	Ar	1.25	0.9	21	SiH	<1.0
3	U015C	500	Ar	2.85	2.05	23	SiH	<0.8
4	270A	575	He	2.5	1.8	14	SiH	<1.0
5	270C	510	He	3.1	2.2	16	SiH	<0.1
6	271A	500	He	2.3	2.5	16	SiH	<1.0
7	300A	375	He	2.7	2.0	16	SiH_2,SiH	<0.1
8	304A	300	He	3.0	1.15	33	$(SiH_2)_n$	<0.1
9[a]	D025C	500	Ar	2.1	0.35	15	SiH	<0.1
10[b]	U005C	540	Ar	3.3	2.06	6.5	SiH_2	<0.9
11[b]	U005A	540	Ar	1.8	1.1	17	SiH_2	<0.9
12[b]	302C	525	He	3.6	8.3	15	SiH	<0.1
13[c]	S-212	475	Ar/H_2				SiH	<0.5

a) p-type sample prepared with 100 ppm B_2H_6

b) prepared in presence of an air leak

c) sputtered sample of Harvard University

of the preparation conditions and sample properties. The deposition temperatures T_s cover a range from 575°K to 300°K. Sample No.1 was prepared with pure SiH_4, for the others the silane was diluted with Ar or He at a ratio of 1:9. The deposition rate usually is larger for cathode than for anode samples. These are identified by the letters C and A, respectively, in the sample code name. The anode plate is grounded, whereas the cathode plate is connected to the r.f. power supply. The sample thickness d was determined from the thin film interference pattern at photon energies below the absorption edge. The hydrogen content C_H (at.%) was calculated from the integrated absorption strength of the Si-H stretching mode using the conversion factor 1.3 eV cm^{-1}/at.%[5].

Table I lists also the bonding configuration of hydrogen. A bond stretching mode at 2000 cm^{-1} without any bond bending mode between 800 and 900 cm^{-1} is characteristic of monohydride SiH. Bond stretching at 2090 cm^{-1} and a bending mode at 880 cm^{-1} signifies the presence of dihydride SiH_2. Polysilane groups $(SiH_2)_n$ are identified by bending modes both at 845 cm^{-1} and at 890 cm^{-1} and a stretching mode at 2090 cm^{-1}. These polysilane modes dominate the infrared absorption spectrum of sample No.8 that was deposited at room temperature. Sample No.9 was doped p-type with 100 ppm B_2H_6 in the silane. Samples No.10-12 were prepared with various amounts of air added to the plasma gas. According to Carlson[6] this procedure enhances photodegradation of a-Si:H solar cells as well as the photo-induced reversible changes of the Si-H absorption bands. Sample No.13 is the sputtered a-Si:H film obtained from the Harvard group. The last column of Table I lists for each sample the upper limits for an undetected relative change in absorbance of the stretching (∿2000 cm^{-1}) and rocking modes (∿630 cm^{-1}). The smallest values, $\Delta A/A<0.1$ percent, were obtained after gaining better control of problems such as the phase and baseline drifts mentioned above.

All our samples exhibited a 10-25% decrease of the 78°K photoluminescence intensity at $h\nu=1.3$ eV after exposure to $3X10^{21}$ cm^{-2} photons of energy 2.41 eV. This results from a light-induced increase in dangling bond density and shows that our samples are subject to the Staebler-Wronski effect.

DISCUSSION

We searched for a reversible light-induced change in the absorption of the infrared Si-H vibration modes in a-Si:H films prepared in several different ways. We did not find any reversible change in absorbance with a detection sensitivity that varied between 0.1 and 1 percent.

In contrast, Zhang et al.[2] found a reversible light-induced increase of 27-34% in the absorption coefficients which corresponds to 12-15% in absorbance. Here we assumed that the thickness of their samples was about 2μm. Zhang et al. prepared their samples by argon sputtering at a total pressure of 0.2 Torr. About 30% of plasma gas was hydrogen. From the absorption at $\bar{\nu}$=2000 cm^{-1} and using the conversion factor 1.3 eV cm^{-1}/at.% we estimated that their samples contained 25 at.% infrared active and bonded hydrogen after annealing. After light exposure this number increased by the 27-34% mentioned above. The dark conductivity of their samples decreased with light exposure in a way that is typical for the Staebler-Wronski effect. However the spin density of their samples increased to $3X10^{18}$ cm^{-3} which is about a factor ten more than normally observed in glow-discharge deposited films.

Carlson et al. reported changes in absorbance of a few percent in a-Si:H samples prepared by glow-discharge and in the presence of an air leak. But surprisingly, two of their measurements exhibit effects of opposite sign. Carlson et al.[3] showed an example of a <u>decrease</u> of the absorbance at 2000 cm^{-1} with light exposure, whereas an example of an <u>increase</u> is shown in a later article by Carlson[4]. The porosity of some of these samples and the exposure to moisture and air during annealing and illumination might have caused changes at the internal void surfaces that produced the effects observed by these authors. We kept our samples in an atmosphere of dry nitrogen except for one that was accidentally annealed in air. None of our samples showed a reversible change in absorbance.

We conclude from the present work that light-induced reversible changes in the absorption coefficient of the Si-H infrared absorption bands exceeding 1 percent or even 0.1 percent are not necessarily associated with the Staebler-Wronski effect. Further studies are needed to identify the preparation conditions and to characterize the film structure that have led to the large changes in the infrared absorption bands found recently in other laboratories.

ACKNOWLEDGEMENTS

We wish to thank Dr. D. Topor for preparing many of our samples and for measurements of the photoluminescence fatigue. We thank W. Paul for providing us with a sputtered a-Si:H film. This work was supported by the Materials Preparation Laboratory of the National Science Foundation at the University of Chicago under Grant No. 79-24007.

REFERENCES

1. D.L.Staebler and C.R.Wronski, J. Appl. Phys. <u>51</u> (1980) 3262.
2. P.-X. Zhang, C.-L. Tan, Q.R. Zhu, and S.-Q. Peng, J. Non-Cryst. Solids <u>59-60</u> (1983) 417.
3. D.E.Carlson, A.R.Moore, D.J.Szostak, B.Goldstein, R.W.Smith, P.J.Zunzucchi, and W.R.Frenchu, Solar Cells <u>9</u> (1983) 19.
4. D.E.Carlson, Solar Energy Mat. <u>8</u> (1982) 129.
5. H.Fritzsche, Solar Energy Mat. <u>3</u> (1980) 447.
6. D.E.Carlson, J.Vac. Sci. Technol. <u>20</u> (1982) 290.

THE EFFECTS OF LIGHT SOAKING ON a-Si:H FILMS CONTAINING IMPURITIES

D.E. Carlson, A. Catalano, R.V. D'Aiello
C.R. Dickson and R.S. Oswald
Solarex Thin Film Division, Newtown, PA 18940

ABSTRACT

A series of amorphous silicon films were grown under identical conditions in a dc glow discharge with different impurity gases added to the silane discharge atmosphere. The constant surface photovoltage technique was used to measure both the diffusion length and the space-charge width as a function of the time exposed to 1 sun illumination. In most cases large changes were observed in the first 48 hours of light soaking with much smaller changes observed in the next 452 hours. After 500 hours of light soaking, diffusion lengths >0.24μm were observed only in films made with pure silane or with impurities such as CH_4, SiH_2Cl_2, or SiF_4 present. Films made in the presence of impurity gases such as N_2, C_2H_4, or $(SiH_2)_2O$ exhibited diffusion lengths <0.15μm after 500 hours of 1 sun illumination.

INTRODUCTION

Recent work has shown that both oxygen [1] and carbon [2] impurities can adversely affect the stability of amorphous silicon solar cells. In the case of oxygen contamination, the characteristic activation energy for annealing is about 1.0eV [3]; for carbon contamination, the activation energy for annealing is about 0.4eV [4]. In both studies, the contamination levels were greater than 10^{+20} cm^{-3}.

Light-induced degradation has been observed even in relatively pure a-Si:H films [5] and may be associated with microstructural imperfections such as voids or polymer chains. The present study shows that the addition of certain impurities increases the light-induced degradation.

EXPERIMENTAL PROCEDURE

The a-Si:H films were grown in a dc glow discharge in silane at a substrate temperature of 250°C. The silane was CCD grade obtained from Airco (Riverton, NJ). The samples were prepared by first depositing a thin layer (∿300Å) of phosphorus-doped a-Si:H (∿3 at.% P) on stainless steel. The deposition chamber was then purged for 1 hour with pure silane before depositing 1 to 2μm of bulk a-Si:H. The impurity gases (see Table I) were added to the discharge atmosphere only during the deposition of the bulk a-Si:H layer.

TABLE I

Impurity Content of Amorphous Silicon Films

Discharge Atmosphere	Impurity Concentrations (cm^{-3})				
	$O(X10^{19})$	$C(X10^{19})$	$N(X10^{19})$	$Cl(X10^{17})$	$F(X10^{18})$
SiH_4 (control)	1.7	2.0	0.03	1.2	----
0.1% SiF_4 in SiH_4	2.3	1.6	0.07	0.07	1.6
2% CH_4 in SiH_4	5.2	25.0	----	0.7	----
1% CF_4 in SiH_4	0.56	14.0	0.15	0.84	140
1% CO in SiH_4	8.7	6.9	0.06	0.36	----
0.07% SiH_2Cl_2 in SiH_4	0.3	0.3	0.15	17.0	----
~ 0.2% $(SiH_3)_2O$ in SiH_4	15.0	0.8	1.0	1.6	----
1% C_2H_4 in SiH_4	0.25	42.0	0.26	0.08	0.3
2% N_2 in SiH_4	3.8	0.2	24.0	0.4	----

The diffusion length was measured by means of the surface photovoltage technique[6,7]. The diffusion length (L_1) was measured in the presence of a white light bias of 100mw cm^{-2} so as to simulate the operational conditions of a solar cell in sunlight. L_1 is actually a collection length that is the sum of the ambipolar diffusion length and the depletion width (W_1) under 1 sun illumination at $V = V_{oc}$. W_1 is often negligible especially after several hours of exposure to 1 sun illumination. In the present experiments, 1 sun illumination was simulated with a quartz-halogen lamp adjusted to 100mw cm^{-2} by means of a calibrated single crystal silicon solar cell.

We also measured the collection length without the white light bias; in this case the only light is that provided by the monochrometer ($\sim 10^{-6}$ suns). This collection length (W_0) is roughly equal to the sum of L_1 and the depletion width in the dark. We estimate that the values of L_1 and W_0 are accurate to within ± 15%[8].

The composition of the a-Si:H films was determined by means of secondary mass ion spectrometry (SIMS), and the impurity concentrations were estimated to be accurate to within ± 10%[9].

DIFFUSION LENGTH MEASUREMENTS

As shown in Table I, a variety of impurity gases were added to the silane discharge atmosphere to produce a series of films with known contaminants. In each case, the SIMS analysis showed a strong correlation between the major impurities in the a-Si:H film and those present in the impurity gas.

Table II shows the values of L_1 and W_0 for the same series of films. The values at t = 0 were taken immediately after deposition while the values at t = 500 hours were taken after exposure to 1 sun illumination for that period of time. In some cases, the initial values of L_1 are relatively large and include a contribution from a space-charge width (W_1). This is apparently the situation for the film grown in the presence of SiH_2Cl_2. Chlorine apparently creates deep acceptor levels[10] that tend to compensate the donor-like defects that are normally found in a-Si:H, and thus a large space-charge width results.

Table II

Effects of Impurities and Light Soaking on Diffusion Lengths

	t = 0		t = 500 hours	
Discharge Atmosphere	L_1 (μm)	W_o (μm)	L_1 (μm)	W_o (μm)
SiH_4 (control)	0.63	1.95	0.40	1.17
0.1% SiF_4 in SiH_4	0.38	1.10	0.24	0.71
2% CH_4 in SiH_4	0.32	0.90	0.25	0.78
1% CF_4 in SiH_4	0.30	0.81	0.15	0.91
1% CO in SiH_4	0.35	2	0.22	1.03
0.07% SiH_2Cl_2 in SiH_4	2	2	0.26	1.50
∼ 0.2% $(SiH_3)_2O$ in SiH_4	0.41	0.95	0.10	0.19
1% C_2H_4 in SiH_4	0.27	0.90	0.14	0.68
2% N_2 in SiH_4	0.37	0.96	0.12	0.17

Figure 1 shows the variation in L_1 and W_0 with exposure time to 1 sun illumination for two of the films listed in Tables I and II. The data in Fig. 1 show that a decrease in both L_1 and W_0 occurs even for the film grown in a pure silane discharge. As shown in Table I, the total concentration of contaminants in this film are $<4 \times 10^{+19}$ cm^{-3}. When ∿0.2 vol.% disiloxane was present in the silane discharge, the SIMS analysis showed that the oxygen concentration was increased to $1.5 \times 10^{+20}$ cm^{-3}, and as shown in Fig. 1, both L_1 and W_0 decreased to relatively small values after 48 hours of light soaking. Somewhat similar results were obtained when 2 vol.% N_2 was added to the silane discharge; in this case, the SIMS analysis showed that the nitrogen concentration increased to $2.4 \times 10^{+20}$ cm^{-3}. As is evident from Table II, additions of either 1 vol.% C_2H_4 or CF_4 also produced films with relatively small values of L_1 after 500 hours of light soaking; however, the values of W_0 were significantly larger than those observed for films contaminated with either oxygen or nitrogen. Thus, carbon impurities do not appear to seriously degrade W_0.

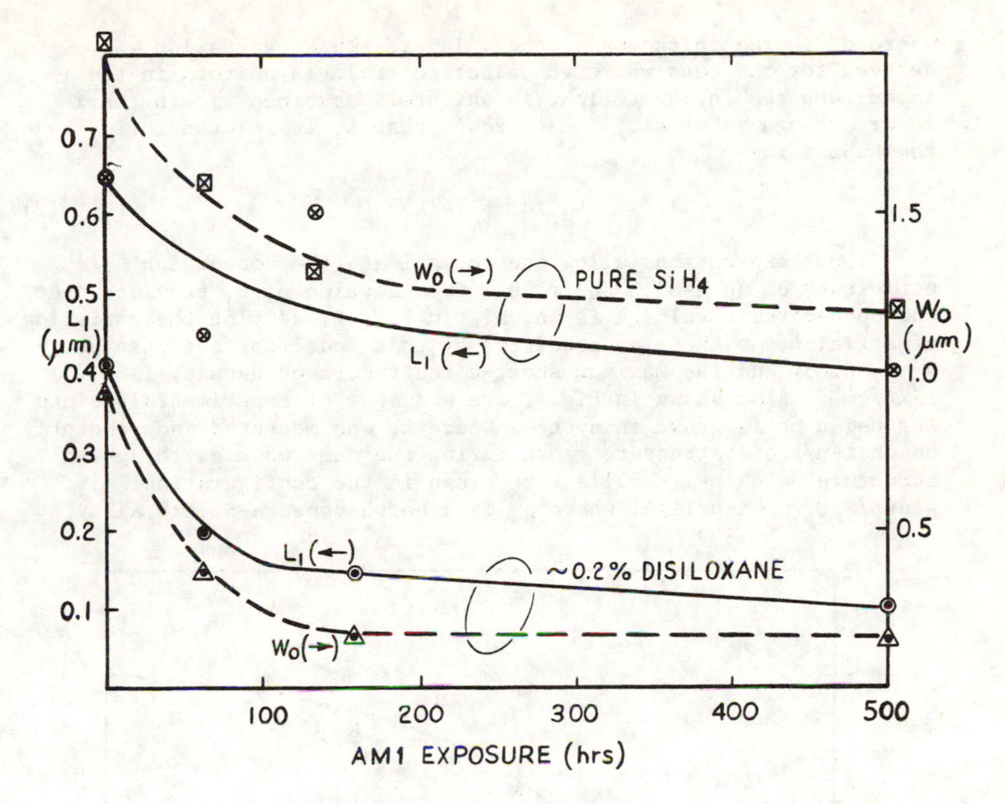

Figure 1
The variation of L_1 and W_0 with exposure time to 1 sun illumination for films grown in discharges of pure silane and silane + 0.2% disiloxane.

Another interesting observation is that oxygen in the form of CO is not as detrimental as in the form of disiloxane. As shown in Table I, oxygen is incorporated into the a-Si:H film more efficiently when present in the silane discharge as disiloxane than when present as CO gas. The CO molecule is strongly bonded and is much less likely to dissociate in the glow discharge. Since the concentrations of oxygen and carbon are comparable in the a-Si:H film in this case, many of the impurities may exist in the film as trapped CO molecules.

STABILITY OF SOLAR CELLS

The decrease in the diffusion length with light soaking can be correlated with a degradation in the conversion efficiency of a-Si:H solar cells. Faughnan and Crandall[11] have developed a simple model for p-i-n cells where the probability of collecting photo-generated carriers is related to the drift length (L_{dr}) by the expression

$$P = (L_{dr}/d_i)\ [1 - \exp(-d_i/L_{dr})] \qquad (1)$$

where d_i is the thickness of the i layer. This expression was derived for the case where the electric field is uniform in the i layer, and the incident light is uniformly absorbed within the i layer. Faughnan et al.[12] have found that L_1 is related to L_{dr} by the expression

$$L_{dr}/d_i = 74(L_1)^2 \quad (2).$$

These expressions allow one to estimate the conversion efficiency of an a-Si:H solar cell from a value of L_1 provided that the open-circuit voltage is known. In Fig. 2, we plot the variation of efficiency with L_1 as predicted by this model for the case where V_{oc} = 0.85V and the maximum short-circuit current density is $15mA/cm^2$. Also shown in Fig. 2 are a number of experimental points for p-i-n cells grown in systems where L_1 was measured independently on an i-n/steel structure grown during the same week as the p-i-n structure. The p-i-n cells were grown in the configuration: glass/SnO_2/p^*-i-n/Ti/Al where p^* is a boron-doped a-Si:C:H alloy.

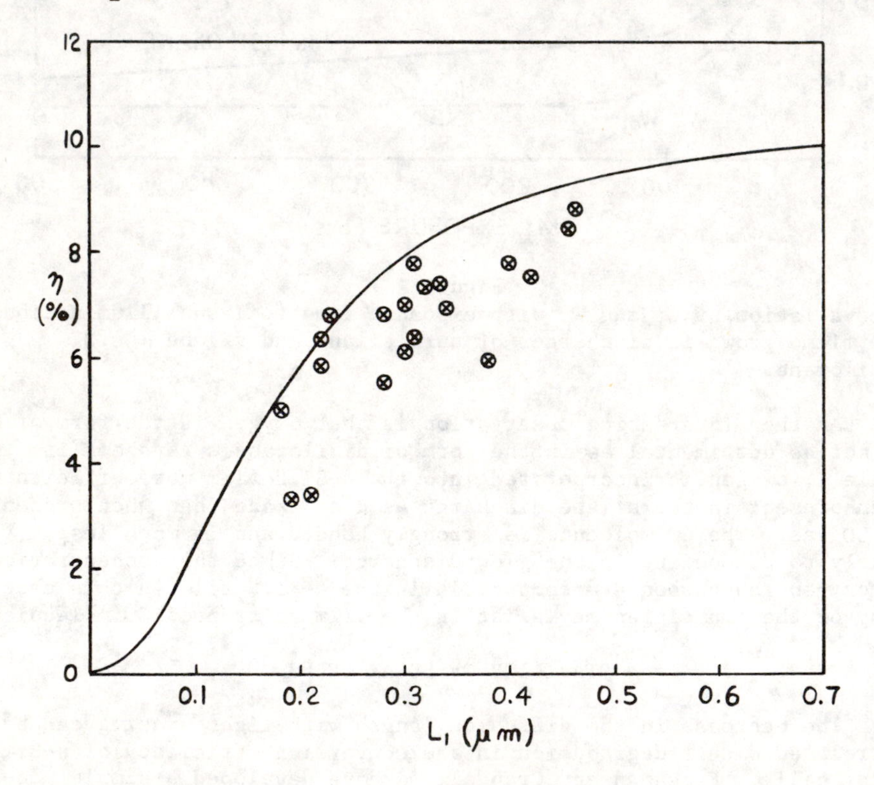

Figure 2
Conversion efficiency vs. L_1 (see text).

The experimental data in Fig. 2 roughly follow the curve predicted by the model. Experimental points would fall below the curve whenever the series resistance becomes significant or the devices are shunted. Also, the electric field is often not uniform in low performance cells. Moreover, the photocurrent will vary somewhat as the thickness and texture of the tin oxide coating varies. However, despite these shortcomings, the model has the advantage of allowing one to directly relate the efficiency of an a-Si:H solar cell to L_1 using a simple empirical relationship.

Another test of the model is shown in Fig. 3 where we plot L_1 as a function of exposure time to 1 sun illumination for an a-Si:H film grown in a discharge system with a small air leak. The films grown in this system at that time contained about 10^{+20} oxygen atoms cm^{-3}. Applying the model to this data, we predict that the conversion efficiency should decrease as shown by the curve in the top portion of Fig. 3. Experimental data for p-i-n cells made in this system are also shown in the top portion of the figure, and the agreement is reasonable.

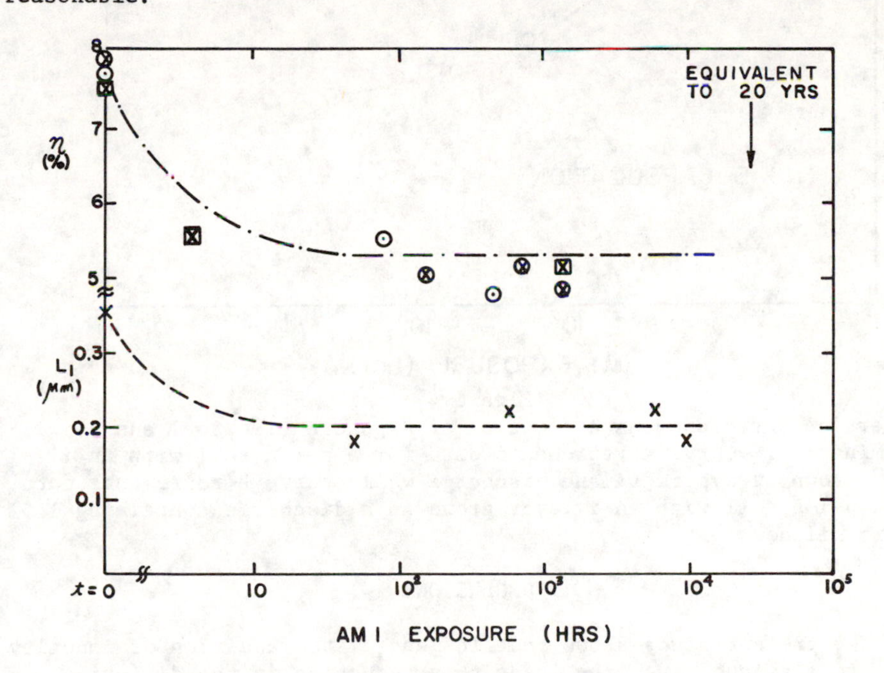

Figure 3
The conversion efficiency and L_1 as a function of exposure time to 1 sun illumination (see text).

The increased degradation caused by carbon impurities is shown in Fig. 4. Curve A follows data for a p-i-n cell where the i layer was made in a pure silane discharge and contained 2.4 X 10^{18} carbon atoms cm^{-3}. Curve B represents data for a cell fabricated under identical conditions except that the i layer was grown in a silane discharge containing 1.5% CH_4. In this case, the i layer contained

1.7×10^{20} carbon atoms cm^{-3}. It is evident from the figure that while the additional carbon caused only a relatively small decrease in the initial efficiency, the decrease in efficiency after 10^{+3} hours of light soaking is much more severe.

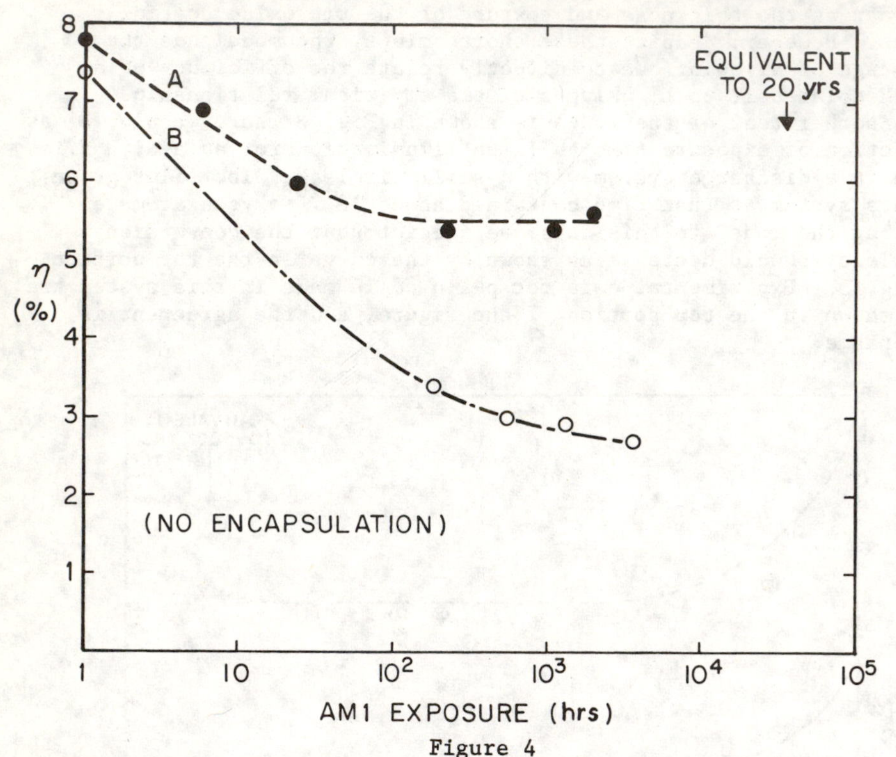

Figure 4
Conversion efficiency as a function of exposure time to 1 sun illumination. Curve A represents data for a p-i-n cell with an i layer grown in a pure silane discharge while curve B represents data for a p-i-n cell with an i layer grown in a discharge containing 1.5% CH_4 in silane.

CONCLUSIONS

The present study shows that in general the addition of impurity gases to a silane discharge leads to a reduction in the diffusion length. In all cases, exposure of the films to 500 hours of 1 sun illumination causes a further reduction in the diffusion length. Most of this light-induced degradation occurs in the first 48 hours. For films where impurity gases such as N_2, CF_4, C_2H_4, or $(SiH_3)_2O$ were added to the silane discharge, the diffusion lengths after light soaking are $\lesssim 0.15 \mu m$. A simple model predicts that such short diffusion lengths would lead to conversion efficiencies <4.5% for p-i-n cells made with such material (see Fig. 2). This model predicts that diffusion lengths must be >0.3μm in order to make p-i-n cells with efficiencies >8%.

In most cases, the space-charge width, W_0, is also reduced by light soaking. However, in three cases, W_0 remains above 1μm even after 500 hours of 1 sun illumination; namely for films grown in a pure silane discharge and where the impurity additives were 1 vol.% CO and 0.07 vol.% SiH_2Cl_2.

While the present work shows that impurities can increase the degradation caused by prolonged illumination, light-induced degradation is still observed for films containing $<4 \times 10^{+19}$ total impurity atoms cm^{-3}. Thus, it seems likely that defects such as microvoids, short-polymer chains or other intrinsic defects [13] are also associated with metastable centers.

ACKNOWLEDGEMENTS

The authors would like to thank A.R. Moore and G. Latham for performing the surface photovoltage measurements, C.W. Magee for performing the SIMS analyses, and H.A. Weakliem for depositing some of the a-Si:H films.

REFERENCES

1. D.E. Carlson, J. Vac. Science and Technology 20, 290 (1982).
2. D.E. Carlson, A. Catalano, R.V. D'Aiello, C.R. Dickson, R.S. Oswald, to be published in the Conf. Record of the 17th IEEE Photovoltaic Specialists Conf. (1984).
3. R.S. Crandall, Phys. Rev. B 24, 7457 (1981).
4. R.S. Crandall, D.E. Carlson, A. Catalano, and H.A. Weakliem, Appl. Phys. Lett. 44, 200 (1984).
5. C.C. Tsai, J.C. Knights, and M.J. Thompson, to be published in Journal of Non-Crystalline Solids (1984).
6. J. Dresner, B. Goldstein, and D. Szostak, Appl. Phys. Lett. 38, 998 (1980).
7. A.R. Moore, Appl. Phys. Lett. 40, 403 (1982).
8. D.E. Carlson, A.R. Moore, and A. Catalano, to be published in Journal of Non-Crystalline Solids (1984).
9. G.J. Clark, C.W. White, D.D. Allred, B.R. Appleton, C.W. Magee, and D.E. Carlson, Appl. Phys. Lett. 31, 582 (1977).
10. A.E. Delahoy and R.W Griffith, Conf. Record of 15th IEEE Photovoltaic Specialists Conf. (IEEE, NY, 1981), p. 704.
11. B.W. Faughnan and R.S. Crandall, Appl. Phys. Lett. 44, 537 (1984).
12. B.W. Faughnan, A.R. Moore, and R.S. Crandall, Appl. Phys. 44, 613 (1984).
13. D. Adler, Journal de Physique 42, Suppl. 10, C4-3 (1981).

THE STAEBLER-WRONSKI EFFECT IN UNDOPED a-Si:H : ITS INTRINSIC NATURE AND THE INFLUENCE OF IMPURITIES

C. C. Tsai, M. Stutzmann and W. B. Jackson
Xerox Palo Alto Research Center, Palo Alto, CA 94304

ABSTRACT

In an attempt to determine the origin of the light-induced metastable defects in undoped a-Si:H, we have examined the purest material available to date, prepared in a UHV plasma deposition system. In addition, the influence of impurities has been studied by investigating a series of SIMS-characterized films in which the oxygen and nitrogen content has been independently varied by nearly four and five orders of magnitude, respectively. We show that, contrary to the earlier indications, impurities do not influence the number of the light-induced defects in a-Si:H if they are below a critical concentration of $\sim 10^{20}$ cm^{-3} for O and $\sim 10^{19}$ cm^{-3} for N. This leads us to conclude that the Staebler-Wronski effect is an intrinsic property of a-Si:H. Also we show that impurities can cause rapid increases in the defect density only when they exceed the critical concentrations. However, in this "alloying" regime where significant modifications of a-Si:H network are expected, the generation of metastable defects may involve a different impurity-related mechanism. A "surface" or "interface" region with a combined thickness of $\sim$ 0.6 μm is observed to exhibit a much higher induced spin density than the "bulk". Thus, surface band bending is believed to play an important role in the defect creation.

INTRODUCTION

Illumination by light can create metastable defects in many semiconductors. Known as the Staebler-Wronski effect (SWE) in hydrogenated amorphous silicon (a-Si:H), the effect was first manifested as a decay in both the photo- and dark conductivity after prolonged illumination.[1] The observed changes are reversible upon annealing to a temperature above ~150 °C. Subsequent work has established that illumination causes an increase in the density of the dangling bonds,[2,3] as well as other changes in the density of gap states.[4,5] Such light-induced metastable changes have important technological implications. Despite extensive studies, however, the nature of the SWE is not well understood.

One important question is the origin of the effect. Previous work suggests that the light-induced metastable defects are related to impurities,[4-6] and thus not an intrinsic property of the material. For example, the formation of a deep trap has been correlated with a simultaneous incorporation of both O and N into the material introduced through an air leak.[4] Also a correlation has been made between excess C and a shallower trap.[5] However, the issue of the origin of the SWE cannot be addressed unless very pure materials are available. It is known that for the conventional a-Si:H material the concentrations of the major impurities, which are O, C, and N, can often be as

0094-243X/84/1200242-08 $3.00

high as 0.1–1 atomic %.[7-10] We have also found that significant amounts of other impurities such as F and Cl can also be present. Moreover, since previous work investigating the effect of impurities utilized materials containing significant concentrations of several impurities, it is difficult to identify the effect due to a particular impurity.

Recently it has been demonstrated that it is possible to reduce impurities to trace levels by employing a UHV plasma deposition system.[9,10] As verified by secondary ion mass spectroscopy (SIMS) measurements, the concentrations of all major impurities can be greatly reduced by over two to four orders of magnitude. The resulting high purity material contains only [N] ~ 5 x 10^{16} cm^{-3}, [C] ~ 4 x 10^{17} cm^{-3}, [O] ~ 2 x 10^{18} cm^{-3}, and no detectable F and Cl.[10] Using the pure material available, we hope to determine the origin of the SWE. Moreover, in the UHV deposition system it is possible to introduce impurities in a controlled way. By incorporating one impurity at a time, and starting from the high purity material, we can identify the influence of a particular impurity.

EXPERIMENT

The details of the UHV glow discharge deposition system and the sample preparation have been published elsewhere.[9] Basically it is a load–locked turbomolecular–pumped system with a base pressure of 5 x 10^{-9} – 2 x 10^{-8} Torr. High purity and low defect density films are deposited under standard deposition conditions: low r.f. power of 2 W, high flow rate of 150 sccm pure silane, and a substrate temperature of 230 °C. Controlled incorporation of O and N impurities has been achieved by adding known amounts of one of the following high purity gas mixtures to the silane discharge each time: O_2 + He, N_2 + He, and NH_3. Thus the effect of a particular impurity can be clearly identified. Both the O and N concentrations in a–Si:H are varied independently over nearly 4 and 5 orders of magnitude, respectively, as verified by SIMS. In addition, samples made in a baked conventional system have also been included for comparison, which have [O] ~ 3–8 x 10^{19} cm^{-3}, and [C], [N] ~ 4–10 x 10^{18} cm^{-3}. Some reductions in the sample impurity content can be made by baking the standard mechanical–pumped system.

The light–induced dangling bonds are observed directly from undoped films on Corning 7059 glass substrates (1 cm x 0.5 cm) by ESR (X–band, dangling bonds, g = 2.0055). The film thicknesses range from 0.1 to 7 μm. The sample was illuminated with unfiltered white light from a tungsten halogen lamp with an incident intensity of 300 mW/cm^2. Alternatively, illumination with homogeneously absorbed monochromatic light from a laser with an incident intensity of 500 mW/cm^2 was used to transform the samples into the saturated light–soaked state (state B). For both cases the heating of the samples during illumination was negligible. Photoconductivity measurements (with bottom Cr gap electrodes) were made to verify that the creation of dangling bonds by illumination does correspond to decreases in the photoconductivity.

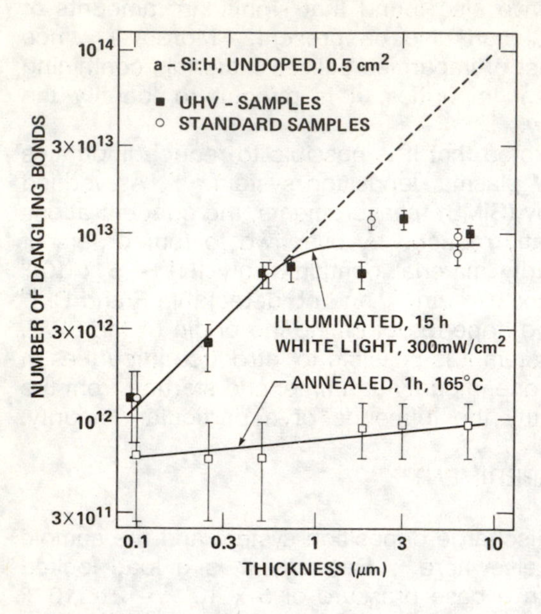

Fig. 1. Total number of dangling bonds in samples prepared in both the UHV system and a baked conventional system as a function of the film thickness in both the light-soaked state A and annealed state B.

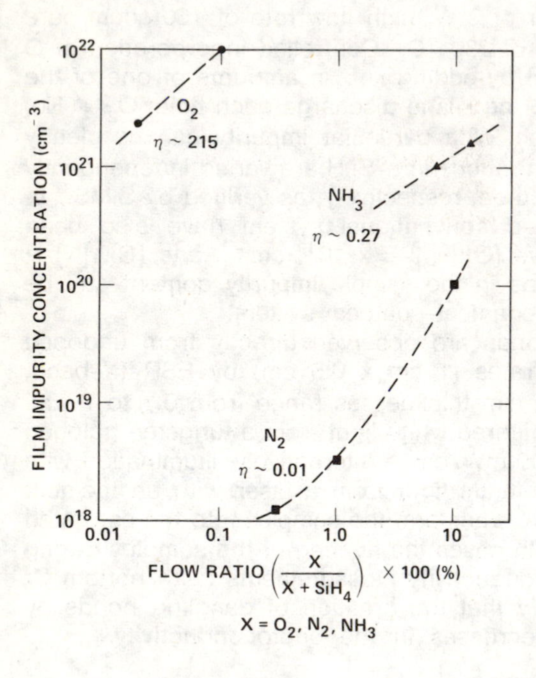

Fig. 2. The O or N impurity concentration in the films as a function of the flow ratio for mixing O_2, NH_3 or N_2 in SiH_4. η is the incorporation efficiency for each impurity gas.

RESULT AND DISCUSSION

We would first like to discuss the SWE in the high purity a–Si:H prepared in the UHV system. Shown in Fig. 1 is the total number of dangling bonds observed in samples in the annealed state A and the light–soaked state B (obtained after a 15 hr illumination with unfiltered white light at 300 mW/cm^2). We have found that metastable defects are created by illumination in all the high purity samples with a wide range of film thicknesses varying from 0.1 to 7 μm. Also included in Fig. 1 are results on samples made in a baked conventional system. Comparison between both types of samples indicates that even though the standard samples have a much higher overall impurity content than the UHV samples, essentially the same magnitudes of changes are observed. These results lead us to conclude that, contrary to the earlier indications,[4-6] the SWE is intrinsic to the material itself.

Having established the SWE in the high purity material, we can now investigate the influence of impurities by controlled additions of impurities to the high purity material. The impurity concentrations in the films, as analyzed by SIMS, are shown in Fig. 2 as a function of the flow ratio, $X / (X + SiH_4)$, where $X = O_2$, N_2, or NH_3. We have found O_2 to be extremely reactive with Si with an incorporation efficiency $\eta \sim 215$, i.e., for just 1 ppm of O_2 molecules in the gas mixture there are ~ 215 ppm of O atoms incorporated in the film. In contrast, it is very difficult to incorporate N from N_2, where η is less than ~ 0.01. It is much more efficient to incorporate N from NH_3, which has $\eta \sim 0.27$. Therefore, NH_3 has been used to make the high nitrogen content films. The large difference in the incorporation efficiency between oxygen and nitrogen can explain in part the relative difficulty in reducing the oxygen content in a–Si:H.

We would first discuss the effect of N impurity on the defect generation. Shown in Fig. 3 (upper part) is the spin density (N_S) in both state A and B as a function of [N]. Typical film thicknesses are between 3–4 μm. For [N] lower than a critical value of ~ 10^{19} cm^{-3}, one always observes ~ 10^{16} $spins/cm^3$ in the state A and ~10^{17} $spins/cm^3$ in state B, independent of the actual nitrogen content. However, for $[N] > 10^{19}$ cm^{-3} the spin density in both state A and B increases. Fig. 4 shows that the light–induced spin density ΔN_s increases strongly with the nitrogen concentration, and reaches ~ 3.5×10^{17} cm^{-3} at [N] ~ 1.5×10^{21} cm^{-3}, a more than 4–fold increase compared to the high purity sample. Whereas for $[N] < 10^{19}$ cm^{-3}, a constant value of $\Delta N_s \sim 8 \times 10^{16}$ cm^{-3} is obtained, independent of [N]. Qualitatively similar results have been obtained with O incorporation, which are shown in Fig. 3 (lower part) and Fig. 4. Again for [O] below a critical value of ~ 10^{20} cm^{-3}, the spin density in both state A and state B is independent of the O concentration, which yields a constant $\Delta N_s \sim 8 \times 10^{16}$ cm^{-3}. But for $[O] > 10^{20}$ cm^{-3}, ΔN_s increases rapidly to ~ 4.2×10^{17} cm^{-3} at [O] ~ 1×10^{22} cm^{-3}. Although no attempt was made to add C to a–Si:H deliberately, it is unlikely that C will influence the SWE since identical results are obtained in samples with very different C contents.

Thus our results strongly suggest the intrinsic nature of the SWE. In

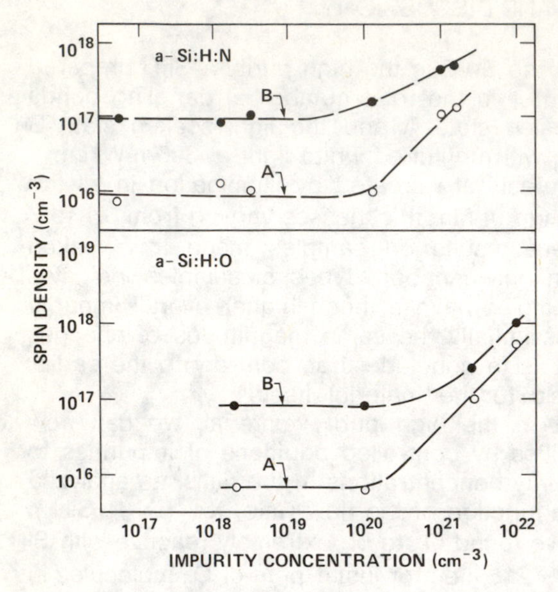

Fig. 3. Dependence of the spin density on the impurity concentration, [N] or [O], for incorporating N (upper part) and O (lower part) into a–Si:H. A and B refer to the annealed and to the light–soaked state, respectively.

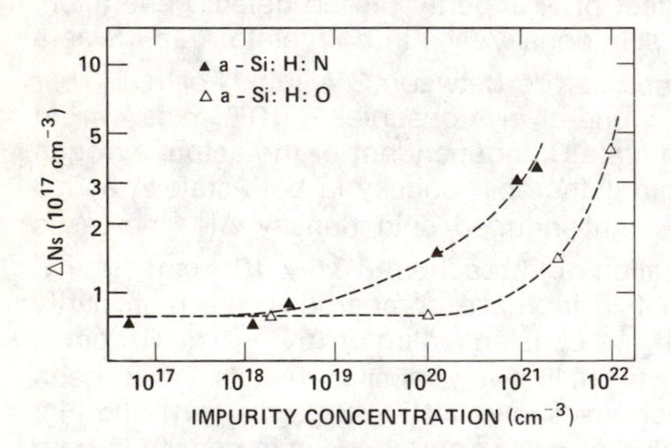

Fig. 4. Light–induced spin density, ΔN_s, as a function of the N and O impurity concentrations for N and O incorporation, respectively.

addition, we have also demonstrated that at high concentrations exceeding the critical values (~ 10^{20} cm^{-3} for O and ~ 10^{19} cm^{-3} for N), either O or N impurity can substantially enhance the defect generation. Both impurities have similar influences, although N appears to be more effective in defect generation by creating one dangling bond for every ~ 3000 N atoms incorporated, while only one dangling bond is created for every ~ 20,000 O atoms incorporated. Nevertheless, neither impurity is very efficient in creating dangling bonds. Since the increase in ΔN_s is always coupled with an increase of N_S in the annealed state A, we believe that the experimental results can be explained by an "alloying" effect. Basically, the presence of a very high concentration of impurities can modify the amorphous Si network significantly, which is supported by the observed increase in the dangling bond density in the state A. The resulting new structures appear to be more defective, and can probably have more metastable defects created after illumination. In the "alloying" regime the generation of metastable defects may involve a different mechanism which is impurity-related.

Since it is not particularly unusual for the conventional a-Si:H material to have a high impurity concentration that exceeds the critical values, the reduction in the impurity content can result in an improved stability under illumination. However, further reduction of impurities for samples already having $[O] < 10^{20} cm^{-3}$ and $[N] < 10^{19} cm^{-3}$ will not eliminate the SWE because of its intrinsic nature.

Consistent with these experimental results, we have proposed a model for the SWE, based on a self-limiting process intrinsic to undoped a-Si:H.[11] A possible kinetic mechanism involves the nonradiative recombination of band tail carriers, and a neighboring Si-H bond may participate in the process. Such a model can explain the time and intensity dependence of the metastable defect generation observed by the ESR and photoconductivity measurements. The details of the model and the experimental results of the kinetics of the SWE are given elsewhere.[11]

One remaining question is the thickness dependence of the induced spins. As indicated in Fig. 1, the total number of dangling bonds in the state B increases linearly with film thickness up to ~ 0.6 μm, and becomes almost saturated above ~ 1 μm. To eliminate the possibility of nonuniform absorption of white light within thick samples, homogeneously absorbed monochromatic light was used to illuminate the samples instead. The thickness dependence is observed to be essentially identical to that in the white light case. The uniformity of the light soaking has been verified by observing that no additional changes are produced if the samples are further illuminated from the substrate side. Also we have estimated the temperature gradient in the samples, and we believe this effect is too small to account for the obseved thickness dependence.

Thus there appears to be a "surface" or "interface" region with a combined thickness of ~ 0.6 μm that exhibits a much higher induced spin density than the "bulk". There are some possible explanations. One is that the difference between surface and bulk contribution is due to the high impurity concentrations present at either the top surface or the substrate interface,[7-10] since we have already shown that the SWE can be enhanced in samples abundant in impurities. Another possibility is that the SWE is related to surface band bending. The first explanation can be easily ruled out since the SIMS results indicate that the surface or interface region with [O] > 10^{20} cm^{-3} or [N] > 10^{19} cm^{-3} is not wider than 0.05 μm in these samples. To test the second hypothesis, we have modified the band bending in a–Si:H by depositing an equally thin layer of heavily doped a–Si:H on both sides of undoped a–Si:H films. Doping has been achieved by mixing 1 % of B_2H_6 or PH_3 in SiH_4. Shown in Fig. 5 is the total number of spins in the light–soaked state B as a function of the doped layer thickness, while the undoped layer has been fixed at a thickness of 2 μm. Indeed we show that the magnitude of the SWE can be substantially changed by altering the band bending. The presence of a 0.05 μm thin layer of B–doped film at both interfaces reduces the SWE by ~ 35 %, while a ~ 40 % enhancement is observed with P–doped films. We do not believe that this effect is due to changes in the bonding or structure of the surfaces which may have been caused by the presence of the adjacent surface layers This is because little change is observed when the doped layer thickness is reduced to 1.2 nm, as shown in the figure. From these preliminary experimental results, it appears that the doped layers change the surface charge density of the undoped a–Si:H, thereby altering the surface band bending. This would imply that surface band bending plays an important role in the SWE. Since a n–type surface layer enhances the SWE, while a p–type layer reduces it, this would also imply that the direction of the band bending in a–Si:H is an electron accumulation layer for either the free surface or the substrate interface. This is in contradiction with other results reporting opposite direction of band bending at the free surface.[12] The additional observation that a 10 nm Al contact modifying one surface increases the SWE by a factor of ~ 3 also is consistent with this band bending hypothesis. One

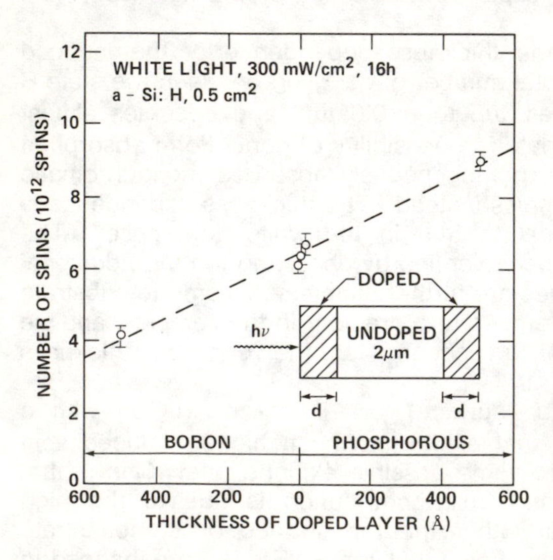

Fig. 5. Total number of spins after light soaking in 2 μm–thick, 1 cm x 0.5 cm undoped a–Si:H samples as a function of the thickness, d, of the doped layer. Both surfaces of the undoped samples have been modified by the presence of either 1 % P–doped or 1 % B–doped a–Si:H thin films with a thickness d.

implication of the result is that samples having different configurations, e.g. p–i–n junctions, coplanar electrodes, Schottky barriers with different barrier heights, etc., or samples having different thicknesses, may exhibit considerable variations in the SWE, since band bending is extremely sensitive to surface conditions. Further work is needed to clearly understand the effect of band bending.

In conclusion, by examining the purest material available to date, we have shown that the Staebler–Wronski effect is an intrinsic property of a–Si:H. It appears that surface band bending can play an important role in the defect generation. Impurities can enhance the defect generation only when they exceed a critical concentration of $\sim 10^{20}$ cm^{-3} for oxygen and 10^{19} cm^{-3} for nitrogen, respectively. The observed enhancement probably originates from an "alloying" effect, however, and a different mechanism related to impurities may be involved. Our results shed new light on the understanding of the origin of the Staebler–Wronski effect, and can also have important implications for device stability since band bending can vary considerably depending on specific device configurations.

ACKNOWLEDGMENT

This work was supported in part by the Solar Energy Research Institute under Contract No. XB–3–03112–1.

REFERENCES

1. D. L. Staebler and C. R. Wronski, Appl. Phys. Lett., 31, 292 (1977).
2. I. Hirabyashi, K. Morigaki and S. Nitta, Jap. J. Appl. Phys., 19, L357 (1980).
3. H. Dersch, J. Stuke and J. Beichler, Appl. Phys. Lett., 38, 456 (1981).
4. R. S. Crandall, Phys. Rev. B, 24, 7457 (1981).
5. R. S. Crandall, D. E. Carlson, A. Catalano and H. A. Weakliem, Appl. Phys. Lett., 44, 200 (1984).
6. D. E. Carlson, J. Vac. Sci. Technol., 20, 290 (1982).
7. A. Delahoy and R. W. Griffith, J. Appl. Phys., 52, 6337 (1981).
8. J. C. Knights in The Physics of Hydrogenated Amorphous Silicon I: Stucture, Preparation, and Devices, ed. J. D. Joannopoulos and G. Lucovsky, (Springer–Verlag, Berlin, 1984), p. 5.
9. C. C. Tsai, J. C. Knights, R. A. Lujan, B. Wacker, B. L. Stafford and M. J. Thompson, J. Non–cryst. Solids, 59 & 60, 731 (1983).
10. C. C. Tsai, J. C. Knights and M. J. Thompson in Proceedings of the International Topical Conference on Transport & Defects in Amorphous Semiconductors, March 22–24, 1984, ed. H. Fritzsche and M. A. Kastner, (J. Non–cryst. Solids, Special Issue).
11. M. Stutzmann, W. B. Jackson and C. C. Tsai, this conference.
12. B. Aker and H. Fritzsche, J. Appl. Phys., 54, 6628 (1983); and R. A. Street, private communications.

GAP STATES DYNAMICS OBSERVED IN LIGHT SOAKED P-DOPED a-Si:H

H.Okushi, M.Itoh*, T.Okuno*, Y.Hosokawa**, S.Yamasaki and K.Tanaka
Electrotechnical Laboratory
Sakura-mura, Ibaraki 305, Japan

ABSTRACT

Gap-state profiles of P-doped a-Si:H before and after the light soaking have been determined using several variations of the isothermal capacitance transient spectroscopy (ICTS) techniques. In parallel with the photo-induced increase in the gap states (D^-) at around 0.5 eV below E_c, a photo-induced decrease in the density of gap states (denoted by X) located at 0.7 eV above E_V has been observed for the first time. It has been demonstrated that X states, originating from "charge coupled dangling bonds ($^*D^-$)", are transformed into D^- states by the light soaking and some of them are relaxed to the original X ($^*D^-$) states even at room temperature. The mechanism of $^*D^-$-D^- transformation is discussed in terms of electron-phonon coupling via $^*D^-$ coupled with P_4^+.

INTRODUCTION

Dark ICTS (isothermal capacitance transient spectroscopy) has provided us much information on the gap states above the midgap in P-doped a-Si:H; density-of-state distribution N(E) of D^- and energy as well as temperature dependence of their electron-capture cross section $\sigma_n(E,T)$.[1-3] In earlier work, through the systematic study of Staebler-Wronski (S-W) effect in P-doped a-Si:H using dark ICTS, we have indicated that the electron-capture cross section at D^- states located at around 0.5 eV below E_C as well as their density increases after the band-gap illumination and is nearly restored to the original state by thermal annealing.[4] Quite recently, we have succeeded in collecting information on the gap states lying near the valence band using photo-ICTS techniques.[3,5]

In this report, we present the detailed experimental results on the gap-state profile of P-doped a-Si:H before and after light soaking under various experimental conditions using dark and photo-ICTS techniques, and discuss the mechanism of "anomalous" gap-state dynamics based on our recent model in terms of "charge-coupled dangling bonds".

EXPERIMENTAL

A. Sample preparation

P-doped (PH_3/SiH_4 = 3×10^{-4}) a-Si:H films used in the present work were deposited on a crystalline Si (n^+, 0.01 ohm-cm) by glow-discharge decomposition under the same condition as described

* Sharp Corporation, Tenri-shi, Nara 632, Japan.
** Kanegafuchi Chemical Industry Co. Ltd., Kobe-shi, Hyogo 652, Japan.

earlier.[6] The optical gap and the activation energy of the dark conductivity were determined to be 1.7 eV and 0.20 eV, respectively. The Schottky diode was fabricated by evaporating Pt on the film with an area of 1×10^{-2} cm^2. Each specimen was subjected to the band gap illumination provided by 500-W Xenon arc through ir cut-off filter (~100 mW/cm^2). Only for the experiment on the excitation spectrum of S-W effect, we used LED (light emitting diode, $h\nu$ = 1.30 eV, ~100 mW/cm^2), semiconductor laser ($h\nu$ = 1.59 eV, ~90 mW/cm^2) and He-Ne laser ($h\nu$= 1.96 eV, ~ 1 mW/cm^2). In photo-ICTS measurement, He-Ne laser ($h\nu$ = 1.96 eV) was used with a low-level light intensity (<400 $\mu W/cm^2$).

B. Dark and photo-ICTS measurements

In the ICTS measurement, a transient capacitance $C(t)$ of the reverse-bias junction is measured, which is induced by applying some perturbation to the diode. In the present work, we have used four different modes of perturbation; (1) a bias-voltage pulse in dark, (2) a step-function light under a constant reverse bias V_R, (3) a light pulse under a constant V_R and (4) a bias-voltage pulse under illumination, which are named as dark, light-ON, light-OFF and under-illumination ICTS, respectively. These four modes of ICTS are schematically given in Fig.1.

For each mode, ICTS signal $S(t)$ is defined as

$$S(t) = t\,df(t)/dt \ , \text{ with } \ f(t) = C^2(t) - C^2(\infty), \qquad (1)$$

where t is the elapsed time after each perturbation is applied. A density of states N(E) is proportional to $S(t)$ and the detailed discussion on their quantitative relationship has been given in the previous paper.[3]

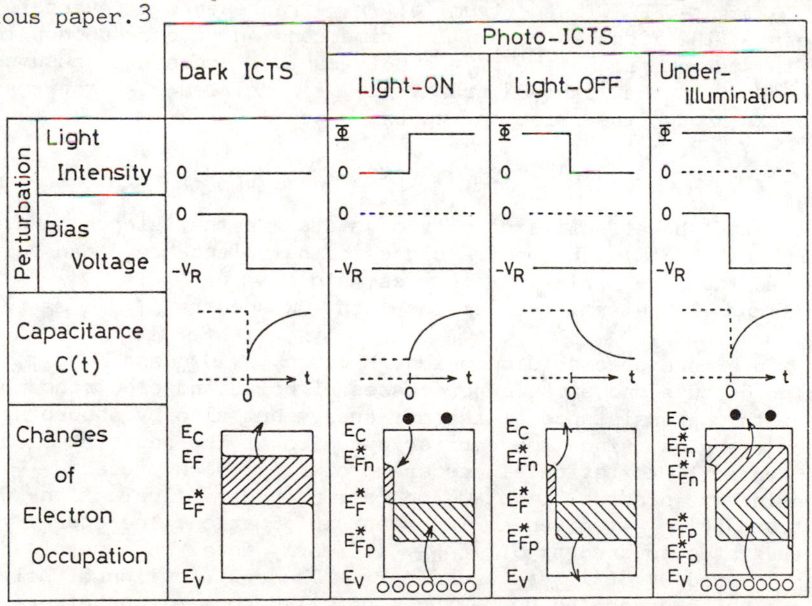

Fig.1. Dark ICTS and three variations of photo-ICTS.

In dark ICTS, a time constant t corresponds to a reciprocal of the electron thermal-emission rate for the case of n-type Schottky diode, and is related with the energy depth measured from E_c through

$$E_c - E = kT \ln\{\nu_n t\}, \quad \text{with } \nu_n = N_c \sigma_n v_{th}, \tag{2}$$

where ν_n is the pre-exponential factor of the thermal-emission rate, N_c the effective density of states in the conduction band, σ_n the electron-capture cross section and v_{th} the thermal velocity of free electrons. Therefore, if ν_n or σ_n is known, exact N(E)-E spectra are estimated from *S(t)-t* curves through Eq.(2). ν_n or σ_n can be experimentally determined from a voltage-pulse-width dependence of the dark ICTS spectrum as reported earlier.[1,2]

In light-ON ICTS, a transient capacitance $C(t)$ originates from a time variation of the electron occupation of the gap states at the depletion region induced by optical excitation. If we assume that the variation of the electron occupation is mainly due to the capture process of excess holes generated by optical excitation, the time constant t is given by

$$t = (p\sigma_p v_{th})^{-1}, \quad \text{with } p = \alpha\Phi d^2/(\mu_p V_R), \tag{3}$$

where α is the optical absorption coefficient, μ_p the hole mobility of the specimen, p the excess hole concentration, Φ the incident photon flux of optical excitation, and d the width of depletion region. Then, if the light-ON ICTS spectrum ($S(t)_{ON}$) has a single peak at t under optical excitation with Φ , we can estimate the hole-capture cross section of gap states mainly contributing to $S(t)_{ON}$ using Eq.(3).

A light-OFF ICTS spectrum ($S(t)_{OFF}$) represents a reverse process of the light-ON mode. The time constant t corresponds to the hole-thermal emission time if the hole-capture process is assumed to be dominant in the light-ON transient; t is related with the energy E above E_v through the relation expressed as

$$E - E_v = kT \ln\{\nu_p t\}, \text{ with } \nu_p = N_v \sigma_p v_{th}, \tag{4}$$

where N_v is the effective density of states in the valence band. Consequently, N(E) - E spectra close to the valence band can be obtained from the $S(t)_{OFF}$ - t curves, since ν_p or σ_p is experimentally determined from the light-ON spectra $S(t)_{ON}$ as mentioned above.

In a more precise discussion, however, $S(t)_{ON}$ and $S(t)_{OFF}$ involve capture and emission processes of holes and electrons not only with the assistance of thermal energy but also by absorbing photon energies, since the carrier dynamics at the gap states are under optical excitation. In order to clarify which process predominates in actual light-ON operation, under-illumination ICTS spectrum $S(t)_{under}$ gives an important information. The energy range of gap states covered by $S(t)_{under}$ is nearly equal to a total energy range covered by $S(t)_{dark}$ and $S(t)_{ON}$. The contribution of holes to $S(t)_{ON}$ can be separated by making a detailed comparison between

$S(t)_{under}$ and a combined spectrum of $(S(t)_{dark} + S(t)_{ON})$, which is summarized in the figure.

RESULTS AND DISCUSSION

A. Gap-state profiles before and after light soaking

Figure 2(a) and (b) show dark and photo-ICTS spectra (light-ON, light-OFF and under-illumination modes) in the specimen before and after light soaking and subsequent room temperature annealing. As described earlier,[1-3] the bump of $S(t)_{dark}$ is originated from the doubly-occupied dangling bond states (D^-) located at around 0.5 eV ~ 0.6 eV below E_c. The energy location of D^- states has been determined by using Eq.(2) with $\nu_n \simeq 10^9$ sec^{-1} or $\sigma_n \simeq 10^{-18}$ cm^2 which were obtained by the voltage-pulse-width dependence of $S(t)_{dark}$.

From photo-ICTS spectra, we have also determined N(E) close to the valence band. The peak of $S(t)_{ON}$ has been found to originate from the hole-capture process. Then, using the peak position of $S(t)_{ON}$, $\alpha = 10^4$ cm^{-1}, $\mu_p \simeq 0.1$ $cm^2/Vsec$ and $\Phi = 1.2 \times 10^{15}$ $cm^{-2} sec^{-1}$, we obtain $p = 5 \times 10^{10}/cm^3$ and $\sigma_p = 2 \times 10^{-16}$ cm^2 from Eq.(3). From $\sigma_p \simeq 10^{-16}$ cm^2 and Eq.(4) it is concluded that the hole-trapping states associated with the light-ON and light-OFF ICTS are located at 0.4 ~ 0.7 eV above E_v.

As is shown in $S(t)_{dark}$ of Fig.2 (a), the bump of $S(t)_{dark}$ increases and its peak position shifts towards into the shorter time constant after light soaking. It means that the density of D^- states as well as their electron-capture cross section increases by the light soaking.

In parallel with the variation of $S(t)_{dark}$, photo-induced decreases of $S(t)_{ON}$ and $S(t)_{OFF}$ are observed, as shown in Fig.2 (b), while no significant changes are observed in $S(t)_{under}$. In contrast

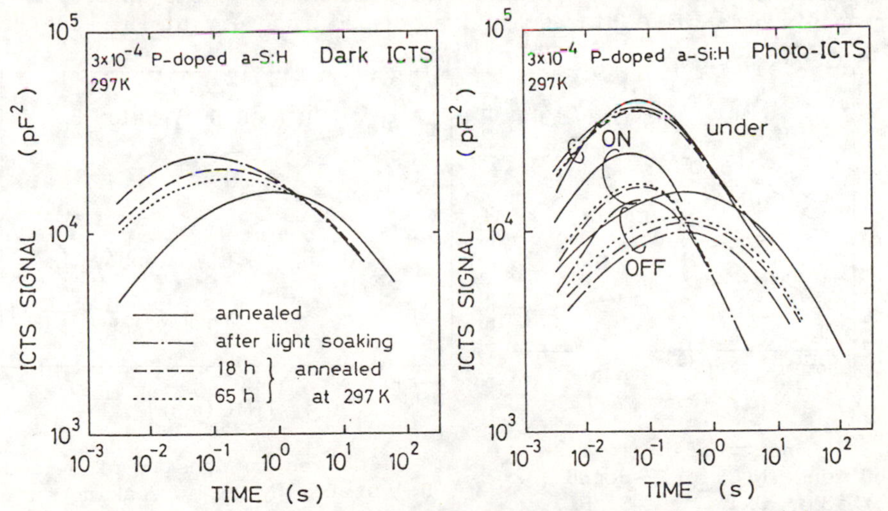

Fig.2. (a) Dark ICTS, and (b) photo-ICTS spectra, $S(t)_{ON}$, $S(t)_{OFF}$, and $S(t)_{under}$, in P-doped ($PH_3/SiH_4 = 3 \times 10^{-4}$) a-Si:H before and after light soaking and subsequent room temperature annealing.

to $S(t)_{dark}$, $S(t)_{ON}$ as well as $S(t)_{OFF}$ in Fig.2(b) reveal a decrease in its peak intensity after the light soaking with a slight peak shift on the t axis. Namely, an increase in the density of D^- states at 0.5 eV below E_c and a decrease in X states at ~0.7 eV above E_V are simultaneously induced by the band-gap illumination.

It should be noted that the decrease in $S(t)_{ON}$ and $S(t)_{OFF}$ is detected only when Φ in photo-ICTS is kept at a small level. In fact, almost no discernible change was observed in $S(t)_{ON}$ and $S(t)_{OFF}$ under strong optical excitation measurement (~100 mW/cm , ~3×10^{17}/cm^{2}sec). Since the quasi-Fermi level for holes approaches closer and closer to the valence band with increase of Φ , it is reasonable to consider that a decrease in the density of X states takes place in a higher energy side (~0.7 eV above E_V) while the density of X states in a lower energy side (~ 0.4 eV above E_V) remains unchanged or rather may increase after the light soaking. This is compatible with PAS measurements.[7] Obtained N(E) of P-doped a-Si:H before and after the light soaking is shown in Fig.3.

The above changes in the density of states N(E), X as well as D^- states, are partially annealed out even at the room temperature(297K), as shown in Fig.4. This complementary relationship between D^- and X states suggests that the defects corresponding to X states are transformed by the light soaking into D^- states located energetically higher than X states and a part of them can be relaxed to the original X states with an assistance of thermal energy.

B. Excitation-energy dependence

Figure 5 shows the dark ICTS spectra before and after light soaking for different excitation energies ($h\nu$ = 1.96 eV, 1.59 eV and 1.30 eV). For each excitation energy we controlled the intensity of excitation so that the absorbed photon flux ($\alpha\Phi \simeq 3\times10^{19}$/cm^{3}sec) was always kept constant. As shown in the figure, the photo-induced increase in N(E) of D^- states and its partial recovery at 297K are

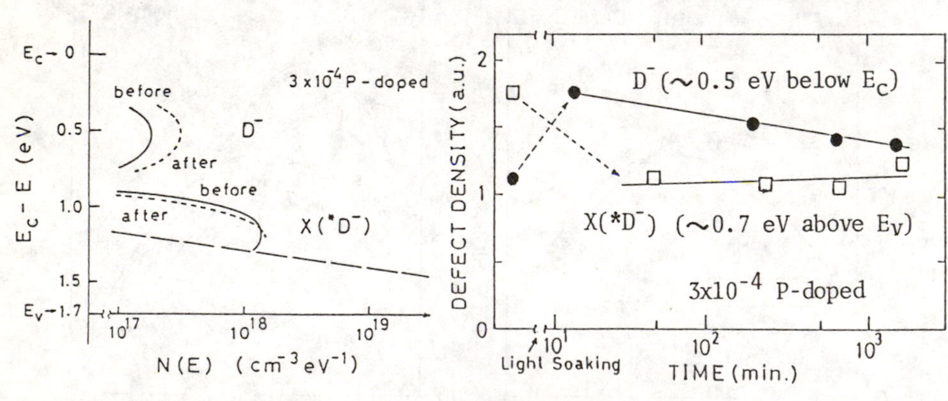

Fig.3. Density-of-state distributions N(E) of P-doped ($PH_3/SiH_4 = 3\times10^{-4}$) a-Si:H before (solid line) and after (dashed line) the light soaking.

Fig.4. Changes in density of D^- and X ($^*D^-$) states by light soaking and subsequent room temperature (297K) annealing.

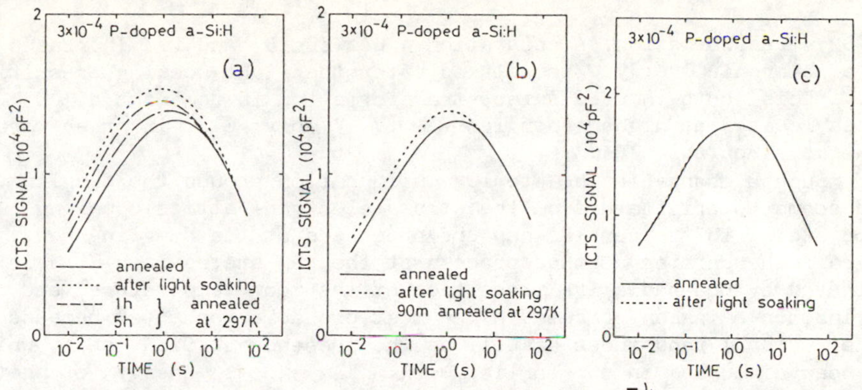

Fig.5 Dark ICTS spectra in P-doped ($PH_3/SiH_4 = 3x10^{-4}$) a-Si:H before and after light soaking and subsequent room temperature annealing for different excitation energies; (a) $h\nu$=1.96 eV, (b) $h\nu$=1.59 eV and (c) $h\nu$=1.30 eV under the constant absorbed photon fulx $\Phi = 3x10^{19}/cm^3sec$.

quite dominant against 1.96-eV excitation, whereas no discernible change was detected against sub-bandgap light (1.30 eV).

Both holes and electrons can be generated by the band-gap excitation (1.96 eV). On the other hand, since the Fermi level is located at 0.20 eV below E_c of the present specimen, only free electrons are excited by 1.30-eV light while free holes are unable to be generated. This discussion leads to an important implication that the capture process of free holes at the gap states plays an essential role in the change of N(E).

C. Mechanism of $^*D^-$-D^- transformation

As described previously, X states located at 1.0 ~ 1.3 eV below E_c increase with a doping level of P atoms and their density as well as the energy location of X states is nearly close to that of a bump of N(E) deduced from the analysis of the below-gap optical absorption determined by PAS.[8] Furthermore, P-doping dependence of a total density of holes captured at X states nearly coincides with that of LESR reported by Street et al.[9] These results have suggested that X states should be attributed to dangling bonds whose energies are modified by P_4^+, or alternatively, other charged defects. This dangling bond strongly coupled with P_4^+, which can be denoted by $^*D^-$ for simplicity, takes a stabler energy compared to D^- because of the Coulombic interaction with P_4^+, and possibly, a strong electron-lattice coupling. This model results in an effective electron-correlation energy U_{eff} at $^*D^-$ states expressed as

$$U_{eff} = U_c - e^2/\varepsilon r - W, \quad (5)$$

where U_c is a pure correlation energy, e the electron charge, ε the dielectric constant, r the $^*D^-$-P_4^+ separation and W a polaronic relaxation energy. Since $^*D^-$ seems to lie below D^o, U_{eff} in Eq.(5) may be negative. The present results of the X-D^- transformation shown in Figs.2-5, therefore, can be interpreted in terms of photo-induced

$^*D^-$-D^- transformation. A separation r between $^*D^-$ and P_4^+ increases and a polaronic energy W is relaxed after the light soaking, resulting in D^-. This photo-induced defect transformation is dominant at $^*D^-$ states located at higher energies (~ 0.7 eV above E_V), in other words, relatively unstable $^*D^-$.

Such a change in the atomic configuration around the dangling bond seems to originate from the strong electron-lattice coupling associated with the carrier-capture processes at the dangling bond. In general, the carrier capture process at the gap states in a-Si:H is dominated by non-radiative transition at the room temperature, and the multiphonon emission process is the most probable mechanism because of the structural randomness as well as the presence of Si:H bonds, which has been discussed in our earlier work.[3] An excess energy of photo excited carrier is released to the lattice through the multiphonon emission when the carrier is captured at the defect states, causing relaxation and/or distortion of local atomic structure.

Judging from the hole-capture cross section of $^*D^-$ states amounting to $10^{-15} \sim 10^{-16} cm^2$ determined from the photo-ICTS, a strong electron- lattice coupling may prevail in this capture process, namely, the hole-capture at $^*D^-$ states is considered to be essential in $^*D^-$-D^- transformation. This is compatible with the experimental observation in Fig.5 that the capture process of free holes at the gap states plays an essential role in the change of N(E).

Figure 6 shows a configuration coordinate diagram of the present model. In the figure, it is indicated that the electron-lattice coupling is strong when hole is captured at the $^*D^-$ state, while it is weak at the D^- state. As described earlier,[1-3] the temperature and the energy dependences of the electron-capture cross section σ_n at D^- states indicate that a multiphonon emission process at weak electron-lattice coupling predominates at the capture process, resulting in W = 0 and positive U_{eff}. Another possible origin of the difference in U_{eff} between $^*D^-$ and D^- may be a difference in electronic configuration of a Si dangling bond as discussed by Adler ; the energetically stable doubly-occupied states ($^*D^-$) may correspond to T_3^- ($s^2 p^3$) while higher states (D^-) to $T_3^0(sp^3)$ + e.[10]

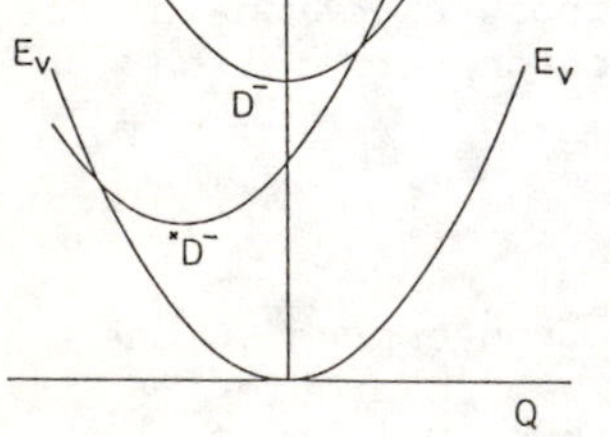

Fig.6 A configuration coordinate diagram for $^*D^-$ and D^-.

It seems that the above mechanism of the $^*D^-$-D^- transformation and the configuration coordinate diagram shown in Fig.6 give insight into the general understanding of the behaviour of dangling bond in a-Si:H associated with the S-W effect. For example, reversible change of the electrical properties of a p-i-n diode by current soaking under forward bias and thermal annealing[11] is well understood by the present mechanism since the $^*D^-$ - D^- transformation is based on the hole-capture process at $^*D^-$ states irrespective of whether holes are provided by optical excitation or current injection.

In the present model, $^*D^-$-D^- transformation is discussed with respect to the presence of P atoms in the network. However, the essence of the present discussion is retained even in undoped a-Si:H, because it is well known that various impurities such as N, O and C are unintentionally incorporated into a-Si:H network during deposition process. It is likely that they produce positively-charged defects coupled with $^{*}D^{-}$.

SUMMARY

We have presented the principle of the photo-ICTS technique (light-ON, light-OFF and under-illumination ICTS) with some details of analyses. Using these techniques as well as the dark ICTS technique, gap-state profiles of P-doped a-Si:H before and after the light soaking have been determined. In parallel with the photo-induced increase in the density of doubly-occupied dangling bond (D^{-}) states located at around 0.5 eV below E_c, a photo-induced decrease in the density of gap states located below midgap (~ 0.7 eV above E_V) has been observed. The origin of the latter has been attributed to the doubly-occupied dangling bond ($^{*}D^{-}$) coupled with P_4^{+}. It has been demonstrated that the higher-energy states (~0.7 eV above E_V) in $^{*}D^{-}$ located at 0.4 ~ 0.7 eV above E_V are transformed into D^{-} states by the light soaking and some of them are relaxed to the original states even at room temperature. The mechanism of the $^{*}D^{-}$-D^{-} transformation has been discussed in terms of electron-lattice coupling in the hole-capture process at $^{*}D^{-}$ states.

The authors would like to thank Prof. K. Morigaki, Dr. I. Hirabayashi and our colleagues of Amorphous Materials Section of ETL for their fruitful discussions.

REFERENCES

1. H. Okushi, Y. Tokumaru, S. Yamasaki, H. Oheda and K. Tanaka, Phys. Rev. B25 4314 (1982).
2. H. Okushi, T. Takahama, Y. Tokumaru, S. Yamasaki, H. Oheda and K. Tanaka, Phys. Rev. B27 5184 (1983).
3. K. Tanaka and H. Okushi, Proc. Int'l Conf. on Transport and Defect in Amorphous Semiconductors (Michigan, 1984), in press.
4. H. Okushi, M. Miyakawa, Y. Tokumaru, S. Yamasaki, H. Oheda and K. Tanaka, Appl. Phys. Lett. 42 895 (1983).
5. H. Okushi, A. Asano, M. Miyakawa, S. Yamasaki and K. Tanaka, J. Non-Cryst. Solids, 59&60 393 (1983).
6. H. Okushi, Y. Tokumaru, S. Yamasaki, H. Oheda and K. Tanaka, Jpn, J. Appl. Phys. 20 L549 (1981).
7. S. Yamasaki (unpublished data); the below-gap optical absorption increases after the light soaking.
8. S. Yamasaki, H. Oheda, A. Matsuda, H. Okushi and K. Tanaka, Jpn, J. Appl. Phys. 21 L539 (1982).
9. R. A. Street, D. K. Biegelsen and J. C. Knights, Phys. Rev. B24 969 (1981).
10. D. Adler, Phys. Rev. Lett. 41 1755 (1978).
11. Y. Uchida, M. Nishiura, H. Sakai and H. Haruki, Solar Cells 9 3 (1983).

PHOTODARKENING AND PHOTOBLEACHING IN a-C:H FILMS

S.Iida, T.Ohtaki and T.Seki

Technological University of Nagaoka, Kamitomioka, Nagaoka
949-54, JAPAN

ABSTRACT

Irreversible photodarkening and photobleaching effects in a-C:H are discussed on the bases of measured excitation spectra and infrared absorptions. From infrared absorption studies, it was found that three absorptions due to CHn, C=O and OH bonds changed with illumination. The CHn absorption decreased for photodarkening as well as photobleaching, while C=O and OH absorptions increased only for photobleaching. From excitation spectra, photodarkening effects were found to be strongly dependent on illuminating photon energy. The knee at about 3.6 eV of the excitation spectrum appears to correspond to the dissociation energy of the CH bond. These results seem to indicate that the photo-dissociation of the CH bond plays an essential role. The temperature dependence of photodarkening seems to imply that thermal energy also participates the process. The estimated quantum efficiency of this dissociation is relatively high, for example, about 0.2 for 365 nm illumination at room temperature.

INTRODUCTION

Photoinduced absorption edge shifts towards higher and lower energies, i.e. photobleaching and photodarkening were reported to occur in a-Si:C:H[1]. Photodarkening is induced by band gap illumination in vacuum and photobleaching by illumination in air. These effects are remarkable for C rich films. The photobleaching is considered to be related to the formation of C=O bonds which appear in the infrared absorption. In order to obtain a more quantitative grasp of the effects experimental data are presented on a-C:H films which exhibit the largest changes and discussions are given on the mechanism of the effects based on these experiments. The main experimental results described are the dependence of the optical gap shifts on illuminating photon energy for photodarkening and infrared absorption measurements for photodarkening as well as photobleaching in a-C:H films.

EXPERIMENTAL

A. Sample Preparation

The a-C:H films were prepared by glow discharge decomposition of C_3H_8 with capacitively coupled rf power input. The chamber was evacuated to below 10^{-5} Torr before C_3H_8 was introduced. The pressure of C_3H_8 was ~40 mTorr and the input rf power was 30-40 W. The films were grown on slide glass or high resistivity silicon substrates at room temperature. The films on silicon substrates were used for IR absorption measurements and the films on glass substrates for

0094-243X/84/1200258-08 $3.00

visible region absorption measurements.

B. Photon Energy Dependence of Photodarkening

Figure 1 shows a change of optical absorption at room temperature appearing in $\sqrt{\alpha h\nu}$ vs $h\nu$ plots for photodarkening of an a-C:H film. The film was placed in a transparent Dewar being evacuated by a rotary pump and illuminated by photons with photon energy of 3.40 eV (365 nm) obtained from a super high pressure Hg lamp by passing through a monochromator. The power density of illumination was about 5 mW/cm^2 and the time of illumination was 5 hours. The main feature of these absorption spectra is a shift to lower energies after illumination. After illumination, the tail in absorption spectrum increased in this case, causing an decrease in the slope, i.e. a proportionality constant in $\sqrt{\alpha h\nu}$ vs $h\nu$ plot. In generally, however, this is not the case.

For convinience, we define a shift ΔE, the energy difference at $\sqrt{\alpha h\nu}$ = 150 $cm^{-1/2}eV^{1/2}$ in $\sqrt{\alpha h\nu}$ vs $h\nu$ plot. This difference is considered to be a relatively good estimate of the optical absorption edge shift, as the values of corresponding absorption coefficients are 8000~10000 cm^{-1} at these photon energies. In the case of Fig. 1, ΔE is about 140 meV. The shift ΔE may be expressed in terms of real absorption and the photon energy as

$$\Delta E \propto \eta \frac{I(1-T)}{h\nu}. \qquad (1)$$

Here, I is the intensity of illumination, T the transmittance of the film, $h\nu$ the illuminating photon energy. $I(1-T)/h\nu$ represents the number of photons absorbed in unit time, and η is a constant proportional to the quantum efficiency of

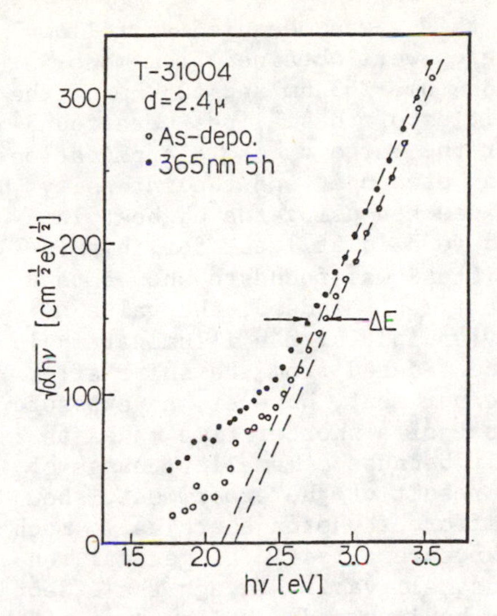

Fig. 1. Spectral change in optical absorption of an a-C:H film at room temperature by light illumination in vacuum (photodarkening). For the definition of the shift ΔE, see the text.

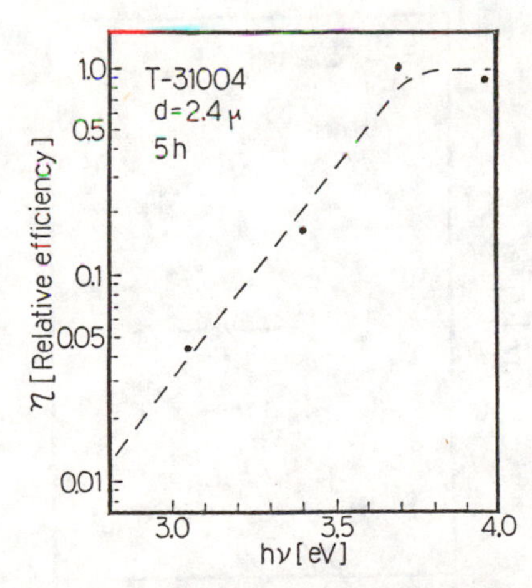

Fig. 2. Relative efficiency η of photodarkening as a function of illuminating photon energy. η is defined in the text. The dotted curve shows the approximate tendency of η vs $h\nu$ relation.

the edge shift.

The photon energy dependence of this η was measured as follows. Photons having different photon energies were obtained from monochromatic emission lines at 313 nm, 334 nm, 365 nm and 405 nm of the super high pressure Hg lamp. Intensities of these lines were found to differ as large as seven times for the largest case. Application of eq. (1) requires a linear relation between ΔE and the intensity of illumination. This linearity was checked using the highest intensity 365 nm line and was confirmed to hold at least for this intensity difference. However, the shift ΔE was found to show a tendency of saturation for long exposure times. Indeed, the value η deduced from the observed shift ΔE by an initial 2 h illumination is almost two times larger than the value deduced from the shift after 5 h illumination. In the following experiment, however, an exposure time of 5 hours was chosen, for the use of a shorter time makes an accurate determination of ΔE difficult because of smaller change of the spectrum. As can be seen in the result of the experiment, shown in Fig. 2, the difference in η for different photon energies is much larger than the difference for the exposure times. The general tendency of η vs hν relation of Fig. 2 is believable, though the effect of the saturation mentioned above may be involved. η increases with increasing photon energy, showing a tendency of saturation above photon energy about 3.6 eV. This effect is discussed later with infrared absorption change.

C. Temperature Effect on Photodarkening

In this section, an investigation was made on the effect of the film temperature on the photodarkening. The experimental results described below are yet preliminary, but seem to be essential for interpretation of the photodarkening phenomenon.

Figure 3 shows absorption spectra before and after optical exposure at room temperature and at liquid nitrogen temperature. The low temperature spectrum was measured at room temperature after optical exposure of the film at 77 K. In this case, the film was immersed in liquid nitrogen contained in a transparent quartz Dewar. The illuminating photon energy was 3.40 eV and the power density and exposure time were also similar as in the case of Fig. 1. From Fig. 3, the shift ΔE at room temperature is 110 meV, while ΔE at 77 K is 30 meV. The experimentally measured absorption spectrum did not show an

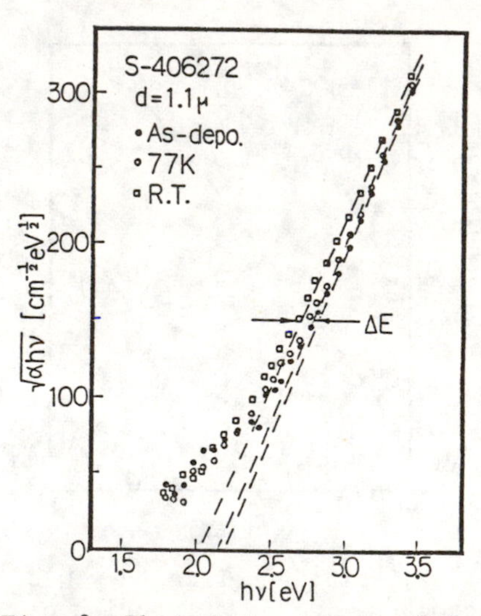

Fig. 3. Photodarkening at different temperatures. At 77 K, the film was immersed in liquid nitrogen and illuminated by 365 nm radiation from an Hg lamp.

appreciable change between 77 K and room temperature. Therefore, the absorbed photon numbers were almost the same at these temperatures. This fact can be interpreted as the efficiency for photodarkening, i.e. η at 77 K being low compared to room temerature.

D. Infrared Absorption Changes in Photodarkening and Photobleaching

A double beam infrared spectrometer (Hitachi Model 260-10) was used for infrared absorption measurements. A single crystaline silicon wafer similar to the one used for the substrate was inserted in the reference pass of the spectrometer.

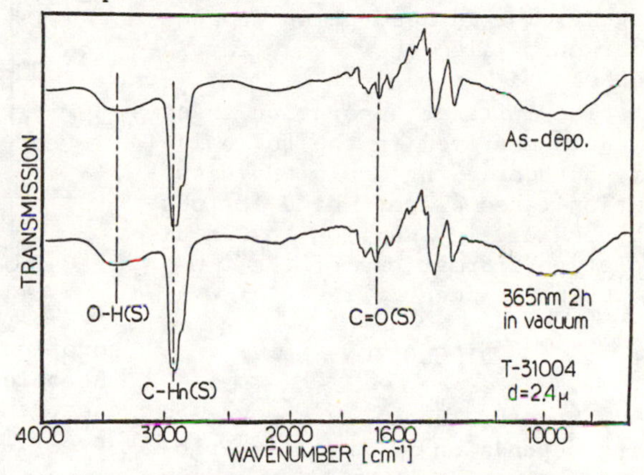

Fig. 4. Infrared absorption change for photodarkening of an a-C:H film. Vertical scale of transmission spectra is same relative units (but shifted).

Figure 4 shows infrared absorption spectra of an a-C:H film before and after photodarkening induced by the 3.40 eV photons. In this experiment, the power density of illumination was 0.4 mW/cm^2, much lower than the case of Fig. 1, taking into account a requirement of larger illuminated area suitable for the IR measurement. Main absorptions seen in the figure are OH at 3400 cm^{-1}(S)[2], CHn at 2900 cm^{-1}(S)[3], C=O at 1700 cm^{-1}(S)[4], CH_2 at 1450 cm^{-1}(B)[5] and CH_3 at 1350 cm^{-1}(B)[6,7] bands. If one makes a close comparison of the CHn absorptions, one can find a decrease after illumination. In order to examine this change in more detail, the peak absorption coefficient of the CHn bond was measured as a function of illuminating time, together with the OH and C=O absorptions. Figure 5 shows the measured CHn, OH and C=O absorptions as a function of illuminating time for the same film as in Fig. 4. Similar measurements for photobleaching were also performed in comparison. The result is shown in Fig. 6. In this case, illuminating source was the 488 nm line of Ar+ laser and the power density used was 20 mW/cm^2. Comparison of Figs. 5 and 6 reveals clear differences between these

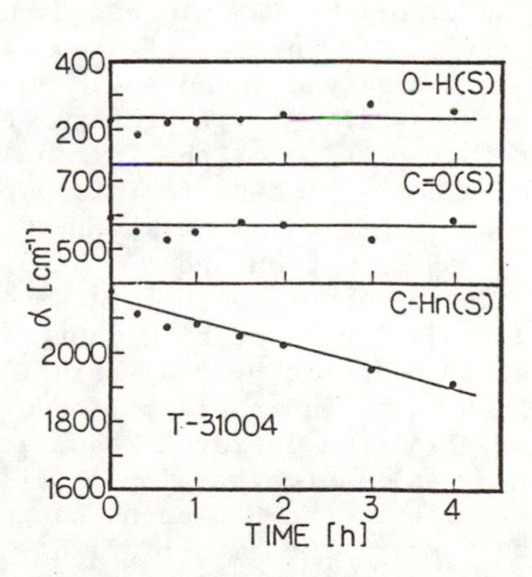

Fig. 5. Changes of OH, C=O and CHn infrared absorption bands as a function of illuminating time for photodarkening of the same film as in Fig. 4.

effects. The CHn absorption decreases linealy with time for photodarkening as well as photobleaching. The C=O and OH absorptions for photodarkening can be regarded almost constant until exposure time of at least 4 h. While in the photobleaching the C=O absorption increases with time rapidly at early stages and more slowly at later stages. The OH absorption in photodarkening also shows an clear increase.

The infrared absorption changes may be summarized in the following. For photodarkening, only CHn dissociation takes place, while for photobleaching, in addition to CHn dissociation, formations of the C=O and OH bonds occur.

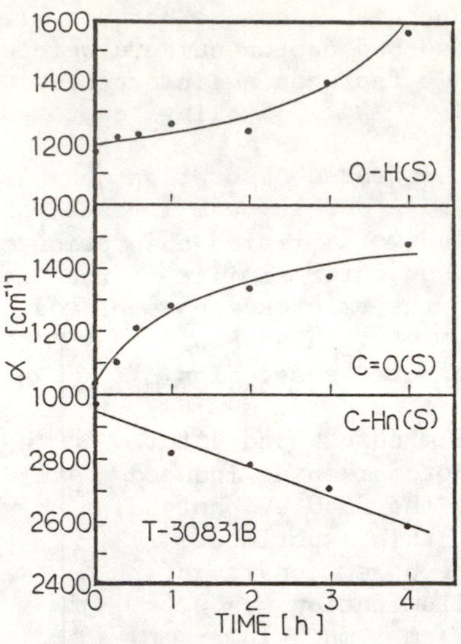

Fig. 6. Changes of the same absorption bands as in Fig. 5 for photobleaching in air. In this case, the light source used was an Ar+ laser.

DISCUSSION

The efficiency for photodarkening depends on illuminating photon energy as described in section B. Apart from the saturation of the shift with time which effectively decreases η compared to the unsaturated case, inhomogeneous optical absorption also affects the photon energy dependence of η. The absorption coefficient α for 3.4 eV photons which were used mostly for the illuminating source is ~17000 cm^{-1} for the film of Fig. 1. The inverse of absorption coefficient α^{-1}, a measure of absorption depth, is only 0.6 μm which is small compared with film thickness 2.4 μm. Therefore, the darkening is thought to occur only at thin portion of the film near the surface. Accordingly the real shift corresponding to this portion near the surface is considered to be larger than the observed shift in Fig. 1. Similarly α^{-1} at 3.97 eV is 0.2 μm and α^{-1} at 3.06 eV is 1.2 μm. The difference between the observed shift and the real shift corresponding to α^{-1} depth is larger at high energy region than at low energy region. The tendency in η vs $h\nu$ relation in Fig. 2 is considered to be much enhanced in the real portion of the film where photodarkening occurs.

From the infrared measurements, a common change for photodarkening and photobleaching is dissociation of the CH bonds. If we assume that the C=O bond change is correlated with CHn bond change, and compare the amount of the C=O bond change in Fig. 3 of the previous report[1] with the changes of C=O and CHn bonds of Fig. 6, corresponding transmission spectrum change at CHn bond in Fig. 3 of the previous report is expected to be only 2-3 %. It is not unreasonable that this small change could not be detected before. As reported before[1], these effects are not due to a thermally activated process.

It is reported[8] that hydrocarbons exhibit photo-decomposition by ultraviolet radiation. The reported dissociation energy for the CH bond is 3.77 eV[9] or 3.51 eV[10]. The energy at which η ceases to increase in Fig. 2 is about 3.6 eV and quite close to the reported dissociation energy of the CH bond. This result also indicates that the CH bond dissociation plays an key role in the photoinduced effects of a-C:H films and supports the conclusion from the infrared results. However, η vs $h\nu$ relation does not show an abrupt drop below 3.6 eV. This fact seems to imply that below 3.6 eV thermal energy may supply an additional energy increment to CH bond dissociation. The result of temperature dependence of the shift ΔE shown in Fig. 3 is at least qualitatively consistent with this idea. Using photons with larger photon energy than the CH dissociation energy, similar experiment as in Fig. 3 should be performed to be compared with the result of Fig. 3, since in this case participation of thermal energy is not expected.

The photo-dissociation efficiency γ may be defined as the ratio of numbers of dissociated bonds and absorbed photons. From Fig. 5, a peak absorption change $\Delta\alpha$ is 260 cm^{-1} after 4 h illumination of 365 nm radiation. The number of dissociated CH bonds ΔN is estimated from the absorption change $\Delta\alpha$ by the following approximate relation[11].

$$\Delta N = As \frac{\Delta\alpha}{\nu} \Delta\nu, \qquad (2)$$

where ν is the wavenumber at absorption peak, $\Delta\nu$ a half width of absorption band, and As a proportionality constant. The number of absorbed photons can be calculated from the real power density of 0.4 mW/cm^2 and the absorption coefficient 17000 cm^{-1} at 365 nm. From these values, and the proportionaly constant As of 8.7×10^{20} cm^{-2} reported by Shimizu et al.[12], the photo-dissociation efficiency γ for the CH bond is estimated to be 0.2 under 365 nm illumination. This value is not unreasonable compared with a reported value of 0.32[13] for hydrogen creation efficiency from gaseous CH_4 by ultraviolet radiation.

In the case of photobleaching, as reported before[1], the C=O bond formation was observed and the change with time is seen in Fig. 6. We interpret this fact as oxidization of the carbon atoms dissociated from CH bond at the surface of the film. Dissociated carbons away from the surface cannot react with oxgen to form C=O bond. This could be the reason of the saturation of C=O bond absorption in Fig. 6. It should be noted that C=O and OH bond formations occur as an aging effect of the film in air kept in the dark in a time scale much longer than the case under illumination[14]. For example, a few hour illumination of Ar+ laser with similar power level as in Fig. 6 corresponds to the change after a few tens of day as the aging effect. It can be considered that the illumination of light to the film in air effictively accelerate the aging effect.

In photodarkening, the compositional change found is only the dissociation of the CH bonds. Therefore, the shift of absorption edge to lower energy as shown in Fig. 1 is considered to correspond to the decrease of hydrogen concentration in the film. Watanabe et al.[15] reported the preparation of a-C:H films at different substrate temperatures by glow discharge decomposition of C_2H_4. The data in their report show that the films prepared at different substrate temperatures have different hydrogen concentrations and that the optical gap changes to lower energy with decrease of hydrogen concentration. The shift of optical gap to lower energy with illumination i.e. photodarkening is at least qualitatively consistent with this data. For quantitative comparison, systematic study is necessary on the optical gap change with hydrogen concentration in the film. In photobleaching, the absorption edge shift to higher energy is presumably related to the oxidization of the film. Systematic study is also required in this case.

The photodarkening in a-C:H films should be compared with the reversible photodarkening in films of chalcogenide glasses[16,17]. In chalcogenides, the edge shift by illumination increases with a decrease in the film temperature and this shift can be recovered by thermal annealing of the films at elevated temperatures. In contrast, the edge shift in a-C:H shows opposite dependence on the film temperature as appeared in Fig. 3 and the shift cannot be recovered by thermal annealing. The process reported here is essentially irreversible, although an apparent reversible gap shift between photodarkening and photobleaching was reported[1].

On the other hand, as reported before[1] there is a similarity between the photobleaching of a-C:H and irreversible photobleaching in amorphous chalcogenide glass $As_{2+x}S_3$ $(x>0)$[18]. In both cases, photo-decomposition as the first step plays an essential role in the effects. But there is a difference in the second processes. In the case of a-C:H, most hydrogen atoms probably remove from the film and carbon atoms become oxidized at the surface, while in $As_{2+x}S_3$, photodissociated As atoms makes crystalline As_2O_3 at the surface and optical absorption is determined by a-As_2S_3 network.

CONCLUSION

Photodarkening and photobleaching effects in a-C:H films were discussed with experimental data on the excitation spectrum, infrared absorptions and temperature dependence of the effect. In the case of photodarkening, it was shown, directly from infrared absorption spectra of the film and indirectly from the excitation spectrum of the effect, that CH bond dissociation and resulting decrease of hydrogen concentration are induced by optical exposure of the film. The temperature dependence of photodarkening seems to imply that thermal energy also participates the process. The efficiency of this CH bond dissociation is relatively high, for example, 0.2 by 365 nm illumination.

In the case of photobleaching, in addition to this CH bond dissociation, the C=O and OH bond formations are seen. Infrared measurements showed that photobleaching has a feature of light-accelerated aging of the film in air.

For further clarification of these photoinduced effects, quantitative comparisons are required between the optical gap shifts and the concentrations of hydrogen and oxygen in the films.

ACKNOWLEDGEMENTS

This work was partly supported by Grant-in-Aid for Scientific Research from the Ministry of Education of Japan.

REFERENCES

1. S.Iida and S.Ohki, Japn, J. Appl. Phys. 21, L62 (1982).
2. T.Imura, K.Ushita and A.Hiraki, Japn. J. Appl. Phys. 19, L65 (1980).
3. R.S.Sussmann and R.Ogden, Philos. Mag. B 44, 137 (1981).
4. For example, R.M.Silverstein, G.C.Bassier and T.C.Morill, Spectrometric Identification of Organic Compounds (John Wiley & Sons Inc., New York, 1974) 3rd ed.
5. H.Kobayashi, A.T.Bell and M.Shen, J. Appl. Polym. Sci. 17, 885 (1973).
6. G.Lucovsky, R.J.Nemanich and J.C.Knights, Phys. Rev. B 19, 2064 (1979).
7. C.Janford, Physical Chemistry of Macromolecules (Wiley, New York, 1961) p.63.
8. Kagaku Binran Kisohen (in Japanese) (ed. Chemical Soc. Japan, Maruzen, Tokyo, 1975) p.1100.
9. Tables of Interatomic Distances and Configuration in Molecules and Ions (ed. The Chemical Soc., London, 1958).
10. The same book as ref.8, p.977.
11. A.Guivarc'h, J.Richard, and M.Le Contellec, E.Ligeon and J.Fontenille, J. Appl. Phys. 51 2167 (1980).
12. T.Shimizu, K.Kumeda and Y.Kiriyama, Proc. Intern. Conf. Tetrahedrally Bonded Amorphous Semiconductors, Carefee, Arizona, 1981(AIP Conf. Proc. No.73, AIP, New York, 1981) p.171.
13. The same book as ref.8, p.1101.
14. T.Ohtaki and S.Iida, unpublished.
15. I.Watanabe, S.Hasegawa and Y.Kurata, Japn. J. Appl. Phys. 21, 856 (1982).
16. K.Tanaka and M.Kikuchi, Appl. Phys. Letters 26, 243 (1975).
17. H.Hamanaka, K.Tanaka and S.Iizima, Solid State Commun. 33, 355 (1980).
18. K.Tanaka and M.Kikuchi, Amorphous and Liquid Semiconductors, eds. J.Stuke and W.Brenig (Taylor & Frances, London, 1974) vol.1, p.439.

LIGHT-INDUCED EFFECT IN a-Si_xC_{1-x}:H FILMS*

Chen Guanghua, Zhang Fangqing, Wang Yinyue, Wang Huisheng and Xu Xixiang
Physics Department of Lanzhou University, Lanzhou, Gansu, China

T. Shimizu
Department of Electronics, Kanazawa University, Kanazawa 920, Japan

ABSTRACT

The paper presents the light-induced effect of n- and p-type GD a-Si_xC_{1-x}:H films by the use of MOSFET structure and coplanar contacts. It was found that, after prolonged light illumination, the gap state density increases and the photoconductance shows a slight decrease. The dark conductance of B state is related to E_A (the A state activation energy).

1. INTRODUCTION

In recent years, Staebler-Wronski effect[1] has attracted a great deal of attention. Lately, Tanielian, Goodman and Fritzsche[2] observed the reverse-effect for some undoped a-Si:H films and also found the increase in the gap state density by field effect measurement. The increase of spin density after prolonged intensive illumination and the temperature dependence of light-induced conductivity in a-Si:H films were reported by Derch et al.[3] and Jin Jang et al.[4], respectively. Recently, Morimoto et al.[5] observed the increase in ESR spin density, the decrease in photoconductivity and the fatigue in photoluminescence on samples of a-Si:H, a-Si_xN_{1-x}:H and a-Si_xC_{1-x}:H after prolonged illumination at room temperature. In this work, we report the light-induced increase in the gap state density studied by the field effect method and the photo- and darkconductance changes for n- and p-type GD a-Si_xC_{1-x}:H films.

2. EXPERIMENTAL AND RESULTS

The MOSFET structure was used for field effect mea-

* Projects Supported by the Science Fund of the Chinese Academy of Sciences

surement. A 1μm layer of high-quality SiO_2 was first deposited on n^+-c-Si (crystalline silicon) substrate; then a 0.25μm thick a-Si_xC_{1-x}:H (1-x=0.1) film was deposited by glow-discharge decomposing SiH_4 and CH_4 gas mixture. Through consequent light-etching and vacuum-evaporating, the source (S) and drain (D) electrodes (the spacing between S and D electrodes was 100μm) were made. The gate-electrode (G) was obtained from evaporating Al on the other side of c-Si substrate.

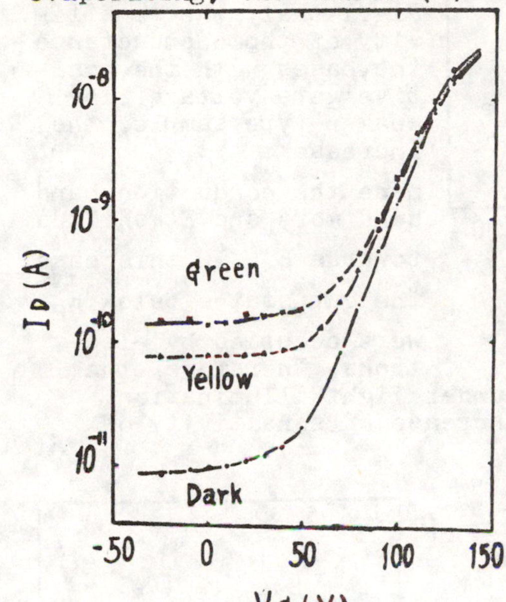

Fig.1. The curves of I_D vs. V_G under different light or in dark.

Coplanar contacts were used for conductance measurement. Al was deposited on the a-SiC:H film as the electrode and the spacing between two electrodes was 1mm. The curve of I-V was measured before the conductance measurement to examine the Ohmic contact. The sample was first annealed (in vacuum, at about 170 °C) for half an hour, i.e. sample being in A state. After prolonged illumination, the sample was in B state.

Figure 1 shows the I_D vs. V_G curves of a-Si_xC_{1-x}:H film without illumination and with illumination (0.2 mW/cm^2) respectively. From the dark-field-effect curve, we calculated the distribution of gap-state density N(E) (see Fig.2). The density of states around Fermi level $N(E_F)$ is about $7\times10^{17}/cm^3ev$ and is larger than that in a-Si:H film (under similar preparing condition), which was resulted from the defect states induced by carbon introduction. But for low-carbon-content samples, the distribution of DOS is approximately similar to that of a-Si:H.

We can see from Fig.1 that the photoconductance of sample is more sensitive to green light (λ=5400Å) than to yellow light (λ=5900Å). When T=330K, extented state conduction is dominant in sample, and for low-carbon-content sample, either yellow light or green light can excite intrinsic photoconductance. The experimental

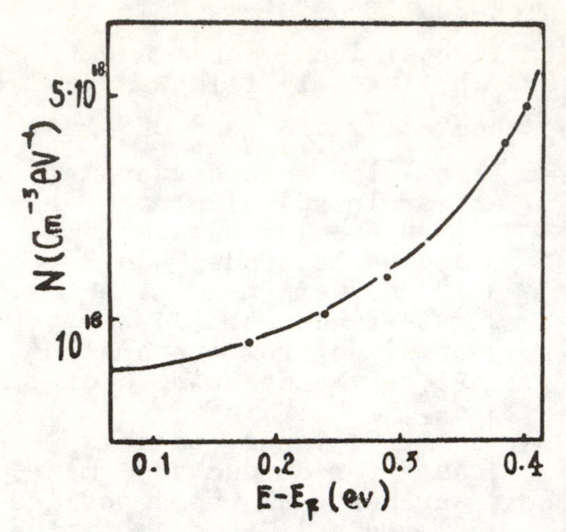

Fig.2. DOS N(E) from dark field effect measurement.

results show that the sample's absorption coefficient in green light region is larger than in yellow light region.

It can be seen from Fig.1 that, the sensitivity of photoconductance increases with the positive gate voltage. For weak n-type sample, the increase of V_G made the conduction band bend more and E_F shifted towards E_c. In this case, the gap states below E_F were occupied by electrons. Therefore, the generation rate increased under light illumination, and this resulted in the increase of sensitivity of photoconductance.

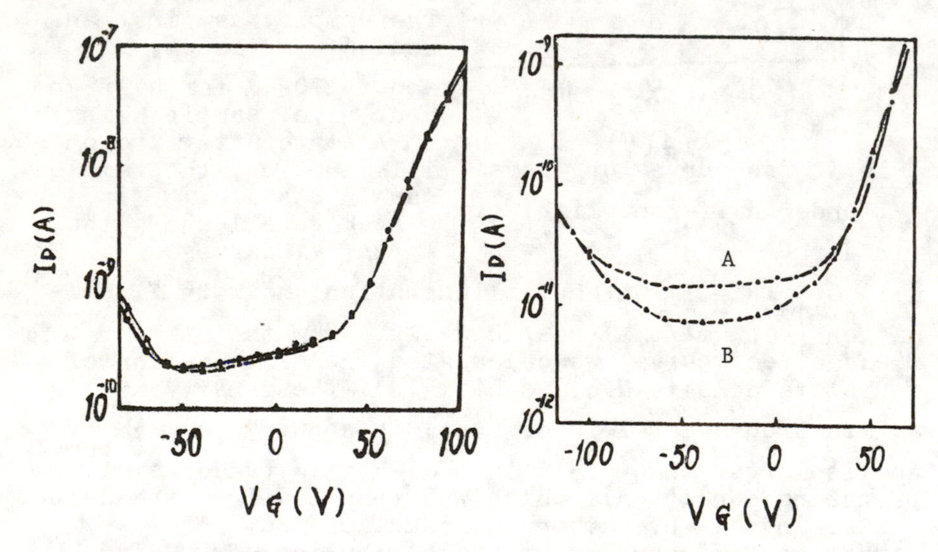

Fig.3. I_D vs. V_G curve of a-SiC:H film with high DOS.
● before illumination
▲ after illumination for 4 hours (200mW/cm^2)

Fig.4. I_D vs. V_G curve of a-SiC:H with lower DOS. The symbols and condition are the same as Fig.3.

The results of dark-field-effect measurement of different sample (before and after illumination) are shown in Fig.3 and Fig.4 respectively. The DOS of sample in Fig.3 is high ($N(E_F) \sim 10^{18}/cm^3ev$). No apparent change of dark-field-effect curve can be found between state A and state B. A probable explanation is that the change of defect state density caused by light illumination is lower than the DOS of sample before illumination so that influence of illumination on the DOS of this sample is not so obvious.

The sample illustrated in Fig.4 had a better light sensitivity, its DOS (in A state) $N(E_F)$ was $7\times10^{17}/cm^3ev$. When being in B state, its conductance near zero gate voltage became smaller than in A state and showed a positive Staebler-Wronski effect. $N(E_F)$ in B state can be obtained by calculation, $N(E_F)$ is about $10^{18}/cm^3ev$ under $200mW/cm^2$ illumination for 4 hours. This approximately showed the magnitude of new gap states resulted from prolonged illumination. Annealing can remove the light-induced effect.

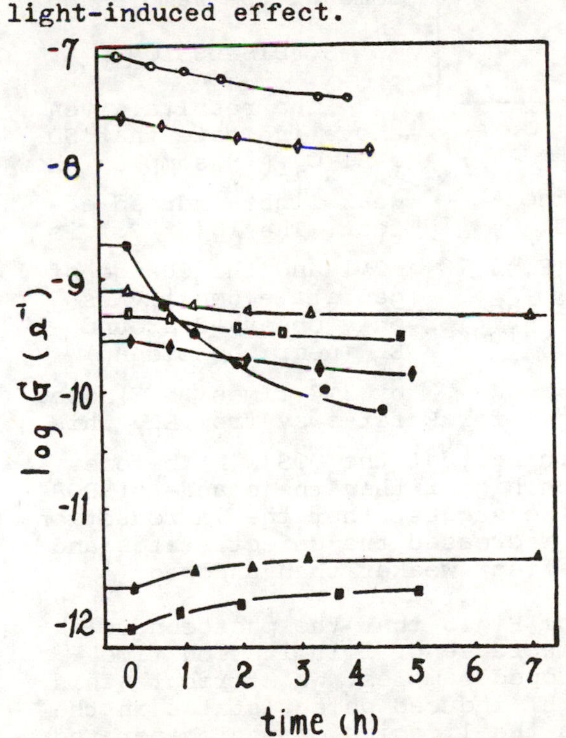

Fig.5. Photo- and dark conductance of n-type a-SiC:H vs. illumination time.

E_A(ev)	dark	photo
0.56	●	○
0.58	♦	◊
0.76	▲	△
0.81	■	□

Figure 5 shows the dependence of photo- and dark-conductances on illumination time t for n-type samples with activation energies of 0.81, 0.76, 0.58 and 0.51ev, respectively. The samples were kept at room temperature during the exposure to light ($100mW/cm^2$). It is obvious that the curves for all samples go down slowly with t and then reach constants. It is also seen that for samples with E_A=0.51 and 0.58ev, the dark conductance decreases with t, i.e. $G_B/G_A<1$, which presents a normal S-W effect. On the other hand, the samples with E_A=0.76 and 0.81ev, $G_B/G_A>1$, show a reverse S-W effect.

In Fig.6, we give the dependence of photo- and dark conductances on t for a p-type sample with E_A=0.78ev. The illumination condition is the same as mentioned before. Obviously it also shows a reverse effect as some n-type samples do.

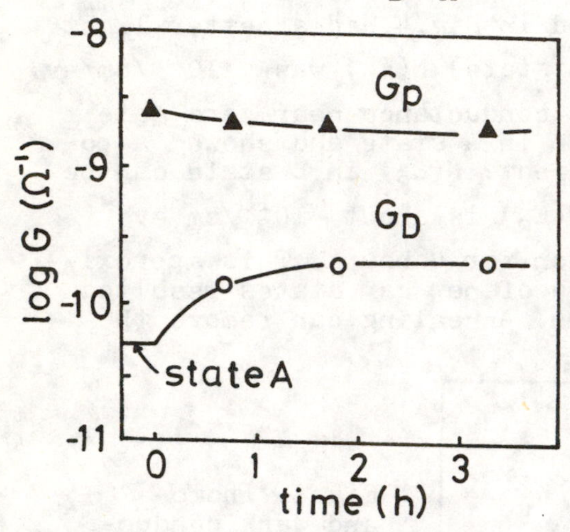

Fig.6. Photo- and darkconductance of p-type a-Si_xC_{1-x}:H vs. exposure time. (E_A=0.78 ev). "▲", the values of photoconductance; "o", the values of darkconductance.

3. DISCUSSION

The results given above indicate that GD a-Si_xC_{1-x}:H samples show weak light-induced effect. The $lg(G_B/G_A)$ is -2∿1 and the change of gap state density is only observed around E_F (near the midgap) by 2∿4 times; N(E) does not show an obvious change for E far away from E_F. This can be explained by noticing that the DOS N(E) before exposure to light is much higher than the change of DOS caused by illumination. We suggest that the introduction of carbon atoms in film increased the defect states and made the light induced effect weaker than undoped a-Si:H film.

It is also seen from Fig.5 that the photoconductances decrease as the increase of illumination time t for both undoped and doped samples. We attribute this to the generation of light-induced defect states in the band gap which decreases the lifetime of photogenerated carriers. This is agreeable to the model suggested by Staebler and Wronski[1].

The observed normal and reverse S-W effects shown in Fig.5 and Fig.6 are found to be related to the position of the Fermi level E_F. The effects can tentatively be explained by the two-phase model in which the a-Si_xC_{1-x}:H film can be considered as the combination of Si, C disorder network terminated by H atoms and the hydrogen-rich trapping region. When E_A is small, the position of E_F is near conduction band edge for n-type samples and a large number of electrons accumulate in the disorder network. These electrons transfer into the trapping region under light exposure and hence cause decrease in dark conductance, i.e. the normal light-induced effect. While E_A is larger, E_F locates near the midgap under illumination, those electrons trapped in the trapping region transfer into the disorder network and result in an increase in dark conductance, i.e., the reverse-effect. The effects can be recovered by annealing in vacuum. However, we can not exclude the possibility that these effects are explained by a bond breaking caused by a prolonged illumination[5].

Because of the complexity of the distribution of gap states, the microstructure of gap defects is far from clear up to date. Therefore S-W effect worths researching further.

ACKNOWLEDGEMENT

The authors greatly thank Mr. Wu Xiangchen, Zhang Qingchun and Miss Fan Yaju for partial experiment.

REFERENCE

1. D. L. Staebler and C. R. Wronski, Appl. Phys. Lett. 31 (1977) 292; J. Appl. Phys. 51 (1980) 3262.
2. M. H. Tanielian, N. B. Goodman and Fritzsche, J. Phys. Paris C4 Suppl 42 (1981) 375.
3. H. Dersh, J. Stuke and J. Beichler, Appl. Phys. Lett. 38 (1981) 465.
4. Jin Jang and Choochon Lee, J. Appl. Phys. 54 (1983) 3943.
5. A. Morimoto, H. Yokomichi, T. Atoji, M. Kumeda, I. Watanabe and T. Shimizu, this conference.

PHOTOCONDUCTIVITY DETECTED S-W RELATED DEFECT LEVELS IN a-Si:H

A. Tříska, M. Vaněček, O. Štika, J. Stuchlík,
A. Kosarev † and J. Kočka
Institute of Physics, Czechoslovak Academy of Sciences
180 40 Prague 8, Czechoslovakia

ABSTRACT

A constant photocurrent method was used to study the Staebler-Wronski effect on a-Si:H samples with a coplanar configuration of electrodes on Schottky diodes. Results on the coplanar samples are partly influenced by surface (interface) band bending but both samples show a higher value of optical absorption coefficient α in an absorption shoulder after prolonged illumination. Some experimental details of the method are discussed. A model for the density of states in p-type a-Si:H material is presented in more detail. A main feature is the existence of a minimum in the density of states about 0.6 - 0.7 eV above the valence band. Some problems connected with deconvolution of α are outlined. Preliminary measurements of the spectral dependence of photocapacitance are presented.

INTRODUCTION

Reversible photoinduced changes in the properties of a-Si:H, the so-called Staebler-Wronski effect[1,2] have attracted much attention but the exact mechanism which is responsible for this effect still needs to be elucidated.

Many methods have been used to study the S-W effect. The important conclusion that silicon dangling bonds are created by illumination has been demonstrated by ESR[3] and LESR.[4] We have demonstrated by our Constant Photocurrent Method (CPM)[5] that after prolonged illumination the optical absorption coefficient α corresponding to the gap states absorption increases[6] and that an annealing at 160°C reversibly restores the original value. Samples with coplanar configuration of electrodes were used. Independently Amer et al.[7] have demonstrated by PDS (Photothermal Deflection Spectroscopy) the same effect.

Jackson and Thompson have shown[8] the influence of the band bending at the surface on photoconductivity of a-Si:H samples with coplanar configuration of electrodes. To eliminate the influence of the band bending we have prepared in the same run undoped samples (thickness 1.7 μm) with coplanar geometry of electrodes and Pt Schottky diodes on n^+ crystalline Si substrate. This substrate was used to exclude cross contamination from the phosphine doping in the usual (glass, NiCr, n^+ a-Si:H, intrinsic a-Si:H, semitransparent Pt) structure. The samples have been prepared at standard conditions

†On leave from FTI, Academy of Sciences, Leningrad, USSR.

0094-243X/84/1200272-08 $3.00

(T_s = 250°C, pressure 0.2 mbar, flow rate 30 sccm of pure SiH_4) by r.f. capacitive glow discharge decomposition.

The results of the S-W effect are presented in the first part, then the results of PDS are compared with our CPM results. A more detailed density of states on p-type a-Si:H is presented, and in the end some comments and conclusions are made.

STAEBLER-WRONSKI EFFECT

In Fig. 1 is shown the spectral dependence of the optical absorption coefficient found by low frequency (4 Hz) ac CPM[5] on an undoped a-Si:H sample with a coplanar geometry of electrodes. The sample is in a state A (annealed at 160°C for 1 hour in air) and after prolonged illumination--state B. The changes were reversible. In Fig. 2, a similar change of the optical absorption coefficient is displayed for undoped (forward biased) Shottky diode on n^+ crystalline Si substrate in state A and B. A few remarks are necessary.

1) The smaller effect on Schottky diodes can be explained by the fact that transmission of the Pt semitransparent contact was less than 50%, so that the effective illumination dose was smaller than that for the coplanar sample.

2) Owing to matching the index of refraction of the crystalline Si substrate with that of the amorphous Si layer, the interference fringes are missing for Schottky diode samples. This is a good way of eliminating the interference fringes.

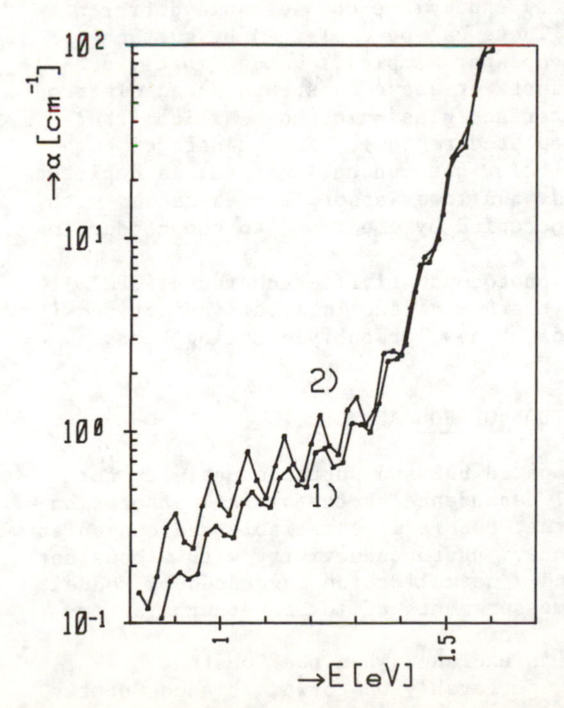

Fig. 1. Spectral dependence of the optical absorption coefficient found by ac CPM on the undoped coplanar a-Si:H sample: (1) State A (1 hour at 160°C), State B (10 hours AM1 illumination).

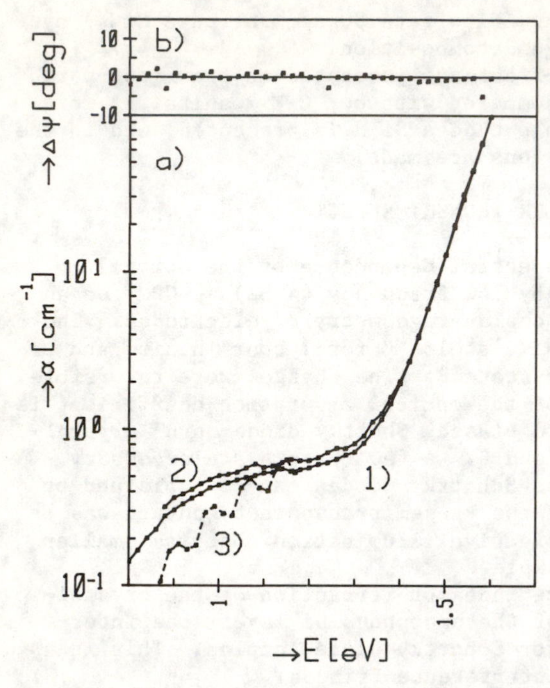

Fig. 2a. Spectral dependence of the optical absorption coefficient found by ac CPM (4 Hz) on the undoped a-Si:H Schottky diode: 1) State A (1 hour at 160°C); 2) State B (10 hours AM1 illumination, see text). Dashed curve 3) - state A of coplanar sample from Fig. 1 is shown for comparison.

Fig. 2b. The changes of phase shift of the ac CPM signal, corresponding to curve 1) in Fig. 2a.

3) When we compare typical results for coplanar and Schottky diode samples (Fig. 2, curve 3) and 1)) we can see some differences below approximately 1.1 eV. These can be explained by surface (interface) band bending on coplanar samples. We use mostly predeposited NiCr strips on a quartz or glass substrate as an electrode, so we suppose that at the interface glass-amorphous silicon there is an upward band bending (depleted region). Thus a part of an initially occupied gap state below the conduction band) is depleted of electrons and this results in a lower absorption connected with transitions from gap states occupied by electrons to the conduction band.

We can conclude that by photoconductivity techniques (CPM) it is clearly demonstrated that the S-W effect is a bulk effect and it is connected with the creation of new, probably dangling bond, states.

COMPARISON OF PDS AND CPM

Jackson et al.[9] have compared PDS and photoconductivity for determination of the spectral dependence of the optical absorption coefficient α and concluded that PDS is more reliable and convenient. They have used standard secondary photoconductivity with a constant number of incident photons and a normalization suggested by Moddel et al.[10] The complementary measurements of lux-ampere characteristics are necessary.

This tedious procedure[9] is excluded when the Constant Photocurrent Method[5] is used, perferably on forward biased Shottky diodes.[11] In this method the optical absorption coefficient is obtained as an inverse of the number of photons necessary to keep the photocurrent constant. Holding the photocurrent constant means the stabilization of the quasi-Fermi level. This is a necessary but not always a sufficient condition in CPM. To stablize the recombination traffic (and so the lifetime or the more easily experimentally accessible response time), when scanning a probe light beam from infrared to the visible part of the spectra, we need to keep a quasiequilibrium in occupation of all gap states. It can be done by additional strong infrared light or by the strong red light (but the latter moves the quasi-Fermi levels far apart).

We can check for a change of the response time by measurement of the spectral dependence of a phase shift of the photocurrent. We can also directly measure the spectral dependence of the response time.[12] It is usually constant at room temperature but the situation at low temperatures can be more complicated.[12]

For the samples investigated here (Figs. 1 and 2) we have checked that there is no change in the phase shift of the ac photocurrent. It was measured by EG+G 5206 lock-in-analyzer, and it is shown in Fig. 2b.

Can we say whether PDS is better than CPM or vice versa? PDS measures total absorption, and we get a correct value of the optical absorption coefficient α. But on the basis of PDS we cannot decide what kinds of optical transitions correspond to the measured α. On the other hand, the absorption coefficient α derived from the photoconductivity technique involves only optical transitions leading to the conducting state (directly or by additional thermal excitation). The question whether the photoconductivity is due to the electrons in the conduction band or holes in the valence band can be answered, e.g., from the knowledge of the sign of photothermopower.[13]

So we believe that both PDS and CPM are useful and one very well complements the other. The systematic difference between PDS and CPM below 0.9 eV is caused sometimes by surface band bending or by a spectral dependence of the lifetime. But if we carefully exclude the above effects the remaining difference must be explained by the fact that PDS also measures localized-to-localized state transitions, not seen in CPM.[11]

DENSITY OF STATES IN P-TYPE a-Si:H

In Fig. 3 the spectral dependence of the optical absorption coefficient (found by dc CPM, on a 100 ppm B_2H_6 doped a-Si:H coplanar sample in state A) is shown.[13] Prolonged illumination again raises the value of α in the absorption shoulder (not shown in Fig. 3).

The measurements were done at three different temperatures. Because the sign of photothermopower in the absorption shoulder is positive,[13] we did the deconvolution of a Guassian deep level with the valence band. Density of states deduced from the 300 K curve of Fig. 3 by deconvolution of α[11] is shown in Fig. 4. For comparison the density of states found on similar samples by Beichler et al.[14] is shown in Fig. 4. Their results, found by transient capacitance on Schottky diodes, support our results measured on coplanar samples.

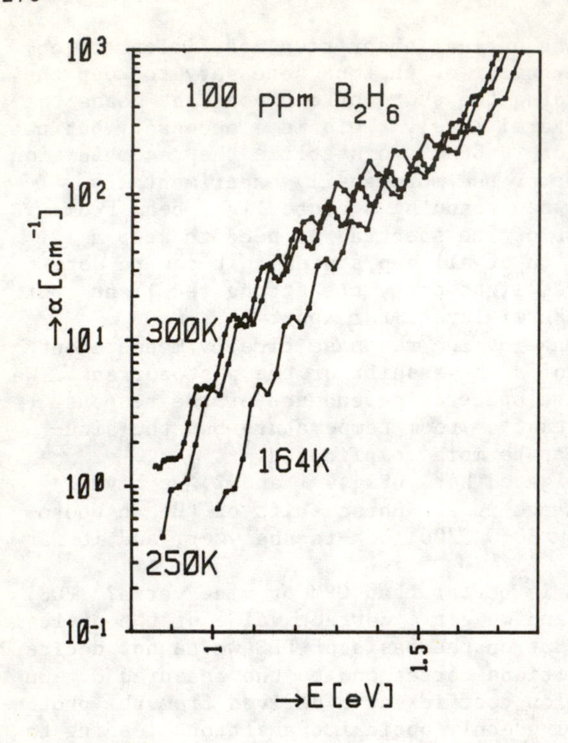

Fig. 3. The spectral dependence of the optical absorption coefficient, measured by dc CPM on a B_2H_6 doped coplanar a-Si:H sample at different temperatures.

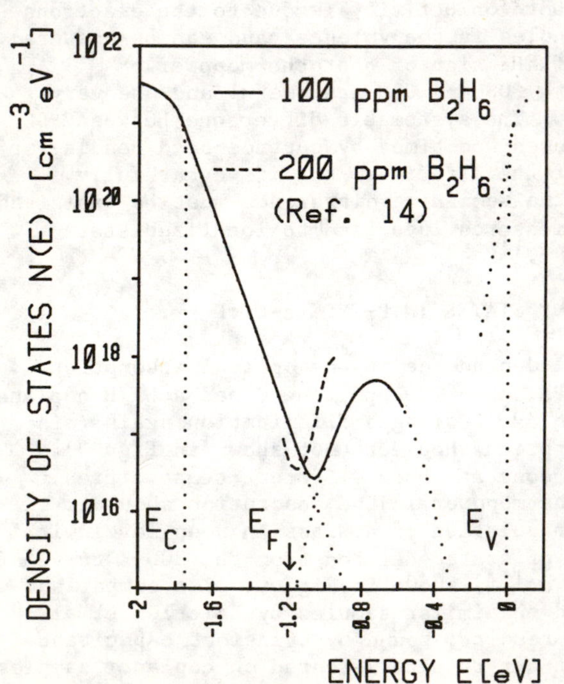

Fig. 4. Density of states deduced from the 300 K curve of Fig. 3 for 100 ppm B_2H_6 doped a-Si:H (full curve). For comparison the resulting density of states of 200 ppm B_2H_6 doped a-Si:H from ref. 14 is shown (dashed curve).

The results by Tiedje et al.[15] on a 100 ppm B_2H_6 doped a-Si:H sandwich structure strongly support our photothermopower conclusion that holes are the main photocarriers in a shoulder region below 1.4 eV.

The temperature dependence of α in Fig. 3 can be explained if we consider not only the transitions of holes from a deep level to the parabolic valence band but also transitions to the exponential valence band tail (relatively broad, compared to the sharp conduction band tail) with subsequent thermal emission of holes below the valence band mobility edge. We are modifying our convolution procedure for p-type material to take this effect into account.

SOME COMMENTS

1. At present, general agreement is growing about the total number of the deep states in a-Si:H found by several different methods, but the energy dependence of the density of states in the gap is still uncertain. The precise position of peaks in the density of states in undoped and doped material, in state A or B, deduced by deconvolution of α should be very important and it could help us to decide, e.g., which of the models[16,17,18] is more suitable. Basic assumptions of the deconvolutions of α[11] (constant optical gap), limit the possiblity to discover more details in the density of gap states. Further improvements could be reached in the near future by direct experimental verification or modification of the above assumptions.

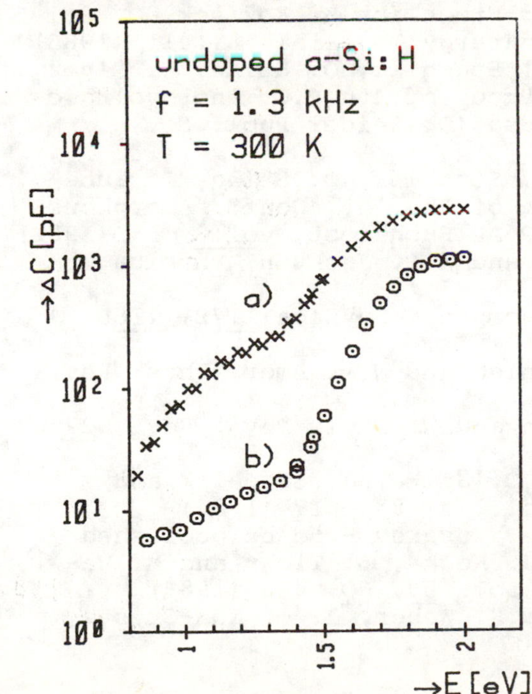

Fig. 5. The spectral dependence of the photocapacitance ΔC of the undoped a-Si:H Schottky diode. Curve a) without and b) with the white light bias illumination.

2. For CPM measurements we prefer Schottky diodes because the influence of the surface can be controlled, but then we have to substitute the measurement of photothermopower (done on coplanar samples). We have tried to use photocapacitance for this purpose.

The first results of the spectral dependence of the photocapacitance ΔC on an undoped a-Si:H Schottky diode are shown in Fig. 5. The method is sensitive so that not only the sign but the whole spectral dependence of ΔC can be measured.

To get correct information, bias illumination is necessary,[19] first to increase the conductivity of the neutral bulk region and also for other reasons.[20] The ΔC is positive as it should be for undoped a-Si:H, where the photocarriers are electrons. Before definitive conclusions based on the photocapacitance results can be drawn, measurements on more n-type and mainly on p-type a-Si:H Schottky diodes have to be done.

REFERENCES

1. D.L. Staebler and C.R. Wronski, Appl. Phys. Lett. 31, 292 (1977)
2. D.L. Staebler and C.R. Wronski, J. Appl. Phys. 51, 3262 (1980)
3. H. Dersch, J. Stuke and J. Beichler, Appl. Phys. Lett. 38, 456 (1981)
4. R.A. Street, P.K. Biegelsen and J.C. Knights, Phys. Rev. B24, 969 (1981)
5. M. Vaněček, J. Kočka, J. Stuchlík, Z. Kožíšek, O. Štika and A. Tříska, Solar Energy Materials 8, 411 (1983) J. Kočka, M. Vaněček, J. Stuchlík, O. Štika, E. Šípek, H.T. Ha and A. Tříska, Proc. of 4th E.C. Photovoltaic Solar Energy Conf., Stresa (D. Reidel Publ. Co., Dordrech, 1982), p. 443
6. J. Kočka, M. Vaněček, J. Stuchlík, O. Štika, I. Kubelík and A. Tříska, Proc. of the Int. Conf. "Amorphous Semiconductors '82", CIP-AP Bucharest, vol.3, 150(1982)
7. N.M. Amer, A. Skumanich and W.B. Jackson, Physica 117B, 118B, 897 (1983)
8. W.B. Jackson and M.J. Thompson, Physica 117B, 118B, 883 (1983)
9. W.B. Jackson, R.J. Nemanich and N.M. Amer, Phys. Rev. B27, 4861 (1983)
10. G. Moddel, D.A. Anderson and W. Paul, Phys. Rev. B22, 1918 (1980)
11. M. Vaněček, A. Abrahám, O. Štika, J. Stuchlík and J. Kočka, phys. stat. sol. (a) 83, 617 (1984)
12. J. Kočka, O. Štika and A. Kosarev - to be published
13. A. Tříska, I. Shimizu, J. Kočka, L. Tichý and M. Vaněček, J. of Non-Cryst. Sol. 59, 60, 493 (1983)
14. J. Beichler, H. Mell and K. Weber, J. of Non-Cryst. Sol. 59, 60, 257 (1983)

15. T. Tiedje, B. Abeles and J.M. Cebulka, Sol. St. Comm. 47, 493 (1983)
16. I. Chen and F. Jansen, Phys. Rev. B29, 3759 (1984)
17. R.A. Street, Phys. Rev. Lett. 49, 1187 (1982)
18. J. Robertson, J. Phys. C - Solid State Phys. 17, L349 (1984)
19. A.M. White, P.J. Dean and P. Porteous, J. of Appl. Phys. 47, 3230 (1976)
20. J. Kočka et al - to be published

LIGHT INDUCED CHANGE IN PHOTOLUMINESCENCE INTENSITY OF HYDROGENATED AMORPHOUS SILICON*

Jin Jang
Dept. of Physics, Kyung Hee University, Dongdaemoon-ku
Seoul 131, Korea

Choochon Lee
Dept. of Physics, Korea Advanced Institute of Science and Technology, P.O. Box 150 Chongyangri, Seoul, Korea

ABSTRACT

The photoluminescence intensity has been measured as a function of time for a-Si:H films deposited by varying substrate temperature, doping concentration, and discharge geometry. The ratio of luminescence intensity measured after light exposure to that before exposure increases monotonously with the substrate temperature between 130 and 350°C and decreases as the hydrogen concentration is increased from 4 to 19 atomic % in the films deposited at 310°C. The luminescence fatigue time decreases as the substrate temperature is lowered, but is independent of the hydrogen and doping concentrations when the substrate temperature is fixed. The results indicate that the mechanism to interpret the light induced effects should involve the hydrogen.

INTRODUCTION

Dark and photoconductivity changes after light exposure have been reported in hydrogenated amorphous silicon(a-Si:H) by Staebler and Wronski[1,2]. They[2] reported that the ratio of conductance(G_A) measured before the exposure to that(G_B) after the exposure increases with the substrate temperature up to ~250°C, and decreases as the substrate temperature is increased above ~250°C. Thereafter, many authors[3,4] measured the conductance ratio(G_A/G_B) to represent the magnitute of light induced effect in a-Si:H film. Accordingly, they reported that the samples showing no conductance change did not have light induced effect.

Light induced changes in spin density[5], photoluminescence intensity[6], Si-H vibrational absorption coeffient[7], diffusion

*This research was supported in part by Korea Research Foundation under '83 project.

length[8], photovoltaic properties[9], and the density of states in the gap[10] have been reported.

In this work, the photoluminescence intensity and conductivity changes with time during light illumination are investigated in a-Si:H films prepared by varying substrate temperature, doping concentration and discharge geometry. Discharge geometry was varied to obtain a-Si:H samples with various hydrogen concentration even when the substrate temperature was fixed[11]. Since there is apparent link between the luminescence intensity and the spin density, the measurement of photoluminescence intensity with time is more convenient way to study the change of spin density by light illumination.

EXPERIMENTS

The samples were prepared by glow discharge decomposition of silane, and silane phosphine mixtures. The rf discharge was sustained between two disk electrodes, and the substrates(Corning 7059 or c-Si wafers) were placed on the cathode. The surface of glass substrate was roughened before deposition to avoid interference effects in the luminescence intensity. The gas pressure during preparation was ~0.5 torr, and the discharge power density 0.1 W/cm^2. The thicknesses of all investigated samples were about 1 μm. The dc samples were also deposited at 310°C. The preparation conditions and the characteristics of the investigated dc samples appeared elsewhere[11]. Optical exposure to investigate the conductance change was done with a light from a tungsten lamp source, and the wavelength of the incident light was in the range 6000 to 9000 Å.

The photoluminescence was excited with a 4 mW output of He-Ne laser operating at 632.8 nm. This beam was focused on to the specimen which was mounted on a cold finger at 77K. The detector was a cooled PbS photoconductor. The photocurrent in a detector was measured by a standard Lock-in technique. The output voltage of Lock-in Amp. was recorded in a strip chart recorder.

RESULTS AND DISCUSSIONS

Fig. 1 shows a representative shape of photoluminescence intensity with time during illumination. The incident light intensity was ~300 mW/cm^2, and the time constant of signal average (Lock-in Amp.) was taken to be 1 second. All data of investigated samples were obtained in an identical condition,i.e., the laser beam size, measuring temperature, and the time constant. The total intensity of photoluminescence decreases, rapidly at first, and gradually saturates with time.

The fatigue time is defined as the time required to decrease

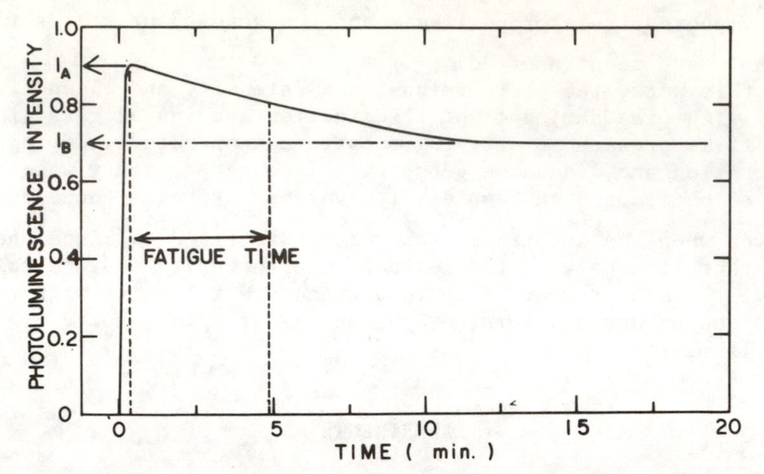

Fig. 1. Relative photoluminescence intensity as a function of illumination time. The fatigue time is defined as the time required to decrease the luminescence intensity(I_A) to $(I_A+I_B)/2$.

the luminescence intensity (I_A) to $(I_A+I_B)/2$, where I_A and I_B are the luminescence intensities at maximum(15-20 seconds after photoexcitation) and at saturation(30 minutes after photoexcitation), respectively. The decrease of luminescence intensity after the exposure was completely recovered by annealing the sample at 180°C for 30 minutes in a-Si:H films deposited at above 180°C. The luminescence was measured after annealing at 180°C for high substrate temperature samples, and right after the preparation for the films deposited at below 180°C.

Fig.2 shows the substrate temperature dependence of I_B/I_A and G_A/G_B. It is well known that the G_A/G_B increases with the substrate temperature, and has a maximun around 250°C, and then decreases as the substrate temperature is raised above ~250°C. It is not clarified yet whether the low temperature samples deposited at below 150°C do not have light induced effects or they have light induced effects even though no longer showing conductance change.

As the substrate temperature is increased up to ~250°C, the I_B/I_A and G_A/G_B increase. The conductance of a sample deposited at 130°C decreases little after the exposure, while the luminescence intensity of that sample changes by 25 % of its original value. Therefore, even though a number of metastable spins are created in low temperature samples, the conductance changes little because the density of states around the Fermi level for these samples are much higher than that of the films deposited around 250°C.

Another interesting point is a relation with the hydrogen content. It is well established that in glow discharge a-Si:H films the hydrogen concentration is decreased as the substrate

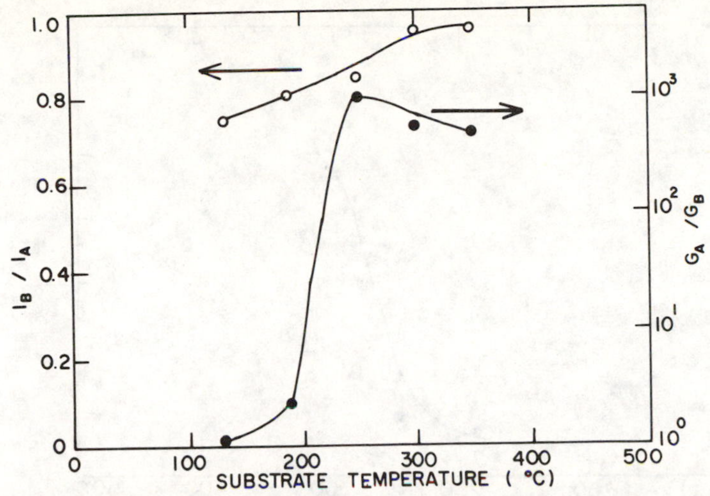

Fig. 2. The G_A/G_B and I_B/I_A as a function of substrate temperature for a-Si:H films.

temperature is raised[12]. The dependence of I_B/I_A on the substrate temperature is similar to that of hydrogen. Therefore, it is expected that the density of light induced metastable centers increases with the hydrogen concentration.

The spin density for good quality sample was found to increase by 1.1×10^{16} cm^{-3} after intense light exposure[8]. On the other hand, Zhang et al[13] reported that the increase of spin density was 4.3×10^{18} cm^{-3} for the reactively sputtered a-Si:H containing high hydrogen concentration. It is apparent from the work of Street[14] that there is a direct link between the luminescence intensity and the spin density. The link appears to be sufficiently strong that the spin density can be inferred from the luminescence intensity and vice versa, with reasonable accuracy.

Since the decrease of photoluminescence intensity by light exposure is large in the sample deposited at 130°C, the density of created spins should be high. However, in the sample deposited at above 300°C the luminescence intensity change is small, indicating the density of created metastable spins being low. Therefore, it is clear that the a-Si:H films containing high hydrogen concentration show large increase of spin density after light exposure.

Fig.3 shows the substrate temperature dependence of fatigue time. As can be seen from this figure, the fatigue time increases with the substrate temperature. Accordingly, the creation of light induced metastable spins is accelerated as the substrate temperature is lowered. It is nearly established that the light induced centers are created by the non-radiative recombination of electron

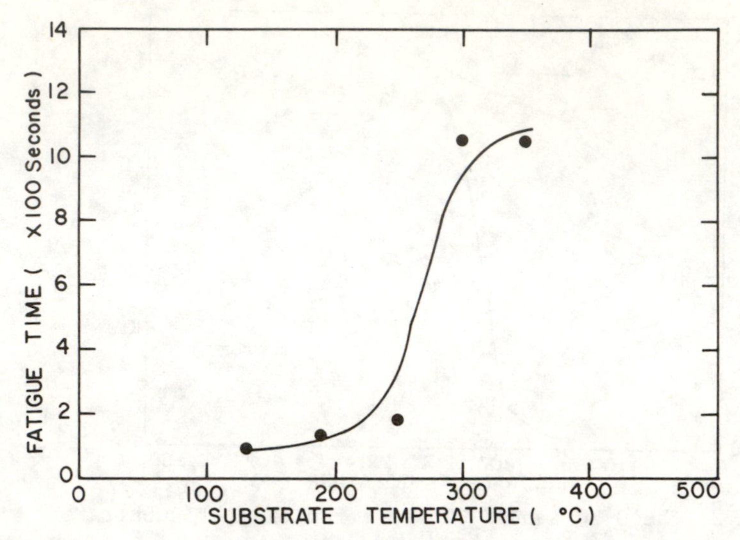

Fig. 3. Fatigue time plotted against the substrate temperature.

and hole[15]. Since the radiative luminescence quantum efficiency for good quality sample is less than 30 % at 72K[16], the difference of non-radiative recombination rate can not give a large difference of fatigue time between high and low temperature samples, shown in Fig.3.

In the samples deposited at low temperatures, the structure may be more flexible four-fold coordination network of a-Si:H containing a large number of weakly bonded hydrogens. The structure appears to be related with the speed of light induced creation of spins in a-Si:H.

Fig.4 shows the doping dependence of I_B/I_A and fatigue time. The I_B/I_A increases with the doping concentration, however, the fatigue time is independent of doping.

The model to explain the quenching of luminescence intensity by doping is that dopants introduce additional defect states and that these act as non-radiative recombination centers. Recent optical absorption measurements in subband gap energy range confirm this model. Using a photothermal deflection technique, the increase in the extrinsic absorption due to defect transitions in doped material has been observed[17].

If we believe that the hydrogen content is not changed by doping up to 100 ppm, and that the created spin density depends on the hydrogen concentration, the I_B/I_A will be increased with increasing of doping concentration; assuming the creation of constant spin density in all doped samples, the decrease of luminescence intensity will be large in low defect density material (undoped) and will be small in high defect density one(doped).

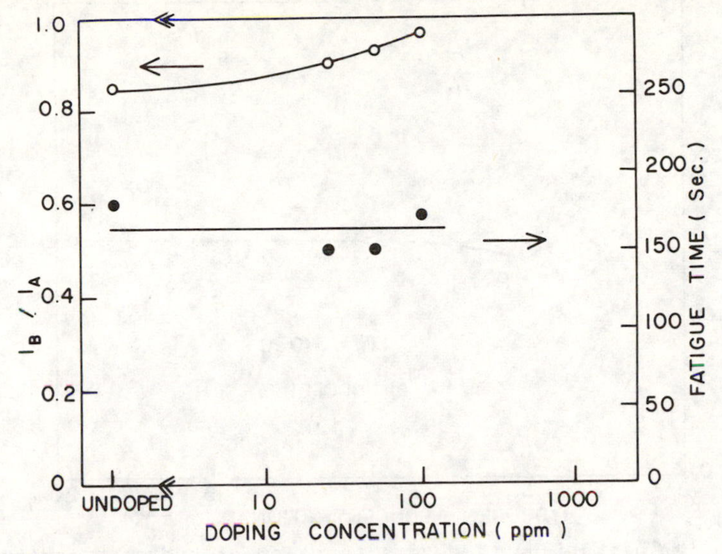

Fig. 4. The I_B/I_A and fatigue time as a function of doping concentration in a-Si:H films deposited at 250°C.

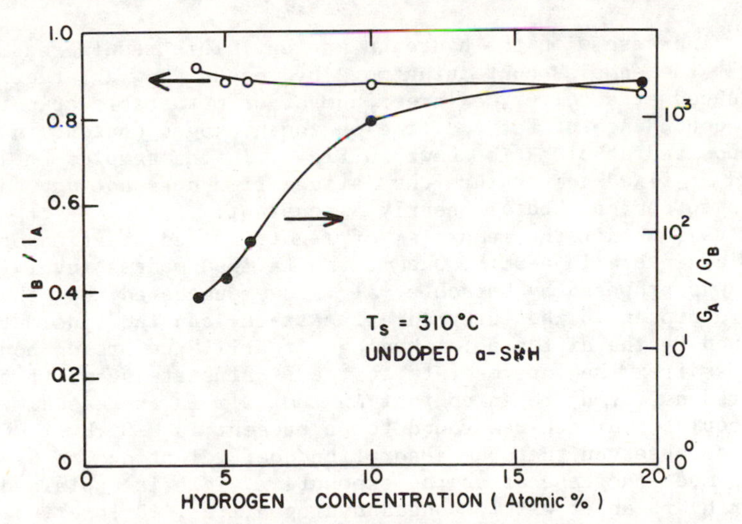

Fig. 5. The I_B/I_A and G_A/G_B as a function of hydrogen concentration for a-Si:H films deposited at 310°C.

The fatigue time is not changed by doping because small amount of doping does not seem to change the flexibility of the structure.

The luminescence intensity and conductance changes after the exposure for undoped a-Si:H films are shown in Fig.5. The samples were prepared by dc glow discharge decomposition. The G_A/G_B

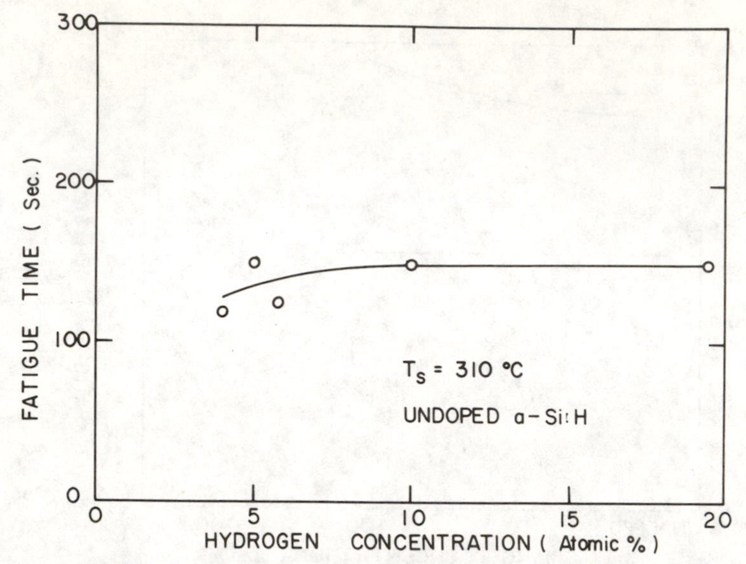

Fig. 6. The fatigue time as a function of hydrogen content in the undoped a-Si:H films deposited by varying discharge geometry.

and I_A/I_B increase with the hydrogen content. This result reconfirms that the samples containing more hydrogens show a larger light induced effect,i.e., a larger number of metastable spins.

The dependence of fatigue time on the hydrogen content in the films deposited at 310° C is shown in Fig.6. In the samples deposited at fixed temperature the fatigue time does not depend on the hydrogen content, but is nearly a constant.

Let us consider the mechanism of light induced creation of metastable centers in a-Si:H. One of the fatigue models involving hydrogen was proposed by Dersch et al[5]. They suggested that the weak bonds were broken in the intermediative state, and the final state was reached by the diffusion of hydrogen to fill the broken bonds. The process from the intermediate to the final state seems to be accelerated as the hydrogen content in the film is increased.

Molecular hydrogen was found to be present in a-Si:H films[18]. It was also observed that the absorption coefficient due to Si-H vibration mode increase after the exposure[13]. This is explained if molecular hydrogen or weakly bonded hydrogen,i.e., IR inactive hydrogen, are bonded to silicon atoms during light exposure, and then these hydrogens change their places to the original positions after annealing.

Another mechanism involving hydrogen was proposed by us[19]. The transfer of majority carriers between the bulk and the trapping regions gives rise to the light induced effect. This model is based on the trapping, but, there are some results indicating that the effect should be caused by recombination[15]. Anyway, the present results can be explained by this model. As the hydrogen concentration increases, the trapping regions will expand, resulting in the increase of metastable spin density.

SUMMARY

From the study of light induced changes in photoluminescence intensity for the a-Si:H films deposited by varying preparation conditions, the following observations are obtained. (1) The density of created spins after light exposure increases with the hydrogen concentration, which is not compatible with the conductivity changes. (2) The speed of the creation of the metastable spins by light exposure increases as the substrate temperature is lowered. (3) The decrease of photoluminescence intensity is small in doped films because the created spin density in doped films is nearly the same as that of undoped film. The experimental results are compared with the existing models involving hydrogen.

REFERENCES

1. D.L. Staebler and C.R. Wronski, Appl. Phys. Lett. 31, 292(1977).
2. D.L. Staebler and C.R. Wronski, J. Appl. Phys. 51, 3262(1980).
3. B.A. Scott, J.A. Reimer, R.M. Plecenik, E.E. Simonyi and W. Reuter, Appl. Phys. Lett. 40, 973(1983).
4. A. Akhtar, V.L. Dalal, K.R. Ramaprasad, S. Gau and J.A. Cambridge, Appl. Phys. Lett. 41, 1146(1982).
5. H. Dersch, J. Stuke and J. Beichler, Appl. Phys. Lett. 38, 465(1981).
6. K. Morigaki, I. Hirabayashi, M. Nakayama, S. Nitta and K. Shimakawa, Solid St. Commun. 33, 851(1980).
7. A. Skumanish et al., Bull. Am. Phys. Soc. 27, 146(1982).
8. J. Dresner, B. Goldstein and D. Szostak, Appl. Phys. Lett. 38, 998(1980).
9. D.L. Staebler, R.S. Crandall and R. Williams, Appl. Phys. Lett. 39, 733(1981).
10. D.V. Lang, J.D. Cohen, J.P. Harbison and A.M. Sergent, Appl. Phys. Lett. 40, 474(1982).
11. J. Jang and C. Lee, Solar Energy Mater. 7, 377(1982).
12. H. Fritzsche, Solar Energy Mater. 3, 477(1980).
13. P.X. Zhang, C.L. Tan, Q.R. Zhu and S.Q. Peng, J. Non-Cryst. Solids 59&60, 417(1983).
14. R.A. Street, Adv. Phys. 30, 593(1981).
15. S. Guha, J. Yang, W. Czubatyi, S.J. Hudgens and H. Hack, Appl. Phys. Lett. 42, 588(1983).
16. W.B. Jackson and R.J. Nemanich, J. Non-Cryst. Solids 59&60, 353(1983).
17. W.B. Jackson and N.M. Amer, Phys. Rev. B25, 5559(1982).
18. H.v. Loehneysen, H.J. Schink and W. Beyer, Phys. Rev. Lett. 52, 549(1984).
19. J. Jang and C. Lee, J. Appl. Phys. 54, 3943(1983).

OPTICALLY INDUCED CHANGES IN PHOTOLUMINESCENCE IN AMORPHOUS Si:H

P. K. Bhat, T. M. Searle, I. G. Austin
Department of Physics, University of Sheffield, Sheffield S3 7RH, UK

ABSTRACT

New two beam photoluminescence (PL) measurements are reported, in which the second beam modifies the PL efficiency. In earlier work, a subgap beam lowered the efficiency of PL excited by above gap light. Here we report subgap excitation experiments, where a second above gap beam enhances PL by up to 50%. These effects are interpreted in terms of optical saturation of non-radiative centres introduced by subgap illumination.

INTRODUCTION

The quantum efficiency (η) of the photoluminescence (PL) and of photoconductivity (PC) depends on the rate of recombination of photogenerated electron-hole pairs. The efficiencies for these two effects differ because in PC any recombination process lowers the density of photogenerated carriers, whereas PL, which directly monitors a radiative recombination process, requires a non-radiative process to reduce its efficiency. In the absence of any non radiative process, PL and PC are directly related, with $\eta_{PC} = 1 - \eta_{PL}$. When non radiative processes compete for photocarriers, the connection between PL and PC is less direct.

Much more is known about radiative than non radiative processes, since the former are more directly accessible via optical spectroscopic techniques. Thus any optical method of probing non radiative processes is valuable. Such methods work by changing the occupancy of a particular centre, usually by changing it from a neutral to a charged state and producing a centre with a large capture cross section for carriers of the opposite charge. These changes in occupation may sometimes be described by a movement of the quasi-Fermi level, but more generally will depend on capture and recombination rates as well as the energy of the non radiative defect.

Recently, several groups have begun to look at such effects in a-Si:H. Persans et al[1,2] used two optical beams of different frequencies to examine the effect of photo induced changes of population of non radiative centres on PC. Their work clearly demonstrated both the existence and complexity of such effects, showing both quenching and enhancing of the photocurrent depending on both temperature and the frequency of the second optical excitation. Two beam measurements of PL have also been reported by two groups[3,4]. In these studies, both used above gap light (> 1.9 eV) to excite the luminescence but used very different photon energies to change the efficiency. The Sheffield group[3] produced a photo quenching effect on the PL with light close to but slightly below the gap (1.65 - 1.55 eV), whereas Carius et al[4] used light of ~ 0.8 eV to produce photo quenching. There were also differences in the intensity dependence reported by these two groups. In the present work the emphasis will be on the intensity dependence rather than the spectral dependence of these effects, though the latter probe the energy dependence of the non radiative centres and will be reported elsewhere.

Intensity effects are commonly seen in PC, where the photocurrent may depend on the excitation density G as G^{ν}, with ν ranging between 1 and 0.5. This change in kinetics can occur without any non radiative process, merely reflecting the freezing out of one of the two types of carrier. In PL, intensity dependent efficiencies η_{PL} require intensity dependent changes of the ratio of radiative to non-radiative centres, for otherwise a constant fraction of the electron-hole pairs must decay by photon emission. Non-linear effects are well known in a-Si:H, even for single beam experiments in which only above gap excitation is used. For example in the temperature range 10K - 50K the efficiency decreases with excitation density, an effect ascribed by Street et al[5] to an Auger process, and by Collins and Paul[6] to radiative centre saturation. At higher temperatures the efficiency of low defect density material can increase still further with increasing excitation intensity.

More complex behaviour is seen with subgap (< 1.7 eV) excitation of Si:H. We have shown (7) that as the excitation energy is lowered below the gap, the efficiency changes from a value independent of the excitation intensity G to one which varies as

$$\eta_{PL} \propto 1/\sqrt{G} \qquad (1)$$

This result implies the production of non radiative centres by subgap light with a density proportional to $\sqrt{G}$. If light produces the conversion of two neutral centres to a charged pair, one of which acts as the dominant non radiative centre, this behaviour is understandable.

The introduction of these non radiative centres by subgap light can also explain the results of two beam PL experiments[3], where a second subgap beam produces photoquenching such that the fractional decrease in PL intensity I is

$$-\Delta I/I \propto \sqrt{G_{subgap}} \qquad (2)$$

In these experiments the second beam always produced a decrease in PL intensity, as it did in the experiment of Carius et al[4]. However, more recently we have observed that the sign of the change of intensity depends upon the photon energies of the two beams, as it does also in PC. In particular, it is possible to observe photoenhancement of the efficiency by up to 50%. In what follows we present data on the photo enhancing effect, with a discussion of the different experimental configurations which produce enhancing and quenching effects. A phenomenological model is given in which the enhancing effects are explained as the saturation of the non radiative centres introduced by subgap light. Finally, microscopic models of these effects are discussed in the context of the dangling bond centre believed to dominate non radiative processes in a-Si:H.

EXPERIMENTAL

The method uses a conventional PL system, with the addition of a second light source which illuminates the same sample area as the excitation source. In principle either light source could be a laser or an incoherent source used in conjunction with filters and/or a monochromator. In practice, laser sources have several advantages. Most common lasers have Gaussian beams of about 1 mm diameter, and these can easily be superposed. An incoherent source usually produces a larger and less well defined spot. Though a dye laser was not used in the present measurements, it is probably the ideal source. If an incoherent source is used, then it should be followed either by a monochromator or by narrow band filters. Wide band filters may well give misleading results, particularly when illuminating near the band edge. Even though such a filter might pass above the gap only a small amount of the power incident at longer wavelengths, the absorbed powers above and below gap might be comparable because of the much higher absorption coefficients for the former. Since, as we discuss below, the two beam kinetics are very different for above and below gap light, it is particularly important to know the relationship of the two photon energies with Eg. Persans' work[1,2] on two beam PC shows how there too the sign of the effect depends critically on photon energy. Some PC studies have used broad band sources, and for the reasons given above, this makes a detailed comparison with other work difficult. For the results presented here, the lines available from an He-Ne and a Kr^+ laser were used.

It is obviously central to the idea of a two beam experiment that one of them can be tagged as the excitation beam which produces PL directly (beam 1) and the other as the beam which changes the centre populations (beam 2). We distinguish the effects caused by the two laser beams by chopping them at two different frequencies f_1 and f_2, say. Then if the PL signal is synchronously detected at f_1, it is beam 1 which is the exciting source, as in a single beam experiment. As discussed further below, we have not seen marked frequency effects in the region $0 < f_1, f_2 < 300$ Hz, and so most of our measurements were made with $f_1 = 40$ Hz and $f_2 = 0$, ie the second beam was unmodulated. We therefore use the subscrips M and C for beams 1 and 2 respectively, 1 being Modulated and 2 Continuous.

In any experiment involving optical excitation one must be careful to avoid heating effects which could give rise to spurious non-linearities. An exchange gas cooled He continous flow cryostat minimised these problems by cooling the sample face on which the laser beams were incident. The power absorbed depends strongly on the wavelength, because of the rapid rise in absorption coefficient near the edge, and so higher powers (up to 50 mW) can be used with the 1.55 eV and 1.65 eV Kr^+ lines than with the remaining shorter wavelength Kr^+ and He-Ne lines. With the latter lines the power was kept below about 1 mW; with such powers the single beam He-Ne efficiency is independent of G.

The spectra were dispersed with a 1/4 m grating monochromator and the PL detected with a cooled Ge p-i-n diode. All spectra are corrected for system response.

RESULTS

General Comments

The two classes of two beam measurements presented below are those in which the exciting (modulated) beam is Above the gap (type A), and those where excitation is Below the gap (type B). Before discussing them separately it is useful to make some general comments which apply to both, and relate to the time response. Though we have not yet made detailed lifetime measurements, we find that the change in intensity produced by the second beam occurs within 10 ms. When the unmodulated beam is removed, the intensity returns to its single beam value equally quickly, with no fatigue effects. Thus we are not observing the creation of metastable centres as in the Staebler-Wronski effect.

Because of the fast response, it is possible to do type A and B experiments simultaneously, by using two lock in amplifiers to measure the components at f_1 and f_2. Such experiments give exactly the same photoinduced changes as with our more normal modulated/unmodulated beam configuration for chopping frequencies up to 300 Hz.

Type A experiments

These have been presented earlier, and so will only be briefly reviewed here. A modulated above gap He-Ne laser excited the PL, and lines from a Kr^+ laser used to change the PL efficiency. The main conclusions so far are

i) no effects are seen with the above gap Kr^+ lines at 2.18 and 1.83 eV,

ii) the lines at 1.65 eV and 1.55 eV produce a photoquenching, the dependence of which on the unmodulated beam intensity G_c($\equiv G_{subgap}$) is given by eqn. (2) as shown in figure 1,

iii) The two beam experiment separates the spectral and intensity effects of the subgap beam. Single beam experiments with subgap light produce a shift to the red at low temperatures[7] of about 70 meV.

iv) The temperature dependence of $\Delta I/I$ is similar to that of the PL efficiency, remaining roughly constant from 10K to about 80K, then falling.

Type B experiments

In this configuration the roles of the two lasers are reversed. The above gap He-Ne is unmodulated, and the modulated Kr^+ lines at E_c= 1.65 eV and 1.55 eV are used to excite PL. The results are more complex than those for type A experiments, particularly since $\Delta I/I$ depends on both G_c and G_m, rather than on G_c alone. The simplest results are obtained when the power absorbed from the modulated beam is low, a condition most easily achieved with the weakly absorbed 1.55 eV line. Typical results are shown in figure 2. Under these conditions, $\Delta I/I > 0$.

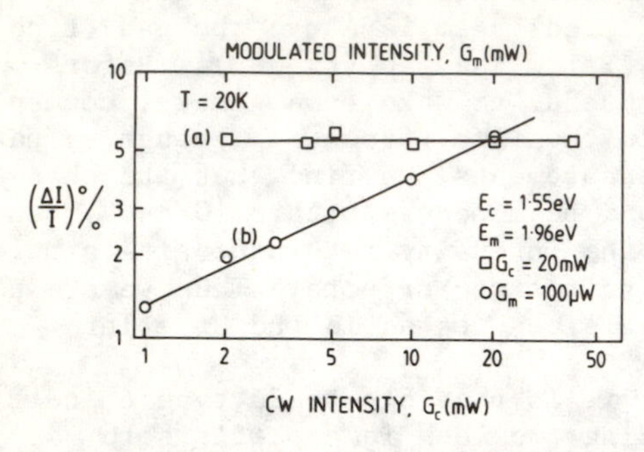

Fig. 1 Type A experiments - intensity dependence of ΔI/I on (a) G_m (b) G_c for a SP film (from ref.3).

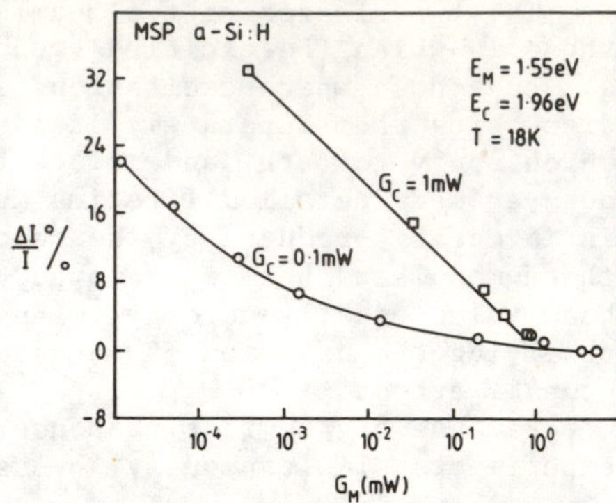

Fig. 2 Type B experiment, with low density 1.55 eV excitation producing only photoenhancement, showing G_m dependence at two unmodulated 1.96 eV beam powers.

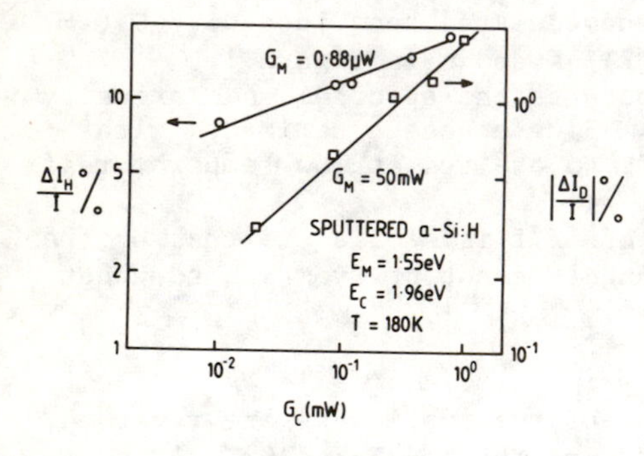

Fig. 3 Type B experiment, showing the dependence of ΔI/I on G_c (1.96 eV) at fixed G_m in both the quenching and enhancing regions.

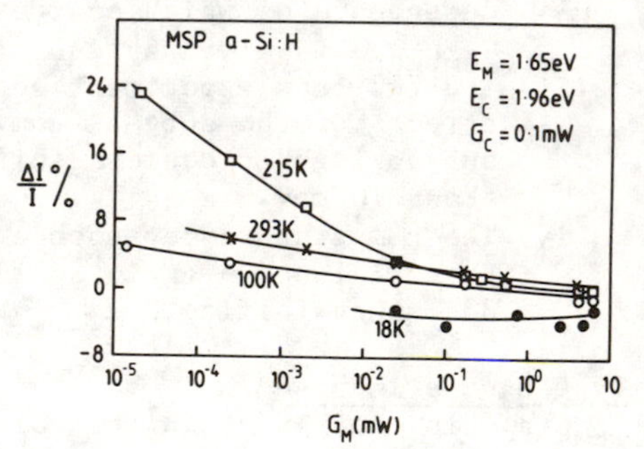

Fig. 4 Type B experiment, with the higher excitation density produced by the 1.65 eV line and showing both enhancing and quenching for a range of temperatures.

The observation of such a photoenhancement in what with $G_c = 0$ is a single beam subgap excitation experiment is less surprising when one recalls that such measurements yield an efficiency decreasing with increasing G_m (equation 1). This behaviour is due to light induced non-radiative centres, and any process which decreases their efficiency will increase η_{PL}. Figure 2 shows that the enhancement can be quite large, and we have observed $\Delta I/I$ as large as 50%. The largest values are for small G_m, i.e. low densities of light induced non radiative centres. We have examined a range of samples prepared by different methods, including glow discharge (GD), sputtered (SP) and magnetron sputtered (MSP) films. Whilst the magnitude of $\Delta I/I$ is sample dependent, the same qualitative behaviour is always seen.

Some properties seem to be sample independent. For example just as in the type A experiments the photoquenching always depended on $\sqrt{G_c}$, in type B experiments at fixed G_m the enhancement

$$\Delta I/I \propto G_c^{\,p}, \tag{3}$$

with p = 0.25 ± 0.05 independent of film or temperature. Figure 3(squares) shows an example of data from a GD sample at 180K. Here the photoenhancing effect is small, because of the high temperature.

Figure 4 shows that at higher excitation powers G_m the behaviour of $\Delta I/I$ becomes more complex. These data are taken with the 1.65 eV Kr^+ line, which is more strongly absorbed than that at 1.55 eV. Above a temperature dependent value of G_m, $\Delta I/I$ changes sign and photoquenching is seen as in a type A experiment. The dependence of this type B photoquenching also obeys equation 1 at fixed G_m as shown by the circles in figure 3. At low tempterature very low powers are sufficient to cause quenching, but as in the type A experiment the magnitude of the quenching falls with increasing temperature.

Figure 4 also shows in contrast that the magnitude of the photoenhancing increases with temperature. At higher temperatures $\Delta I/I$ vs temperature goes through a maximum then begins to fall. The temperature of this maximum is sample dependent.

Another difference between the observations of A and B type experiments is that whereas there are no spectral shifts in the former[3], there are in the latter. In the absence of the second beam, subgap excitation at low temperature shifts the PL peak some 70 meV to the red, though at higher temperature (⩾ 200 K) the shift is to the blue. When the second above gap beam is added we observe that the PL peak moves to the red if one is in the photoquenching region, and to the blue if there is photoenhancing. In both cases the shifts are small, ⩽ 50 meV.

DISCUSSION

We have already argued[3] that the single beam subgap and two beam type A results of equations (1) and (2) can be understood if subgap light introduces non-radiative centres with a density proportional to $\sqrt{G_{subgap}}$. This power dependence is immediately understandable if subgap light produces the ionization of a T_3° neutral dangling bond

$$2T_3^\circ \rightarrow T_3^+ + T_3^-$$

and if one of the two charged centres on the right acts as a much more effective carrier trap than the other. The dependence of the reaction on photon energy could result from excitation into the T_3° absorption band. The absence of any photoquenching with above gap photons would require that the cross section falls rapidly above the gap. In an alternative model, photons are selectively absorbed into tail states near the T_3°. The latter model would imply higher tail state densities near dangling bonds, and that carriers excited above the mobility edge diffuse rapidly away from these centres.

The observation that the photoenhancing effects occur in the type B configuration and when the subgap intensity is low suggests a simple explanation in terms of optical saturation of these non-radiative centres. In a B type experiment the quantum efficiency is always less than unity because the exciting beam (G_m) itself introduces non-radiative centres, excitation occurring at subgap energies. When G_m is low, the density of these centres is small, and they are readily saturated. As the density increases with increasing G_m, $\Delta I/I$ falls as a smaller fraction of the centres are saturated. A simple quantitative model which will be discussed elsewhere leads to at least qualitative agreement with the data of figure 2.

Such a model can also describe the temperature effects seen in figure 4 and discussed above. Saturation of the non-radiative centres will occur at lower powers if the radiative centres are being thermally ionized, increasing the free carrier concentration. Thus we ascribe the initial increase of $\Delta I/I$ with temperature to the same ionization mechanism which determines η. The subsequent decrease in $\Delta I/I$ can be understood if at higher temperature the non-radiative centres are also being ionized. The data therefore suggest that the radiative excited state is nearer the mobility edge than that of the non-radiative centre.

The observed spectral shifts mentioned above provide further indications of saturation effects which include the radiative centre. Shifts of PL bands with excitation intensity are well known in crystalline donor-acceptor pair systems, and have been reported by Collins and Paul[6] in a-Si. They occur because of a correlation of lifetime and emission energy within the band; if both increase together then saturation of the low energy long lifetime centres occurs first, moving the band to the blue. Photoenhancement, resulting from saturated non-radiative centres is accompanied by an increase in carrier concentration, mimicing an increase in excitation density and producing the observed blue shift. Likewise photoquenching is accompanied by a red shift. The radiative centre populations being observed in A and B type experiments are different, as indicated by the red shift in subgap single beam excitation. It is for this reason that corresponding spectral shifts are not seen in type A experiments. The existence of two radiative channels, one monitored but of low efficiency, the other efficient but not detected because not modulated in type B probably explains the observation of photoquenching with 1.96 eV light. Since our earlier measurements showed no photoquenching in type A experiments at this energy, we do not expect such photons to be able to produce significant densities of non-radiative centres. We note however that the photoquenching has the same intensity dependence in A and B type experiments (figure 3, equation 2), which might again suggest non-radiative centre production. It may be that type B experiments have greater sensitivity, since the radiative centre density is much lower than that excited by above gap photons.

SUMMARY

We have presented new two beam PL measurements, and in particular reported the existence of photoenhancing effects up to 50%. The studies have clarified the kinetics of these processes, particularly the relationship between A and B type experiments, and between the photoquenching and photoenhancing effects.

REFERENCES

1. P. D. Persans, Phil. Mag. B46, 435 (1982).
2. P. D. Persans and H. Fritzsche, J. Phys., Paris 42, 597 (1981).
3. P. K. Bhat, D. J. Dunstan, I. G. Austin and T. M. Searle, J. Non-Cryst. Solids 59 and 60, 349 (1983).
4. R. Carius, W. Fuhs and M. Hoheisel, J. Non. Cryst. Solids (to be published).
5. R. A. Street, Adv. Phys. 30, 593 (1981).
6. R. W. Collins and W. Paul, J. Non Cryst. Solids 59 and 60, 369 (1983).
7. P. K. Bhat, T. M. Searle, I. G. Austin, R. A. Gibson and J. Allison, Solid State Comm. 45, 481 (1983).

THE EFFECT OF EXTENDED LIGHT EXPOSURE ON THE URBACH TAIL AND THE SUBBANDGAP ABSORPTION IN a-Si:H

Daxing Han* and H. Fritzsche
James Franck Institute, The University of Chicago
Chicago, IL 60637

ABSTRACT

We present a study of the effect of light exposure at 160°K and 300°K and of annealing on the spectral dependence of the photoconductivity of hydrogenated amorphous silicon (a-Si:H) prepared by d.c. glow-discharge deposition. We find no evidence for a change in the Urbach tail of the optical absorption curve larger than the experimental accuracy of 3 percent. We find that samples prepared either by r.f. or by d.c. glow-discharge deposition show evidence that two kinds of centers are produced by extended light exposure. These centers are distinguished by their annealing kinetics and the temperature at which the samples are exposed to light.

INTRODUCTION

Recent experiments indicate[1] that prolonged light exposure, the Staebler-Wronski effect,[2] produced at least two but perhaps even a spectrum of defects in hydrogenated amorphous silicon (a-Si:H). These defects differ for instance in the way they affect the recombination lifetime and the subbandgap absorption. They also differ in their annealing kinetics, and the kinds that are produced depend on the temperature at which the sample is exposed to light.

One part of this paper explores the question of whether these observations depend sensitively on preparation conditions. In this paper we present data on a sample prepared in a different laboratory and by d.c. instead of r.f. plasma deposition.

Another part of this paper deals with the question whether the slope of the Urbach edge of optical absorption is affected by prolonged light exposure. One would expect that the slope should be changed because of two independent studies. One is the observation of Cody et al.[3,4] that the slope of the exponential part of the optical absorption curve (Urbach edge) decreases with an increase in disorder. The disorder parameter, according to their analysis, can be increased by raising the temperature or by increasing the concentration of defects. The other study is by Hauschildt et al.[5]. These authors found that the difference in activation energies of the conductivity and of the thermopower increases as a result of prolonged light exposure. This difference in turn has been associated with internal potential fluctuations in the material[6]. An increase of these potential fluctuations by light exposure is therefore an additional

* Present address: Institute of Physics, Chinese Academy of Sciences, Beijing, China.

reason for expecting an effect on the slope of the Urbach edge.

EXPERIMENTAL DETAILS

We used an undoped a-Si:H sample that was prepared at 300°C in a d.c. glow-discharge system from undiluted silane gas. The d.c. power was 20W. The 0.7 μm thick film contained about 15 at.% hydrogen. We used either NiCr or carbon paint contacts separated by 0.1cm in a coplanar geometry. Both kinds of electrodes provided ohmic contacts up to 400V, the highest voltage applied. All measurements were carried out in a stainless steel chamber that was evacuated to about 10^{-4} Torr by a liquid-nitrogen cooled sorption pump. The sample could be exposed to heat-filtered light of intensity F_o=100 mW/cm^2s from a 300W tungsten-halogen lamp (AM1 light) or to monochromatized light having photon energies between $h\nu$=0.4 and 3.4 eV.

RESULTS

Figure 1 presents an overview of the changes in the conductivity as the sample is first annealed for 40 min at 430°K (state A). The sample is then exposed for 3h to AM1 light at 300°K. One observed a decrease of the photoconductivity σ_p during exposure because of the creation of new recombination centers. The state after 300°K light exposure is called B1. Annealing for 40 min at a higher temperature of 480°K yields the state called A*, and light exposure at

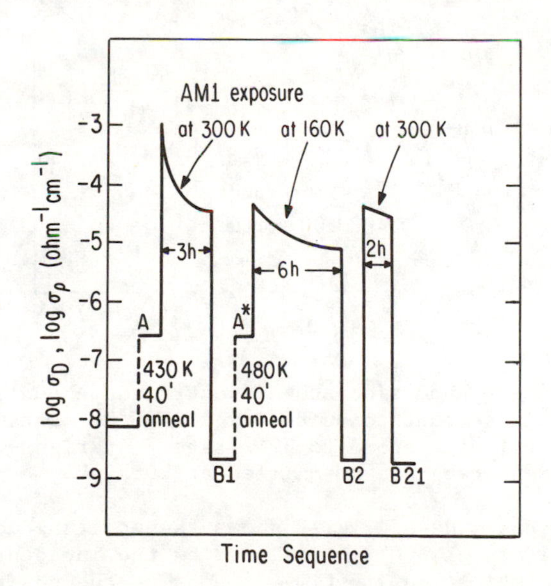

Fig.1. Time sequence of annealing and light exposures. The time dependences of the photoconductivity during exposure are shown.

160°K leaves the sample in state B2. When the sample is subsequently exposed again but this time at 300°K we call the new state B21. One finds the dark conductivities σ_D of states A and A* are the same and so are the σ_D values of states B1, B2, and B21. The different treatments are given and distinguished because they produced very different photoconductivities in the r.f. plasma deposited samples described previously. All states shown in Fig. 1 are reversible and reproducible except of course the unlabelled original state at the far left of the figure. At the beginning the properties of the sample depend on the exposure to moisture and air prior to mounting the sample in the measuring chamber[7].

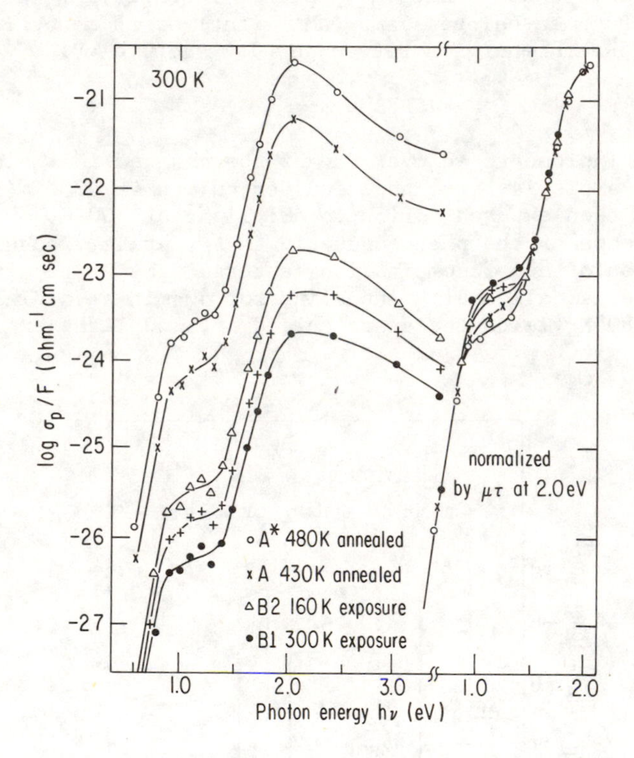

Fig.2. Left-hand side shows spectral dependence of normalized photoconductivity at 300°K. Data symbols (+) correspond to B21 state. At right-hand side curves are normalized at $h\nu$=2 eV.

The left-hand side of Fig. 2 shows the spectral dependence of the photoconductivity at 300°K normalized by the incident light flux F. Since the recombination lifetime depends on the position of the quasi-Fermi level these measurements were carried out at a constant value of the photoconductivity, $\sigma_p=3.5\times10^{-7}$ ohm^{-1}cm^{-1} for states A and A* and $\sigma_p=3.5\times10^{-9}$ ohm^{-1}cm^{-1} for the three B states. These values yield a total conductivity that is just a factor 2 larger than σ_D. At these low light intensities there is no danger of disturbing the states A and A* by creating a noticeable number of defects during the measurements.

The major difference in the curves at the left-hand side of Fig. 2 is a vertical shift that corresponds to a change in the recombination lifetime τ. In order to reveal changes in other features of the photoconductivity spectra, we shifted the curves vertically and made them coincide at $h\nu$=2.0 eV. The curves thus normalized at $h\nu$=2.0 eV are replotted on the right-hand side of Fig. 2. One now observes differences in the relative absorption between $h\nu$=1.0 and 1.4 eV. In this photon energy range absorption is governed by transitions between localized states and extended band states. It is generally believed that transitions from occupied gap states below the Fermi level E_F to the conduction band contribute much more to the subbandgap absorption than transitions from the valence band to unoccupied localized states above E_F because of the larger density of states below midgap. If this is correct, then the absorption and photoconductivity below $h\nu$=1.4 eV may be taken as a measure of the number of defect states below E_F[8].

The magnitude of the defect absorption below $h\nu$=1.4 eV in this sample is least in the well annealed state A* and strongest after light-soaking at room temperature (state B1). In the r.f. glow discharge sample previously examined, light exposure at 160°K (state B2) did not increase the defect absorption whereas exposure at 300°K (state B1) did. Exposure at either temperature on the other hand strongly reduced the recombination lifetime. In the d.c. glow discharge sample of Fig. 2, the different behavior after 160°K and 300°K exposure is also noticeable but in a less extreme manner: there are fewer defects created by light exposure at 160°K for 6 hours (state B2) than at 300°K for 3 hours (state B1) even though the recombination lifetime τ is decreased by a similar factor.

Figure 3 is essentially equivalent to Fig. 2 except that the data of Fig. 3 are taken at 130°K. Here the difference between light exposure at 300°K (state B1) and at 160°K (state B2) is more evident. Although the reduction in σ_p due to a decrease in τ is similar for B1 and B2 as seen on the left-hand side of Fig. 3, the increase in the defect absorption below $h\nu$=1.4 eV is much less for B2 (low temperature exposure) than for B1 (room temperature exposure). These spectra were also measured at constant photoconductivity values, $\sigma_p = 3.5\times10^{-10}$ and 3.5×10^{-11} $\text{ohm}^{-1}\text{cm}^{-1}$ for the states A,A* and for the states B1,B2, respectively.

We now address the question whether extended light exposure affects the slope of the exponential rise of the absorption curve. Figure 4 shows that there is no detectable change in slope after annealing and after exposure either at 300°K or 160°K. This statement is not affected by the qualifying remarks that we have to make before discussing other aspects of the results shown in Fig. 4.

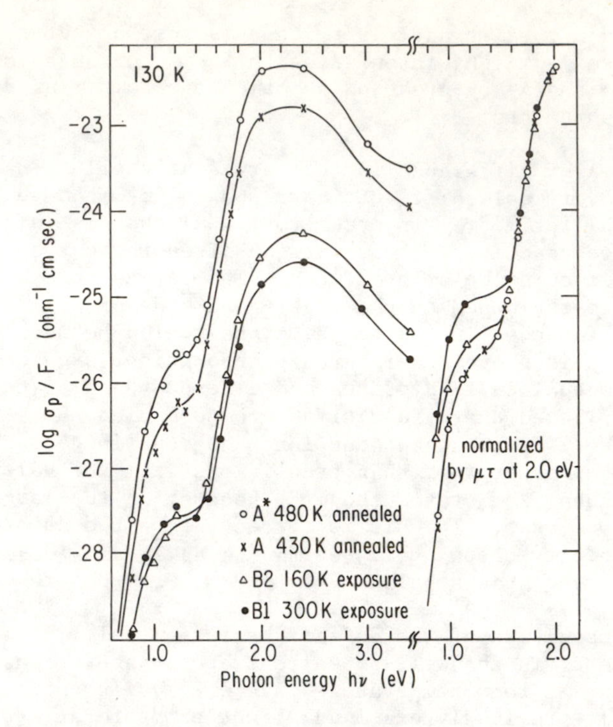

Fig.3. Left-hand side shows normalized photoconductivity at 130°K. The curves at right-hand side are normalized at hν=2 eV.

The absorption coefficient is obtained from the photoconductivity by matching the curves to direct measurements of the absorption coefficient in the range $\alpha > 10^3 cm^{-1}$ where these values can be obtained from optical transmission data. The photo-conductivity curves represent then $\alpha(h\nu)$ if the quantum efficiency is independent of photon energy and temperature. One expects that the quantum efficiency η drops below unity as the temperature and the photon energy is decreased because the photoexcited electrons can no longer escape the Coulomb attraction of the holes before recombination. Carasco et al.[9] found that $\eta \simeq 1$ efficiency above hν=1.5 eV and that η drops by about a factor ten below that energy. This energy is close to the onset of the defect absorption. The smaller apparent defect absorption observed at 130°K compared to that at 300°K in Fig. 4 is probably caused by a decrease of the quantum efficiency with temperature. We believe that the slopes of the exponential regime of the absorption curves are not affected by changes in η. We find a slope of E_o=62m eV at 130°K and E_o=82m eV at 300°K. The broadening of the Urbach edge with increasing temperature agrees with the observations of Cody et al.[3,4]. The magnitudes of E_o are, however, about 30 percent larger than those observed in r.f. glow-discharge samples prepared at a substrate temperature of T_s=260°C.

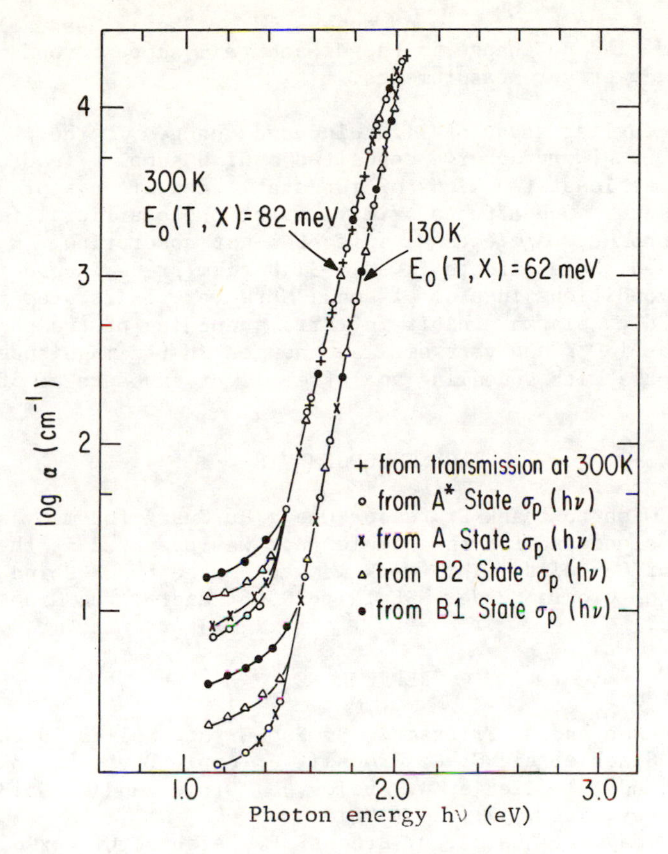

Fig.4. Absorption coefficient α obtained from optical transmission and photoconductivity at 300°K and 130°K. $E_0(T,X)$ is the slope of the exponential part and the Urbach edge parameter of reference 4.

DISCUSSION AND CONCLUSIONS

In agreement with Cody et al.[3] we find that the slope of the optical absorption curve (Urbach tail) depends on the preparation conditions and on temperature. The Urbach tail broadens with temperature and is wider in the present a-Si:H sample prepared by dc glow discharge than in high quality r.f. glow discharge samples. Because of the sensitivity of the slope of the Urbach tail to a generalized disorder parameter one might expect that the Staebler-Wronski effect would produce a change in the slope. A broadening would be expected if defects created by light exposure increase the internal potential fluctuations as suggested by conductivity and thermopower measurements[5]. On the other hand, a sharpening might have been observed if the Staebler-Wronski effect could be interpreted as a photo-induced sharpening of the band tails as suggested by Shapiro and Adler[10]. Instead, we find no change in the Urbach tail slope beyond the 3 percent accuracy of our measurements.

An earlier study of light-induced changes in the photoconductivity of r.f. glow discharge deposited a-Si:H samples lead us to conclude that at least two kinds of metastable centers are produced by light exposure. One affects primarily the photocarrier lifetime and the other produces an increase in subbandgap absorption. Here we have observed these effects in an a-Si:H film prepared under very different conditions in a dc plasma. Moreover, films prepared in a r.f. and a d.c. plasma exhibit infrared quenching of the photoconductivity at 130°K and very similar changes in the magnitude of the quench spectra with annealing and after light exposure at 300°K and at 160°K [1].

ACKNOWLEDGEMENTS

We wish to thank Professor Cheng Ru-Guang for many stimulating discussions and for the sample that was prepared at the Shanghai Institute of Ceramics. This work was supported by US-China cooperative research program under NSF INT 8203084 and by the Grant NSF DMR 8009225.

REFERENCES

1. Daxing Han and H. Fritzsche, J. Non-Cryst. Solids 59-60,397(1983).
2. D. L. Staebler and C. R. Wronski, J. Appl. Phys. 51, 3262 (1980).
3. G. D. Cody, T. Tiedje, B. Abeles, B. Brooks and Y. Goldstein, Phys. Rev. Lett. 47,1480 (1981).
4. G. D. Cody, B. Abeles, B. Brooks, P. Persans, C. Roxlo, A. Ruppert and C. Wronski, J. Non-Cryst. Solids 59-60, 385 (1983).
5. D. Hauschildt, W. Fuhs and H. Mell, Phys. Stat. Solidi(b) 111,171 (1982).
6. H. Overhof, J. Non-Cryst. Solids 66, 261 (1984).
7. M. Tanielian, Phil. Mag. B45, 435 (1982).
8. W. B. Jackson and N. M. Amer, Phys. Rev. B25, 5559 (1982).
9. F. Carasco and W. E. Spear. Phil. Mag. B47, 495 (1983).
10. F. R. Shapiro and D. Adler, J. Non-Cryst. Solids 66, 303 (1984).

DIFFERENCE IN THE INFLUENCE OF OXYGEN AND NITROGEN ON THE LIGHT INDUCED EFFECT

N. Nakamura, S. Tsuda, T. Takahama, M. Nishikuni,
K. Watanabe, M. Ohnishi and Y. Kuwano

Research Center, SANYO Electric Co., Ltd.
1-18-13, Hashiridani, Hirakata City, Osaka, Japan

ABSTRACT

The influences of impurities such as oxygen and nitrogen on the light induced effect were studied. Although both oxygen and nitrogen had a strong influence on the light induced effect, there were large differences in their influences. The oxygen doped film had a longer time constant for the light induced change in the conductivity and gap state density than the nitrogen doped film. It was also found that oxygen was associated with the degradation of a-Si solar cells. It is considered that the light induced effect is associated with the hole traps from the results of the degradation in n-i-n and p-i-p diodes. The degradation in a 2-kW a-Si power generating system was also analyzed.

INTRODUCTION

The light induced effect in hydrogenated amorphous silicon (a-Si:H) was first reported by Staebler and Wronski in 1977.[1] Prolonged illumination affects the dark conductivity, photoconductivity,[1] electron spin density,[2] density of gap states[3] and solar cell performance.[4] Although the prevention of the light induced effect is very important for the application of a-Si solar cells, the origin of the phenomenon is not clear.

In this paper, the influences of impurities such as oxygen and nitrogen on the light induced effect of glow discharged a-Si:H films and solar cells are discussed, and the stability of a 2-kW a-Si power generating system is also presented.

EXPERIMENT

a-Si films were deposited on glass substrates and crystalline silicon (c-Si) wafers by a capacitively coupled r.f. glow discharge reaction in $SiH_4 + O_2$ or $SiH_4 + N_2$ gas. The fundamental conditions in the plasma reaction are shown in Table 1. The content of oxygen and nitrogen in the a-Si films was measured by IMA (Ion Microprobe Analysis).

Table 1. Reaction conditions

R.F.power	22 mW/cm^2
Frequency	13.56 MHz
Substrate temp.	250 °C
Gas pressure	0.3 Torr
Flow rate	40 cc/min
Deposition rate	1 μm/hour

The conductivity, space charge density and gap state density were measured in order to estimate the light induced effect. The space charge density was obtained from the $1/C^2$-V plot of an Au/ a-Si:H/n^+ c-Si Schottky diode. The thickness of the a-Si:H film was about 1~2 μm and the modulation frequency was 0.01 Hz. The gap state density was measured by employing the ICTS (Isothermal Capacitance Transient Spectroscopy) method.[5] In this case, 100ppm phosphine doped a-Si films were used for the Schottky diode in order to eliminate the influence of the series resistance on the capacitance measurement. As the irradiation sources, a solar simulator (AM-1) and He-Ne laser (6328A) were used.

In the p-i-p and n-i-n diodes, the heavily doped p-layer (B_2H_6/SiH_4 = 1%, 0.1 μm) and n-layer (PH_3/SiH_4 = 1%, 0.1 μm) were used. The thickness of the i-layer was about 1 μm.

RESULTS AND DISCUSSIONS

(1) Conductivity

The conductivity of a-Si films decreases by light exposure, and it is recovered by thermal annealing.[1] But it was found that the change in the conductivity by light exposure and thermal annealing depends on the oxygen and nitrogen concentrations.

The changes in the dark conductivity (σd) by light exposure (AM-1, $100mW/cm^2$) and thermal annealing (150°C) are shown in Fig. 1 for the nitrogen (~1%) doped, oxygen (~1%) doped, and undoped a-Si films. The longitudinal axis denotes the dark conductivity normalized at the initial value. The initial value of σd increased slightly by the nitrogen doping, and decreased slightly by the oxygen doping. The decrease ratio of σd by light exposure increased by the oxygen or nitrogen doping. But the time constant of the decrease in the σd was different for the oxygen doped film and the nitrogen doped film. In the case of the nitrogen doped film, the σd decreased rapidly by light exposure, and then became stable. On the other hand, the σd of the oxygen doped film decreased more slowly by light exposure in comparison with the nitrogen doped films. Also, the σd was hard to recover by thermal annealing in the case of the oxygen doped film. It is believed that the mechanism of the light induced effect is different for the oxygen doped film and the nitrogen doped film.

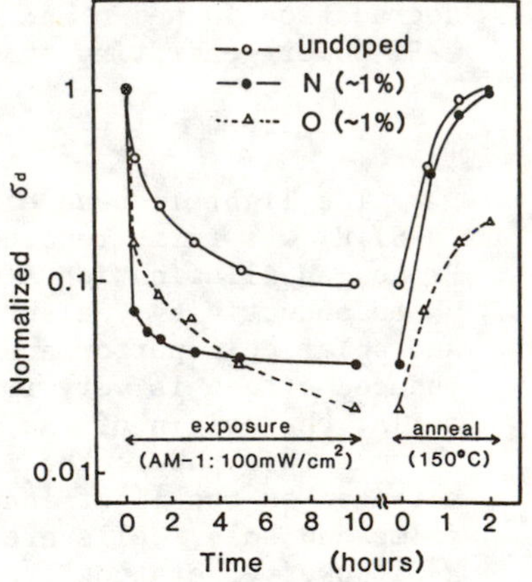

Fig. 1. The change in the σd by light exposure and thermal annealing

(2) Space charge density

The change in the space charge density of an Au/a-Si:H/n^+ c-Si Schottky diode is shown in Fig. 2 as a function of exposure time to He-Ne laser light (50mW/cm^2). The ΔN_i denotes the difference in the space charge density before and after light exposure. The ΔN_i increased rapidly at the beginning of light exposure, and then the increase became slower. The ΔN_i can be expressed by the following equation as the first approximation.

$$\Delta N_i = N_o \ (1-e^{-t/\tau})$$

N_o is a constant; τ is the time constant; and t is the exposure time. In this sample, N_o was 1.5 x 10^{16} cm^{-3} and τ was 50 minutes. This value of N_o corresponds to the result of the electron spin density measurement[2].

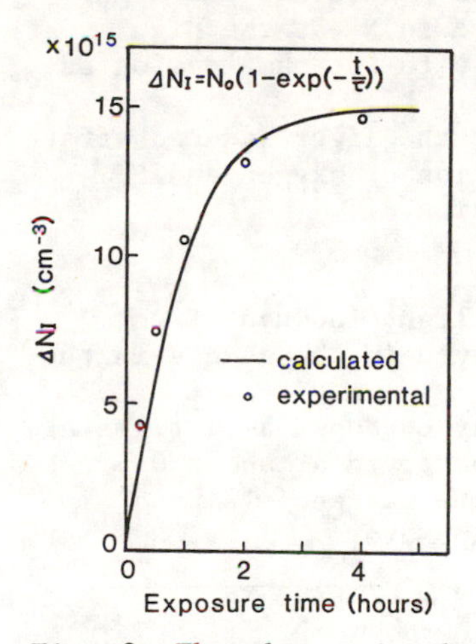

Fig. 2. The change in the space charge density by light exposure

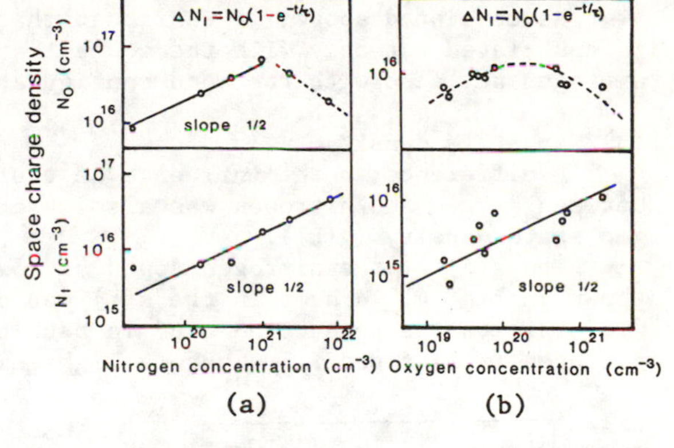

Fig. 3. The change in the N_I and N_o as a function of (a) nitrogen concentration and (b) oxygen concentration

The initial values of space charge density (N_i) and N_o are shown in Fig. 3(a) as a function of the nitrogen concentration in a-Si films. It was confirmed that a change in the oxygen concentration in the a-Si films did not occur when the nitrogen concentration was varied. As shown in the lower part of Fig. 3(a), N_i increased in proportion to the square root of the nitrogen concentration in the a-Si films. This result can be explained by the law of mass action the same as in the case of phosphorus doping.[6] In fact, the σd increased slightly by the nitrogen doping. The following reaction can be considered in the growth of nitrogen doped films.

$$N_3^0 + T_4^0 \rightleftarrows N_4^+ + D^-$$

where N_3^0, T_4^0, N_4^+ and D^- are the neutral nitrogen, neutral silicon, four-fold coordinated nitrogen and doubly occupied silicon, respectively. Therefore, the nitrogen atoms induce the negatively charged dangling bonds, and increase the space charge density.

As shown in the upper part of Fig. 3(a), N_o also increased in proportion to the square root of the nitrogen concentration. This result indicates that the nitrogen atoms accelerate the light induced effect in a-Si films. Also, it is considered that the decrease in N_o at the high nitrogen concentration region is due to the change in the bond configuration.

The oxygen also had a strong influence on the film properties and light induced effect as shown in Fig. 3(b). In this experiment the change in the nitrogen concentration in a-Si films was not observed. The N_i increased as the oxygen concentration in a-Si films increased. Because N_o increased at the low oxygen concentration region, the oxygen atoms also accelerate the light induced effect in a-Si films. The decrease in N_o at the high oxygen concentration region seems to be due to the same reason as in the case of nitrogen doping.

As mentioned above, it was found that the light induced effect is associated not only with the concentration of oxygen and nitrogen but also with the bond configuration.

(3) Gap state density

A difference in the influence on the light induced effect between oxygen and nitrogen was also observed in the change in the gap state density (g(E)).

The g(E) of the nitrogen doped and oxygen doped a-Si films are shown in Fig. 4. A bump in the g(E) was observed at about 0.5 ~ 0.6 eV below the conduction band in both the oxygen (~1%) and nitrogen (~1%) doped films before light exposure.

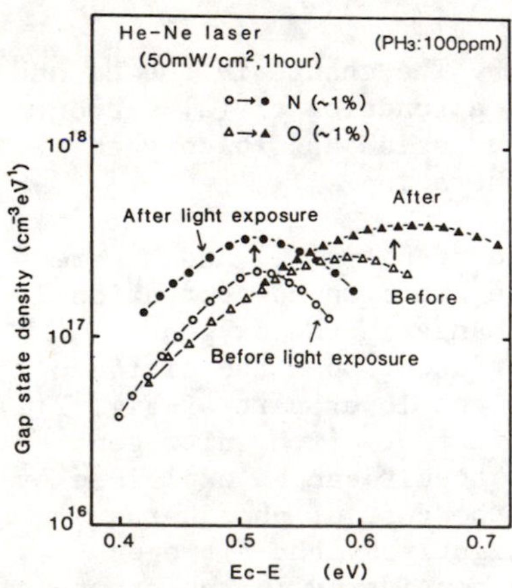

Fig. 4. The gap state density of the nitrogen and oxygen doped films before and after light exposure

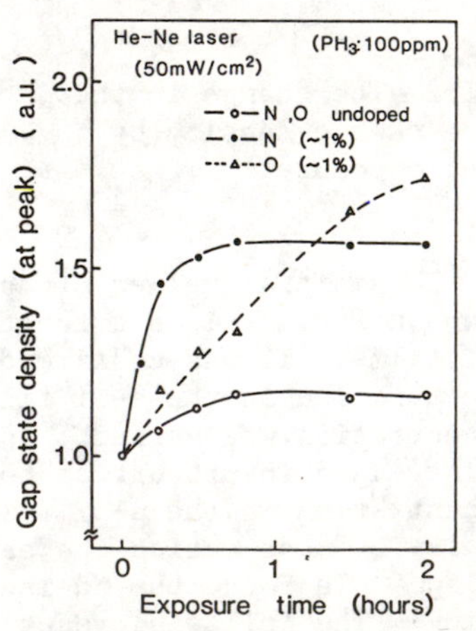

Fig. 5. The change in the gap state density at peak as a function of light exposure time

But the peak position of the bump in the oxygen doped film was slightly deeper than that of the nitrogen doped film, and its position shifted to the deeper energy level by light exposure. The bump of the g(E) is associated with the dangling bonds.[7] So, it is considered that the energy level of the dangling bond generated by light exposure in the oxygen doped film is different from that in the nitrogen doped film.

The changes in the g(E) at peak by light exposure are shown in Fig. 5. The longitudinal axis denotes the g(E) normalized at the initial value. As shown in Fig. 5, the increase ratio of the g(E) by light exposure increased by the oxygen or nitrogen doping, and the oxygen doped film had a longer time constant in the increase of the g(E) by light exposure than the nitrogen doped film. These results correspond to the results of the space charge density and dark conductivity.

Furthermore the electron-capture cross section (σ_n) of the oxygen doped film increased more than one order of magnitude by light exposure as shown in Fig. 6. In the case of the nitrogen doped film, the σ_n increased only several factors. Because the increase of the σ_n directly means a decrease of the carrier lifetime, it seems that the oxygen has a stronger influence on the light induced degradation of a-Si solar cells than the nitrogen.

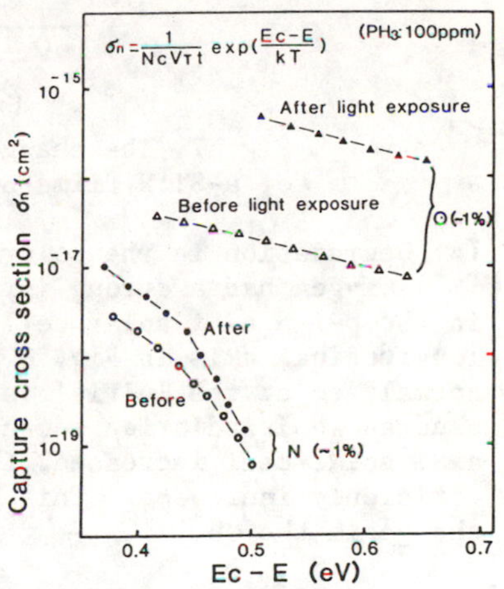

Fig. 6. The σ_n of the nitrogen and oxygen doped films before and after light exposure

As mentioned above, there are many differences in the influence on the light induced effect between oxygen and nitrogen, such as the time constant of degradation, annealing behavior and σ_n. It is considered from the above mentioned results that defects such as dangling bonds are generated in a different manner by light exposure and that their properties are different between oxygen and nitrogen doped films. And it seems that these differences between oxygen and nitrogen are associated with differences in the coordination number, the bonding energy and the electronegativity.

In the above mentioned experiments, oxygen was intentionally doped into the a-Si films. Then the light induced effect at the lower oxygen concentration region was also investigated. The changes in the photoconductivity by light exposure are shown in Fig. 7 for the a-Si films with various oxygen concentration. The longitudinal axis denotes the photoconductivity normalized at the initial value. As the oxygen concentration in the a-Si film decreased from 2×10^{19} cm^{-3} to 3.5×10^{18} cm^{-3}, the degradation ratio of the photoconductivity by light exposure also decreased. It is considered that the light induced effect can also be reduced more and more by decreasing the number of oxygen atoms in the lower concentration region.

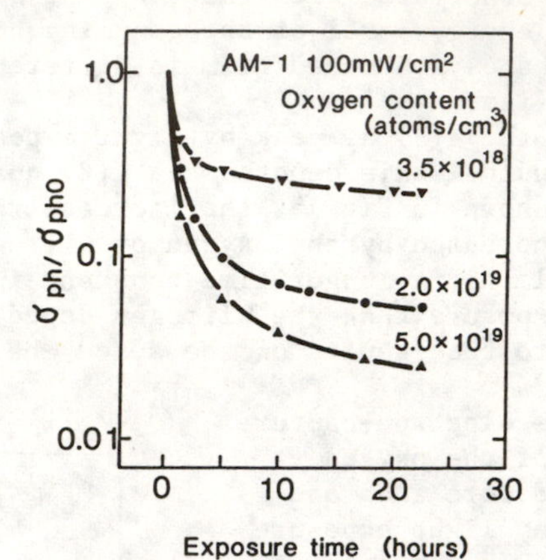

Fig. 7. The changes in the photoconductivity of a-Si:H films by light exposure

(4) Degradation in the solar cell

Oxygen had a strong influence on the light induced degradation in the p-i-n a-Si solar cells, as shown in Fig. 8. The longitudinal axis in Fig. 8 denotes the conversion efficiency normalized at the initial value, and AM-1 ($100mW/cm^2$) light was used as an irradiation source. As the oxygen concentration in the a-Si solar cell increased, the degradation in the conversion efficiency increased. This result corresponds to the result of the change in the σn.

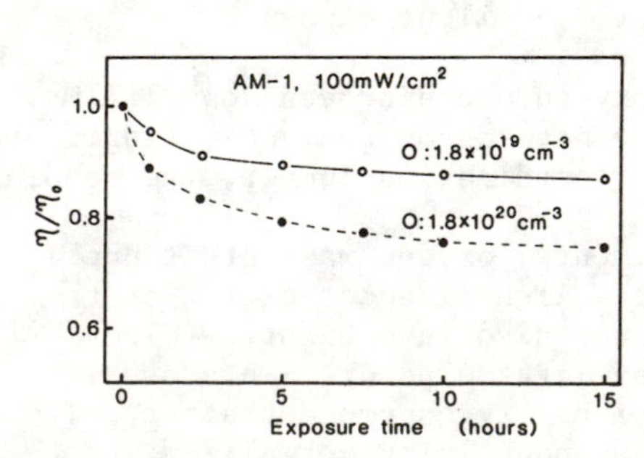

Fig. 8. The degradation of pin a-Si solar cells for various oxygen content

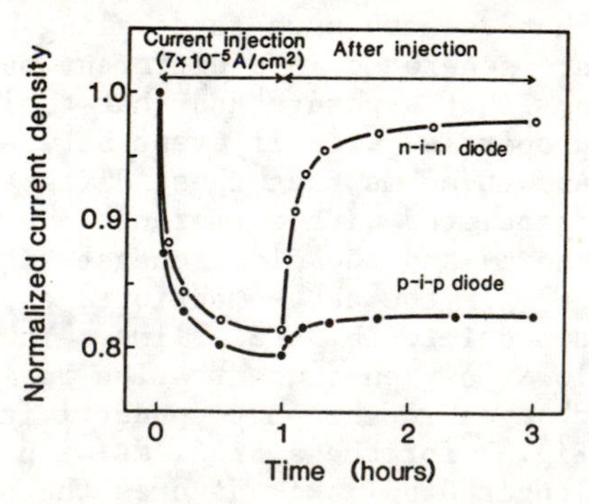

Fig. 9. The change in the current density of the p-i-p and n-i-n diodes by current injection

The light induced degradation of the a-Si solar cell is associated with the recombination and trapping of excess carriers [8]. So, the degradations in p-i-p and n-i-n diodes by current injection were also studied to distinguish the recombination from the trapping. The blocking structures make it possible to reduce recombination in these diodes. The changes in the current density of the p-i-p diode and the n-i-n diode are shown in Fig. 9. A decrease in the current density was observed in both the p-i-p diode and the n-i-n diode by current injection. But, the current density of the n-i-n diode was rapidly restored at room temperature after ceasing current injection, as shown in Fig. 9. In the case of the p-i-p diode, the degraded state was almost stable at room temperature, and was recovered by thermal annealing. These results suggest that the hole trapping is associated with degradation by current injection. It is considered that the light induced effect is mainly associated with the hole trapping because the degradation by current injection seems to be the same as the degradation by light exposure [8]. It seems that oxygen atoms induce the hole trapping centers.

(5) A 2-kW a-Si power generating system

A long term investigation has also been performed for a 2-kW a-Si solar cell power generating system since April, 1981. As shown in Fig. 10, about a 10% degradation in the output performance was observed in the initial period of one month, and then the degradation ratio became small. After one month, the conversion efficiency decreased in proportion to the equation of $\exp(-t/\tau)$ as the first approximation. The value of the time constant(τ) was about 410 months in this case.

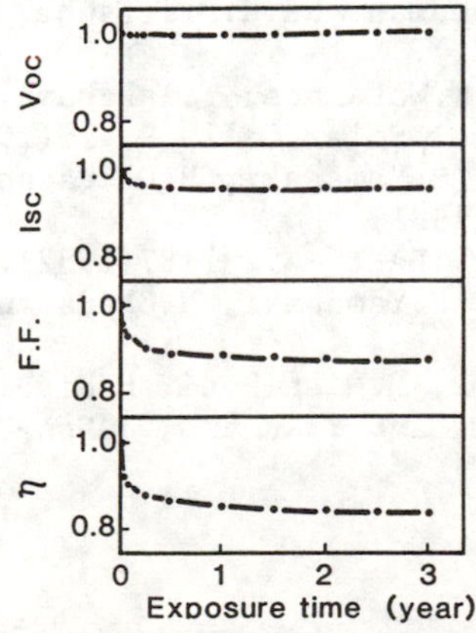

Fig. 10. The degradation of the output performance for a 2-kW a-Si power generating system

SUMMARY

The influences of oxygen and nitrogen on the light induced effect of a-Si films were reported. Both oxygen and nitrogen had a strong influence on the light induced effect, but there were large differences in their influences. The oxygen doped film had a longer time constant for the degradation than the nitrogen doped film. The dark conductivity decreased by light exposure was hard to recover by thermal annealing in the oxygen doped film. And it was also found that oxygen had a bad influence on the light induced degradation of a-Si solar cells. From the results of the degradation in the n-i-n and p-i-p diodes, it was considered that the light induced effect is mainly associated with the hole traps. The results of a 3-year experiment on a 2-kW a-Si power generating system were also presented.

ACKNOWLEDGMENTS

The authors wish to express their sincere appreciation to Professor Y. Hamakawa for his kind guidance and to Dr. M. Yamano, the executive managing director of Sanyo Electric Co. Ltd., for their encouragement and advice.

This work is supported in part by NEDO (New Energy Development Organization) as a part of the Sunshine Project under the Ministry of International Trade and Industry.

REFERENCES

1. D.L.Staebler and C.R.Wronski, Appl. Phys. Lett. 31, 292(1977).
2. H.Dersch, J.Stuke and J.Beichler, Appl. Phys. Lett. 38, 456(1981).
3. M.H.Tanielian, N.B.Goodman and H.Fritzsche, J. de Phys. 42, suppl. 10, C4-375(1981).
4. S.Tsuda, N.Nakamura, K.Watanabe, T.Takahama, H.Nishiwaki, M.Ohnishi and Y.Kuwano, Solar Cells 9, 25(1983).
5. H.Okushi, Y.Tokumaru, S.Yamasaki, H.Oheda and K.Tanaka, Jpn. J. Appl. Phys. 20, L549(1981).
6. R.A.Street, Phys. Rev. Lett. 49, 1187(1982).
7. H.Okushi, Y.Tokumaru, S.Yamasaki, H.Oheda and K.Tanaka, Phys. Rev. B25, 4313(1982).
8. N.Nakamura, K.Watanabe, M.Nishikuni, H.Hishikawa, S.Tsuda, H.Nishiwaki, M.Ohnishi and Y.Kuwano, J.Non-Cryst. Solids 59&60, 1139(1983).

INFLUENCE OF BIAS AND PHOTO STRESS ON a-Si:H-DIODES WITH nin- AND pip-STRUCTURES*)

W. Krühler, H. Pfleiderer, R. Plättner and W. Stetter
Research Laboratories of Siemens AG, D-8000 Muenchen 83

ABSTRACT

We observed the current-voltage characteristics of a-Si:H diodes having nin- and pip-structures after annealing, bias stressing and photo stressing. Both diode types show photo degradation, the pip diodes also bias degradation. The degradation is reversible in all cases by annealing. Furthermore, photo and bias degradations proceed independently of each other. We conclude that recombination and hole trapping induce different metastable defects.

INTRODUCTION

The reversible (by annealing at 150 °C) photodegradation of amorphous silicon (a-Si:H) films (reduction of dark conductance and photoconductance) is known as the Staebler-Wronski (SW) effect /1/. Pin diodes serving as solar cells suffer from an analogous degradation (loss in efficiency) that can be caused by prolonged photo exposure (P stress) or forward-bias application (B stress) /2,3/. Both stress conditions stimulate excess electrons and holes. Their recombination occurs, to a small degree, directly between tail states. The release of energy due to tail-tail recombination creates SW defects (dangling bonds) that are believed to be responsible for the degradation phenomena /4/. Consistently with this view the P degradation of nin diode characteristics showing electron injection is considerable, while their B degradation may be viewed as insignificant /5/. But pip diodes considerably degrade under B stress as well /3,6/. Therefore degradation (of conductance) is observed as a consequence not only of double injection (P stress, forward biased pin diodes) but, beyond that, also of one-carrier injection, especially of holes (B stressed pip diodes). Thus both (tail-tail) recombination and trapping (of holes) stimulate (reversible) degradation by the introduction of R and T centers, respectively.

Are the R and T centers identical? We attempt to answer this question here by observing some properties

*) This work was supported under the Technological Program of the Federal Department of Research and Technology of the FRG. The authors alone are responsible for the contents.

of nin, pip, and pπp diodes. (An i layer is not intentionally doped and behaves as weakly n-doped. A π layer is weakly p-doped.) The diode characteristics depend on the bias polarity (i.e. are not symmetric). The rectification ratio attains in the injection region values up to about 10. All the conductances shown below represent mean values with respect to the two bias polarities.

THE nin DIODE

Our diodes are structured as follows: NiCr coated Corning glass (substrate), glow-discharge deposited a-Si:H films (about 0.7 μm thick), ITO islands (6 mm^2) on top, thickness of the highly doped n layers about 30 nm. Fig. 1 shows measured σ(U) functions plotted in double logarithmic scales. σ represents the mean conductance (normalized to express the specific conductance), U the applied voltage. For U → 0, the conductance tends to become constant (ohmic current). The increase of σ with U exhibits one-carrier injection (electrons). The 3 curves were measured at room temperature and correspond to different states imparted by certain pretreatments. Curve A represents the annealed state (after 1/2 h at 180 °C), B the bias-stressed state (after 64 h at U = 2 V), P the light-soaked state (after 64 h of AM1 illumination). The B and P treatments were performed after restoration of state A. The

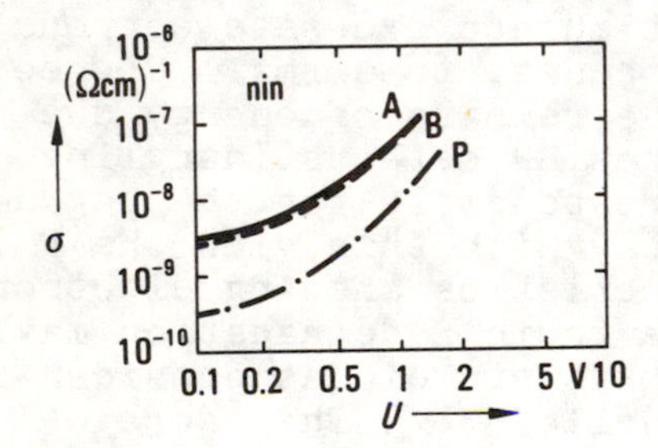

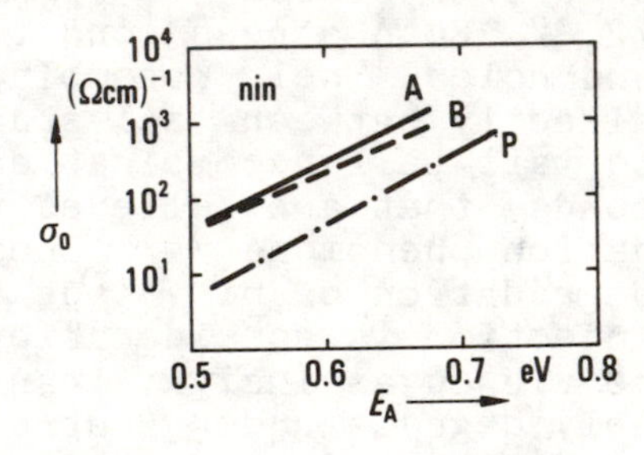

Fig. 1: Conductance σ of an nin diode versus voltage U in states A, B and P (see text).

Fig. 2: Conductance prefactor σ_o versus activation energy E_A.

stress voltage was chosen to yield a current density of about 10 mA/cm^2. The B and P conductances are lower than the A conductance (degradation). The B degradation remains relatively small. We measured σ at 3 temperatures (20,50, 80 °C) and found $\sigma(T) = \sigma_o \exp(-E_A/kT)$. Both, the prefactor σ_o and the activation energy E_A depend on U. Fig. 2 shows semilogarithmic $\sigma_o(E_A)$ plots. Any point on the curves corresponds to a definite voltage value. The voltage parameter increases from the right end of the curves to the left one through the interval $0.1\ V \leq U \leq 2\ V$. Thus E_A decreases when U increases. This behavior is to be expected for space-charge-limited currents (SCLC). Furthermore, the activation plots of Fig. 2 having the form of straight lines demonstrate the Meyer-Neldel rule (MNR) /7/ for the SCLC of our nin diode. Figs. 1 and 2 show the σ and E_A pictures of degradation.

An SCLC characteristic in a-Si:H depends on the density of gap states (DGS), or else a measured characteristic can be transformed into a DGS distribution near the Fermi level E_F. Fig. 3 shows this distribution for the A and P states. The DGS for the A and B states differ negligibly. The abscissa of Fig. 3 represents the gap ener-

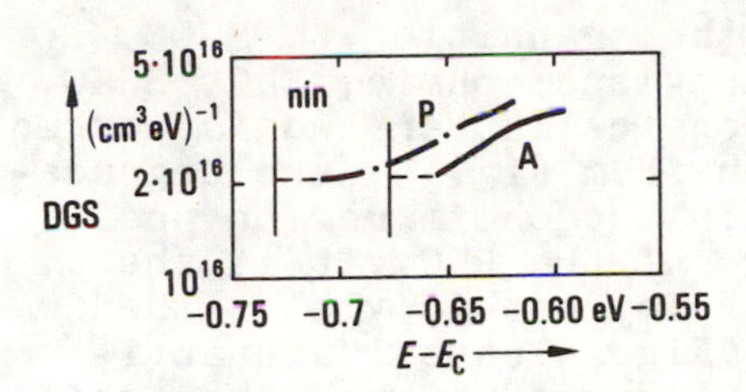

Fig. 3:
Density of gap states (DGS) versus gap energy E minus conductance-band edge E_c. The Fermi level $E = E_F$ is indicated by a vertical line for both curves (A and P).

gy E referred to the conduction-band edge E_c. The E_F positions were inferred from the activation energy for $U = 0.1\ V$. In the immediate vicinity of $E = E_F$, the DGS evaluation becomes uncertain (dashed lines). The σ degradation $A \rightarrow P$ has 2 causes: a Fermi-level shift (activation increase in the ohmic region) and emergence of an excess DGS (SW centers). The DGS evaluation was accomplished by the scheme /8/ being identical to the later proposal /9/.

THE pip DIODE

The pip diode has the same structure and dimensions as the nin diode, but with highly doped p layers. Fig. 4 shows the conductance in states A, B and P. A significant B degradation becomes visible. The limited degradation in the ohmic region points to a minimum conductance $\sigma \approx 10^{-10}\ (\Omega\ cm)^{-1}$. The hooked activation plot in Fig. 5 reveals a change-over of the injection-current mechanism. The activation energy first increases with the voltage and then decreases. On the low injection level we assume a recombination-limited current (RLC). The ma-

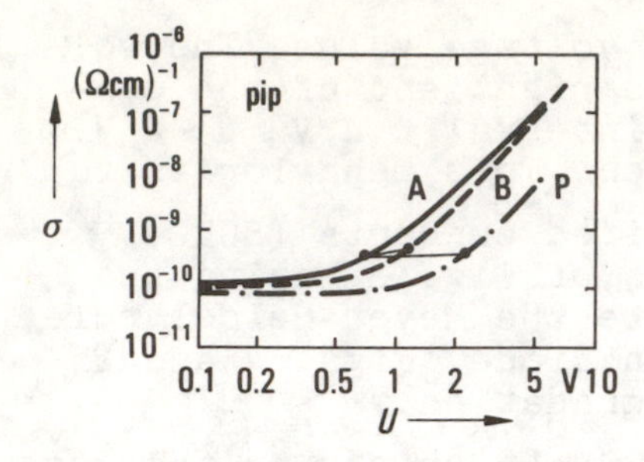

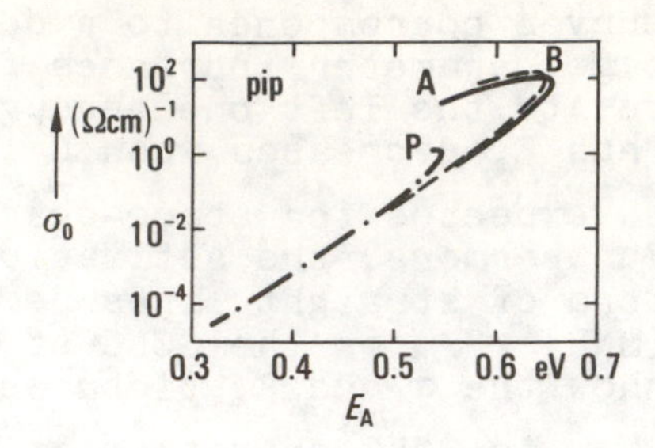

Fig. 4: $\sigma(U)$ for a pip diode in the annealed state A, in the state B after 64 h at 5 V and in the state P after 64 h AM1 illumination.

Fig. 5: $\sigma_o(E_A)$ for the states A, B and P. Parameter U from 0.1 V (lowest σ_o values) to 2.5 V.

ximum activation energy marks the transition to SCLC. Thus we distinguish 3 current regions: ohmic, RLC, SCLC. In Fig. 4 the transition between the latter two is marked by thick points. As can be seen from Fig. 5, the degradation decreases the activation in the ohmic region and thus enhances the voltage range of the RLC. It is the injection of holes into the "i" layer that yields the RLC. The conductance curves in Fig. 4 reveal a quantitative difference of the B and P degradations, the activation energy in Fig. 5, however, is a qualitative one. Both RLC and SCLC satisfy the MNR.

THE p π_1 p DIODE

The high-resistance layer of this diode was deposited through glow-discharge decomposition of silane containing 1 vppm diborane. The conduction and activation plots, Figs. 6, 7a and 7b, look similar to those for the pip diode. The new feature shown now is alternate degradation: B stress is immediately followed by a P stress without an annealing step in between, resulting in states B and BP, respectively. The interchange of the stress sequence yields the states P and PB. According to Figs.

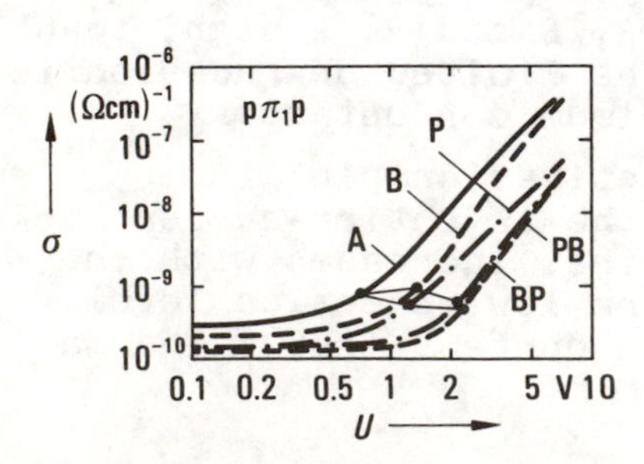

Fig. 6: $\sigma(U)$ for a pπ_1p diode in the annealed state A, in the state B (24 h at 6 V), in the state P (24 h illumination) and in the alternately stressed states BP and PB.

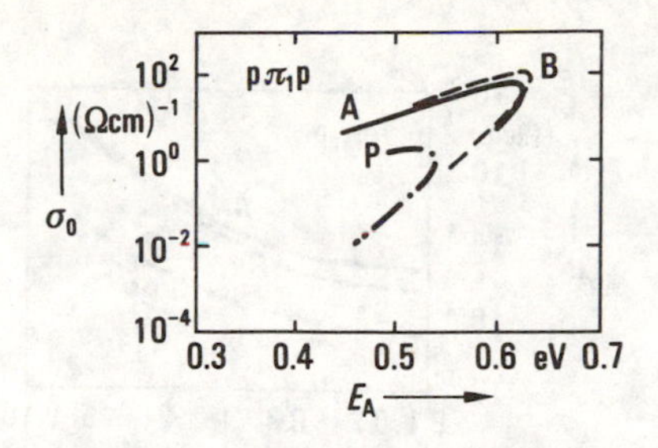

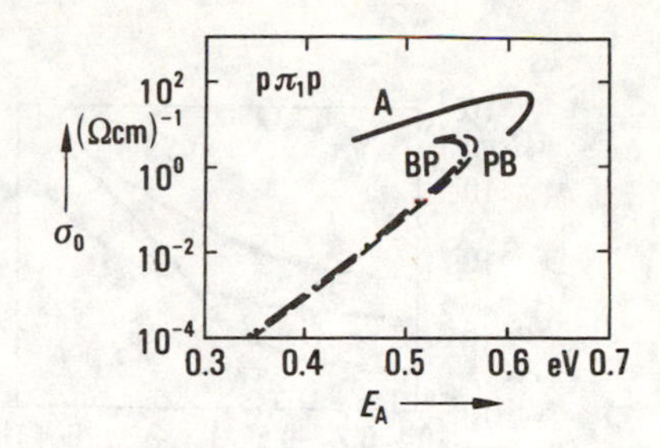

Fig. 7a: $\sigma_o(E_A)$ supplementing Fig. 6

Fig. 7b: $\sigma_o(E_A)$ supplementing also Fig. 6.

6, 7a and 7b the B and P stresses lead to cumulative degradations; i.e. the conductance and activation plots of states BP and PB fall close together, or else the B and P degradations proceed independently of each other. Therefore recombination and trapping events lead to different defects, called R and T centers, respectively.

THE p π_6 p DIODE

In order to achieve a hole injecting diode (showing SCLC without RLC) a highly resistive π_6 layer was deposited from a mixture of silane with 6 vppm diborane. Fig. 8 shows the conductance in states A_o (as-grown) and A (annealed), and in the corresponding photo-stressed states P_o, P. The difference between A_o and A is peculiar to the hole diode p π_6 p (and does not occur with the electron diode nin). A_o appears as severely degraded. Figs. 9 and 10 give the plots $\sigma(U)$ and $\sigma_o(E_A)$ for the sequence A, B and BP. From Fig. 9 together with the curve P from Fig. 8 we infer that B and P stresses induce cumulative degradation effects also in the case of our p π_6 p diode. Fig. 10 exhibits straight lines (MNR). The voltage parameter increases from the right to the left. Therefore the p π_6 p diode shows SCLC, at least in states A and B. The B stress increases the activation energy (in the ohmic region), while the P stress decreases it. The latter phenomena and the difference between A and A_o casts some doubt on the assumption of a pure SCLC mechanism in states A_o and P. (Further observations are necessary in order to clarify this point). Therefore the results of DGS evaluations are given in Fig. 11 for the states A und B only. The valence-band edge now serves as reference ener-

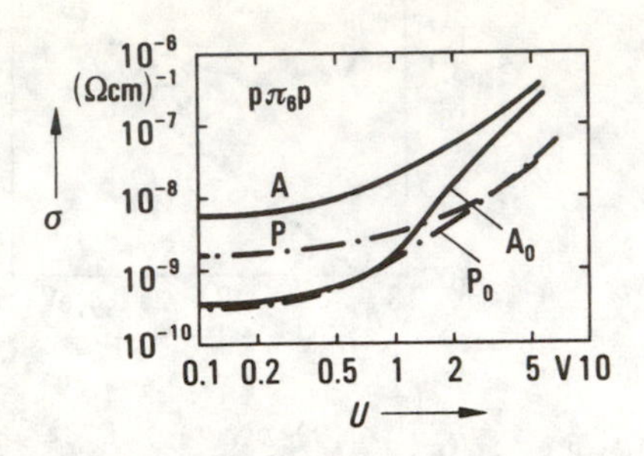

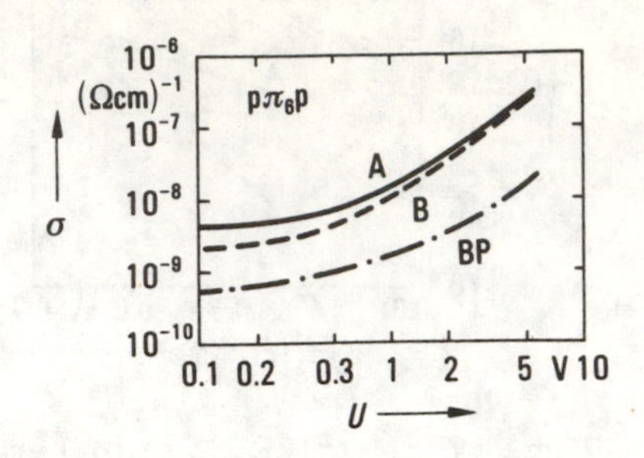

Fig. 8: $\sigma(U)$ of a $p\pi_6 p$ diode in the states A_o, P_o, and A, P (see text).
Fig. 9: $\sigma(U)$ of a $p\pi_6 p$ diode in the states A, B (64 h at 4,0 V) and in the state BP (alternately stressed).

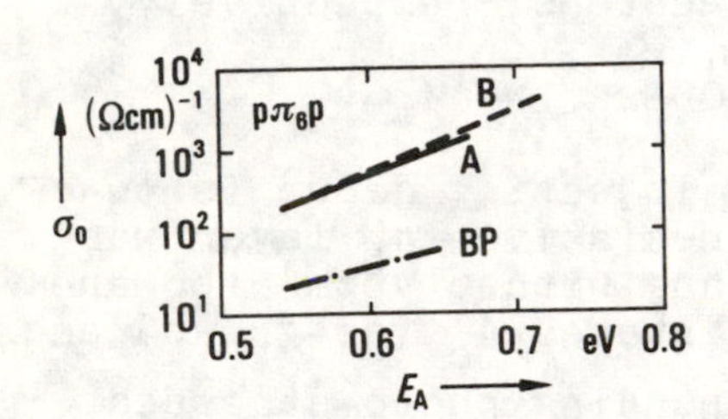

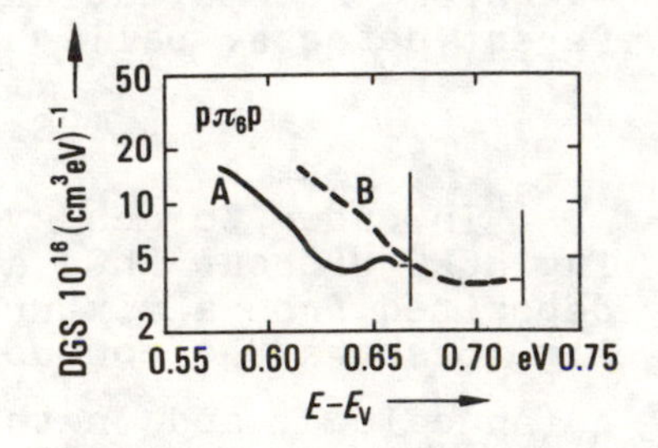

Fig. 10: $\sigma_o(E_A)$ supplementing Fig. 9.
Fig. 11: Density of gap states versus gap energy E minus valence-band edge E_V.

gy. Considering Figs. 11 and 3 the DGS increases more steeply for the hole diode ($p\pi_6 p$) than for the electron diode (nin), which is consistent with a smaller $\sigma(U)$ slope; compare Figs. 9 and 1.

ANNEALING BEHAVIOR

Further evidence for the existence of different R and T centers is provided by the temperature dependence of the annealing process of P and B stressed diodes. After each annealing step (1/2 h at the elevated temperature) we measured the I(U)-characteristic at 20 °C. Plotting the conductance at a fixed voltage versus the annealing temperature shows a steep rise in the conductance at a critical temperature T_A, at which the state A is reached.

Table 1 shows the collection of the T_A values of the different diodes. Note, that P-stressed diodes show about 20 °C higher T_A than the B-stressed diodes.

Table I: Annealing temperature T_A (°C)

Diode	P-stressed	B-stressed
nin	160	-
pip	160	140
p π_1 p	160	140
p π_6 p	115	90

CONCLUSION

While the observations presented are by no means complete, they prompt us to draw a tentative conclusion: in addition to SW defects introduced by recombination (R centers), the trapping especially of holes yields other (SW-like) defects (T centers). The two centers influence the conductance as well as the activation characteristics in a cumulative manner, i.e. act independently of each other. The activation behavior allows a distinction to be drawn between recombination and space-charge-limited excess carrier transport in the wake of one-carrier injection. The R centers support (in particular) the first transport mechanism. A hole diode (with boron-compensated base) severely degrades already during preparation. And hole injection is more active in creating T centers than electron injection.

REFERENCES

1. D.L. Staebler and C.R. Wronski; J. Appl. Phys. 51 (1980) 3262
2. D.L. Staebler, R.S. Crandall, and R. Williams; Appl. Phys. Lett. 39 (1981) 733
3. N. Nakamura, K. Watanabe, M. Nishikuni, Y. Hishikawa, S. Tsuda, H. Nishiwaki, M. Ohnishi, and Y. Kuwano; J. Non Crystalline Sol. 59/60 (1983) 1139
4. M. Stutzmann, W.B. Jackson, and C.C. Tsai; to be published
5. H. Pfleiderer, W. Kusian, and W. Krühler; Sol. St. Comm. 49 (1984) 493
6. W. den Boer, M.J. Geerts, M. Ondris, and H.M. Wentinck; Intern. Conf. on Transport and Defects in Amorphous Semiconductors, Detroit 1984
7. P. Irsigler, D. Wagner, and D.J. Dunstan; J. Phys. C16 (1983) 6605
8. F. Stöckmann; phys. stat. sol. (a) 64 (1981) 475
9. R.L. Weisfield; J. Appl. Phys. 54 (1983) 6401

THE EFFECT OF DOPANTS ON THE STABILITY OF a-Si SOLAR CELLS

H.Sakai, A.Asano, M.Nishiura, M.Kamiyama, Y.Uchida, and H.Haruki
Fuji Electric Corporate Research and Development, Ltd.
2-2-1, Nagasaka, Yokosuka City, 240-01, Japan

ABSTRACT

This paper describes the effect of doping boron atoms to the intrinsic(i-) layer on the stability of metal/n-i-p/ITO amorphous silicon solar cells. The stability is improved as the amount of doped boron atoms increases. We have explained the improvement in cell performance and its stability by the counterdoping of phosphorus with boron on the basis of the photo-induced change of the electrical properties of slightly doped a-Si:H films and the decrease of gap state density.

INTRODUCTION

We have studied the stability of p-i-n a-Si solar cells with relatively high conversion efficiency.

It was shown that the photovoltaic performance of this type of solar cell and its stability to sunlight were improved by slightly doping boron during the i-layer deposition[1],[2]. However the dependence of the stability on the boron-doping level and the role of doped boron have not been clarified yet. Recently, we have improved the conversion efficiency of metal/n-i-p/ITO a-Si solar cell up to 9.5%(area:$1cm^2$) by employing back electrode with high reflection. In this type of solar cell, we have found phosphorus atoms of more than $8x10^{16}$ atoms/cc were auto-doped in the i-layer by the secondary ion mass spectrometry(SIMS). This phosphorus atom incorporation stimulated our interest in the effect of doping boron on the photovoltaic performance and its stability.

In this paper, we report the stability of two types of a-Si solar cell with emphasis on the role of the dopants in the i-layer. One solar cell with relatively high efficiency of more than 8% possessed a metal/n-i-p/ITO structure and another formed on the heavily doped n-type single crystalline silicon possessed n(c-Si)/i-p/ITO structure. Since the i-layer of the latter type solar cell is not auto-doped with phosphorus atoms, we can compare the photovoltaic properties and their stability of undoped a-Si cells with those of slightly doped a-Si cells. In addition, light soaking effect of singly- and counter- doped a-Si:H films are studied. The effect of counterdoping with boron on the silicon dangling bond defects will be also discussed on the basis of the experimental results obtained by the isothermal capacitance transient spectroscopy[3].

1. PERFORMANCE STABILITY OF metal/n-i-p/ITO SOLAR CELL

The a-Si film used were deposited in a conventional RF(13.56 MHz) glow discharge apparatus. An 1% phosphorus-doped n-layer was formed on metal substrates. Then an intrinsic layer and an 1% boron-doped p-layer were deposited successively. Both doped layers were microcrystallized to decrease the light absorption loss and to form a good ohmic contact to the electrodes[4]. The i-layer in most of the cells was slightly doped with boron with B_2H_6-SiH_4 gas mixture. Finally indium-tin-oxide(ITO) transparent electrode and metal grid electrode were evaporated by the electron-beam technique. The solar cells used in this study had an area of $1cm^2$ and their initial conversion efficiencies distributed between 8% and 9.5%.

The solar cells were continuously exposed to simulated AM1 ($100mW/cm^2$) light. We investigated the changes of photovoltaic performance by exposing the cells for a relatively short period of 4 hours because the cells degrade in the early stage of the exposure and then tend to stabilize. The exposure was conducted under open-circuit condition: the severest loading condition [1].

Figure 1 shows the conversion efficiency after 4 hours exposure to the light of AM1($100mW/cm^2$) normalized to its initial value for metal/n-i-p/ITO cells with 0-1vppm boron-doped i-layer. The cells tend to stabilize as the concentration of boron in the i-layer increases. Excess boron-doping of more than 1vppm caused the decline in initial conversion efficiency.

Figure 2 shows the depth profiles of boron and phosphorus atoms in the i-layer of the metal/n-i-p/ITO solar cell measured by SIMS. This cell had i-layer doped with 0.5vppm boron. As shown in the figure, phosphorus atoms of more than $8x10^{16}$ atoms/cc were unintentionally incorporated in the i-layer.

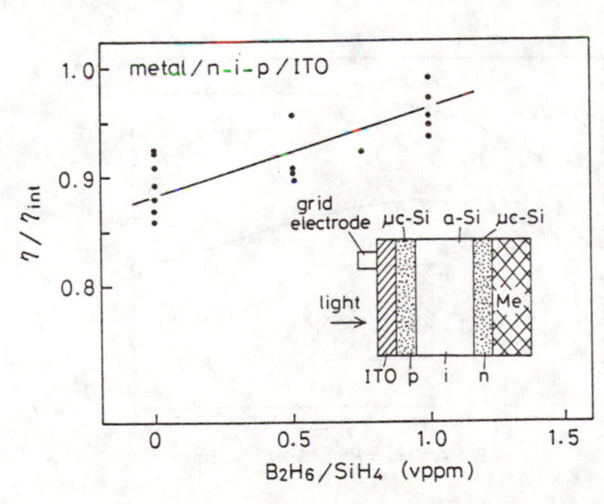

Figure 1
Dependence of performance stability on the boron-doping level in the i-layer for metal/n-i-p/ITO solar cells.

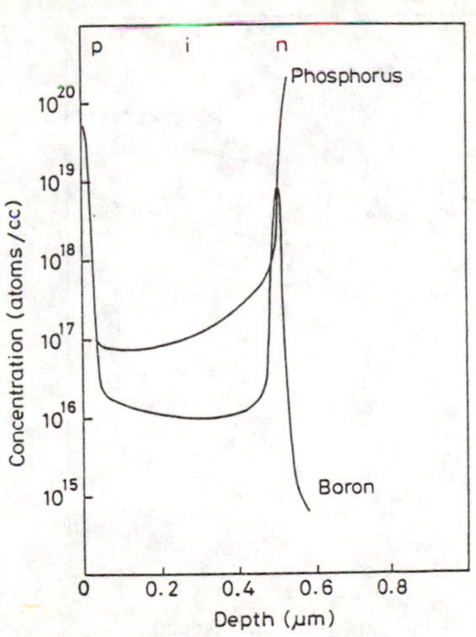

Figure 2
Depth profiles of dopants for metal/n-i-p/ITO solar cells.

Therefore, there are two possible effects of boron-doping in the i-layer. One is that the boron-doping effect originates from boron atoms only. Another is that the i-layer auto-doped with phosphorus atoms is counterdoped with boron atoms. We will discuss the latter possibility in next section.

2. COUNTERDOPING EFFECT ON THE PERFORMANCE STABILITY OF n(c-Si)/i-p/ITO SOLAR CELL

The incorporation of phosphorus atoms in the i-layer is unavoidable in metal/n-i-p/ITO solar cells as shown in Fig.2. To clarify the effect of counterdoping on the photovoltaic performance stability, we adopted n(c-Si)/i(a-Si)/p(μc-Si)/ITO structure with heavily doped n-type($\rho<0.018\Omega cm$) crystalline silicon substrate. The fabrication procedure was almost the same as that of the metal /n-i-p/ITO solar cell. The cells, fabricated with the doping gas mixture of $PH_3/SiH_4=B_2H_6/SiH_4$=0-2vppm, had the i-layer with a thickness of about 0.5μm.

Figure 3 shows the relation between the gas flow ratio of PH_3 or B_2H_6 to SiH_4 and the concentration of dopants measured by SIMS. Even if PH_3/SiH_4 was equal to B_2H_6/SiH_4,phosphorus atoms doped into the a-Si:H films were several times as many as boron atoms. In this figure, phosphorus concentration for PH_3/SiH_4=0vppm is about $5x10^{16}$ atoms/cc: the detection limit in our SIMS measurement. Some of the data represented by the closed circles were obtained in a-Si films doped under the $PH_3/SiH_4=B_2H_6/SiH_4$ condition. The other data represented by the open circles in Fig.3 shows the results of singly-doped a-Si films. Since the two series of data exhibited

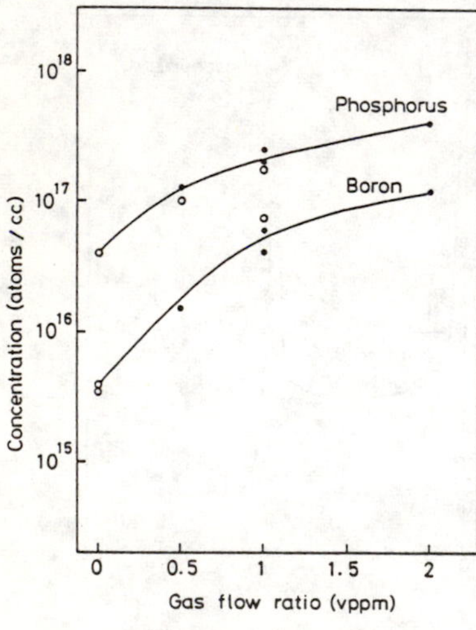

Figure 3
Relation between the gas flow ratio and the concentration of dopants.

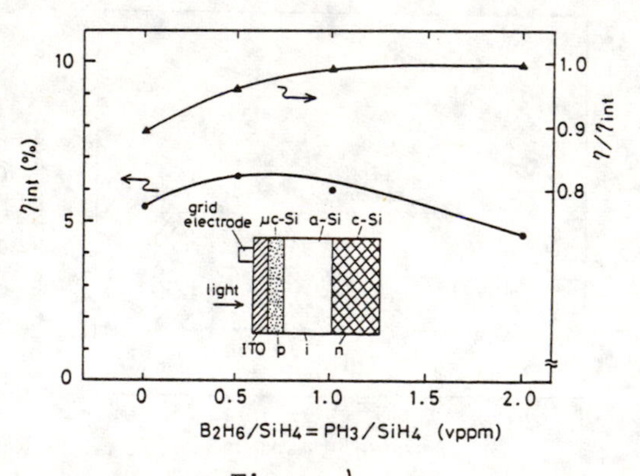

Figure 4
Dependence of photovoltaic performance and its stability on the level of counterdoping for n(c-Si)/i-p/ITO solar cell.

almost the same doping tendency, the abnormal doping[5] did not occur in the present experiment.

Figure 4 shows the dependences of photovoltaic performance and its stability on the level of counterdoping in the i-layer. Initial conversion efficiency shows maximum when $PH_3/SiH_4(=B_2H_6/SiH_4)$ is equal to 0.5vppm, which corresponds to the concentration of about $1x10^{17}$ atoms/cc and $2x10^{16}$ atoms/cc for phosphorus and boron, respectively. This indicates that truly nondoped a-Si:H is not always the best as the i-layer material of p-i-n devices. Counterdoping of phosphorus with boron can improve the performance of the solar cell. A maximum efficiency of about 6.5% is inferior to that of the metal substrate solar cell(9.5%). The reflection at the back contact of the c-Si substrate solar cell is almost zero, which causes the decreases of spectral response in the longer wavelength region, resulting in the lowering of photocurrent under the light of AM1.

The stability is improved as the level of counterdoping becomes higher. We find the same tendency in the boron-doping of metal substrate solar cell(see Fig.1).

We have found that the counterdoping of a small amount of boron and phosphorus during the i-layer deposition is effective to obtain the stable a-Si solar cell. To investigate the effect of counterdoping on the stability of p-i-n a-Si solar cell, we studied the properties of counterdoped a-Si:H films. First we discuss the electrical properties and their light soaking changes of slightly doped a-Si:H films. Afterward, the mechanism of counterdoping with boron to the phosphorus-doped a-Si:H films is discussed in relation to the density of gap states. Here, we measured the density of doubly-occupied silicon dangling bond defects(D^-) by the isothermal capacitance transient spectroscopy(ICTS)[3]. The change of D^- by counterdoping with boron and by light exposure are studied.

3. ELECTRICAL PROPERTIES OF SLIGHTLY DOPED a-Si:H FILMS

First, we deposited a-Si:H films of about 0.5μm thick onto Corning 7059 substrates. The doping level was 0-2vppm for phosphorus and boron. Then coplanar, magnesium electrodes[6] were evaporated on these films and annealed at 420K for 2 hours in air(annealed state). Dark- and photo- (under AM1; $100mW/cm^2$) conductivities of these films were measured. Then the films were exposed to the light of AM1 for 4 hours(light soaked state) and the change in both conductivities were measured.

Figure 5 shows the dark- and photo- conductivities in the annealed state. Phosphorus single doping increases the photoconductivity, but it also raises the dark conductivity more greatly. These results indicate that the electrical properties of the film were not necessarily improved by phosphorus-doping. In contrast, single doping with boron decreases both conductivities. The decrease of dark conductivity corresponds to the shift of Fermi level to the mid gap; electronic compensation occurs. And the decrease of photoconductivity arises from the decrease of the mobility-lifetime product of electrons by boron-doping[7],[8]. The dashed curves in Fig.5 were obtained under the full counterdoping conditions: $PH_3/SiH_4=B_2H_6/SiH_4$. As the full counterdoping level becomes higher, the photoconductivity drops but the dark conductivity remains almost constant. The decrease of photoconductivity by the full counterdoping suggests the decrease of mobility-lifetime products of electrons. Considering that the sum of mobility-lifetime products of both carriers is increased by the full counterdoping as shown in Fig.4, we know that the full counterdoping increases the mobility-lifetime product of holes.

Light-induced changes of photoconductivity in these a-Si films are shown in Fig.6. This figure exhibits the ratio of photoconductivity in annealed state to that in light soaked state. The change of photoconductivity increases as the doping level of phosphorus becomes high and it decreases by counterdoping with boron. The dashed line represents the doping condition of $PH_3/SiH_4=B_2H_6/SiH_4$. The light-induced change is reduced as the counterdoping level increases. This inclination is similar to that of the performance stability of c-Si substrate solar cell. This suggests the possibility of explaining the stabilization of photovoltaic performance by counterdoping, for the reduction of photoconductivity change by the light soaking leads to stabilization of the electron mobility-lifetime product.

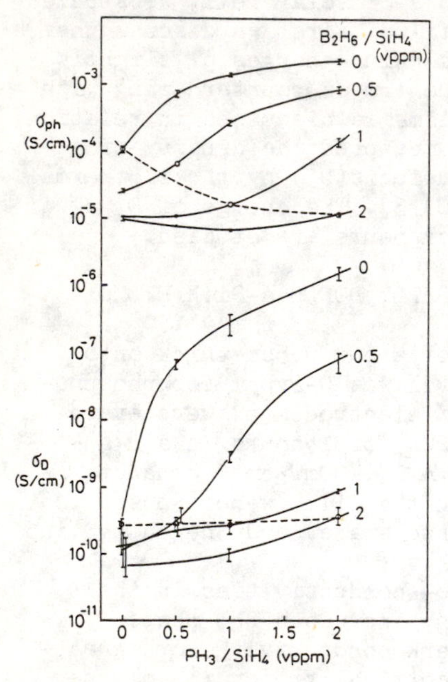

Figure 5
Dark- and photo- conductivities in annealed state of singly- and counter- doped a-Si:H.

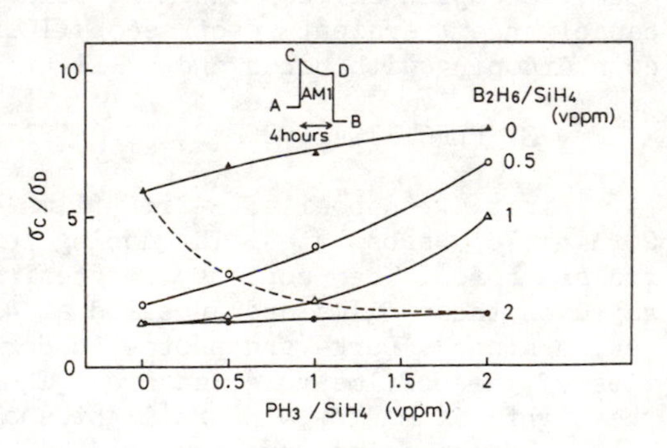

Figure 6
Ratio of photoconductivity in annealed state to that in light soaked state of singly- and counter- doped a-Si:H films.

4. EFFECT OF COUNTERDOPING WITH BORON ON THE DANGLING BOND DEFECTS

To discuss the effect of counterdoping with boron in detail, the doubly-occupied dangling bond defects(D^-) induced by phosphorus-doping[9], was measured by the ICTS technique. The samples were prepared as follows. First a thin, heavily(1%) phosphorus-doped microcrystalline Si(μc-Si) was deposited onto stainless steel substrates. Then a-Si films(PH_3/SiH_4=300vppm, B_2H_6/SiH_4=0-50vppm) to be measured were deposited onto the microcrystalline silicon. The doping level were chosen so as not to make the film too resistive for the fast capacitance measurement. The a-Si films thus prepared were annealed at 420K for 2 hours and Au top electrode with an area of $1mm^2$ was evaporated(Fig.7). In this way, we fabricated Schottky diodes with the diode quality factor n of 1.08-1.14 and dark reverse saturation current Jsat of less than 10^{-7} A/mm^2 at -1 volt bias. These values reduced, namely improved, as the amount of counterdoped boron atoms increased.

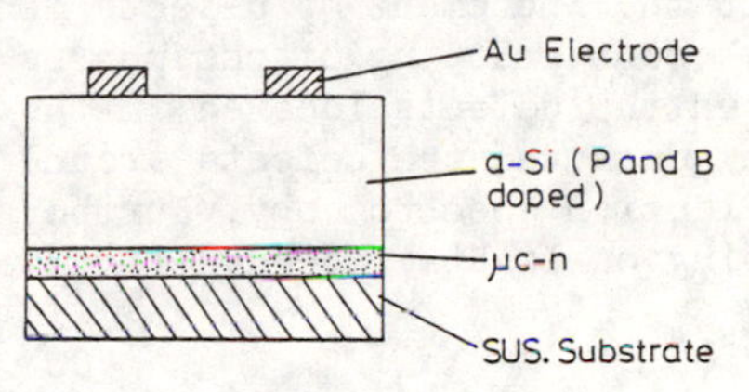

Figure 7
Structure of Schottky diode prepared for ICTS measurement.

Curve A-D in Fig.8 show the density distribution of gap states in a-Si:H doped with 300 vppm PH_3 and 0-50vppm B_2H_6. The energy scale for A-D(0.22-0.38eV below the Fermi level) was determined by the trap filling experiment(pulse width 200ns-1s) [10]. The density of gap states has the bump near 0.35eV below the Fermi level, which is assigned to D^-. Doubly-occupied dangling bonds decrease significantly by counterdoping with boron without changing the emission energy of trapped electrons to the conduction band. These decreases of D^- greatly improves the transport of holes, which corresponds to the effect of boron-doping to the i-layer on the photovoltaic performance of p-i-n a-Si solar cells.

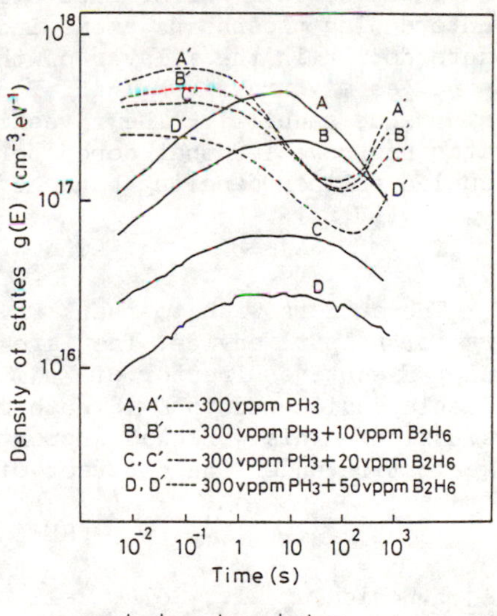

Figure 8
Density-of-state distributions for a-Si:H counterdoped with boron(A-D). And their increase by light soaking(A'-D').

To measure the dependence of photo-induced change of D^- on the level of boron-doping, we exposed the annealed a-Si films to the light of AM1($100mW/cm^2$) for 4 hours. Then Au top electrodes were evaporated. This exposure brought about slight increase of n and Jsat , which indicates the increase of recombination at the junction. The dashed line A'-D' of Fig.8 shows the density of gap states after exposure to the light of AM1. If we compare the light soaked state with the annealed state, counterdoping with boron do not affect the increment of D^- so significantly. Though, our ICTS results contradicts the results of the light soaking experiments shown in Fig.4 and Fig.6, we employed a relatively high doping with phosphorus in ICTS study and a relatively low doping in the light soaking experiments. Amer et al. investigated the effect of compensation on light-induced defects in a-Si:H films by the photothermal deflection spectroscopy[11]. They reported that the increment of defects density after light soaking decreases with counterdoping of phosphorus with boron. We deduce that we measured the defects located in the mid gap in ICTS experiment, while they measured the defects around the band tail by the photothermal deflection spectroscopy. Further studies on the effects of counterdoping on light-induced defects are in progress.

CONCLUSIONS

The stability of two types of a-Si solar cell has been investigated. Boron-doping to the i-layer of the metal/n-i-p/ITO a-Si solar cell improves cell performance stability under illumination by counterdoping phosphorus atoms incorporated in the i-layer. Full counterdoping in the i-layer of the n(c-Si)/i-p/ITO cell with the dopant gas mixture of $PH_3/SiH_4=B_2H_6/SiH_4$ also improves the cell performance stability. The investigation on the density of gap states has revealed that doped boron atoms reduce the doubly-occupied silicon dangling bond defects(D^-) induced by phosphorus.

ACKNOWLEDGMENTS

The authors wish to thank Dr.Okushi of Electrotechnical Laboratory for helpful advice. They are also grateful to Mr.K.Suzuki, Senior Executive Director of Fuji Electric Corporate R&D, for valuable advice and to Dr.S.Hushimi & Dr.Y.Ichikawa for fruitful discussions. This work was supported by the New Energy Development Organization under the contract of the Sunshine Project.

REFERENCES

[1] Y.Uchida et al., Solar Cells 9(1983) 3
[2] H.Haruki et al., Solar Energy Materials 8(1983) 441
[3] H.Okushi et al., J. Phys. (Paris) 42, Suppl.10 (1981) c4-613
[4] Y.Uchida et al., J. Phys. (Paris) 42, Suppl.10 (1981) c4-265
[5] R.A.Street et al., Phys. Rev. B24(1981) 969
[6] H.Matsuura et al., Jpn. J. Appl. Phys. 22(1983) L197
[7] R.A.Street et al., Appl. Phys. Lett. 43(1983) 672
[8] H.Sakai et al., J.Non-Cryst. Solids 59-60(1983) 1151

[9] H.Okushi et al., J.Non-Cryst. Solids 59-60(1983) 437
[10] H.Okushi et al., Jpn. J. Appl. Phys. 21(1982) Suppl.21-1, 447
[11] N.M.Amer et al., J. Non-Cryst. Solids 59-60(1983) 409

OPTICALLY DETECTED ESR OF LUMINESCENCE CENTERS IN a-As_2S_3

T.Tada, H.Suzuki, K.Murayama and T.Ninomiya
Department of Physics, University of Tokyo, Tokyo 113, Japan

ABSTRACT

Optically detected ESR (or optically detected magnetic resonance —— ODMR) was observed in a-As_2S_3 with the microwave of 25GHz and 9.6GHz. Two kinds of signals were observed, which are attributed to the resonance of the triplet exciton and that of the distant electron-hole pair respectively. Analyzing the width for 25GHz and 9.6GHz, we found that the pair resonance had the unresolved hyperfine structure with an As-nucleus.

The pair resonance was excited with the energy below 2.6eV. This can be explained in terms of the mobility edge.

INTRODUCTION

In chalcogenide glasses, Bishop, Strom and Taylor have observed two kinds of signals in light-induced ESR (LESR)[1]. One is the ESR associated with a chalcogen hole center and the other is associated with an arsenic center. The width of the latter is attributed to the hyperfine interaction with the As-nuclear spin (I=3/2).

ODMR studies in a-As_2S_3 have been made by Suzuki et al.[2] and Depinna and Cavenett.[3] They observed two kinds of signals, which were attributed to the resonance of a triplet exciton (the triplet resonance) and a distant electron-hole pair (the pair resonance) respectively. Depinna and Cavenett have shown both the ODMR spectra are common in a-As_2S_3, a-As_2Se_3 and a-P.[4] Suzuki et al. have reported the compositional dependence in a-As_xS_{1-x}.[2] The intensity of the pair resonance increases with increasing x. This suggests that the center of the pair resonance is associated with As.

In this paper, microwave frequency dependence of the width of the pair resonance is shown to be very small and this suggests that the origin of the width is the hyperfine interaction with an As-nuclear spin. We also report the dependence of the ODMR signals on the excitation energy, the emission energy and the microwave chopping frequency.

EXPERIMENTAL PROCEDURE

The optically detected magnetic resonance (ODMR) measurements were made at 1.6K with K-band (25GHz) and X-band (9.6GHz) microwave. An argon ion laser was used for the excitation, and the luminescence was detected by a cooled Ge detector with a response time of 300 nsec (North Coast EO-817P) or a photomultiplier (Hamamatsu Photonics R636).

We made two different types of ODMR measurements.

(i) A usual ODMR method: The microwave is chopped and excitation light

0094-243X/84/1200326-07 $3.00

is kept on continuously. The ESR is obtained as the synchronous change in the luminescence intensity using a lock-in detection method. (ii) Time resolved ODMR : The microwave is chopped and the excitation light is pulsed using an acousto-optic modulator. The microwave chopping frequency should be sufficiently lower than the excitation pulse repetition frequency and asynchronous. The time resolved luminescence is measured with a box-car integrator with a chosen delay time and a gate width. The output from the box-car integrator is lock-in detected,synchronously to the microwave modulation.

Samples of bulk amorphous As_2S_3 were prepared from elemental As and S, each of six-nines purity. The samples were sealed in evacuated quartz ampoules, held in a rocking furnace at 600°C for 24 hours and air-quenched to room temperature.

RESULTS AND DISCUSSION

(i) HYPERFINE STRUCTURE

The ODMR spectra with 25GHz and 9.6GHz are shown in Fig.1. The samples were excited with 2.41eV and we monitored the luminescence in the energy region from 0.69eV to 1.6eV. The spectrum is composed of two kinds of resonance: One is the broad double peaks with $g \simeq 6.0$ and 2.3, the other is the narrower peak with $g \simeq 2.03$. The former is the triplet resonance, and the latter is the pair resonance due to an electron or hole localized at an arsenic site, which are previously mentioned.

The width of the pair was 650 gauss for 9.6GHz, which is consistent with that by Depinna and Cavenett[3], and for 25GHz was 750 gauss. The width of the pair resonance is almost independent of the frequency. This suggests that the width of the pair resonance is due to the unresolved hyperfine structure, because the hyperfine interaction is independent of the external magnetic field. Since the S-nucleus does not have the spin, the hyperfine structure is due to the As-nucleus (I=3/2).

This is consistent with the compositional dependence of ODMR signals in a-As_xS_{1-x}.[2] In the As-rich samples the pair resonance is strong, while in the S-rich samples it is very weak.

The LESR spectrum observed by Bishop et al.[1] in a-As is best fitted with the hyperfine constants of $A_{/\!/}$ = 400 gauss and $A_\perp$ = 57 gauss. Using these values, we analyze the experimental data of the pair resonance. If the ODMR center of the pair resonance in a-As_2S_3 is the dangling bond of an As-atom, the values of the hyperfine constants are expected to be approximately the same as in a-As.

Since the anisotropy in the hyperfine interaction is small compaired with the resonant field (8.9kG for 25GHz), the spins are quantized very close to the direction of the magnetic field. Thus the spin Hamiltonian can be written as:

$$H = g\beta HS_z + aI_zS_z + bI_zS_z\,(3\cos^2\theta - 1) \qquad (1)$$
$$a = 1/3\,(A_{/\!/} + 2A_\perp) \qquad b = 1/3\,(A_{/\!/} - A_\perp)$$

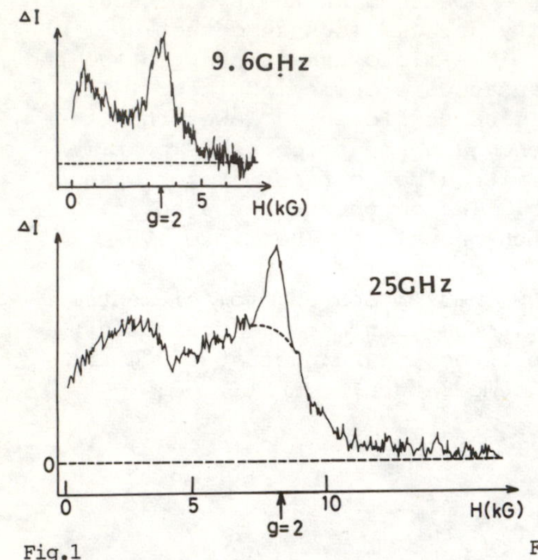

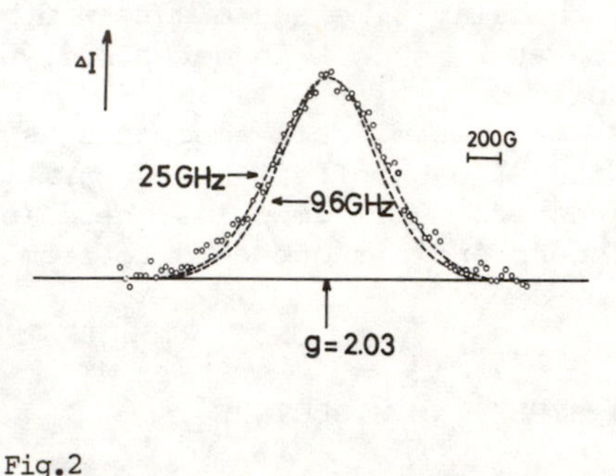

Fig.1
The ODMR spectra for the microwave of 9.6GHz and 25GHz. The widths of the pair component for both the frequencies are almost identical.

Fig.2
Computer calculated ODMR spectra (broken lines) of the pair component for 25GHz and 9.6GHz. Open circles are the experimental data of the pair component(25GHz).

where z is the direction of the magnetic field H and θ is the angle between the magnetic field and the symmetry axis of the hyperfine interaction. β is the Bohr magneton and S=1/2 for the distant pair. In amorphous materials, θ is randomly oriented and the observed spectrum is averaged over all orientations. We introduce the Gaussian broadening with the width σ. This is due to the dipolar interaction, random variation of local environments etc. We decompose this σ into the two components σ_1 and σ_2 . The component σ_1 is independent of the value of the resonant magnetic field (or micro wave frequency) and σ_2 is proportional to the value of the resonant magnetic field, and $\sigma = \sigma_1 + \sigma_2$. The origins of σ_1 are the dipolar interaction, the spin-lattice interaction or the structural disorder etc. and those of σ_2 are the anisotropy of g-value or the distribution of the g-value due to random variations of local environments etc.

In this way we fitted the pair component of 25GHz. The experimental data is the superimposition of the triplet and the pair resonance. We subtracted the triplet resonance from the experimental data, and obtained the pair component. Hence there is a little ambiguity in this subtraction, but it makes little effect on the fitting.

We obtained the following fitting parameters:

$$\sigma_1 = 210 \text{ gauss} \quad \sigma_2 = 110 \text{ gauss (25GHz)}$$

The parameters for 9.6GHz are calculated as follows:

$$\sigma_1{}' = \sigma_1 \qquad \sigma_2' = 9.6/25 \times \sigma_2$$

Fig.2 is the results of the fitting. The open circles are the experimental data of the pair component for 25GHz and the dashed lines are calculated curves for 25GHz and 9.6GHz. The FWHM is 750 gauss (25GHz) and 650 gauss (9.6GHz) respectively. This is consistent with the experimental data.

(ii) EXCITATION ENERGY DEPENDENCE

In the following parts, the ODMR was measured using the micro wave of 25GHz.

In Fig.3, the ODMR spectra excited with 2.41eV and 2.61eV are shown. The pair resonance vanishes when excited at 2.61eV. Depinna and Cavenett have observed the similar dependence using 9GHz micro wave.[3] In Fig.4, ΔI/I for the triplet component and the pair component is plotted against the excitation energy, where we monitored the luminescence ranging from 0.69eV to 1.6eV. As is seen, the pair resonance disappears when excited above 2.6eV, while the triplet resonance is still observed above 2.6eV. This can be explained as follows: Murayama et al. have shown from excitation energy dependence of the peak energy of the luminescence at 0 delay that the energy of excitation to the mobility edge is estimated to be 2.6eV.[6] Then for the excitation energy below 2.6eV, the electron is excited directly to the As-localized center and recombines through the distant electron-hole pair recombination channel. But above 2.6eV the electron is excited to the extended states,

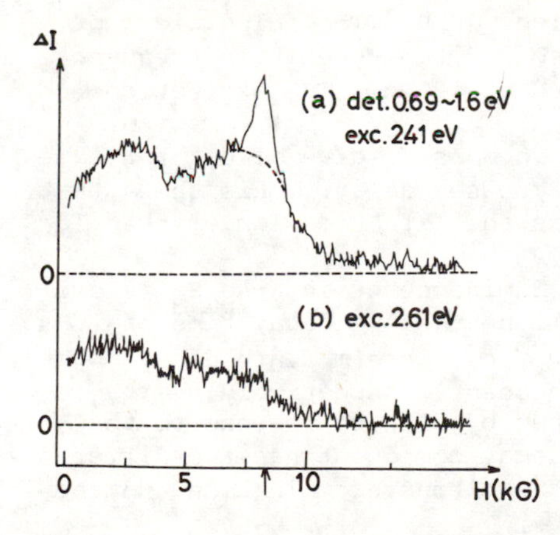

Fig.3
The ODMR spectra excited with 2.41eV and 2.61eV. The detected emission energy region is 0.69eV-1.6eV. The pair resonance vanishes when excited with 2.61eV.

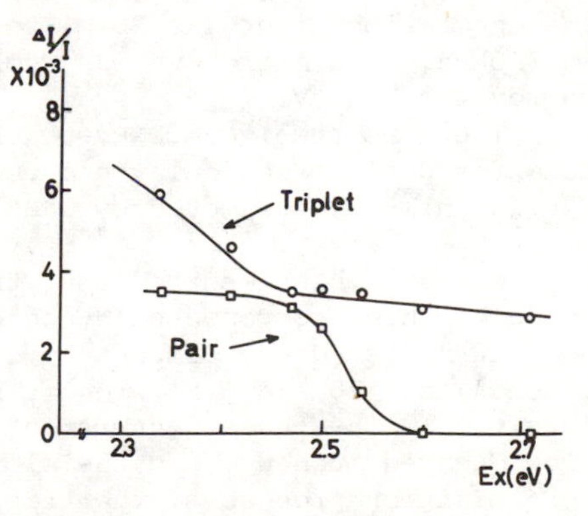

Fig.4
ΔI/I for the pair and the triplet component is plotted against the excitation energies. ΔI/I for the pair component disappears above 2.61eV.

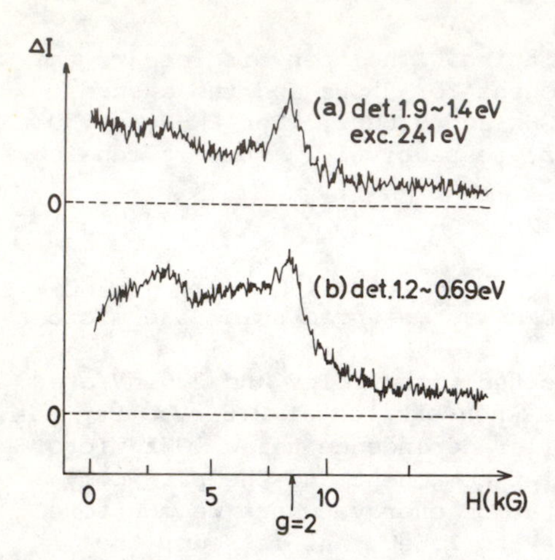

Fig.5
The ODMR spectra for different emission energy ranges: (a) 1.4eV - 1.9eV , (b) 0.69eV - 1.2eV.

recombines through other channels, and the pair resonance is weakened.

(iii) EMISSION ENERGY DEPENDENCE

The luminescence in a-As_2S_3 has the wide emission spectrum ranging from 0.6eV to 1.9eV. The emission energy dependence of the ODMR signals in a-As_2S_3 was studied by monitoring the luminescence in the following three energy ranges:(i)1.9eV-1.4 eV (ii)1.6-0.69eV (iii)1.2-0.69eV, while the excitation energy was always 2.41eV. The results of the cases (i) and (iii) are shown in Fig.5 (a),(b) and the case (ii) is shown in Fig.3 (a). The pair resonance was relatively strong when monitoring the high energy side of the luminescence. On the other hand when monitoring the low energy side of the luminescence, the pair resonance was relatively weak and the triplet resonance was strengthened.

(iv) LIFE TIME OF THE ODMR CENTERS

We also measured the time resolved ODMR. The pulse width of the excitation light was 20 micro sec and the repetition frequency was 10kHz. The microwave was chopped at 120Hz. The excitation energy was 2.41eV.

The time resolved ODMR spectra in a-As_2S_3 give almost the same data for 0 micro sec delay and 50 micro sec delay. This shows that the luminescence detected by ODMR has the life time longer than 10^{-4} sec.

Murayama has reported that the luminescence in a-As_2S_3 is composed of three components.[7] Each component has a decay time of $\sim 10^{-8}$sec (fast decay component), $\sim 10^{-6}$sec (intermediate decay component) and $\sim 10^{-4}$sec (slow decay component) respectively. Thus the luminescence centers detected by ODMR correspond to the slow decay component, and the slow decay component has two different recombination processes: the triplet exciton recombination and the distant pair recombination.

Microwave chopping frequency dependence (90Hz to 2.2kHz) of the ODMR spectra shows that the pair resonance is strongly dependent on the chopping frequency. $(\Delta I)_{pair} / (\Delta I)_{triplet}$ $(\equiv r)$ is plotted against the chopping frequency in Fig.6. The pair resonance is relatively strong at low frequencies, and weak at high frequencies.

The response of the system with life time t for the external

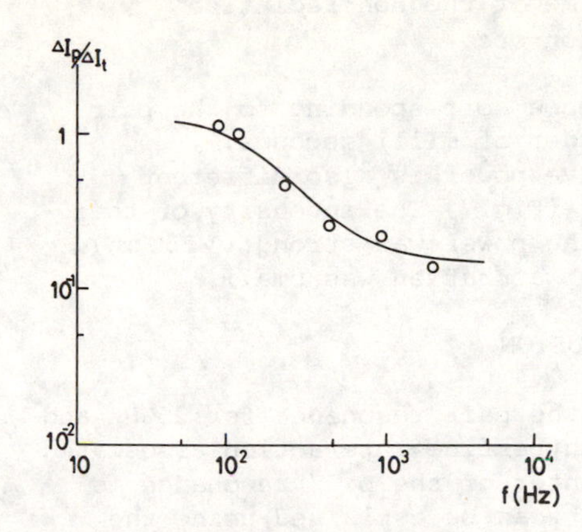

Fig.6
$(\Delta I)_{pair} / (\Delta I)_{triplet}$ is plotted against the chopping frequency. The solid line is the theoretical curve described in the text.

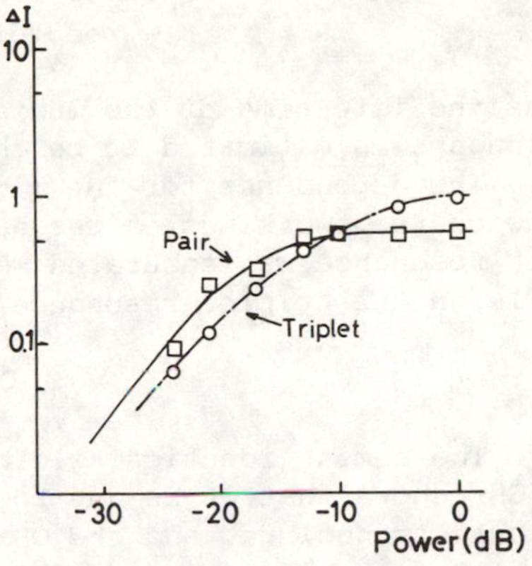

Fig.7
The microwave power dependence of the ODMR spectra. The pair component saturates faster than the triplet component.

modulation with frequency f can be written as[8]:

$$R = Ro / \sqrt{1 + 4\pi^2 f^2 t^2} \quad ; \; Ro = \text{constant} \qquad (2)$$

Thus r obeys the following equation:

$$r = r_o \sqrt{1 + 4\pi^2 f^2 t_1{}^2} / \sqrt{1 + 4\pi^2 f^2 t_2{}^2} \qquad (3)$$

f : the chopping frequency of the microwave
t_1: the effective spin relaxation time for the triplet component.
t_2: the effective spin relaxation time for the pair component.
r_o: constant

We fitted the data using (3) and obtained the fitting parameters:

$$t_1 = 2 \times 10^{-4} \text{ sec}$$
$$t_2 = 2 \times 10^{-3} \text{ sec}$$

The effective spin relaxation time t with the optical pumping cycle can be written as:

$$1/t \sim 1/t_r + 1/T_1 + 1/t_n \qquad (4)$$

t_r: the life time of the radiative recombination
T_1: the spin-lattice relaxation time

t_n: the life time of the non-radiative recombination etc.

Thus the life time of the luminescence corresponding to the pair resonance is estimated to be the order of milli seconds.

The dependence for the microwave power is also different between the two kinds of resonance. (Fig.7) The intensity of the pair resonance was saturated when the power was strong ($\sim$250 mW), while in the triplet resonance, the saturation was small.

CONCLUSION

The almost identical width of the pair resonance for 25GHz and 9.6GHz shows that it is due to the hyperfine interaction associated with an As-nucleus, and the ODMR center of the pair resonance is localized on the As-atom. The width can be explained using the hyper fine constants by Bishop et al.

The pair resonance was preferentially excited below 2.6eV. This is because 2.6eV corresponds to the mobility edge and excitation above the mobility edge makes the electron move freely and recombine through other recombination channels than that localized on the As-atom.

ACKNOWLEDGEMENTS

The authors gratefully acnowledge Prof. K.Morigaki, Dr. I.Hirabayashi and Miss M.Yoshida for the use of their X-band ESR apparatus and helpful advice.

REFERENCES

1. S.G. Bishop, U. Strom and P.C. Taylor, Phys. Rev. B15, 2278, (1977).
2. H. Suzuki, H. Nakata, K. Murayama and T. Ninomiya, to be published.
3. S.P. Depinna and B.C. Cavenett, Phys. Rev. Lett. 48, 556 (1982).
4. S.P. Depinna and B.C. Cavenett, Phil. Mag. B46 71, (1982).
5. G.E. Pake and T.L. Estle, The Physical Principles of Electron Paramagnetic Resonance (W.A. Benjamin, Inc. 1973), p. 214.
6. K. Murayama, H. Suzuki and T. Ninomiya, J. Non-crystalline Solids 35 & 36, 915 (1980).
7. K. Murayama, J. Non-crystalline Solids 59 & 60, 983 (1983).
8. P.W. Kruse, L.D. McGlauchin and R.B. McQuistan, Elements of Infrared Technology (J. Wiley, New York, 1961), p. 274.

OPTICALLY-INDUCED ESR AND ABSORPTION EDGE SHIFT IN GeS_x GLASSES

J. Shirafuji, N. Kumagai* and Y. Inuishi**
Osaka University, Suita, Osaka, 565, Japan

ABSTRACT

The isothermal and isochronal annealing characteristics of the photo-induced ESR spin density and the absorption edge shift have been measured on melt-quenched bulk GeS_x glasses (x = 1.5, 1.8, 2.0 and 2.3). The annealing curves can be explained in terms of the first order reaction with a Gaussian distribution of the activation energies. A configuration-coordinate model taking into account of a strong electron-lattice interaction is proposed to interpret consistently the experimental results. The doping and gamma-ray irradiation effects on the photo-induced ESR intensity give negative support for the bond switching mechanism under band-gap light illumination.

INTRODUCTION

When chalcogenide glasses are illuminated with band-gap light, various phenomena related to the introduction of metastable states such as absorption edge shift,[1,2] optically-induced below-gap absorption[3] and photo-induced ESR[3,4]. These optically-induced effects seem to relate to each other. Bishop et al[3] have observed that the optically-induced paramagnetic centers are located at the mid-gap and induce the below-gap absorption. The optically-induced effects can be annealed thermally or bleached by illumination of sub-band-gap light. Taylor et al[4] have measured the saturated density of ESR centers in As_2Se_3 as a function of temperature and explained under the assumption of a broad flat or a Gaussian distribution for the density of the thermal release energies. They speculated the origin of the distribution of the thermal release energies to be a distribution of hopping distances, rather than trap depths of the localized states. Street and Biegelsen[5] have measured the decay characteristics of the photo-induced ESR centers at various temperatures in $GeSe_2$ and explained their results in terms of thermally-activated tunneling between two neutral centers.

In this paper, we have measured the effects of doping and gamma-ray irradiation, and the isothermal and isochronal annealing characteristics of optically-induced ESR centers in bulk GeS_x glasses. The comparison is made with the annealing of the photo-induced absorption edge shift. The annealing characteristics are satisfactorily analyzed on the basis of simple first-order reaction with a Gaussian distribution of the activation energies.

* Present Address: Fuji Electric Co. Ltd., Hino, Tokyo, 191, Japan
** Present Address: Kinki University, Higashi-Osaka, 577, Japan

EXPERIMENTS

The compositions x of melt-quenched bulk GeS_x glasses were varied from 1.5 to 2.3. For ESR measurement, the granulated glasses were sealed into fused silica capillaries. After the saturation of the ESR spin density was reached at 150 K by the illumination with light from a Xe arc lamp through an infrared-cut filter, isothermal or isochronal annealing process was carried out in the same microwave cavity. All of the ESR measurements were performed at 150 K.

The measurement of the photo-induced absorption edge shift was made on the samples (50 - 100 μm thick) with mirror-finished surfaces which were kept at 150 K.

RESULTS

In GeS_x glasses the ESR signals were observed even at room temperature in as-prepared samples. The shape of the signals changes drastically at the stoichiometric composition of GeS_2.[6] The illumination, gamma-ray irradiation and Ag or Li doping caused a change in the ESR intensity, but nothing in the line shape, indicating that these perturbations change only the number of ESR centers and do not introduce any different kind of paramagnetic centers even by gamma-ray irradiation.

The doping with monovalent elements such as Ag and Li led to a decrease in both inherent and photo-induced ESR intensities as shown in Fig. 1. However, the effect is very small compared with the case of evaporated films.[7] The reduction in the photo-enhancement by the doping suggests that the optically-induced ESR centers are associated with dangling bonds which are terminated by monovalent elements to be electronically inactive. In the case of Li doping, excess Li atoms were locating at interstitial sites and caused a slowly decaying transient current in the dark-conductivity measurement possibly due to ionic conduction of interstitial Li atoms. The magnitude of the optical enhancement of the ESR intensity is roughly proportional to the ESR intensity before illumination (Fig. 1), suggesting that dangling bonds responsible for the photo-induced ESR are inherently existing in as-quenched samples and are not newly produced during illumination by a process, for example, of bond switching.[8]

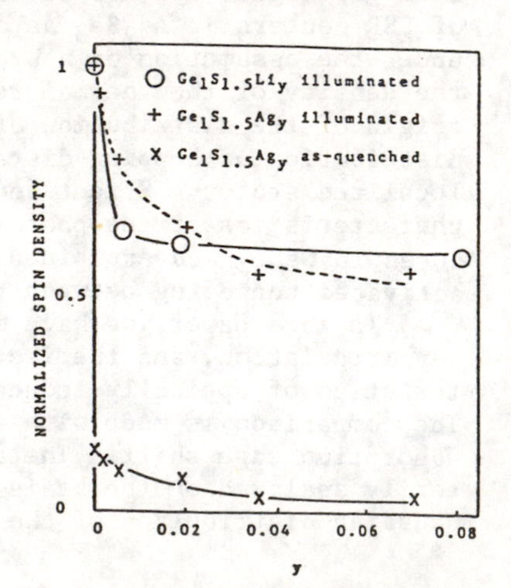

Fig. 1. Effect of doping with Li and Ag.

The ESR signal was remarkably increased by gamma-ray irradiation (^{60}Co) at room temperature (3×10^9 rö/cm^2) and furthermore the photo-enhancement in gamma-ray irradiated samples was much larger than in unirradiated samples as seen in Fig. 2. The effect of the gamma-ray irradiation is not only photoelectric excitation similar to light illumination, but also includes other factors, as supported from the fact that the photo-enhancement of the ESR intensity after gamma-ray irradiation is about an order of magnitude larger than that before irradiation. An increase in the dangling bond density by gamma-ray irradiation effectively enhanced the photo-induced effect. The thermal annealing decreased the excess dangling bonds and thus the photo-enhancement of the ESR intensity (Fig. 2).

These observations give support for the model of the photo-induced ESR in which neutral dangling bonds are produced through the process

$$D^+ \text{ (or } D^-) + e^- \text{ (or } e^+) \longrightarrow D^0$$

and not by bond switching.[8]
It may be possible that the network containing four-coordinated Ge is not so flexible to cause bond switching as in As-Se glasses where As atoms are three-coordinated.

The photo-induced ESR could be degraded by the illumination with sub-band-gap light (for instance, longer wavelength than 540 nm for $GeS_{1.5}$ glass) which brought as well the recovery of the absorption edge shift; this phenomenon gives an insight into the model for the photo-induced ESR.

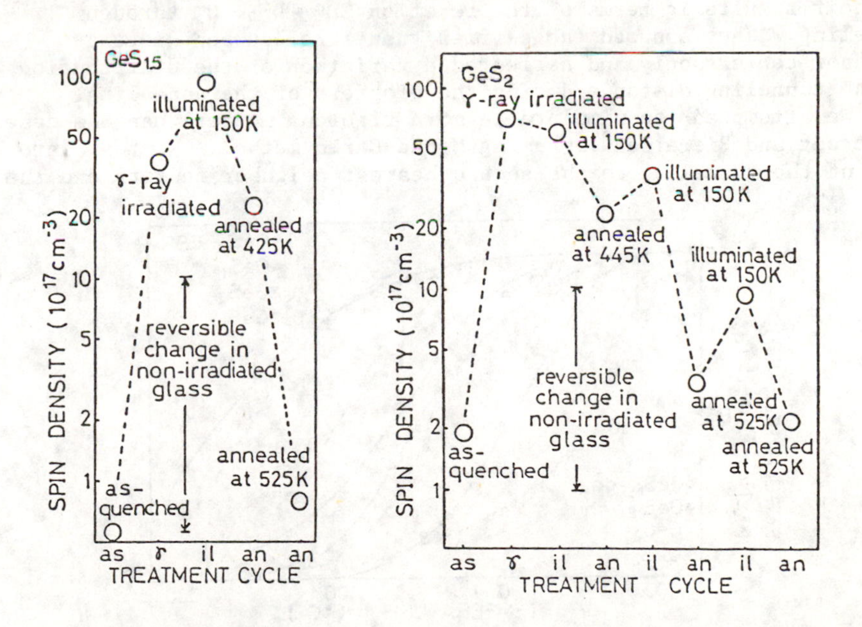

Fig. 2. Photo-enhancement and annealing of the ESR intensity after gamma-ray irradiation in $GeS_{1.5}$ and GeS_2.

Figure 3 shows the isochronal (30 min) annealing characteristics of the photo-induced ESR intensity N_s and the absorption edge shift E_g for glasses of various compositions, where subscripts s, so and a stand for "annealed for 30 min at each temperature", "saturated by illumination" and "fully annealed out," respectively. As seen clearly in Fig. 3, the temeprature of the annealing stage depends critically on the glass composition and becomes maximum at the stoichiometric composition of GeS_2. This behavior is in parallel with the composition dependence of the glass transition temperature, suggesting a role of atomic displacement in the photo-induced effects. Moreover, the similarity of the isochronal annealing behaviors for the photo-induced ESR intensity and the absorption edge shift indicates that these photo-induced effects may be different appearances of a single phenomenon.

Figure 4 shows a typical example of the isothermal annealing characteristics of the photo-induced ESR spin density in $GeS_{1.8}$ glass. These characteristics can not be fitted by simple processes such as first-order reaction with a single activation energy and second-order one with a single reaction constant.

The isochronal annealing experiments have been carried out on As_2Se_3 glass by Taylor et al.[4] and on $GeSe_2$ glass by Street and Biegelsen.[8] Taylor et al.[4] assumed a broad flat distribution for the density of the thermal release energies due to a distribution of hopping distances and, moreover, assumed that all of the states having thermal release energies less than kT anneal out by the annealing at the temperature T. On the other hand, Street and Biegelsen[8] explained their results in terms of the reaction $2D^o \rightarrow D^+ + D^-$ through tunneling. They assumed the carrier tunneling between nearest-neighbor centers only and neglected a variation of the distribution of the tunneling distance during the progress of the annealing.

We attempted[9] to simulate in more rigorous manner than the case of Street and Biegelsen[8] by using Monte-Carlo method. We took into account the tunneling to the second nearest neighbor centers and the

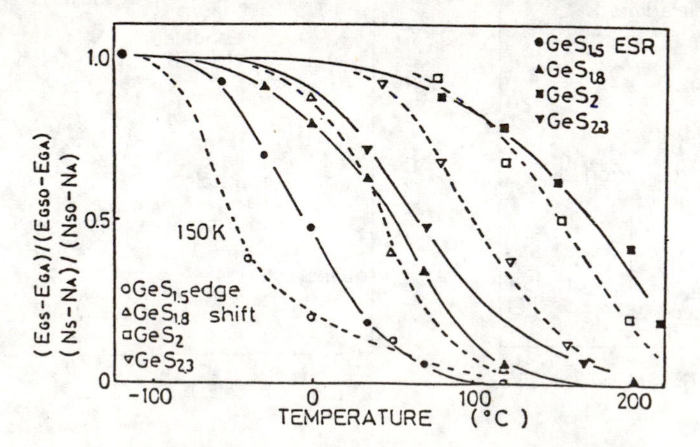

Fig. 3. Isochronal annealing characteristics of the photo-induced ESR intensity and absorption edge shift.

change in the neutral center distribution due to a decrease in neutral center density. Unfortunately, however, the result was quite unsatisfactory as seen in Fig. 5. A set of the adjustable parameters of tunneling factor, activation energy W_0 and frequency factor ν can not fit both of annealing curves taken at different temperatures. This failure may suggest a distribution of the activation energy to be considered. However, this procedure will introduce much more complication into the simulation. Then, we tried a rather simple calculation to interpret consistently the two curves in Fig. 4.

We assumed simple first-order reaction with widely-distributed activation energies of a Gaussian distribution. The adjustable parameters were W_0 (the center value of the activation energies), σ (the deviation) and ν (the frequency factor). Fig. 6 shows the fitting between the experimental points and the calculated curves. The value of $\nu = 10^{13}$ s^{-1} gives a good fit for both curves simultaneously using the common values of W_0 = 1.1 eV and σ = 0.17 eV. The value of ν is quite reasonable in view of the fact that most of the

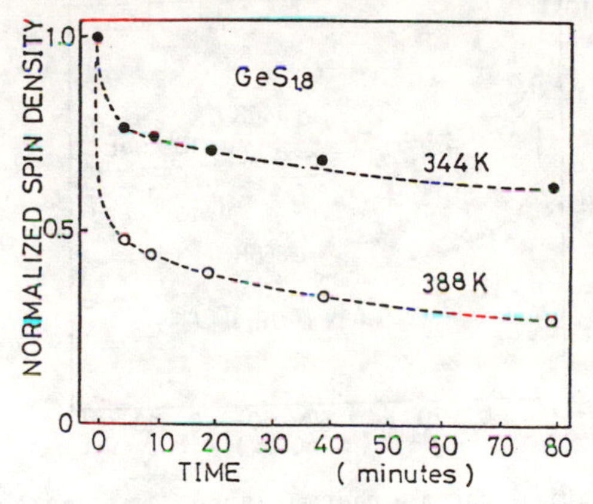

Fig. 4. Experimental result of the annealing characteristics of the photo-induced ESR .

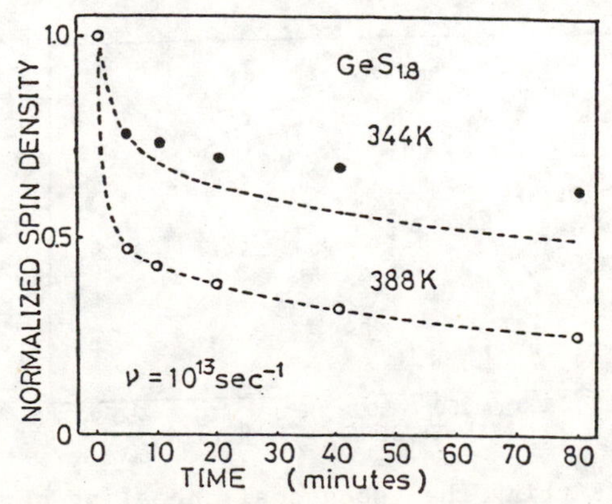

Fig. 5. Monte-Calro fitting of the annealing characteristics.

vibrational modes in GeS_x glasses are close to 300 cm^{-1}. Similar beautiful fitting was also attained by using $\nu = 10^{13}$ s^{-1} as shown in Fig. 7, where the isochronal annealing characteristics of glasses with different compositions are drawn. The parameters used in the fitting are listed in Table I. The numerical figures for $GeS_{1.8}$ are in good agreement with those obtained from the isothermal experiment (Fig. 6).

DISCUSSION

The experimental results obtained from the doping and gamma-ray irradiation effects on the photo-induced ESR lead to a negative conclusion for the bond switching model[8] at least in the case of GeS_x glasses. A plausible configuration-coordinate description based on the Street and Mott model[10] which is consistent with our experimental results is depicted in Fig. 8. Metastable neutral centers (C) are

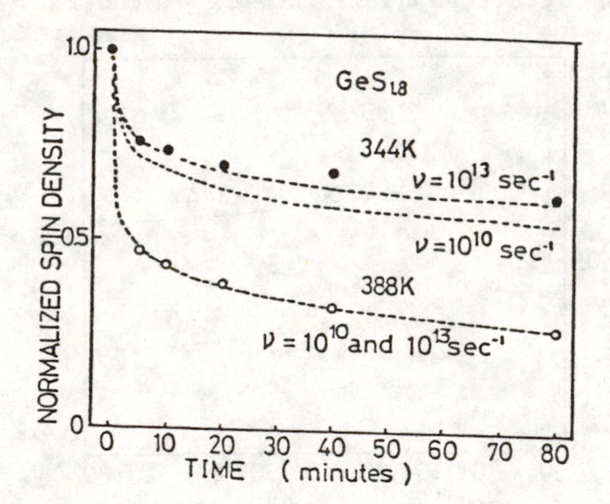

Fig. 6. Fitting the isothermal annealing characteristics by the first-order reaction with a Gaussian distribution of activation energies.

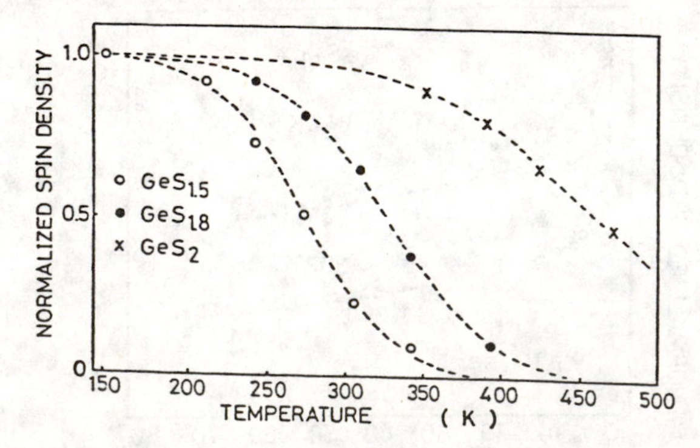

Fig. 7. Fitting the isochronal annealing characteristics by the first-order reaction with a Gaussian distribution of activation energies.

produced through optical excitation (A → B). These neutral centers can be responsible for the photo-induced ESR and possibly for the absorption edge shift by a secondary effect accompanying structural distortion. The model of Fig. 8 is compatible with the fact that **the illumination by sub-band-gap light causes a reduction in the** photo-induced ESR centers and a shift of the absorption edge to higher energies.

The widely-distributed activation energies come from local fluctuations of electron-lattice coupling strengths. Figure 9 (a) and (b) illustrates these situations. When the lattice distortion of 0.3 Å necessary for producing the neutral centers and a parabolic configuration-coordinate are reasonably assumed, the fluctuation of 3 - 5 % for the frequency factor (10^{13} s^{-1}) or for the structural distortion at $\nu = 10^{13}$ s^{-1} can explain the deviation of W_0 of 0.1 to 0.3 eV, which is in good agreement with the values in Table I.

Table I. Estimated parameters by fitting in Fig. 7.

COMPOSITION	W_0 (eV)	σ (eV)	Eopt (eV)
$GeS_{1.5}$	0.90	0.15	2.01
$GeS_{1.8}$	1.08	0.18	2.48
GeS_2	1.53	0.28	2.72

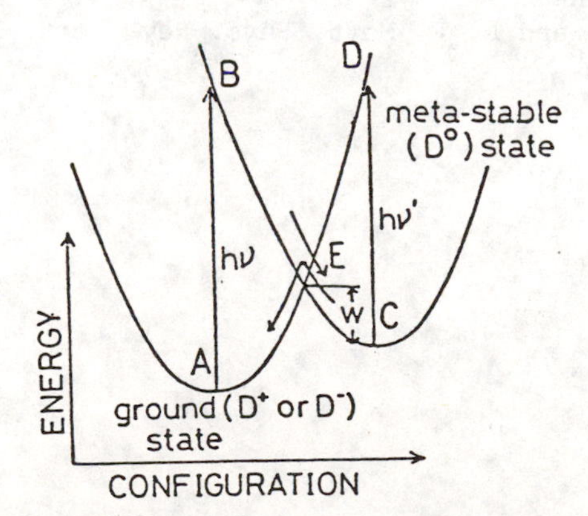

Fig. 8. Configuration-coordinate model including a strong electron-lattice coupling to explain the experimental results consistently.

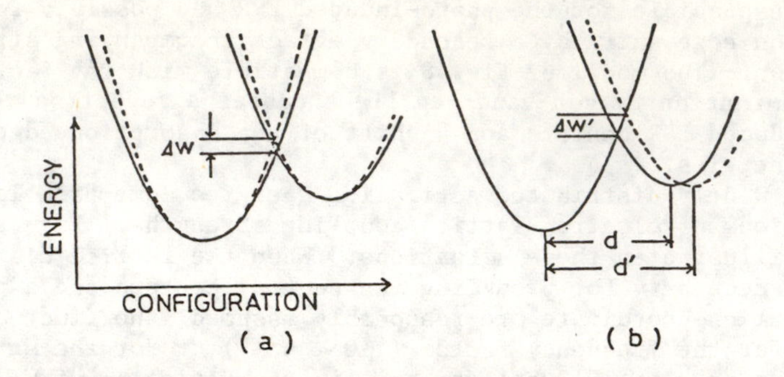

Fig. 9. Schematic presentation of the two kinds of fluctuations of the electron-lattice coupling strength.

REFERENCES

1. K. Oe, Y. Toyoshima and H. Nagai, J. Non-Cryst. Solids 20, 405 (1976).
2. K. Tanaka, Appl. Phys. Lett. 26, 243 (1975).
3. S. G. Bishop, U. Strom and P. C. Taylor, Phys. Rev. B15, 2278 (1977).
4. P. C. Taylor, U. Strom and S. G. Bishop, Philos. Mag. B37, 241 (1978).
5. R. A. Street and D. K. Biegelsen, J. Non-Cryst. Solids 32, 339 (1979).
6. K. Arai and H. Namikawa, Solid State Commun. 13, 1167 (1973).
7. I. Watanabe and T. Shimizu, Solid State Commun. 25, 795 (1978).
8. R. A. Street, Solid State Commun. 24, 363 (1977).
9. N. Kumagai, J. Shirafuji and Y. Inuishi, Jpn. J. Appl. Phys. to be published.
10. R. A. Street and N. F. Mott, Phys. Rev. Lett. 35, 1293 (1975).

THE CONDUCTION BAND OF HYDROGENATED AMORPHOUS SILICON

W. B. Jackson, S. - J. Oh, C. C. Tsai, and J. W. Allen
Xerox Palo Alto Research Center, Palo Alto, CA 94304

ABSTRACT

The valence and conduction band density of states of hydrogenated amorphous silicon (a - Si:H) are determined using x-ray photoemission (XPS) and bremstrahlung isochromat spectroscopy (BIS), respectively. Evidence for the Si - H antibonding orbital of polyhydrides is found near the conduction band edge. Implications of these results for the transport and recombination are discussed. In samples doped with 1% boron, unoccupied gap states are found. Finally, corrections used in photoemission are applied to BIS spectra to determine a more accurate density of states.

INTRODUCTION

Knowledge concerning the conduction band is important for the understanding of the optical and transport properties of a - Si:H. The electrical transport characteristics are determined by the density and capture cross section of the conduction band (CB) states particularly near the band edge. The optical absorption depends on a convolution of the conduction and valence band (VB) states.

Despite its importance, little is known about the CB density of states (DOS). Previous experiments have attempted to determine the CB DOS using core level absorption or threshold yield.[1-3] In these measurements, the spectral dependence of the absorption from the relatively narrow core levels to the CB is measured. Unfortunately, it is a widely accepted fact that such absorption is distorted by the Coulomb attraction between the highly localized core hole and the electron in the conduction band (i.e. exciton formation or many body effects).[4] The result is that core absorption spectra exhibit artificial peaks and shifts of spectral weight not found in the true one - electron CB DOS. There are other problems with the absorption measurements as well. First, a superposition of absorption spectra from core levels spin split by ~.6 eV require a deconvolution of the spectra. Second, the core levels are broadened asymetrically by chemical shifts which are as large as .75 eV for a - Si:H.[5] Third, there are significant distortions due to the different matrix elements for the transitions from p and s core levels. Because of these various problems, the CB DOS for crystalline Si, for example, differs significantly from the calculated DOS.[4] The best estimates of the CB DOS for a - Si:H consist of a broadened step function or a broadened crystalline DOS.

The problems of core level absorption can be overcome using a recently developed technique, bremstrahlung isochromat spectroscopy (BIS), to measure the CB DOS of a-Si:H.[6] A monoenergetic electron beam of variable energy impinges on the sample causing the emission of x-rays. As the incident electron energy is varied, the number of x-rays emitted at a fixed photon energy (1486.6 eV) as the incident electron energy is varied is proportional to the density of unoccupied states. BIS is the time reversed process of x-ray photoemission spectroscopy (XPS). Since there is no core hole, BIS spectra is undistorted by the core hole-electron Coulomb **interaction** and therefore are a more accurate measurement of the CB DOS. Strictly speaking, BIS measures the spectrum for adding an electron to the system which for our purposes is equivalent to the one-electron DOS.

In this paper, both the conduction band and valence band are measured using BIS and XPS, respectively. A method for subtracting the inelastic tails of BIS is presented. A peak near the CB edge which disappears upon annealing is due to the Si-H antibonding orbital. Heavily boron doped silicon exhibits additional gap states above the Fermi level.

Experiment

Undoped and 1% boron doped samples of a-Si:H (100nm thick) were deposited on Mo substrates in a UHV chamber. The undoped samples were deposited using low rf power and 100% silane at 25 °C and at 230 °C.[7] The boron doped films were deposited at 230 °C. XPS with the Al K_{α} line and BIS were performed using an Vacuum Generators ESCALAB, factory-modified for BIS. The Fermi level for both the XPS and BIS measurements were determined by measuring the Fermi level of the metal substrate. The resolution of the measurement was 0.72 eV. The base vacuum of the chamber was 10^{-10} torr or better. Further details of the measurement may be found in Ref. 8.

The samples were transferred from the deposition chamber to the measurement chamber with less than 3 minutes exposure to air. There are a number of reasons why the results were not affected by oxidation. First, the oxidation was found to be less than 0.05 monolayers by measuring the O 1s core levels with XPS. This is consistent with previous results which found that the atomic hydrogen present during deposition not only passivates Si surface states but also interior dangling bonds as deep as 1 μm.[9] The oxidation rate of a-Si:H is at least seven orders of magnitude less that of a cleaved crystalline surface.[10] Second, the measurements are not very sensitive to surface oxidation since the escape depth for 1.4 keV electrons is at least 30 Å. Third, the spectra were not changed by light sputtering or increasing the exposure to air. Fourth, no evidence of an Si-O bonding level was found in the XPS spectra. Because of the above reasons, oxidation is considered to be unimportant.

Results

Fig. 1 presents the BIS spectra of a 25 °C as-deposited sample and the same sample annealed within the measurement chamber, first for an hour at 230 °C and subsequently for another hour at 310 °C. The rise above 1500 eV is due to a plasmon replica of the conduction band. The electron loss function was determined from the core level and valence band XPS spectra. The magnitude and the energy of the rise is consistent with the electron loss function convoluted with the conduction band below 1500 eV. In the region 5 to 16 eV above the CB edge, there are no observable critical point structures. The absence of structure is expected since the delocalized CB states are more sensitive to the absence of long range order than the more localized VB states. On this energy scale the CB edge is fairly abrupt.

The most interesting feature is the ~3 eV wide peak near the CB edge in the as-deposited film with ~ 30% H determined by hydrogen evolution. The peak decreases when the sample is annealed at 230 °C and completely disappears upon further annealing at 300 °C (19 at. % H). Further anneals at temperatures as high as 650 °C do not result in a reappearance of the peak. This behavior indicates that the peak is not due to crystallization of the sample. The peak is not observed in as-deposited 230 °C samples.

A BIS spectrum of the 1% boron doped sample is shown in Fig. 2 superimposed on a spectrum of an undoped 230 °C sample. The boron spectrum was shifted downwards by 0.5 eV to match the CB edge. This shift is consistent with the ~0.6 eV lowering of the Fermi level caused by heavy boron doping. The curves are normalized to coincide at 4 eV. The boron sample has an additional sub-gap defect band which is unoccupied.

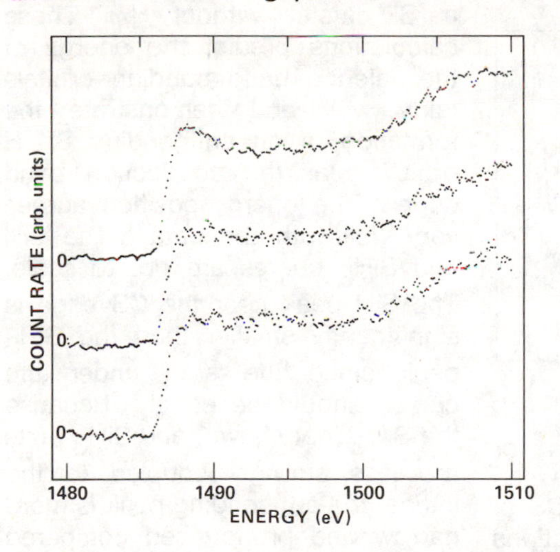

Fig. 1 BIS spectra for undoped a-Si:H deposited at 25 °C (Top curve-as deposited, middle curve-annealed 1 hr at 250 °C, Bottom curve-annealed 1 hr further at 310 °C).

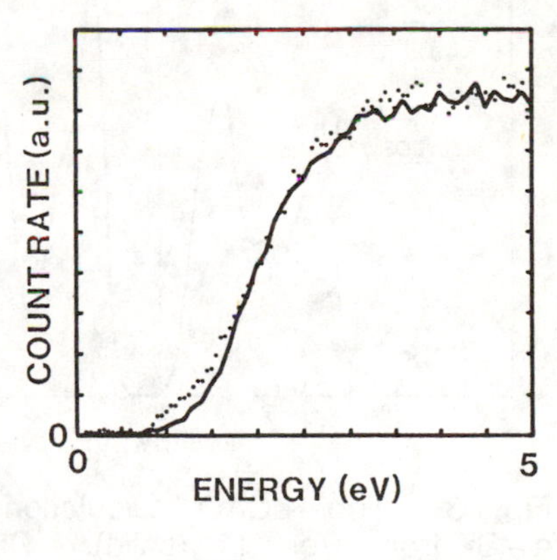

Fig. 2 BIS spectra for 1% boron doped (points) and 230 °C undoped (line) films. The boron curve was shifted relative to the undoped upwards by 0.5 eV. The Fermi level is at 0.7 eV for the boron doped sample.

Discussion

A. Silicon – Hydrogen Antibonding States

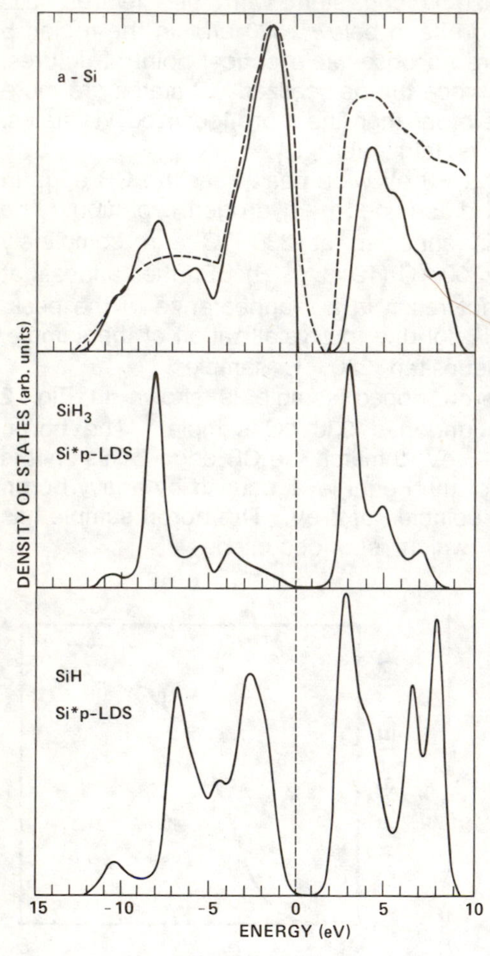

Fig. 3 Top – LCAO calculations for a – Si from Ref. 11 (solid). Dashed curve represents the measured DOS of a 230 °C sample after corrections disussed in the text. The local DOS for the silicon atom bonded to 3 H atoms (middle curve) and to 1 H atom (bottom curve) from Ref. 11. The SiH states are too large relative to the SiH_3 states since the area under the curves should be equal.

From its annealing behavior, magnitude, and energy position, the peak in Fig. 1 can be assigned to the Si – H antibonding orbital. The antibonding states are most likely due to the polyhydride configuration (Si – H_2, Si – H_3, etc.) rather than the monohydride configuration (Si – H) for the following reasons.

The energy position is consistent with LCAO calculations on large clusters shown in Fig. 3.[11] The lower two curves represent the local density of p – like states of the Si atoms bonded to the hydrogen for SiH_3 and SiH. The upper curve denotes the density of states for the a – Si cluster without H. These calculations predict the energy of the valence band bonding orbitals fairly well and demonstrate the formation of an antibonding Si – H orbital near the conduction band edge. The energy position agrees very well with the data. The SiH and SiH_3 curves are not to scale. The SiH peak near the CB edge is significantly smaller than the SiH_3 peak since the area under the curves should be equal. Because the SiH_2 (not shown) and SiH_3 units are less strongly coupled to the lattice, the antibonding peak is more narrow and pronounced compared with the SiH states. Consequently, the polyhydrides are mcre readily observed over the background Si network states than monohydride. Furthermore, the calculations in Fig. 3 show that as the hydrogen concentration increases, the CB moves to slightly lower energy. Analysis of expanded BIS spectra of the CB edge reveals that the edge of the 25 °C sample recedes by ~0.2eV as hydrogen is driven off.

The magnitude of the peak is also consistent with that expected for a SiH antibonding orbital. When the H concentration, X, of a film decreased from 30% to 19%, the number of Si – H bonds/Si atom, X/(1 – X), is reduced by roughly 0.2 states/Si atom. Depending somewhat on how the curves are normalized, the observed area decrease corresponds to a 0.1 – 0.3 states/Si atom change.

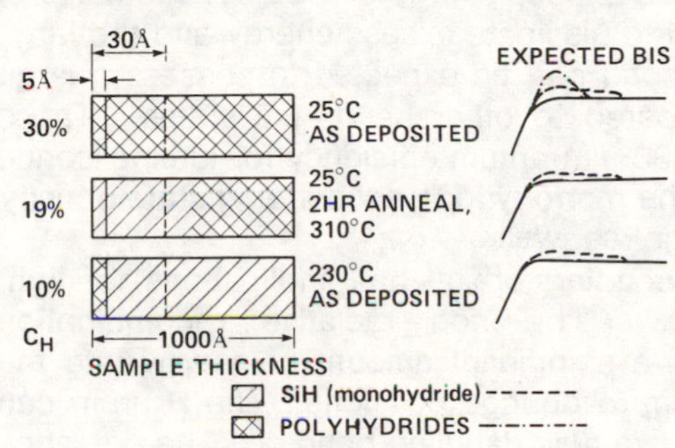

Fig. 4 The H distribution within the sample and the expected contributions to BIS. The escape depth for electrons in UPS ($h\nu = 40.8$ eV) and XPS ($h\nu = 1486$ eV) are 5 and 30 Å, respectively.

Finally, the annealing dependence of the peak is consistent with the change of hydrogen bonding. The distribution of hydrogen for different films and the expected BIS spectra are shown in Fig. 4. The as – deposited 25 °C sample contains significant polyhdride throughout the sample. Because the ratio of Si – H bonds to Si – Si bonds is large and the polyhdride peak is narrow, the BIS spectrum exhibits a well defined peak. UV photoemission spectroscopy (UPS) spectra of the VB show the polyhydride structure.[12] After annealing, nuclear reaction profiling of the hydrogen,[13] and UPS spectra indicate that the near surface region of the sample is depleted of hydrogen.[12] The UPS spectra are characteristic of the monohydride structure. The antibonding states due to the estimated 8% remaining H bonded as SiH is expected to be only 0.02 of the background Si states. Because of this small magnitude and broad energy distribution of the monohydride CB states (almost 8 eV), the corresponding antibonding states are not observable. The as – deposited 230 °C sample contains roughly 8% SiH within the sampling region. Hence, no peak appears in the BIS spectra of these films. The UPS spectra will be characteristic of the polyhydrides since the surface is hydrogen enriched and the escape depth of the UPS electrons is ~5Å at 40.8 eV.[3] Thus, the results indicate that peak is due to the Si – H antibonding states from the polyhydride configurations.

The presence of the Si – H antibonding orbital near the CB edge has important implications for transport in a – Si:H particularly in films containing polyhydride configurations.[14] The antibonding states are close to the bottom of the CB and a portion of the states overlap with the mobility edge. Consequently, the electron can occupy these states during transport at the mobility edge. Unlike most bonding configurations found in tetrahedrally bonded semiconductors, the H atom is not constrained by other bonds. When an electron occupies the Si – H antibonding state, the Si – H bond weakens and the H moves away from the Si atom. The lattice deformation and the electron – lattice coupling can be expected to be quite large. Thus, the **Si – H antibonding orbital is a site where the electron self – traps quite readily.** Such states can significantly reduce the mobility of the electrons particularly if they are comparable to the H density of 10^{20} states/cc or greater.

Not only are the trapping properties unique for the antibonding state but the recombination from these states are also potentially quite interesting. Because of the large phonon energy for the Si – H vibration only ~5 phonons of .25 eV each are required to dissipate an excess energy of 1.25 eV. Because of the small number of phonons required to dissipate excess energy and the large electron – lattice coupling, the Si – H bond can be expected to possess a large nonradiative recombination rate compared to other band – tail states. These states may be a factor in causing the 30% quantum efficiency for luminescence **observed in a – Si:H[15] particularly if the monohydride antibonding state density is found to overlap the mobilty edge as well.**

The presence of the Si – H antibonding orbital has implications for light induced defect formation as well. The non – radiative recombination mechanism at a Si – H bond deposits a significant amount of energy into the Si – H bond and can cause the H atom to change positions. The H atom can break a weak Si – Si bond resulting in two new dangling bonds.[16] The presence of the Si – H antibonding orbital in the CB makes this process much more likely.

B. Boron Induced Gap States

The BIS spectra in Fig. 2 reveal that boron doped samples have a significant number ($2X10^{20}$ cm^{-3}) of states located roughly 0.7 eV above the Fermi level, i.e., ~0.5 eV below the CB edge. The number or shape of the excess state distribution does not change if the film is annealed for 2 hr at 450 °C indicating that the states are not associated with hydrogen.

The origin of these states is unclear. The dangling bond density is estimated to be only $5X10^{18}$ cm^{-3}.[17] A more likely origin is that the boron doping causes significant broadening of the conduction band edge. The edge does seem to be broader in Fig. 2 but there seems to be excess states above the broadening. A final possibility is that the states are due to the unoccupied lone pair orbital of threefold coordinated boron atoms. The lone pair orbital is estimated to lie within the band gap roughly at the observed position.[18] If the states are due to boron lone pair states, the density given above would increase to $1.4X10^{21}$ cm^{-3} since the boron 2p cross section is 0.2 that of the Si 3p cross section.[19] The magnitude agrees well with the estimated number of boron atoms ($1X10^{21}cm^{-3}$) for a 1% doping level. Although the position of the peak agrees with optical data, the magnitude does not.[17,20] Further investigation is required.

C. Transformation of BIS Spectra Into DOS

In this section, we examine processes which affect the BIS spectra and indicate how a more accurate CB DOS may be obtained. Fortunately, since BIS is equivalant to a time reversal of XPS, the factors affecting BIS are the same and are well known.

One factor to consider is inelastic scattering. When a monoenergetic beam of electrons enters or leaves a sample, it loses energy by e – h pair production and plasmons. The plasmon losses are 16.6 eV and are readily detected. The energy loss due to e – h production creates a constant tail (see inset Fig 5) . At an energy of ~1460 eV in Fig. 5, all the counts are due to inelastic scattering. The ratio of the area under the valence band to that caused by inelastic scattering can be determined such that there are no counts at ~1460 eV. This ratio of the inelastic tail to the peak area can be checked with the core level spectra. The background at an energy E is removed by subtracting an amount proportional to the area between E and the Fermi level. This is a commonly used technique.[21]

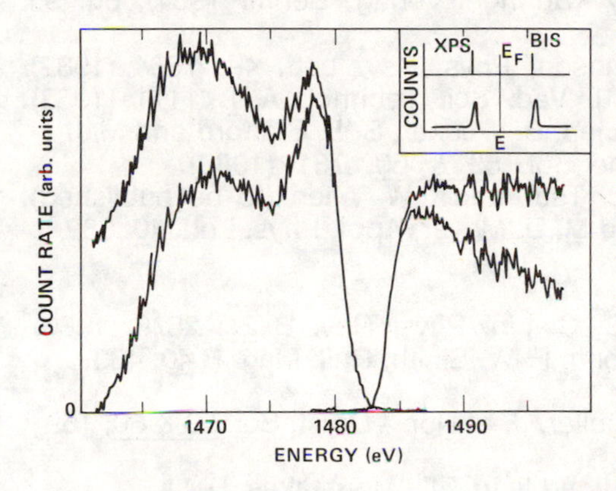

Fig. 5 XPS and BIS spectra uncorrected and corrected for inelastic tailing. The inset shows the XPS and BIS spectra for delta function densities of states. The tail is due to inelastic scattering and the finite width is due to instrumental broadening.

The electron energies for BIS and XPS are within 10-20 eV of each other. Since the XPS and BIS measurements are the time reversed processes of each other, the same correction procedure can be used although in the case of BIS the inelastic tails extend to higher energy. The correction procedure was tested on Ce which has a large f peak just above the Fermi level. Any tailing due to inelastic scattering should be readily observable. The procedure discussed above effectively removed the inelastic scattering of the large peak. The corrected spectra for a – Si:H using this procedure are shown in Fig 5.

The other effect which must be considered is the different cross section for Si 3s and 3p states. The lower peak of the valence band is reduced by a factor of 3.4 since the photoemission cross section for Si 3s states of the lower peak is 3.4 times as large as the Si 3p states of the upper peak. In principle, the same corrections are necessary for the CB. Fortunately, the conduction states are of mixed s and p character as could be determined from comparing c – Si BIS to calculations and comparisions of the K and L absorption edge spectra. Consequently, the correction is probably not necessary. Fig. 3 (dashed curve) shows the final VB and CB DOS for a – Si:H. Both corrections discussed above are widely used in XPS and are well understood; the application to BIS is consequently straightforward.

Acknowledgments

We would like to thank M. Stutzmann and R. Street for helpful discussions. This work was supported in part by Solar Energy Research Institute Contract No. XB-3-03112-1.

References

1. F. C. Brown, R. A. Bachrach, and M. Skibowski, Phys. Rev. B 15, 4781 (1977).
2. C. Senemaud and B. Pitault, Solid State Comm. 43, 483 (1982).
3. L. Ley in The Physics of Hydrogenated Amorphous Silicon II, ed. J.D. Joannopoulos and G. Lucovsky (Springer-Verlag, Berlin, 1984), pg. 95.
4. Ref. 3, pg. 96.
5. L. Ley, J. Reichardt, R. L. Johnson, Phys. Rev. Lett. 49, 1664 (1982).
6. T. Fauster and F. J. Himpsel, J. Vac. Sci. Technol. A 1, 1111 (1983).
7. C. C. Tsai, J. C. Knights, R. A. Lujan, B. Wacker, B. L. Stafford and M.J. Thompson, J. of Non-Crystalline Sol. 59 & 60, 731 (1983).
8. W. B. Jackson, S.-J. Oh, C. C. Tsai, and J.W. Allen (to be published).
9. N. Johnson, D. K. Biegelsen, and M. D. Moyer, Appl. Phys. Lett. 40, 882 (1982).
10. Ref. 3, pg. 120..
11. W. Y. Ching, D. J. Lam, and C. C. Lin, Phys. Rev. B 21, 2378 (1980).
12. B. von Roedern, L. Ley, M. Cardona, F. W. Smith, Phil. Mag. B 40, 433 (1979).
13. M. Reinelt, S. Kalbitzer, and G. Moller, J. of Non-Cryst. Sol. 59 & 60, 169 (1983).
14. An early discussion of the Si-H bond is in T.D. Moustakas, D.A. Anderson, and W. Paul, Solid State Comm. 23, 155 (1977).
15. W.B. Jackson and R.J. Nemanich, J. of Non-Cryst. Sol. 59 & 60, 353 (1983).
16. M. Stutzmann, W.B. Jackson, and C.C. Tsai (to be published in Appl. Phys. Lett.).
17. W. B. Jackson and N. M. Amer, Phys. Rev. B 25, 5559 (1982).
18. J. Robertson, Phy. Rev. B 28, 4666 (1983).
19. S. M. Goldberg, C. S. Fadley, and S. Kone, J. Electron. Spectros. and Rel. Phenon. 21, 285 (1981).
20. A. Triska, I. Shimizu, J. Kocka, L. Tichy, and M. Vanecek, J. of Non-Crystalline Sol. 59 & 60, 493 (1983).
21. E. Antonides, E.C. Janse and G. A. Sawatzky, Phys. Rev. B 15, 1669 (1977).

Optical Properties of a-Ge:H - Structural Disorder and H Alloying

P. D. Persans, A. F. Ruppert, G. D. Cody, B. G. Brooks
Exxon Research and Engineering Company
Annandale, New Jersey 08801

W. Lanford
State University of New York at Albany
Albany, New York 12222

ABSTRACT

We report measurements of the optical absorption edge of pure and hydrogenated amorphous Ge. It is proposed that both network site disorder and pseudo-binary H alloying play important roles in controlling the optical gap.

INTRODUCTION

Over the past few years significant advances have been made in our understanding of the microscopic origins of the hydrogen-induced increase in the band gap of a-Si:H [1-3]. Only a few studies on the optical properties of the second important amorphous tetrahedral semiconductor system, a-Ge:H, have been reported [4,5] although a great deal is known about both structure and optical properties of unhydrogenated material [6,7,8]. A more thorough study of this system may test and clarify our understanding of a-Si:H and lead to a more general understanding of the fundamental limitations and applications of amorphous tetrahedral semiconductors and their alloys.

Hydrogen may affect band edge electronic properties of a-Ge:H in a variety of ways. H incorporated during film growth may decrease Ge network distortions by substituting for overconstrained Ge-Ge bonds; H may also saturate bonds which might otherwise be dangling bonds. A relationship between optical gap E_G and network distortions has been suggested by Cody and collaborators for a-Si:H[1,9] on the basis of optical characterization of the quasi-exponential absorption tail and its temperature dependence. A second mechanism by which H may increase E_G is through an explicit H alloying effect. In the virtual crystal approximation the band gap of an alloy is determined by the volume averaged potential of the alloy components [10]. Since the Ge-H bond is stronger than the Ge-Ge bond the introduction of H may be expected to increase E_G through the formation of a $Ge_{1-y}(GeH)_y$ pseudo-binary alloy. Other more subtle effects may play a role in determining E_G [1,3] but we will confine our discussion here to network distortion and binary alloy effects.

Samples for the present study were deposited on fused SiO_2 and Si substrates by reactive sputtering of Ge in an Ar, H_2 gas mixture and by rf plasma-enhanced chemical vapor deposition from 10% GeH_4 in H_2. Important preparation parameters are included along with optical data in Table 1. Details of sample preparation and optical characterization will be published elsewhere [11].

NETWORK DISORDER EFFECTS

Structural disorder can be linked to the band gap and the quasi-exponential tail slope through theories which include the effects of site energy disorder [12,13]. Although a qualitative connection between network distortions and site disorder for amorphous networks has been made [14] results are not yet quantitative enough for comparison to experiment. A semiempirical model connecting site energy disorder to network distortions has been developed by Cody et al. [9]; we make use of this approach here.

The relationship between network disorder and optical properties can be explored by changing measurement temperature to vary thermal (dynamic) disorder or by annealing or otherwise changing preparation conditions to vary static disorder and associating measured or calculated structural variations with optical properties. In the following we focus on changes in mean square distortions σ^2 in the second neighbor distance (bond angle distortions) and changes in the optical absorption exponential tail width E_o as measures of network disorder. We correlate network disorder with changes in either the extrapolated gap E_G or an isoabsorption energy such as E_{o4} [15].

THERMAL DISORDER

Increasing measurement T leads to an increase in thermal motion and thus in σ^2. In the theory of the T-dependence of E_G for crystals the pseudo-potential is expanded to second order in fluctuations in atomic position [16]. In reference 1 this approach is generalized to include average static and thermal fluctuations in amorphous semiconductors. Assuming that σ^2 is controlled by a single oscillator of energy $k\Theta$ we write by analogy:

$$E_G\,(\sigma^2) = E_G\,(0) - D\,\sigma^2 \qquad (1)$$

where:

$$\sigma^2 = \frac{3\hbar^2}{2Mk\Theta}\left(\frac{1}{e^{\Theta/T}-1} + \frac{1}{2} + X\right) \qquad (2)$$

M is the mass of the oscillator X is the static network disorder, and D is a second-order coupling coefficient.

In Fig. 1 we show $\Delta E_G(T)$ and $\Delta E_{04}(T)$ for film #111B, described in Table 1. The solid line through the data points is for M=72, $k\Theta$ = 22 meV and D ~ 15 eV/Å^2. About 20-30% of $\Delta E_G(T)$ is due to thermal expansion, therefore the effective coupling coefficient D is ~ 10eV/Å^2 for thermal disorder σ^2. We note that this result is in substantial agreement with the interpretation of Cody et al. on the T-dependence of E_G for a-Si:H[1].

STATIC DISORDER

In Fig. 2 we show the effect of annealing on the absorption coefficient of an unhydrogenated sputtered a-Ge sample, series A in Table 1. We observe that E_{04} increases by 0.14 eV as E_o decreases by ~ 20 meV from 145 to 125 meV as illustrated in Fig. 3. Structural measurements on room temperature sputtered and fully annealed a-Ge indicate that σ^2 at room temperature decreases by 0.02 ± 0.01 Å^2 [8,17,18]. Thus for annealing induced changes in σ^2 we find D ~ 8 ± 5 eV/Å^2, in agreement with the thermal studies discussed above.

Recent Raman and optical studies indicate that hydrogenation can also reduce σ^2. In Fig. 4 we show α measured by transmission and photothermal techniques for a series of hydrogenated films in which H content y was varied from 0.02 to 0.2 by varying H_2 partial pressure from 0.04 to 1.0 mT in 5mT of Ar sputtering gas (series B in Table 1). For y <0.15 we find that E_{04} increases from 0.92 eV to 1.16 eV while the acoustic to optical Raman mode intensity ratio drops by ~15%. We infer a decrease in σ^2 of 0.02 + 0.01 Å^2 for an increase in E_{04} of 0.24 eV. This result is consistent with the changes in E_{04} with σ^2 found for the annealing experiment but does not exclude binary alloy effects from playing a significant role.

The relationship between ΔE_G and ΔE_o found by Cody et al. [9] for a-Si:H appears to hold up here for y < 0.15 and for annealed samples, but a systematic deviation appears for y > 0.15. The gap E_{04} continues to increase with increasing y but E_o ceases to decrease. Consideration of data on samples prepared at different deposition temperatures by reactive sputtering or by rf glow discharge suggests that samples with higher H content tend to have higher E_{04} for a given E_o. A summary of this data is given in Table 1. Unless otherwise noted H content was determined by nuclear reaction profiling.

It is thus clear within the (E_o, E_{04}) description that a second mechanism in addition to the network disorder mechanism must be considered to interpret the data.

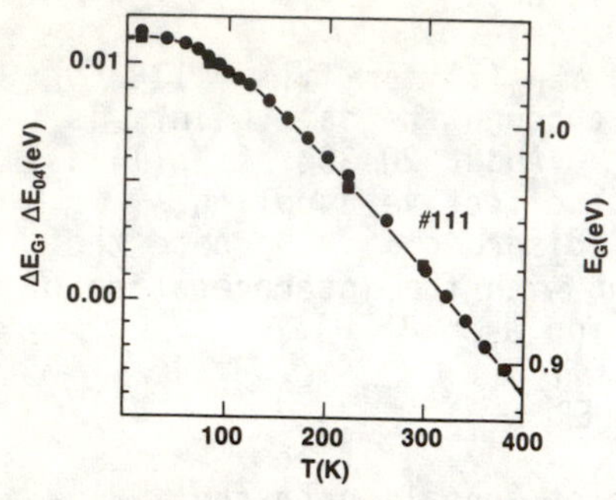

1). Thermal shift in the extrapolated gap E_G and in E_{04} plotted against temperature for a-Ge:H sample #111. Solid line through data points is given by equations 1 and 2.

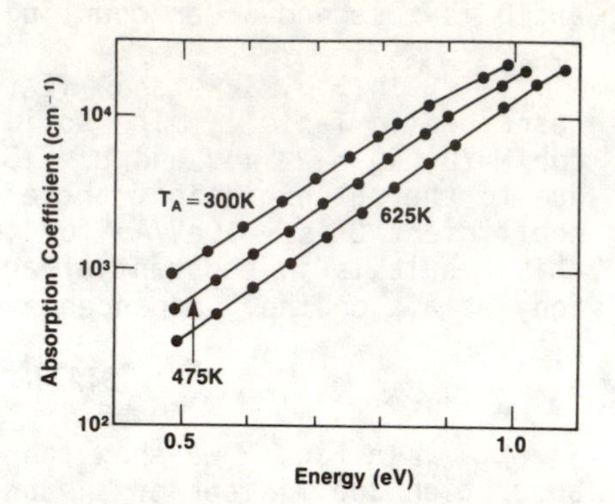

2). Absorption coefficient α plotted against E for an annealing series on a-Ge sample #71 (Series A) described in Table 1.

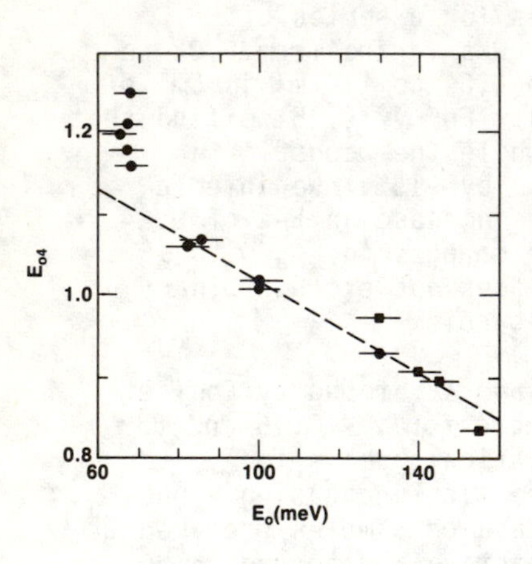

3). Correlation plot of energy gap (E_{04}) plotted against quasi-exponential tail slope E_o for annealing series A and hydrogenation series B.

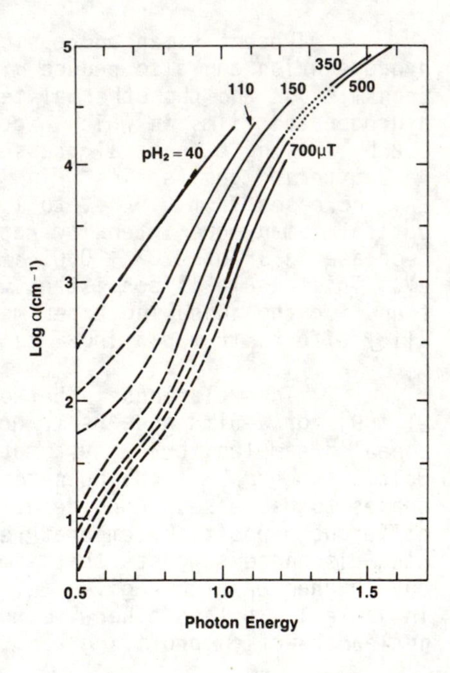

4). α plotted against E for a-Ge:H series B in which H content is varied from $\sim$ 2 to 15 atom %.

TABLE I

Sample #	d (μm)	T_d (K)	T_a (K)	$P[H_2]$ (mT)	y	E_G (eV)	E_{04} (eV)	E_o (meV)
71A	1.18	300	-	0	0	-	0.83	145
71A	1.18	300	475	0	0	-	0.90	140
71A	1.18	300	525	0	0	-	0.92	135
71A	1.18	300	625	0	0	-	0.98	125
93B	1.70	410	-	0.04	0.03	-	0.93	130
97B	1.05	410	-	0.08	0.04*	0.86	1.01	100
96B	1.58	410	-	0.11	0.08*	-	1.02	100
98B	0.95	410	-	0.15	0.09*	0.92	1.07	85
111B	1.73	410	-	0.20	0.09*	-	1.06	82
112B	1.24	410	-	0.35	0.13*	-	1.16	68
81B	1.13	410	-	0.50	0.19	1.07	1.18	67
110B	1.24	410	-	0.50	0.18	1.07	1.20	65
91B	1.31	410	-	0.70	0.21	-	1.21	67
87B	0.98	410	-	1.00	0.23	1.14	1.25	68
77	0.90	300	-	0.50	0.26	1.09	1.20	88
78	1.10	300	-	1.0	0.33	1.18	1.32	84
90	1.09	570	-	0.50	0.10	0.99	1.14	73
105	1.2	350	-	GD	0.11*	1.00	1.19	88
109	1.1	560	-	GD	0.08	1.02	1.14	65

* H content estimated from IR wag mode intensity

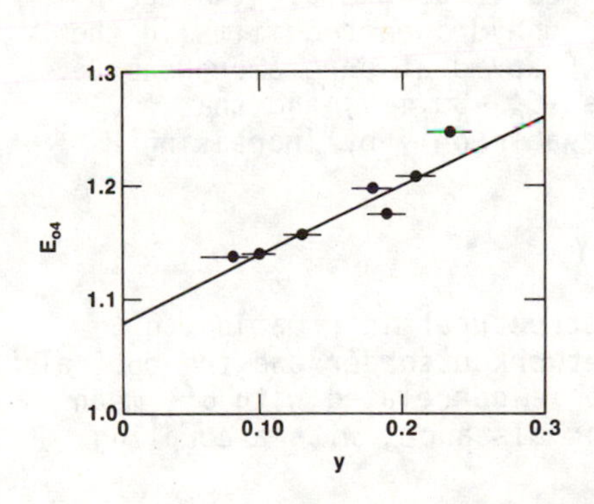

5). Energy gap (E_{04}) plotted versus H composition y in $Ge_{1-y}(GeH)_y$ for all samples with E_o 65 + 5 meV.

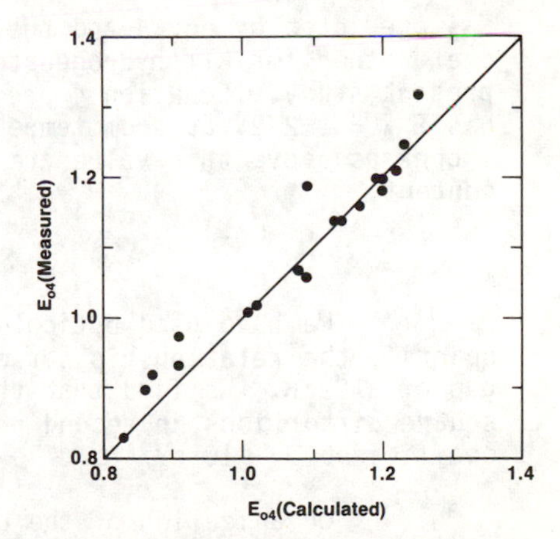

6). Correlation plot of energy gap (E_{04}) plotted against E_o for all samples

HYDROGEN ALLOYING

Increases in E_G due to the change in average alloy potential have been calculated for a-Si:H [3] and shifts in $E_G \sim 2$ times smaller than experimental shifts with H are found. No work on a-Ge:H has been reported.

We experimentally separate network disorder from binary alloy shifts in E_G by plotting E_{04} against y for a set of films for which the measure of disorder, E_o, is the same, In Fig. 6. we plot E_{04} vs y for all films with $E_o = 68 \pm 5$ meV. In the virtual crystal approximation for pseudo-binary alloys the gap is determined by the volume-averaged potential of the alloy components [10], in this case we take Ge and GeH.

$$E_G = E_G\,(Ge)\cdot(1-y) + E_G(GeH)\cdot y \qquad (3)$$

The solid line through the data points in Fig. 6 corresponds to E_G (Ge) = 1.075 eV and E_{04} (GeH) = 1.7 eV. Calculations on Si:H indicate that E_G (SiH) is ~ 1 eV larger than E_G (Si) [3]; the band gap parameters here are therefore quite reasonable.

The network disorder and binary alloy approaches can now be combined to parameterize the band gap of a-Ge:H prepared under a wide variety of conditions. In the absence of more direct structural measurements for all films we use E_o rather than σ^2 to characterize network disorder. The best fit to all the data is given by

$$E_{04}\,(y,E_o) = 1.29 + 0.62y - 3\,E_o \qquad (4)$$

E_{04} predicted by eq. 4 and the measured value are plotted against one another for all hydrogenated and unhydrogenated films in the present study. Equation 4 predicts that ideal pure amorphous Ge has $E_{04} = 1.2$ eV at room temperature ($E_o \sim 30$ meV) and that increases above this value are achievable only by increasing H content.

SUMMARY

We have used optical and structural information to quantify the relationship between network disorder and the optical gap of a-Ge:H. We find that the gap is correlated with σ^2, mean square distortions in second neighbor distance, with a coupling coefficient D ~ 10 eV/Å^2.

Organization of the data in terms of gap E_{04} and quasi-exponential tail slope E_o indicates that both network disorder and explicit H alloying effects must be considered. We suggest that these two mechanisms have comparable effects upon the band edge energy.

REFERENCES

1. G. D. Cody in Hydrogenated Amorphous Silicon II, ed. J. Pankove, (Academic Press, 1984) and references therein.
2. L. Ley in The Physics of Hydrogenated Amorphous Silicon, ed. J. Joanopoulos and G. Lucovsky, (Springer-Verlag, 1984) and references therein.
3. D. DiVincenzo, J. Bernholz and M. Brodsky, Phys. Rev. B28, 3246, (1983).
4. G. A. N. Connell and J. R. Pawlik, Phys. Rev. B13, 787, (1976).
5. P. D. Persans, A. F. Ruppert, S. S. Chan, G. D. Cody, Sol. St. Commun., 54, 203, (1984).
6. G. A. N. Connell, W. Paul and R. J. Temkin, Adv. in PHys. 22, 643, (1973).
7. W. Paul, G. A. N. Connell and R. J. Temkin, Adv. in Phys. 22, 531, (1973).
8. R. J. Temkin, W. Paul and G. A. N. Connell, Adv. in Phys. 22, 581, (1973).
9. G. D. Cody, T. Tiedje, B. Abeles, B. Brooks and Y. Goldstein, Phys. Rev. Lett., 47, 1480, (1981).
10. J. C. Phillips, Bonds and Bands in Semiconductors (Academic Press, New York), 1973.
11. P. D. Persans and A. F. Ruppert (to be published).
12. S. Abe and Y. Toyozawa, J. Phys. Soc. Japan, 50, 2185, (1981).
13. C. M. Soukoulis and M. H. Cohen, this proceedings.
14. F. Yonezawa and M. H. Cohen, in Fundamental Physics of Amorphous Semiconductors, ed. F. Yonezawa (Springer, Berlin, 1981).
15. The extrapolated gap E_G is defined by the relationship $(\alpha \cdot h\nu)^{1/2} = C(E-E_G)$. We use E_{04} (the energy at which $\alpha = 10^4$ cm^{-1}) to measure shifts in E_G when a reliable extrapolation cannot be made. The tail width E_o is defined by $(d \ln \alpha/dE)^{-1}$ for $5 \times 10^2 < \alpha < 5 \times 10^3$ cm^{-1}.
16. P. B. Allen and V. Heine, J. Phys. C, 9, 2305, (1976).
17. R. Tsu, J. G. Hernandez, F. H. Pollak, Proc. of International Topical Conf. on Transport and Defects in Amorphous Semiconductors, (1984).
18. J. E. Yehoda and J. S. Lannin, J. Vac. Si Technol. A1 392, (1983).

ELECTRO-OPTICAL EFFECTS ON THE ABSORPTION EDGE OF a-Si:H

U. Mescheder and G. Weiser
Fachbereich Physik der Philipps-Universität Marburg,
3550 Marburg, F.R. Germany

ABSTRACT

Electroabsorption measurements on doped and undoped glow discharge silicon films in a transverse electrode configuration are reported. The EA signal depends on the polarization of light being about 3 times larger for external field and polarization of light parallel. The results are not consistent with the Franz-Keldysh effect or similar models. In undoped samples the high voltage applied leads to charge injection (electrons) over surprisingly large distances in excess of 100 μm. The injection of space charge is reduced by illumination with white light, reversibly on annealing, and is improved by exposing the sample to a second beam.

INTRODUCTION

Electroabsorption and electroreflectance have provided a very sensitive and selective tool to study electronic states in crystalline semiconductors. Its success induced similar studies on amorphous solids, mainly on chalcogenide glasses which proved that the optical properties of amorphous semiconductors are indeed changed by moderate electric fields in the range of 10 kV/cm. The underlying mechanism, however, it not yet understood which leaves uncertain what knowledge we can gain from those investigations.

Section 2 of the present paper will point out some discrepancies between the current models, like the Franz Keldysh effect and the experimental results on glow discharge silicon. Recent results point to an alternative interpretation of the spectra. Section 3 will employ electroabsorption to measure internal fields in silicon which arise from charge injected into the sample by the high voltage applied for electroabsorption measurements.

The report is restricted to the study of electric field induced changes of the absorption which is the best accessible experimental quantity. All measurements were performed in a transverse configuration where a sinusoidal field of the frequency f (∿ 1 kHz) is applied parallel to the film by two coplanar electrodes with a spacing of 0.5 or 1 mm. The field-induced change of the absorption has been measured at twice the frequency of the field (2f) because for the chosen symmetric configuration, positive and negative halfwave of the field should have the same effect on the electronic states. It will be shown, however, that asymmetries due to injected charge will create a signal at the fundamental frequency as well which is easily separated. The corresponding inhomogeneity of the electroabsorption signal has been investigated by selecting small areas of the homogeneously illuminated gap between the electrodes for the detection of the field-induced change of the transmittance.

ELECTROABSORPTION SPECTRA OF AMORPHOUS SILICON (Si:H)

The electroabsorption spectrum Δα of a-Si:H, shown in fig. 1, is very similar to that of the chalcogenide glasses reflecting the disordered state rather than specific properties of the material. Only transitions near the absorption edge are sensitive to electric fields. Δα increases quadratically with the applied electric field. It rises steeply in the exponential part of the absorption edge, reaches a maximum near 1.85 eV where the absorption constant is about $10^4 cm^{-1}$ and decreases slowly at higher energy. Variation of the hydrogen concentration of the films or doping does not affect the shape of this spectrum; the rather broad peak, however, moves to different energies, following the shift of the absorption edge.

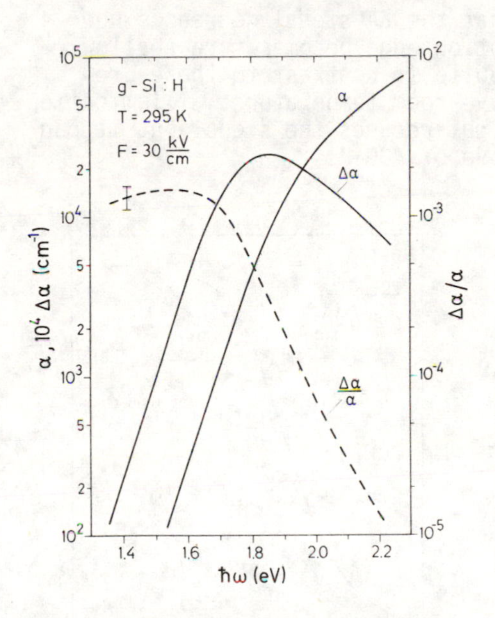

Fig. 1. Absorption and electroabsorption spectra of undoped glow discharge Si.

In the past two models were offered to explain those spectra, the Franz Keldysh effect[1] which is a redistribution in the density of states by field mixing of the free electron states and the Dow-Redfield[2] model which considers the alteration of the Cb-interaction of electron and hole by an external field. Both models propose that large internal fields induced by disorder cause broadening of the sharp band edge of a crystalline network resulting in an exponential absorption edge. An external field leads to further broadening giving rise to the electroabsorption spectrum of amorphous solids. Several inconsistencies have been pointed out for the results obtained on chalcogenide glasses[3] or on amorphous silicon[4].

Both models, based either on a redistribution of states or on field broadening of an exciton require a sign reversal in the spectrum Δα which has not been observed for any amorphous semiconductor. In all cases the absorption increases with an electric field over the whole range of the absorption edge.

Since an external field leads to additional broadening of the absorption tail the relative change of the absorption constant Δα/α should increase with decreasing energy. In silicon and more pronounced in the chalcogenides, however, Δα/α reaches a maximum and decreases again at low energy. Further assumptions on local anisotropy for amorphous Se[5] or on transitions between localized states[6] have been made to account for this behaviour. It should be mentioned, however, that in the low absorption region the accurate determination of α is quite difficult if only thin films are available.

The relative change of absorption $\Delta\alpha/\alpha$ is surprisingly large reaching values of 10^{-3} at moderate fields. Experiments on single crystals show that perturbation of the delocalized states by defects or by phonons rapidly decreases the influence of external fields and thus the electroabsorption signal. Usually good crystals and low temperatures are needed to obtain values of $\Delta\alpha/\alpha$ of 10^{-3}. Scattering of delocalized electrons in an amorphous solid by the disorder related potential fluctuations should be much more severe than by phonons in a single crystal. Therefore, it is very likely that the Franz Keldysh effect on a highly broadened band edge of delocalized states with or without exciton effects remains below of the detection limit.

The models require further that the EA signal decreases when with rising temperature the absorption edge becomes more shallow. As shown in fig. 2 (full curves) quite in contrast to those predictions $\Delta\alpha$ increases strongly above room temperature. Illumination of an undoped sample with white light reduces the signal but it can be restored again by annealing above of 400 K.

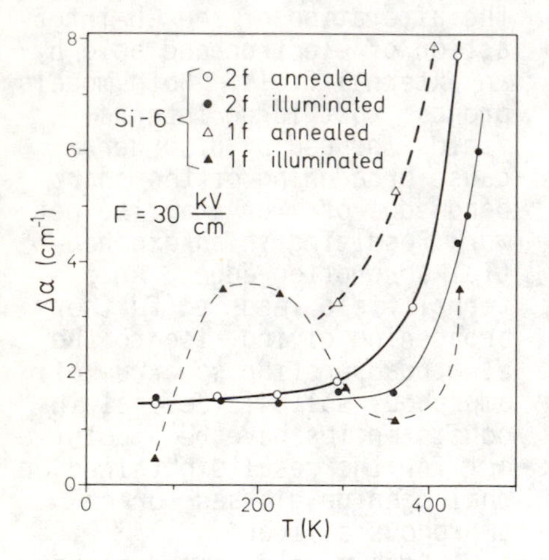

Fig. 2. Temperature dependence of the first and second harmonic of the EA signal in undoped Si and its change by illumination with white light.

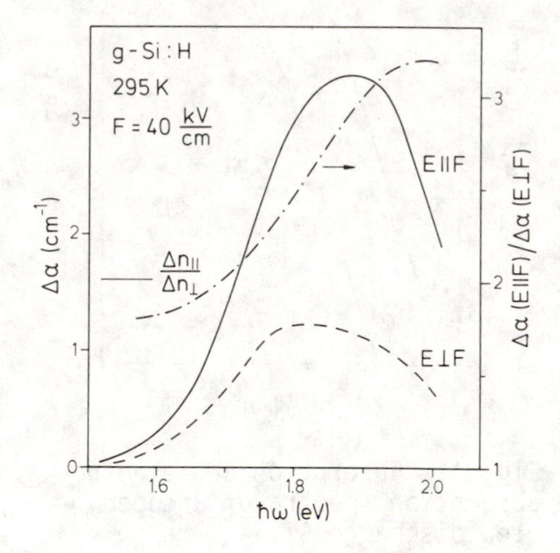

Fig. 3. Polarization dependence of electroabsorption $\Delta\alpha$ and of the changes Δn of the refractive index.

A further result which is not consistent with the Franz Keldysh effect nor with the behaviour of transitions between localized states is presented in fig. 3. Quite unexpected the electroabsorption spectrum in amorphous solids depends on the polarization of light. $\Delta\alpha$ is considerably larger when field F and polarization E of the light are **parallel and this polarization ratio $\Delta\alpha_{\parallel}/\Delta\alpha_{\perp}$ is not constant.** In a-Si:H it is somewhat less than 2 at low energy but it increases to values of more than 3 at higher energy. This anisotropy is also apparent in the refractive index which has a value of $\Delta n_{\parallel}/\Delta n_{\perp} = 2$ at low energy. The anisotropy of the electroabsorption spectrum is different for different materials. It is smaller in evaporated Si (~ 1.5) whereas in selenium even higher values are observed. Such field-induced anisotropy in an isotropic medium cannot be understood in the frame of the Franz Keldysh effect or of the exciton model. Optical excitation of a charge to another site (charge transfer transition) results in a large dipole moment and to some polarization dependence of the EA-signal if this signal originates from the energy shift of this dipole in an electric field. In case of an isotropic medium, however, $\Delta\alpha_{\parallel}/\Delta\alpha_{\perp}$ cannot exceed a value of 1.5 which has been observed in disordered films of organic molecules.[7] Those transitions yield also a different shape of the EA spectrum and it appears unlikely that the transition probability for optical excitation between spatially separated localized states is sufficiently large to account for absorption constants of $10^4 cm^{-1}$.

At the present state we can exclude density of state effects as the origin of the electroabsorption spectrum in amorphous solids. Both, the large polarization of $\Delta\alpha$ and the observation that electric fields always enhance the absorption points to matrix element effects. It is assumed that the electroabsorption spectrum covers the range where optical transitions between localized tail states and delocalized band states contribute to optical absorption. The corresponding matrix elements should become small with rising localization of the tail states. Field mixing of the localized states reduces the localization along the direction of the field which enhances the transition probability to the extended states, for polarization of light parallel to the field. Phonons which also mix the localized states should assist the field mixing rather than competing, which leads to the unexpected temperature dependence of electroabsorption. In this model which needs more detailed studies of optical matrix elements for transitions between localized and extended states the electroabsorption spectrum is centered near the mobility edge. The decrease of $\Delta\alpha/\alpha$ towards higher energy is assigned to the onset of transitions between delocalized states which are considered as unsensitive to the external field because of the severe broadening of these states by disorder.

CHARGE INJECTION BY HIGH ELECTRIC FIELDS IN ELECTROABSORPTION

Even if we do not really understand the mechanism of electroabsorption in amorphous solids we can use this method to study internal fields and the related charge distribution. Using symmetric electrode configuration, sinusoidal fields cause field induced changes of the optical constants only at twice the frequency of the field, unless internal fields are present or created by the applied voltage. Injection of charge for instance creates inhomogeneous fields and, if injection of electrons and holes is different, also an asymmetry which produces a signal at the fundamental frequency f of the external field.

Fig. 4 shows an example of spatial inhomogeneity of the EA-sig-

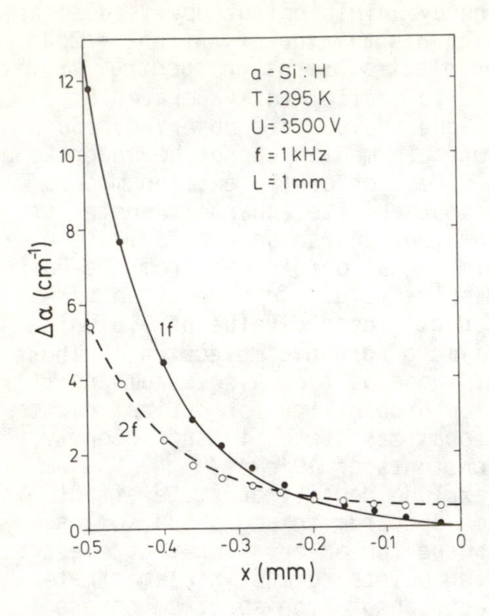

Fig. 4. Spatial inhomogeneity of the first and second harmonic of the EA signal. Signals decrease from the Cr contact (x = -0.5 mm) towards the middle of the sample.

nal across a gap of 1 mm when 3500 V are applied. Near the contact (Cr) in undoped films a very large signal is observed at the fundamental frequency f which disappears in the middle of the sample and appears again on the other contact with an opposite sign. The EA-signal at 2f is also larger near the contacts but it does not change its phase. The 1f signal has the same spectral and polarization dependence as the 2f signal considered before. Its temperature dependence is different (fig. 2, dashed curves) being very small at low temperature. It is also strongly reduced by illumination with white light. The size of the 1f signal is much less reproducible for different samples than the 2f signal. It disappears with boron doping and is not present in evaporated Si.

The inhomogeneity of the EA-signal points to an inhomogeneous field. The large extension of this inhomogeneity excludes contact potentials as a possible source. We assume therefore that the inhomogeneity results from space charge injected from the contacts. The occurance of a 1f signal indicates that injection is not symmetric for electrons and holes. Assuming exponential spatial decay of the injected charge one obtains a time dependent charge density:

$$\rho = \rho_s + \rho_m \cos\omega t \ , \ \rho_s = \rho_{os} \cosh(x/a) \ , \ \rho_m = \rho_{om} \sinh(x/a) \ ,$$
$$-L/2 < x < L/2 \qquad (1)$$

where the static part ρ_s results from the asymmetry of electron and hole injection and the modulated part ρ_m accounts for the modulation of the space charge with the external field. L is the spacing of the electrodes and a is the injection depth. The space charge related fields are

$$F_s = (\rho_{os}/\varepsilon\varepsilon_o)\ a \sinh(x/a) \ , \ F_m = (\rho_{om}/\varepsilon\varepsilon_o)\ a \cosh(x/a) \qquad (2)$$

and the corresponding voltages are

$$U_s = -(\rho_{os}/\varepsilon\varepsilon_o)\ a^2(\cosh(x/a)-1) \ , \ U_m = -(\rho_{om}/\varepsilon\varepsilon_o)\ a^2 \sinh(x/a) \qquad (3)$$

Since the externally applied ac voltage U must provide the modulation of the space charge, the homogeneous part of the field is given by

$$F_h = (U - 2\,U_m(L/2))/L \qquad (4)$$

The quadratic dependence of the EA signal on the total field $F_S = F_s + (F_m + F_h)\cos\omega t$ leads to a component S_{1f} at the fundamental frequency and S_{2f} at twice that frequency:

$$S_{1f} \propto 2\,F_s(F_m + F_h) \quad , \quad S_{2f} \propto (F_m + F_h)^2/2 \qquad (5)$$

The fit to the experimental data in fig. 4 yields the space charges given in fig. 5, the related fields and the injection depth a. The injection depth is 160 μm, surprisingly large but typical for undoped samples. The space charge related fields are in the same range as the homogeneous field. About half of the applied voltage is needed to modulate the space charge. Nevertheless, owing to the large injection length the space charge density remains small with values near 10^{13}e cm^{-3} even at the contacts. Such small concentration of trapped charge will not move the Fermi level sufficiently to affect the current voltage characteristic of the sample which remains linear.

Quite unexpected is the large penetration depth of the charge. We assume that injection occurs via the extended states or shallow tail states. The space charge is built up by injected carriers which become trapped deeper in the gap. With this assumption we expect a correlation between the μτ-product of photoconductivity and the occurrence of a spatially inhomogeneous EA signal. Since undoped samples have values of $\mu\tau \approx 10^{-6} cm^2/V$ fields of 10 kV/cm move the charge over a distance of 100 μm before the carriers are trapped.[8] the product μτ is much smaller at low temperature or boron doping which is consistent with the reduction of the 1f signal at low temperature or its disappearance in boron doped samples.

The space charge related part of electroabsorption is reduced when the samples are illuminated by white light and can be restored by annealing very similar to the Staebler Wronski effect of the conductivity[9]. An example is given in fig. 6.

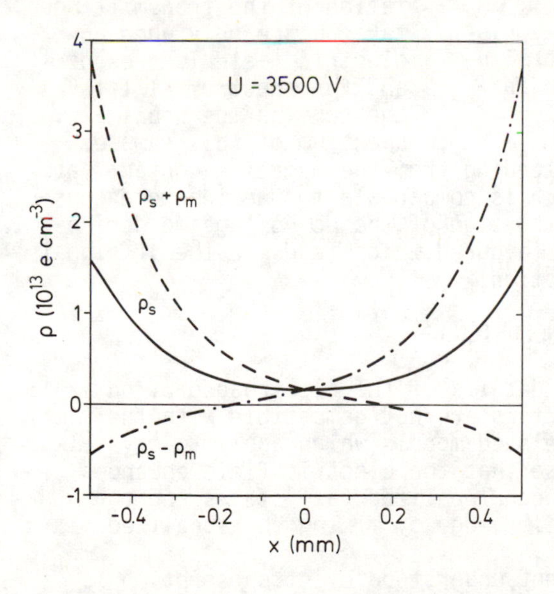

Fig. 5. Distribution of the injected space charge $\rho_s \pm \rho_m$ as derived from the results in fig. 4.

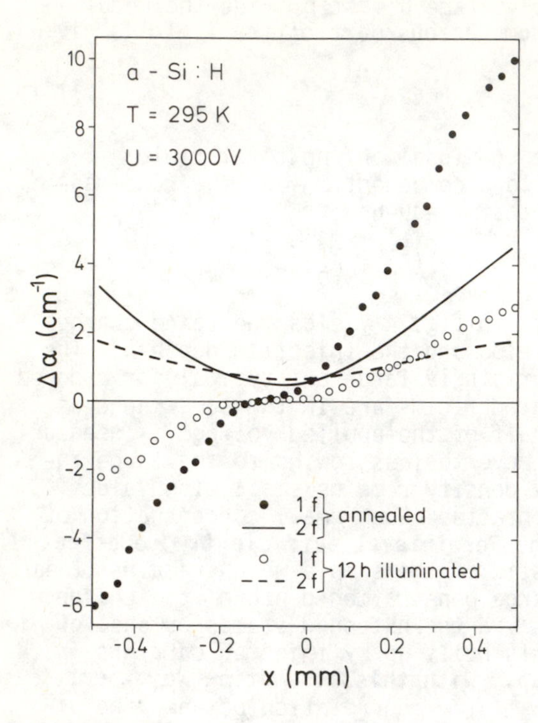

Fig. 6. Variation of the spatial distribution of the EA signal with illumination (white light) and with annealing.

After illumination the EA signal is more homogeneous. Analysis shows that the injection depth has not changed but the charge density is reduced. Since white light creates defects in Si which can act as recombination centers the corresponding faster recombination rate may contribute to the reduction of the trapped charge density.

The charge density can also be modified by additional light as shown in the two beam experiment sketched in fig. 7. Different from the common EA experiment a dc field is applied to the sample and the transmittance is measured as before. This transmittance is modulated by a second beam (a He-Ne-Laser) chopped with a frequency of 350 Hz. The modulation of the transmittance takes place only when an electric field is present and the spectrum $\Delta I/I$ is the same as the usual EA spectrum of this sample. The signal is inhomogeneous extending from the negative contact into the sample over a distance which is comparable to the inhomogeneous part of the EA- signal. From the sign of the dc voltage we derive that injection of electrons is favoured which leads to the 1f signal in the transverse electroabsorption.

CONCLUSIONS

We have outlined that the details of the electroabsorption spectra of amorphous semiconductors are not compatible with the Franz Keldysh effect or with related models which apply to crystalline semiconductors. We propose that the electric field enhances the transition probability between localized tail states and the extended states above the mobility edge by mixing the localized states along the external field.

Although the details are not understood electroabsorption is useful to monitor internal fields and the related space charge. We have shown that in undoped glow discharge silicon high voltage leads to injection of electrons over surprisingly large distances in excess of 100 μm. This injection varies with field, doping and tem-

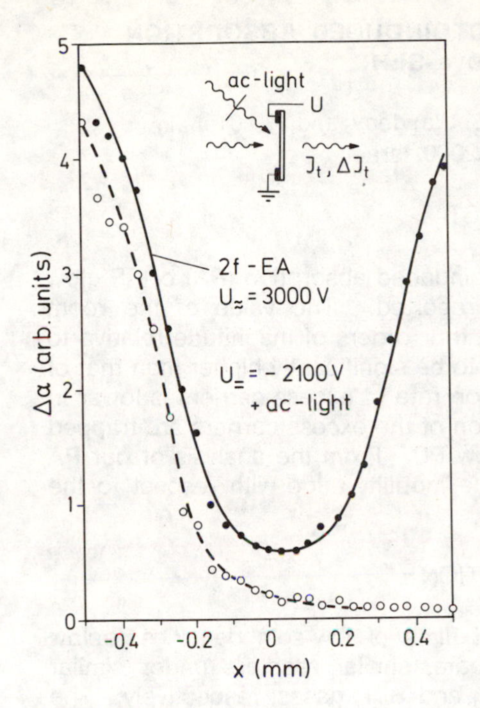

Fig. 7. Comparison of the spatial distribution of 2f EA-signal and of the modulation of the absorption by a second beam (He-Ne-Laser, 350 Hz) in presence of a dc electric field. Injection occurs from the negatively biased contact.

perature. The stationary space charge density (time scale of the experiment is in the range of 1 ms) is reduced by illumination with white light and restored again by annealing in a manner similar to the Staebler Wronski effect. Since electroabsorption is sensitive to the total charge density including immobile charge in traps it may provide a complementary tool to transport studies. Such injection of charge may also take place in EA experiments using a sandwich configuration[10]. In this case the internal fields derived may not be due solely to built-in potentials like Schottky barriers but may partially arise from injected charge.

This work has been supported by the Bundesministerium für Forschung und Technologie (BMFT).

REFERENCES

1. B. Esser, phys.stat.sol. (b) 51, 735 (1972)
2. J.D. Dow and D. Redfield, Phys.Rev. B 5, 594 (1972)
3. R.A. Street, T.M. Searle, I.G. Austin, and R.S. Sussmann, J. Phys. C 7, 1582 (1974)
4. S. Al Jalali and G. Weiser, J. Non-Cryst. Solids 41, 1 (1980)
5. R.S. Sussmann, I.G. Austin, and T.M. Searle, J. Phys. C 8, L182 (1975)
6. B. Esser and P. Kleinert, Proc. 7th Int. Conf. on Amorphous and Liquid Semiconductors, Edinburgh 1977, p. 244
7. L. Sebastian, G. Weiser, and H. Bässler, Chem. Phys. 61, 125 (1981)
8. M. Hoheisel and W. Fuhs, to be published
9. D.L. Staebler and C.R. Wronski, J. Appl. Phys. 51, 3262 (1980)
10. S. Nonomura, H. Okamoto, and Y. Hamakawa, Appl. Phys. A 32, 31 (1983)

PHOTOCONDUCTIVITY AND PHOTOINDUCED ABSORPTION IN a-Si:F AND a-Si:H

M. Janai,* R. Weil, B. Pratt, Z. Vardeny and M. Olshaker
Technion, Haifa 32000, Israel

ABSTRACT

The photoconductivity (PC) and photoinduced absorption (PA) of RF glow discharge a-Si:F and a-Si:H samples is reported. The value of the room temperature PC of a-Si:F is lower by some four orders of magnitude relative to that of a-Si:H, but the PA of a-Si:F is found to be significantly higher than that of a-Si:H. This indicates that the recombination rate of excess carriers is lower in a-Si:F than it is in a-Si:H, yet a larger fraction of the excess carriers are trapped in the band tails of a-Si:F, leading to its low PC. From the analysis of our PA data, we also determine the position of the mobility edge with respect to the extrapolated band edge in a-Si:H.

INTRODUCTION

RF glow-discharge a-Si:F and a-Si:H films of low-spin densities (below $3\times10^{16}cm^{-3}$) were prepared in two separate, similar reactors under similar plasma conditions from the plasma of SiF_2 and SiH_4 gases, respectively. The deposition temperatures were 225°C for the a-Si:F and 250°C for the a-Si:H samples. The a-Si:H films had similar optical and electronic transport properties to those of non-doped, high quality GD a-Si:H films reported in the literature.[2] The a-Si:F films contained about 8 at.% fluorine, bonded predominantly in the monofluoride (Si-F) form.[1,3] The optical band-gap was 1.75±0.1 eV, the dark conductivity at room temperature (RT) was about 10^{-10} $(\Omega\cdot cm)^{-1}$, and the activiation energy of the dark conductivity above RT was 0.78±0.04 eV for the a–Si:F films. The undoped a-Si:F films were n-type and could be doped by either phosphorous or boron, indicating that there is no large density of mid-gap defect states pinning the Fermi level.

While the optical properties and the equilibrium carrier transport properties of the RF GD a-Si:F samples were similar to those of the a-Si:H samples, we found significant differences in the excess carrier transport properties of both types of films.

Below, we present a comparative study of the steady state excess carrier behaviour of a-Si:F and a-Si:H, as derived from PC and PA measurements of non-doped films. The results are used to estimate the difference in the density of tail states of both types of materials.

*Present address: Xerox PARC, 3333 Coyote Hill Rd., Palo Alto, CA 94304.

PHOTOCONDUCTIVITY (PC)

The steady state PC was measured in co-planar electrode geometry with an interelectrode spacing of 0.3 mm, and an applied voltage of 20V. The light intensity was $25 mW/cm^2$, the wavelength was 5461 Å, and the chopping frequency was 170Hz.

Figure 1 shows the PC of one of the a-Si:H samples (# H-5) and four of the a-Si:F samples (# F-191, F-174, F-176 and F-170). The PC is presented in terms of the product $\eta\mu\tau$, where η is the quantum yield, μ the mobility, and τ the mean carrier lifetime.[4] The figure also indicates the spin density of the samples, N_s. The spin density of samples F-176 and F-191 was higher than that of the other samples as a result of postdeposition thermal anneals, at 325°C and 425°C, respectively.[1] There are a few interesting points presented in figure 1. First, the room temperature PC of the a-Si:F samples appears to be some four orders of magnitude lower than that of the a-Si:H sample. Second, there seems to be no correlation between the magnitude of the PC of the a-Si:F samples and their spin density, unlike the results reported for a-Si:H.[2] We received similar values for the PC of the a-Si:F samples up to spin densities of $10^{18} cm^{-3}$. Third, there seems to be a significant difference in the magnitude of the temperature dependence of the PC of both types of samples above RT.

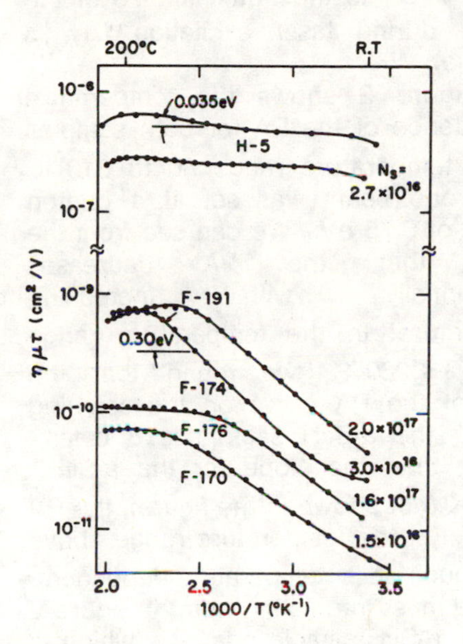

Fig. 1: PC of a-Si:H and a-SiF vs. temperature.

PHOTOINDUCED ABSORPTION (PA)

In order to understand the origin of the difference in the excess carrier transport properties of the a-Si:F and a-Si:H samples of equal spin densities, we performed PA measurements. PA measurements give information on the trapped fraction of the photocarriers, while PC measures the mobile photocarriers, so the two experiments are complementary. The results reported below were obtained on samples H-5 and F-174, which had similar values of spin density and equilibrium (dark) transport properties.

Our experimental set-up was similar to that used by O'Connor and Tauc for steady-state PA measurements.[5,6] Excitation of photocarriers (pumping) was done with an Ar^+ laser at a power of $0.1 W/cm^2$, and at a wavelength of 5145 Å. The laser beam was chopped at a frequency of 160 Hz. The films were thick enough to completely absorb the pump beam. The PA was probed with ir light coming from a monochomatized tungsten lamp at a power of 50 μW/cm^2. The wavelength varied between 1 and 4 μm. The samples for the PA experiment were deposited on crystalline silicon substrates, so that interference effects in the probe beam were eliminated. The amorphous structure of these samples was verified by Raman spectroscopy. The PA results are presented in terms of $\Delta\tau/\tau$, where τ is the transmittance of the samples at the probe-beam wavelength in the absence of pumping, and $\tau + \Delta\tau$ is the transmittance of the sample during laser excitation ($\Delta\tau$ is negative).

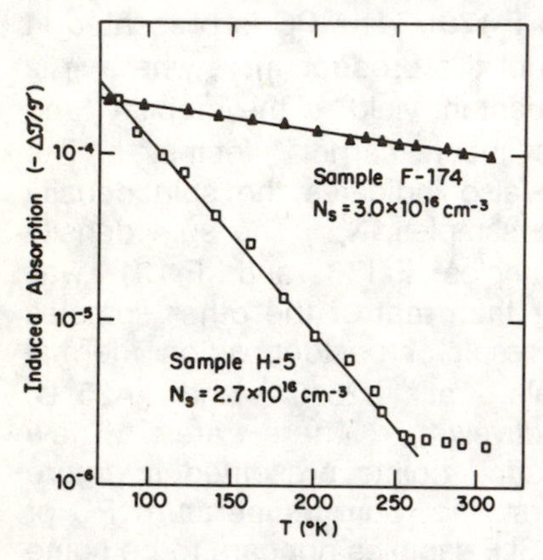

Fig. 2: PA vs. temperature

Figure 2 shows the temperature dependence of the PA for both samples in the temperature range 80 to 310°K. The probe beam was set at a photon energy of 0.75 eV. We can see from the figure that the PA decreases exponentially with increasing temperature. In the temperature range 60°K$\leq$T$\leq$260°K, we found that the slope of the PA curve on the semi-log plot for the a-Si:H sample is 8 times steeper than the slope for the a-Si:F sample. In the temperature range below 60°K (not shown in the figure), the PA of the a-Si:H sample reached a steady value. In the temperature range above 270°K, the PA of a-Si:H became, again, much less temperature-dependent.

For our present discussion, perhaps the most interesting point in figure 2 is the absolute value of the PA signal of the a-Si:F sample near RT, which is almost two orders of magnitude higher than that of the a-Si:H sample. We note that this result was obtained for the two samples measured with the same pump and probe beam intensities, respectively. Figure 3 shows plots of $[(\Delta\tau/\tau)h\nu]^2$ vs. the photon energy of the probe beam, $h\nu$, at various temperatures. The solid lines in the figure are the fits of the linear portion of the data to the function

$$(\Delta\tau/\tau)h\nu = A(h\nu - h\nu_0)^{1/2}, \qquad (1)$$

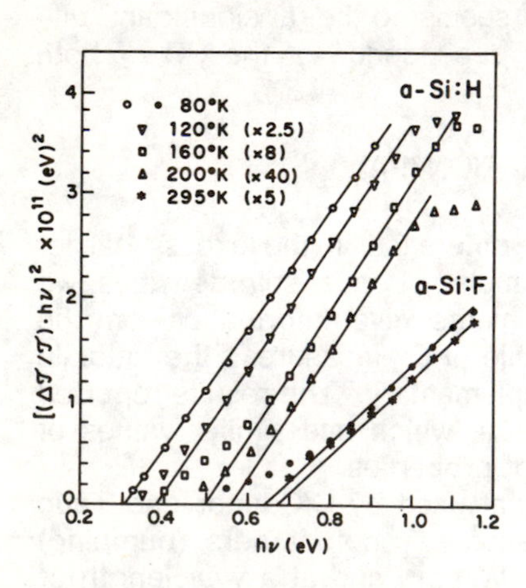

Fig. 3: PA vs. photon energy

where $h\nu_0$ is the (extrapolated) threshold photon energy for the PA process. As discussed by Tauc,[6] eq. (1) is consistent with the assumption that the major contribution to the PA signal in the corresponding range of photon energies is due to the transitions of trapped photocarriers from a narrow energy region in the band tail, into the parabolic energy band. If the density of states in the energy band away from its edge is given by

$$D_u(E) = N_u|E - E_u|^{1/2}, \qquad (2)$$

and if the trapped photocarriers are excited from the quasi-Fermi level E_{qF} into the band,[6] then $h\nu_0$ is given by

$$h\nu_0 = |E_u - E_{qF}|, \qquad (3)$$

where E_u is the band-edge energy (see fig. 5). E_u should not be confused with the mobility edge $E_{\mu u}$, which lies lower in the band tail, as we shall show in the discussion below. The subscript u in eqs. (2) and (3) may denote either the conduction or the valence band; we can not determine which band tail makes the dominant contribution to the PA from PA measurements alone.[7]

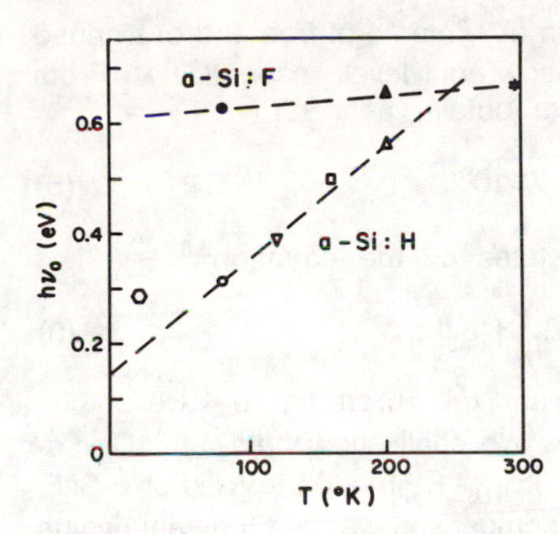

Fig. 4: $h\nu_0$ vs. temperature

Figure 4 presents the values of $h\nu_0$ for a-Si:F and a-Si:H as a function of temperature. In the case of a-Si:H, $h\nu_0$ has a relatively strong dependence on T. It indicates that, with increasing temperature, the quasi-Fermi level sinks rapidly down into the band tail. For a-Si:F, on the other hand, the quasi-Fermi level lies deep in the tail at all temperatures. Under our experimental conditions, E_{qF} of the a-Si:F sample was about 0.65 eV in the tail away from E_u.

DISCUSSION

The low PC of a-Si:F with respect to a-Si:H could be explained by one or more of the following reasons:

(a) Low microscopic mobility (that is, low value of the mobility of the carriers at energy states above the mobility edge),

(b) Low macroscopic mobility (that is, a relatively large fraction of photocarriers are trapped in the band tails), and

(c) Fast recombination rate, which would cause a low steady-state value of the density of photocarriers in a-Si:F.

As to (a), the close similarity of the values of the dark conductivity of a-Si:F and a-Si:H over five decades[1] suggests that their microscopic mobilities are, most likely, the same. The third possibility, that a shorter photocarrier lifetime in a-Si:F is the reason for the low PC of this material, is unlikely. We saw, above, that the value of the PA in a-Si:F is higher than in a-Si:H. With the assumption of similar matrix elements for optical transitions from the tail into the band in both types of samples, the higher PA of a-Si:F near RT implies a higher density of trapped photocarriers in this sample. If the trapped photocarriers at E_{qF} are in thermal equilibrium with the free photocarriers, the total number of steady-state photocarriers must be higher in a-Si:F, and hence their average lifetime must be longer.

We conclude, then, that the lower PC of a-Si:F is due to the larger density of trapping states at the band tails of this material. For an order-of-magnitude calculation of the difference in density of tail states between a-Si:F and a-Si:H, we write the ratio of free photocarriers, n_f, to trapped photocarriers, n_t, as

$$n_f/n_t \approx [D_f(E_{\mu u})/D_t(E_{qF})] \exp\left[-|E_{\mu u}-E_{qF}|/kT\right] , \quad (4)$$

where D_f and D_t are the densities of states (in $eV^{-1}cm^{-3}$) of free and of trapped carriers at the mobility edge and at the quasi-Fermi level, respectively. From the PA and PC data presented above, we obtain, near RT:

$$[n_f/n_t]_{a\text{-}Si:H} \,/\, [n_f/n_t]_{a\text{-}Si:F} \approx 10^6 . \quad (5)$$

Let us describe the density of tail states by the equation[6,8]

$$D_t(E) \approx 10^{21} \exp[-|E-E_{\mu u}|/E_0] , \quad (6)$$

where E_0 is the characteristic width of the band tail. From figure 4 we obtain, near RT, $|E_u - E_{qF}| \approx 0.65$ for both materials. We shall show below that for a-Si:H, $|E_u - E_{\mu u}| = 0.15$ eV, so that, near RT, $|E_{\mu u} - E_{qF}| = 0.5$ eV. (For a-Si:F, $|E_u - E_{\mu u}|$ may be larger, but for the sake of comparison we take the same value for both materials.) We further assume that $D_f(E_{\mu u})$, for both a-Si:F and a=Si:H, is about the same. If we take the value $E_0 = 0.03$ eV for a-Si:H, then from eqs. (4), (5), and (6) we get $E_0 = 0.18$ eV for a-Si:F. As we discuss elsewhere,[9] this large value of E_0 for a-Si:F in eq. (6) is inconsistent with the low density of spins in our a-Si:F samples; with $E_0 = 0.18$ eV, the predicted density of overlapping tail states alone comes out to the order of 10^{18} cm^{-3}. This inconsistency can be explained in several ways. As we show elsewhere,[9] one of the possibilities is that the density of tail states is given by a Gaussian distribution, rather than by eq. (6). A Gaussian distribution will have an exponential appearance near $E_{\mu u}$,[9] but it decays much faster towards mid-gap, and gives consistent results with our observed spin density.

Our model for the PA of a-Si:H and a-Si:F is schematically presented in figure 5, below. In a-Si:H, at the temperature range $60°K \leq T \leq 270°K$, the multiple-trapping model may be applied.[6,10,11] Then, the dominant mechanism by which the band tails get depleted of photocarriers is thermal excitation to the mobility edge, and the subsequent drift of the carriers to recombination centers.

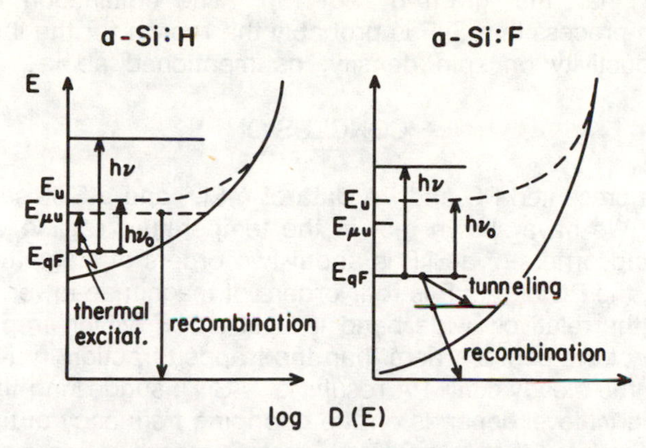

Fig. 5: Schematic diagram of the density of tail states and the dominant transitions in the PA of a-Si:H and a-Si:F

In this case, the quasi-Fermi level stabilizes at an energy level given by

$$|E_{qF} - E_{\mu u}| = kT \ln \nu t \,, \tag{7}$$

where ν is the "attempt to escape" frequency[6,10,11] ($\nu = 10^{12}$ sec^{-1}) and t is the time scale of the experiment. The value of t should be equal to the chopping period, since states deeper than E_{qF} given by eq. (7) would not empty during the dark part of the chopping period.

From eqs. (3) and (7), $h\nu_0$ should be given by

$$h\nu_0 = |E_u - E_{\mu u}| + kT \ln \nu t \,. \tag{8}$$

Indeed, the slope of the curve $h\nu_0(T)$ of the a-Si:H sample (fig. 4) is 2×10^{-3} eV·°K^{-1}, in very good agreement with the predicted value k ln νt = 8.5×10^{-5} eV°K^{-1} x ln(10^{12}/160) = 1.9×10^{-3} eV·°K^{-1}.[12] From the figure, we also find that, at T = 0, $|E_u - E_{\mu u}|$ = 0.15 eV. Hence, we find that, in a-Si:H, the mobility edge lies 0.15 eV in the tail relative to the extrapolated band edge E_u. (As mentioned before, we can not say whether the value we found refers to the conduction or to the valence band, or to both.) We should point out that if one assumes a linear dependence of $D_u(E)$ in eq. (2),and the data of figure 3 is fitted accordingly, the last result is modified, and we get $|E_u - E_{\mu u}| \simeq 0$. The rest of our analysis remains unaffected. For a-Si:H, however, particularly at low T, where the PA signal was large and clear, we found that the fit to a parabolic density of states holds over a wider energy range than the fit to a linear density of energy states.

In a-Si:F (as well as in a-Si:H at temperatures below 60°K), the dominant mechanism of tail depletion is probably tunneling and direct tail-to-tail recombination. The temperature dependence of E_{qF} is then expected to be much weaker than that given by eq. (8). The domination of a tail-to-tail recombination process in a-Si:F is probably the reason for the independence of its photoconductivity on spin density, as mentioned above.

CONCLUSION

We have presented PC and PA data of a-Si:F and a-Si:H samples of equal spin density. We have shown that in the temperature regime where the total density of photocarriers of a-Si:F is about two orders of magnitude higher than that of a-Si:H, the PC of a-Si:F is four orders of magnitude lower. We attributed this result to the relatively wide band tails of a-Si:F, which trap a significantly higher faction of the photocarriers than the trapped fraction in a-Si:H. We have re-interpreted the steady state PA results of a-Si:H, suggesting that the position of the quasi-Fermi level depends on the chopping frequency of the pump beam. We have also claimed that in a-Si:F the major recombination mechanism is not via thermal release of the trapped carriers and their subsequent drift to recombination centers, but tunelling and recombination through tail states. Such a tunneling mechanism is consistent with the large density of tail states found in this material. Finally, our results indicate that fluorine can eliminate deep spin centers in a-Si, but it does not eliminate the tail states. We thus conclude that the unique property of hydrogen in a-Si:H is not the elimination of spin centers (which flourine does, too), but the elimination of tail states, which gives the material its good photoconductive quality.

This research was supported by grants from the National Council for Research and Development, Israel, and K.F.A., Juelich, Germany. The technical assistance of R. Hida, A. Kesel, L. Patlagin and P. Ron is greatly appreciated.

REFERENCES

1. M. Janai, R. Weil, B. Pratt, submitted to Phys. Rev. B.
2. H. Fritzsche, Solar Energy Mater. 3, 447 (1980).
3. M. Janai, L. Frey, R. Weil and B. Pratt, Solid State Commun. 48, 521 (1983).
4. W.E. Spear and P.G. LeComber, in "Photoconductivity and Related Phenonema," ed. J. Mort and D.M. Pai (Elsevier, NY 1976) p. 185.
5. P. O'Connor and J. Tauc, Phys. Rev. B25, 2748 (1982).
6. J. Tauc, in "Advances in Solid State Physics," ed. P. Grosse Aachen, 22, (Vieweg, Braunschweig, 1982) p. 85.
7. In Ref. 6 it is suggested that the PA is due to holes trapped in the valence band tail.
8. T. Tiedje, T.M. Celbulka, D.L. Morel and B. Abeles, Phys. Rev. Lett. 46, 1425 (1981).
9. M. Janai, submitted to Phys. Rev. Lett.
10. T. Tiedje and A. Rose, Solid State Commun. 37, 49 (1981).
11. J. Orenstein and M.A. Kastner, Phys. Rev. Lett. 43, 161 (1981).
12. Our interpretation to fig. 4, which is based on eq. (8), differs from that given in ref. 6.

INTERBAND OPTICAL ABSORPTION IN AMORPHOUS SEMICONDUCTORS

Morrel H. Cohen, Costas M. Soukoulis[+] and E. N. Economou[*]
Exxon Research and Engineering Company
Route 22 East, Clinton Township
Annandale, New Jersey 08801

ABSTRACT

We present a simple, analytic derivation of the exponential tails universally observed in the optical absorption of three-dimensional disordered systems based on our recent analysis of exponential band tails. These tails contain states localized inside relatively large regions within which the local average of the potential falls below its overall average. The physical extent of a tail state is governed by the fluctuation size and not the localization length. Strong anticorrelation in the fluctuations of the valence and conduction band edges implies that the Urbach tail arises from valence tail to conduction tail transitions. Otherwise, the Urbach tail arises from a superposition of valence tail to conduction extended and valence extended to conduction tail state transitions. The wider band tail then dominates the Urbach tail, as in the case, e.g., for a-Si:H, where the valence band tail dominates.

The interband optical absorption of amorphous(a) semiconductors displays several remarkable regularities, explored in detail for a-SiH_x by Cody[1] and recently reviewed by him[2]. First, as originally observed by Tauc[3], the single main absorption peak in amorphous silicon, or, e.g., germanium appears to be a broadened version of that of the crystal. From the broadening, one can estimate that the r.m.s. value of the potential fluctuations introduced by structural and thermal disorder into a SiH_x is of order several eV[4], a typical value. Second, on the low-energy side of the peak, the optical absorption α is well fitted by the Tauc relation[3,5] as modified by Cody[1,2],

$$\left\{\frac{\alpha}{\hbar\omega}\right\}^{1/2} = B[\hbar\omega - E_{Go}], \qquad (1.)$$

*Permanent Address: University of Crete and Research Center of Crete, Heraklio, Crete, Greece.
+Present Address: Department of Physics, Iowa State University, Ames, Iowa 50011.

down to α values of order 10^4. Third, after a transition region between $\alpha \sim 10^3$ and $\alpha \sim 10^4$, an exponential Urbach tail sets in with width E_{oo}. Fourth, Cody has found a linear relation between the apparent optical gap E_{Go} and E_{oo},

$$E_{Go} = E_G^o - C\,E_{oo}, \qquad (2.)$$

where E_G^o is the limiting value of the optical gap for vanishing tail width, Fifth, there exists an Urbach focus at $\hbar\omega \cong E_G^o$. That is, as the sample temperature or hydrogen content is changed, the changing Urbach tails nevertheless intersect at a single point on extrapolation to higher frequencies. Sixth, the optical matrix element between states in the valence and conduction bands appears to be constant in these regions, independent of the initial and final states, with the value expected from a sum rule analysis of the crystalline material[1,2]. Seventh, the values of E_{oo} are about two orders of magnitude smaller than those of W.

Our present purpose is to derive or at least interpret the above observations. We start with the general expression for the imaginary part of the dielectric constant following from linear response theory[6],

$$\varepsilon_2 = \frac{4\pi^2 e^2}{\Omega} \sum_{nn'} |X_{nv,n'c}|^2 \delta(E_{n'c} - E_{nv} - \hbar\omega). \qquad (3.)$$

In (3.) Ω is the sample volume, $X_{nv,n'c}$ is the matrix element of the x coordinate between states n in the valence band v and n' in the conduction band c, and E_{nv} and $E_{n'c}$ are the corresponding energies. Eq. (3.) can be rewritten as

$$\varepsilon_2 = \frac{4\pi^2 e^2}{\Omega} \iint dE\,dE'\; M^2(E,E') N_v(E) N_c(E')\; \delta(E' - E - \hbar\omega) \qquad (4.)$$

where

$$M^2(E,E') = \frac{\sum_{nn'} |X_{nv,n'c}|^2 \delta(E_{nv}-E)\delta(E_{n'c}-E')}{N_v(E)\ N_c(E')} \qquad (5.)$$

and $N_v(E)$,$N_c(E')$ are the total densities of states. The double sum on n and n'in (5.) is tantamount to taking an ensemble average of $|X_{nv,n'c}|^2$ over all equivalent structures of the amorphous material with the energies E_{nv} fixed at E and $E_{n'c}$ fixed at E',

$$M^2(E,E') = \overline{|X_{nv,n'c}|^2}^{E,E'} . \qquad (6.)$$

The eigenfunctions entering $X_{nv,n'c}$, ψ_{nv} and $\psi_{n'c}$ can be written in tight binding form as

$$\psi_{nv}(\vec{r}) = \sum_{\ell\beta} a_{\ell\beta}^{nv} \phi_\beta(\vec{r}-\vec{r}_\ell), \qquad (7.)$$

where the $\phi_\beta(\vec{r}-\vec{r}_\ell)$ is a pseudoatomic orbital of type β on the ℓth site. s,s*, p_x, p_y and p_z orbitals would give an accurate representation of the valence band and the bottom of the conduction band. We now make the simplification that the actual linear combination of $\phi_\beta(\vec{r}-\vec{r}_\ell)$ on each site can be replaced by a typical combination $\phi_v(\vec{r}-\vec{r}_\ell)$ or $\phi_c(\vec{r}-\vec{r}_\ell)$, the same for each site in the energy regions under consideration,

$$\psi_{nv} = \sum_\ell a_\ell^{nv} \phi_v(\vec{r}-\vec{r}_\ell), \qquad (8.)$$

so that

$$M^2(E,E') = \overline{d^2 \sum_{\ell\ell'} (a_\ell^{nv})^* a_{\ell'}^{nv} (a_{\ell'}^{n'c})^* a_\ell^{n'c}}^{E,E'} \qquad (9a.)$$

where
$$d = (\phi_{v}, \times \phi_c). \qquad (9b.)$$

We now make two assumptions: (1.) the phase coherence length for the a_ℓ is of order of or smaller than an interatomic separation;

(2.) there is no statistical correlation in amplitude or phase between the a_ℓ^{nv} and the $a_{\ell'}^{n'c}$. The first assumption is justified by the fact that we are dealing with strongly disordered systems in the general vicinity of the mobility edges. The second assumption is justified by the fact that there are many elements of structural disorder in covalent materials, bond-length disorder, bond-angle disorder, dihedral angle disorder, odd rings, etc.[7] Each of these affects the top of the valence band and the bottom of the conduction band in quite different ways, presenting a totally different random potential to each[7]. We therefore expect no correlation in the amplitude fluctuations. With these assumptions (9.) becomes

$$M^2(E,E') = d^2 \sum_\ell \overline{|a_\ell^{nv}|^2}^{\,E} \; \overline{|a_\ell^{n'c}|^2}^{\,E'} \qquad (10.)$$

After the ensemble averaging, the $\overline{|a_\ell|^2}$ become translation invariant, and normalization requires that

$$\overline{|a_\ell^{nv}|^2}^{\,E} = \overline{|a_\ell^{n'c}|^2}^{\,E'} = \frac{1}{N} \qquad (11.)$$

where N is the number of sites. Eq. (11.) is true irrespective or whether the states involved are localized or extended, uniform in amplitude or undergoing violent amplitude fluctuations. Similarly, the phase incoherence argument applies equally well to real localized states and complex extended states. Thus we arrive at the constant matrix element approximation by a more generally valid argument than that of Tauc and Velicky, which allowed neither for localized states nor amplitude fluctuations in extended states, and in precisely the form used by Cody[2],

$$M^2(E,E') = d^2/N \qquad (12.)$$

Inserting (12.) into (4.) gives

$$\varepsilon_2 = (2\pi ed)^2 \, v_a \int dE n_v(E) n_c(E+\hbar\omega), \tag{13.}$$

where v_a is the atomic volume and $n_v(E)$ is the density of states per unit volume.

We have shown[8] that the density of states of each band has exponential tails of width $E_{ov}(E_{oc})$ above (below) an energy E_{1v} (E_{1c}). The transition energy E_{1v} (E_{1c}) is not sharply defined but lies in the region of localized states above (below) the mobility edge E_v (E_c). We have related E_{ov} and $E_{1v}-E_v$ explicitly to the disorder W and find $E_{ov} \simeq 0.1$ W as is required by the data[1,2]. The density of states is continuous across the mobility edge into the extended states. A form for the density of states consistent with all of the above is

$$n_v(E) = A_v \, (E_{1v}-E_{ov}/2 - E)^{1/2}, \; E<E_{1v} \tag{14a.}$$

$$= A_v \, (E_{ov}/2)^{1/2} \, e^{(E_{1v}-E)/E_{ov}}, \; E > E_{1v} \tag{14b.}$$

and similarly for the conduction band. We now have three classes of states for each band: extended ($E<E_v$); localized with their spatial extent given by the localization length ($E_v>E>E_{1v}$), the near tail; localized with their size given by the spatial extent of a large-scale potential fluctuation ($E_{1v}<E$), the far tail. There are thus nine possible kinds of transition.

The four kinds among the extended and near-tail states give rise to the Tauc region with

$$E_{Go} = [E_{1c}-E_{1v}-(E_{ov} + E_{oc})/2] \tag{15a.}$$

$$B = \pi^3 e^2 d^2 v_a A_v A_c / 2n, \tag{15b.}$$

where n is the refractive index at $\hbar\omega \simeq E_{Go}$. We see that the apparent optical gap is, to about 5%, equal to the separation between the edges of the near tails of the valence and conduction bands.

The Urbach tail is dominated by transitions from the wider far tail to the near tail of the other band. ε_2 in that region has the form

$$\varepsilon_2(\text{URBACH}) = (2\pi ed)^2 v_a n_v(E_{1v}) n_c(E_{1c})$$
$$\times \left[\left(\frac{2E_{oo}}{E_<}\right)^{1/2} + 1\right] E_{oo} e^{\{\hbar\omega - (E_{1c} - E_{1v} - E_</2)\}/E_{oo}} \qquad (16a.)$$

where $$E_{oo} = \sup\,(E_{ov}, E_{oc}) \qquad (16b.)$$

$$E_< = \inf\,(E_{ov}, E_{oc})$$

under the assumption that E_{oo} is significantly greater than $E_<$.

Provided that (1.) the disorder potential W is substantially smaller than either band width, (2.) the characteristic lengths associated with amplitude variation of the wavefunctions are longer than interatomic separations, (3.) the probability distribution of the random potential possesses a second moment, and (4.) the correlation length of the potential is of atomic size, one can show that there are characteristic units of energy and length for each band, E_v and E_c, ℓ_v and ℓ_c[9]. The E's are monotomically increasing functions of disorder and the ℓ's are monotomically decreasing functions. Assuming, as we have, a single overall measure of disorder W for both the valence and conduction band, we have in fact supposed there to be a single energy unit E in terms of which all characteristic energies E_γ can be expressed,

$$E_\gamma = \varepsilon_\gamma\, E \qquad (17.)$$

For example, we have

$$E_v = E_v^o + \varepsilon_v\, E$$
$$E_v - E_{1v} = \varepsilon_{1v} E \qquad (18.)$$
$$E_{ov} = \varepsilon_{ov}\, E$$

in the valence band and similarly for the conduction band. In (18.), E_V^o is the limit of the mobility edge for vanishing disorder, and ε_V is found to be very much smaller than ε_{1V} by numerical analysis[10].

Cody's linear correlation (2.) between E_{Go} and E_{oo} follows immediately from (15.), (16.), and (18.), with

$$E_G^o = E_C^o - E_V^o \qquad (19a.)$$

$$C = \frac{\varepsilon_c + \varepsilon_v + \varepsilon_{1c} + \varepsilon_{1v} + (\varepsilon_{oc} + \varepsilon_{ov})/2}{\sup(\varepsilon_{oc}, \varepsilon_{ov})} \qquad (19b.)$$

For $aSiH_x$ we estimate[8] from (19b.) that C is between 5 and 10. Cody[1,2] finds a value of 6.2. The existence of an Urbach focus at E_G^o with E_G^o given by (19a.) follows from (16.) and (18.). The importance of these two results is that optical determination of E_G^o either from the correlation of E_{Go} and E_{oo} or from the Urbach focus gives the limiting value of the mobility gap $E_c^o - E_v^o$. Moreover, since ε_c and ε_v are small[10], the mobility gap remains very close to $E_c^o - E_v^o$ as disorder is introduced into the structure.

For $aSiH_x$, Cody[1,2] finds E_G^o to be 2.1eV. The unperturbed mobility gap is thus about one eV larger than the energy gap in pure crystalline silicon. It has already been shown that the Lifshitz limits to the valence and conduction band are given by the crystalline band edges in the absence of band-length and band-angle disorder in pure a Si[7]. It has also been shown that the mobility edges are very close to the CPA band edges[10]. Thus, the unperturbed mobility gap $E_G{}^o$ corresponds to the band gap of the actual disordered material treated as an effective medium. The effects of the dihedral-angle disorder and the bond-angle disorder push the valence band edge down by about 0.5 eV, and the effects of odd rings and bond-angle disorder push the conduction band edge up a comparable amount. These effects provide an explanation of Cody's observation[1,2] that the product A_vA_c is about 100 times the free-electron-mass value.[11] The separately observed shapes of the top of the valence band[11] and bottom of the conduction band[12] are also consistent with such large values of A_v and A_c.

E_{Go} has frequently been confused with the mobility gap E_c-E_v in the past. Because E_{Go} shows a marked temperature dependence, it has been supposed that E_c-E_v is comparably temperature dependent. However, E_{Go} is essentially E_{1c}-E_{1v}, the temperature-dependence of which derives primarily from thermal disorder, while E_c-E_v is only weakly sensitive to thermal disorder. Such arguments grossly overestimate the temperature dependence of the mobility gap.

The primary theoretical problem that remains is understanding why the square root density of states holds down below the mobility edge or even why it holds at all.

We are grateful to Dr. G. Cody for informative discussions.

REFERENCES

1. G. D. Cody, T. Tiedje, B. Abeles, B. Brooks and Y. Goldstein, Phys. Rev. Lett. 47, 1480 (1981).
2. G. D. Cody, "The Optical Absorption Edge of α Si:H_x in Amorphous Silicon Hydride", ed. by J. Pankove, (Vol. 21B of Semiconductors and Semimetals, Academic Press 1984) p. 11.
3. J. Tauc, in The Optical Properties of Solids, ed. by B. Abeles, North-Holland, Amsterdam 1970, p. 277; and in Amorphous and Liquid Semiconductors, ed. by J. Tauc, Plenum Press, New York (1974).
4. T. Tiedje, J. M. Cebulka, D. L. Morel and B. Abeles, Phys. Rev. Lett., 46, 1425 (1981).
5. N. F. Mott and E. A. Davis, Electronic Processes in Non-Crystalline Materials, Clarendon Press, Oxford, (1979).
6. See e.g. J. Callaway, Quantum Theory of the Solid State, Academic Press, New York (1976).
7. F. Yonezawa and M. H. Cohen, in Fundamental Physics of Amorphous Semiconductors, ed. by F. Yonezawa, Springer Series in Solid State Sciences, 25, 119 (1981).
8. C. M. Soukoulis, M. H. Cohen and E. N. Economou, submitted to Phys. Rev. Lett.
9. M. H. Cohen, C. M. Soukoulis, and E. N. Economou, unpublished.
10. E. N. Economou, C. M. Soukoulis and A. D. Zdetsis, Phys. Rev. B. (1984).
11. L. Ley, S. Kawalczyk, R. Pollak, and D. A. Shirley, Phys. Rev. Lett. 29, 1088 (1972).
12. F. C. Brown and O. P. Rustgi, Phys. Rev. Lett. 28, 497 (1972).

ELECTRONIC PROPERTIES OF STRAINED BONDS IN AMORPHOUS SILICON: THE ORIGIN OF THE BAND-TAIL STATES?

L. Schweitzer and M. Scheffler
Physikalisch-Technische Bundesanstalt, Bundesallee 100,
D-3300 Braunschweig, Federal Republic of Germany

ABSTRACT

The electronic properties of a 'distorted-bonds defect' embedded in a macroscopic solid have been studied by means of a self-consistent Green's-function technique. The defect consists of a central silicon atom and its four nearest neighbours, connected either by streched or compressed bonds. We have calculated the change of the density of states and the charge density due to the variation of the bond length. We find strong resonances, and bound states within the gap of the host silicon. The density of states is strongly enhanced at the band edges by bond streching, whereas compression leads only to minor changes within that energy range. We consider the possibility of these states to be the origin of the localised band-tail states of amorphous silicon, which strongly influence the optical and transport processes in this material.

INTRODUCTION

In recent years the knowlege about the gap states in hydrogenated amorphous silicon (a-Si:H) has steadily grown. It is now generally agreed that in this material the so called dangling-bond is the major defect, which mainly governs the photovoltaic properties by introducing localised states into the gap. These states act as effective non-radiative recombination centers. Depending on the inter-electronic repulsion of electrons occupying these levels and on the magnitude of the accompanying lattice distortion, the effective correlation energy determines whether this defect may be neutral, positively or negatively charged. Although the microscopic details of its structure, which exhibits an ESR-resonance with g = 2.0055 when singly occupied, are not completely resolved, there is at present no doubt that a somehow relaxed threefold coordinated silicon atom is responsible for this defect.

On the other hand, there exists a big confusion in the literature and absolutely no agreement on further defects and imperfections that might give rise to states in the gap, such as dihedral angle variations, streched bonds, vacancies and divacancies, and other point defects involving hydrogen or dopant atoms (see e.g. Mott 1980)[1].

For undoped a-Si:H a model has been recently suggested[2] which, besides dangling-bonds and localised tail states, did not involve any additional defect. This model was successfully applied to describe various recombination processes observed in this material[3,4]. One of the proposals of the model was to assign the E_y peak, commonly observed in field-effect and DLTS studies, to doubly occupied valence band-tail states with a corresponding effective correlation energy

0094-243X/84/1200379-07 $3.00

U of about .25 eV. According to the model the energy dependence of U(E) is the reason for the pronounced shoulder and the fact that the valence band-tail is not so steep as the conduction band-tail for which U amounts only ∿.01 eV.

But what is now the microscopic origin of the localised tail states? There exist several possible explanations, and the one most frequently given is that in terms of stretched silicon bonds. Other suggestions concerning the nature of the tail states are: variations in the dihedral angles[5], the twofold coordinated silicon atom (T_2^0)[6], and the silicon hydrogen bond (Si-H)[7,8]. A three center Si-H-Si bond[9] with a corresponding negative effective correlation energy has also been proposed to give rise to localised gap states. The Si-H_2-Si bond has also been considered[10].

It has been shown both by experimental[11] and by theoretical investigations[8] that incorporation of hydrogen strongly influences the electronic states in the gap region. It not only saturates the dangling bonds but also removes gap states from the valence band edge and simultaneously increases the optical gap. Hence, all models for the band-tail states involving hydrogen are conceivable. However, their exist also band-tails in evaporated and sputtered amorphous silicon containing no or only a negligible amount of hydrogen. Furthermore, even for a-Si:H it is not obvious how hydrogen removes gap states and likewise creates localised band-tail states. It seems, therefore, that 'strained bonds' is the more natural explanation. In particular one could imagine that in the presence of hydrogen their exists a critical maximal bond length[1] for which it is energetically more favourable to include one or two hydrogen atoms into a streched bond. This mechanism would then, by hydrogenating the material, only remove deeper tail states and leave steeper band-tails composed of less strained bonds.

It is now the purpose of the present work to investigate theoretically whether electronic states due to streched or compressed bonds could be the origin of the band-tail states and to what extent these states would act as electron or hole traps. We further need information about the effective correlation energies associated with the different occupancy of the distorted bonds.

As a first approach we consider a tetrahedron with stretched or compressed bonds embedded in an otherwise perfect, infinite crystal. We perform self-consistent Green's-function calculations in order to determine the changes in the local density of states, and the charge redistribution, which results from the local distortion due to bond-length variations. Furthermore, the electronic correlation energy of conduction band derived states is evaluated.

MODEL AND THEORY

Our 'distorted-bonds defect' consists of a central silicon atom and its four nearest neighbours, which have been deliberately moved out- or inwards along their bond axes. The choice of this defect, which embedded in an otherwise perfect silicon crystal, permits a thorough investigation of the effects caused only by local bond distortions, which are probably blurred in a real amorphous Si:H network by dihedral-angle variations, Si-H bonds, impurities, and perhaps odd membered rings. We note that the density of states of amorphous silicon is similar to that of a Si-crystal. We therefore assume that the embedding in a crystal is not a severe approximation for our distortion study.

The electronic structure of this defect is calculated for various distortions using the self-consistent Green's-function method[12] in the formulation of ref. 13. The host crystal Hamiltonian, h^0, is written as

$$h^0 = -\nabla^2 + \sum_R v_{ion}(\vec{r}-\vec{R}) + 1/2 \int \rho^0(\vec{r}')/|\vec{r}-\vec{r}'|\, d^3\vec{r} + v_{xc}(\rho^0), \quad (1)$$

where $\vec{R}$ are atomic sites and $v_{ion}(\vec{r}-\vec{R})$ are the electrostatic potentials describing the host ions[14]. $\rho^0(\vec{r})$ is the valence electron charge density and $v_{xc}(\rho^0)$ is the exchange-correlation potential in the local-density approximation. The band structure calculation is performed using a plane wave basis set and the charge density $\rho^0(\vec{r})$ is determined self-consistently.

The Hamiltonian, h, of the embedded 'distorted-bonds defect' is given by an expression analogous to (1), just with different positions of the ions, $\vec{R}^*$, and a different charge density, $\rho(\vec{r})$. The perfect crystal Greens-function is

$$G^0(E) = \lim_{\eta \to o} (E + i\eta - h^0)^{-1} \quad (2)$$

and the electronic structure of the embedded defect is given by solving self-consistently the equations

$$G(E) = G^0(E) + G^0(E)\,[h(\rho) - h^0(\rho^0)]\,G(E) \quad (3)$$

and

$$\rho(\vec{r}) = -2/\pi \int_{-\infty}^{E_F} \mathrm{Im}\, G(E, \vec{r}, \vec{r})\, dE \quad (4)$$

For this calculation, which depends only on the local change of the electronic charge, induced by the distorted bonds, we use localised basis orbitals centered at the central Si atom and at the 16 atoms of the two neighbouring shells. The density of states is given by

$$N(E) = -2/\pi\ \mathrm{Tr}\ \mathrm{Im}\, G(E). \quad (5)$$

RESULTS AND DISCUSSION

The results of our calculations are shown in fig. 1. The upper part displays the gap region of the density of states (DOS) of crystalline silicon $N^0(E)$. The change of the density of states, $\Delta N(E) = N(E) - N^0(E)$, due to a local distortion of +10%, which means that the four nearest neighbour atoms have been moved 10% away from the central silicon atom and consequently increased

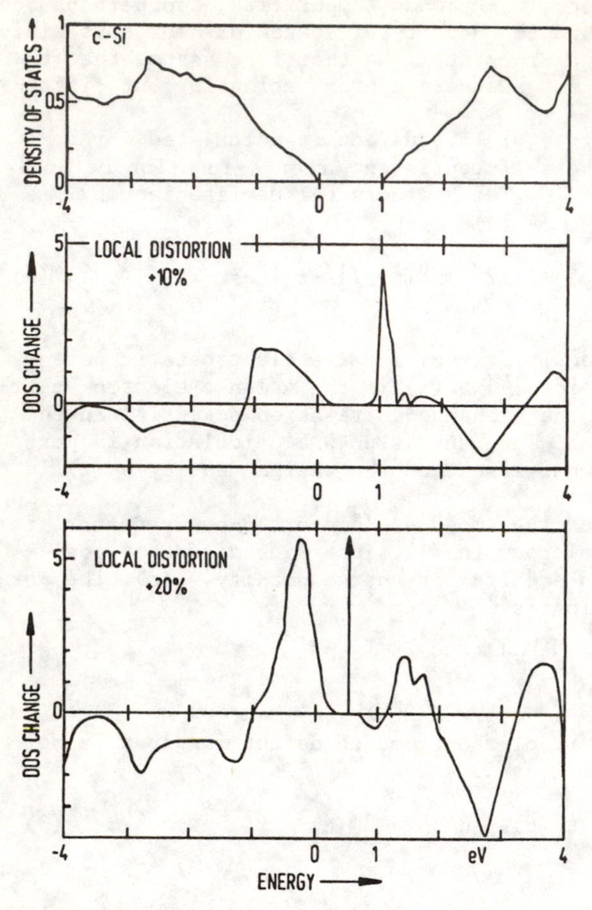

Fig. 1. Density of states (DOS) of c-Si.

Change of the DOS due to a 10% outward relaxation.

Change of the DOS due to a 20% outward relaxation.

the bond length from 2.35 Å to 2.59 Å, is shown in the middle panel. A strong enhancement of the density of states at the valence band edge and an even more pronounced increase at the bottom of the conduction band are clearly visible. After a further distortion of altogether +20% the conduction band peak splits off the band and appears as a bound state in the gap, at E = .54 eV. This is shown in the lower panel of fig. 1. Also, at the top of the valance band a strong peak of additional states has piled up.

In fig. 2 the results of a -10% distortion (inward relaxation) are shown. Here the full energy range is displayed and it is seen that there are only small changes of the DOS in the gap region. The major changes appear in the lower part of the valence band, where states are shifted to lower energies by the inward relaxation of the nearest neighbour atoms. The sharp resonance at the bottom of the valence band indicates a strong increase of the s-like bonding character. As there is no appreciable change in the DOS at the band edges, we will no longer discuss the inward relaxed model defect in

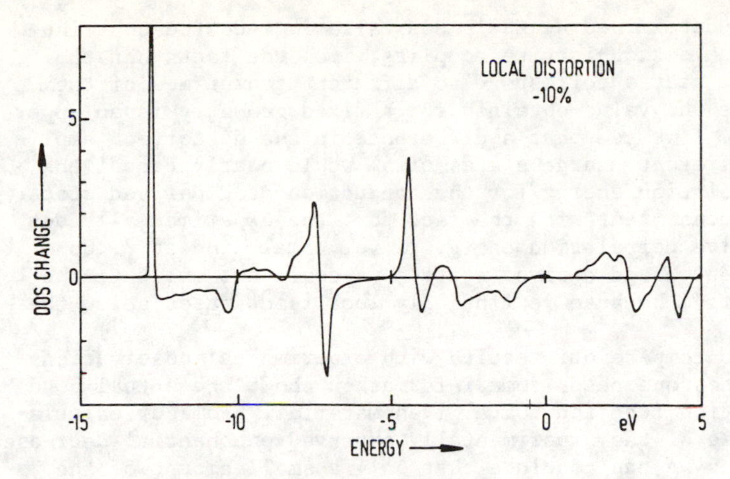

Fig. 2. Change of the DOS due to a 10% inward relaxation.

this paper. Instead, we will take a closer look at the outward relaxation process itself. In moving the four nearest neighbour atoms 10% out of their normal position, not only the bond length is increased by .24 Å, but also the tetrahedral angle is decreased by ∿5%. We believe that both variations significantly effect the electronic structure, while the corresponding reduction of the bond length between the moved atoms and the next nearest neighbours of 0.67 Å (∿2.85%) should not be so important. We note that in a real amorphous system it is usually not possible to vary only the bond length without changing the tetrahedral angle. Therefore, a distinction between the two effects would be artificial.

For the +20% distorted bonds we have also studied the electronic correlation energy by adding one or two electrons to the system (i.e. changing the Fermi energy; see equ. (4)). It turns out that in doing so the peak of the DOS change at the valence band edge (fig. 1, lower panel) splits off the band, becoming also a bound state in the gap (see fig. 3). From fig. 3 we extract the electronic correlation energy for the conduction band derived states as U_{CD} ∿0.3 eV.

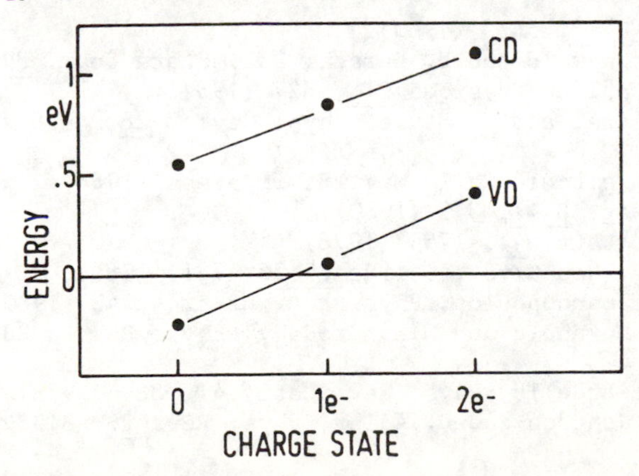

Fig. 3. Single particle energies as a function of the charge state. The higher level (conduction band derived) is filled with 0,1 and 2 electrons. The lower (valence band derived) level is fully occupied. The local distortion is +20%.

The effective U determined by ESR[15] does also include the contribution due to the electron-lattice coupling, i.e. the fact that the different charge states correspond to different geometries of the atoms. Therefore the value obtained for a fixed geometry is an upper bound. From fig. 1 we see that a difference in the distortion amplitude for two different charge states of 5% would nearly cancel the electronic correlation energy for the conduction-band derived state. This were then consistent with the fact that the experimentally determined effective correlation energy is very small indeed[15]. Concerning the valence-band derived states, we could not evaluate the correlation energy, because we find only localised states below the Fermi level.

In order to compare our results with experimental observations of the tail states one needs some information about the actual bond-length distribution function for a given material. From our calculations and because of the experimentally observed exponential decrease of the band edges we can conclude that only a small amount of the bonds would have to be strongly distorted. It is often inferred from the radial distribution function (RDF) that the bond length fluctuations must be less than 1%. For the localised band-tail states, however, it would suffice if only 0.1% of the total number of bonds were strongly distorted.

Since hydrogenation removes states from both band edges[8], only such bonds would remain which exhibit a small distortion or are not accessible to hydrogen during deposition. The observed inhomogeneity of the material points to the latter possibility.

In conclusion we presented self-consistent Green's-function calculations of the electronic properties of distorted silicon bonds in order to elucidate the extend that these states are responsible for the localised tail states in amorphous silicon. It is found that bond streching gives rise to an enhancement in the DOS at both band edges and for very strong distortions even bound states within the gap[16]. Our results support the view that strained bonds give rise to electron and hole traps in a-Si:H.

REFERENCES

1. N.F. Mott, J. Phys. C 13, 5433 (1980).
2. L. Schweitzer, M. Grünewald and H. Dersch, Sol. State Comm. 39, 355 (1981), and Journal de Physique C 4, 827 (1981).
3. H. Dersch, L. Schweitzer and J. Stuke, Phys. Rev. B 28, 4678 (1983).
4. H. Dersch and L. Schweitzer, Phil. Mag. B, in press (1984).
5. J.H. Davies, Phil. Mag. B 41, 373 (1980)
6. D. Adler, Phys. Rev. Lett. 41, 1755 (1978)
7. J.D. Joannopoulos, J. Non-Cryst. Solids 35-36, 781 (1980), D.C. Allan and J.D. Joannopoulos, Phys. Rev. Lett.44, 43 (1980)
8. D.P. DiVincenzo, J. Bernholc and M.H. Brodsky, Phys. Rev. B 28, 3246 (1983)
9. R. Fisch and D. Licciardello, Phys. Rev. Lett. 41, 889 (1978)
10. M.E. Eberhart, K.H. Johnson and D. Adler, Phys. Rev. 26, 3138 (1982)

11. J. Reichardt, L. Ley, and R.L. Johnson, J. Non Cryst. Sol. 59 & 60, 329 (1983)
12. G.A. Baraff and M. Schlüter, Phys. Rev. B 19, 4965 (1979), and J. Bernholc, N.O. Lipari and S.T. Pantelides, Phys. Rev. B 21, 3545 (1980)
13. M. Scheffler, J. Bernholc, N.P. Lipari and S.T. Pantelides, Phys. Rev. B 29, 3269 (1984)
14. We use ionic pseudopotentials of M. Schlüter, J.R. Chelikowsky, S.G. Louie and M.L. Cohen, Phys. Rev. B 12, 4200 (1975)
15. H. Dersch, J. Stuke and J. Beichler, phys. stat. sol. (b) 105, 265 (1981)
16. J.D. Joannopoulos and D.C. Allan, Advances in Solid State Physics 21, 167 (1981)

STRUCTURAL ORDER AND THE URBACH SLOPE IN AMORPHOUS PHOSPHORUS*

L. J. Pilione, R. J. Pomian† and J. S. Lannin, Department of Physics,
The Pennsylvania State University, University Park, PA 16802

ABSTRACT

Extensive x-ray, Raman and optical absorption measurements have been performed on thin and thick rf sputtered a-P films. Low angle x-ray diffraction and Raman spectra on as-deposited films of variable pressure and temperature, as well as annealed films, indicate that changes in intermediate and short range order influence the optical gap. However, the Urbach slope is found to correspond to a range of optical gap values implying that this parameter is in general not a measure of network structural order.

INTRODUCTION

As an elemental noncrystalline solid amorphous phosphorus plays a special role in structural and physical property studies. This is based on the observation that a-P may be prepared in thin film form with an extensive range of degrees of disorder.[1-3] These studies, as well as related work[4-6] in a-As and a-Si and a-Ge, have demonstrated that the amorphous state is in general not characterized by a single local network order. Previous measurements in a-P have indicated that substantial variations of intermediate range order (IRO) are possible as functions of thin film sputter deposition conditions.[1] The presence of such IRO, whose presence and variability has been observed in x-ray diffraction and Raman scattering studies,[1,3] has been discussed in the context of a quasi-two dimensional, local layer-like model. While a detailed confirmation of a layer-like picture has not been fully demonstrated, the results to date imply some form of anisotropic structure that allows, to first order, a distinction between interunit and intraunit interactions and their associated responses.

It has recently been suggested[7] that a tubular model, similar to that originally suggested by Krebs and Gruber,[8] may also be applicable to bulk a-P. One difficulty with this model is the considerable angular and ring statistics constraints of a tubular unit relative to that of other amorphous solid models. In particular, it is somewhat difficult to see how the apparent continuium of structures obtained in thin films may be obtained in a simple tubular array model. A more natural model appears to be one with layer correlations which evolves in the limit of maximum disorder to continuous random network models that appear appropriate for disordered a-As and a-Sb.

Previous thin and thick film structural and optical property measurements in a-P involved films prepared by higher pressure argon rf sputtering. In these studies changes in substrate

*Supported by NSF Grant DMR-8402894
†Present address: Eaton Corporation, Melville, NY 11747

temperature were the predominant means of varying structural order and associated physical properties. More recent[3] Raman scattering measurements performed on rf sputtered, low pressure films indicate that changes in short range structural order may further occur due to bombardment effects in the sputtering process. The present work is concerned with the physical properties of films prepared under such conditions and their relation to more ordered higher pressure films. Of particular interest is the relationship of changes in structural order to the optical properties of a-P. The results presented below suggest that while the optical gap is related to network order, the Urbach slope parameter is not simply related to structural order as has been suggested for selected a-$Si_{1-x}H_x$ films.[9]

EXPERIMENTAL

Amorphous phosphorus films were prepared in a Materials Research Corporation rf-diode sputtering system at power-voltage settings of 100W-1100V. A 2" inch crystalline black-P target was used with a target to substrate distance of ~2.5" and argon gas pressure of 10 μ and 60 μ. Films were deposited onto polished single crystal silicon wafers for Raman, x-ray and infrared measurements and onto microscope slides for optical measurements. Substrate temperatures, T_s, ranging from -2°C to 177°C were determined directly with a Ag-Al thin film thermocouple. Film thicknesses (D = 0.5 to 15 μ) were measured with a Taylor-Hobson Talysurf. X-ray diffraction experiments were performed on the thick films with a Rigaku vertical goniometer from k = >0.5 to 5.5 $Å^{-1}$. The Si substrate background was subtracted to obtain corrected intensities, I(k). Raman scattering experiments were performed on the thick films at ~80°K utilizing the Kr 6764 Å laser line.

Optical measurements (400 nm to 1500 nm) were performed at room temperature with a Cary 14 double beam spectrometer (%T) and a Beckman DK-2A integrating sphere spectrometer (%R). The optical absorption coefficient (α) and the optical gap (E_0) were determined in a manner described in a previous paper.[2] Data measured on the thick films (D = 8 -15 μ) were used to find α from 10^2 cm^{-1} to 5 x 10^3 cm^{-1} and the thinner films (D < 1 μ) for values of α >5 x 10^3 cm^{-1}. The Urbach slope (Γ) was found by plotting ln α vs $h\nu$ for the thicker films deposited at the same argon pressure at different substrate temperatures or the same film at different annealing temperatures. The curves were linear over a substantial range and extrapolated to a common point (ln α_u,E_u) whose coordinates were used to find the value of the slope by: $\Gamma = \ln(\alpha/\alpha_u)/(E-E_u)$.

Upon completion of optical, Raman and x-ray measurements the thin and thick 10 μ samples were annealed in vacuum (10^{-5} T) up to a maximum temperature T_a = 250°C. This upper limit was dictated by the physical appearance of the films. The thicker films began peeling at ~200°C, the thinner films deteriorated at ~250°C.

RESULTS AND DISCUSSION

Substantial modification in structural order in a-P films is noted by large changes with deposition conditions of the low wave-vector diffraction spectra.[1] Indicated in Fig. 1 are the temperature dependence of I_1/I_2, which represents the ratio of the low angle diffraction peaks at 1.1 A^{-1} and 2.14 A^{-1}, respectively. Also shown is the variation of $1/\Delta k$, the full width at half maximum of the low angle peak. The X symbols represent the results on high pressure sputtered films, while the circles and triangles indicate low pressure data; the open symbols refer to the as deposited films, the solid symbols correspond to annealed films. The high pressure films indicate enhanced order for a fixed deposition temperature as indicated by larger I_1/I_2 and $1/\Delta k$ values. The rate of increase of $1/\Delta k$ with T_a of the low pressure films is similar, however, to that of high pressure films with T_s. In contrast, I_1/I_2 increases more rapidly, particularly above ~100 C, in as-deposited high pressure films. This suggests that intraunit intermediate range order, which I_1/I_2 is sensitive to, is more dependent on deposition conditions. Subsequent annealing thus requires greater structural relaxation to achieve comparable order. The similar behavior of $1/\Delta k$ with T_s or T_a may indicate that interunit (layer) correlations are more readily relaxed. This is reasonable, given weak interlayer coupling relative to more strongly covalently bonded nearest neighbors in a-P. It is useful to note that a more substantial reduction of order relative to the results in Fig. 1 has been observed in dc sputtered a-P films deposited at ~90 K.[3]

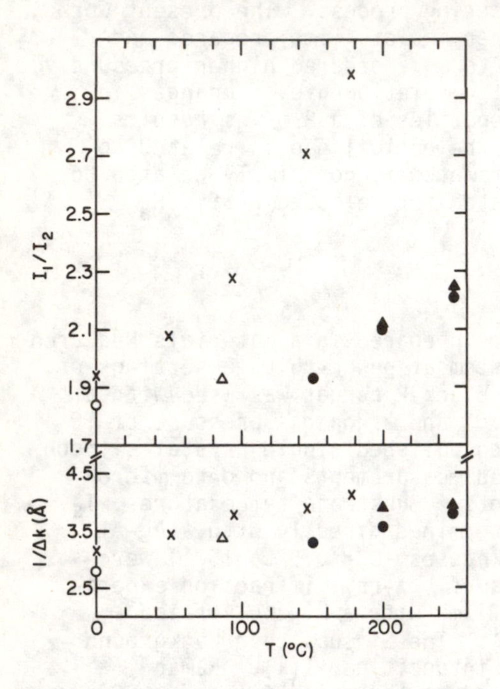

Fig. 1. Temperature dependence of the low k diffraction peak intensity, I_1/I_2 and width, $1/\Delta k$. X represents 60 μ film variation with T_s, open symbols represent 10 μ film T_s data and solid symbols indicate T_a data on 10 μ films of a-P.

Raman scattering measurements have also been shown to be highly sensitive to changes in network order in a-P *within* structural units. Previous studies[1,3] have noted changes in the depolarized and polarized spectra, I_{VH} and I_{HH}, which have been attributed to variations in intermediate as well as short range order. Figure 2 indicates an example of the very large range of variations of HH spectra that have been observed between highly ordered, high temperature and pressure rf sputtered films and those prepared at lower temperatures or pressure. It has been suggested[3] that in the former changes in intermediate range order may account for variations in the Raman spectra, while in the latter modifications of short range order have also occurred. Such changes in short range order are attributed to the increased width of the bond angle distribution. In the limit of maximum disorder, observed in low T_s dc sputtered films, the broad, relatively structureless spectra qualitatively resembles that of sputtered a-As and a-Sb.[4] This suggests the absence of any substantial IRO and the possible description of the structure using a CRN model similar to that developed for a-As.[10] The intermediate character of the Raman spectra of low pressure sputtered films as well as high pressure films deposited at low temperatures suggests that changes in intermediate and short range order may both influence the Raman spectra.[3]

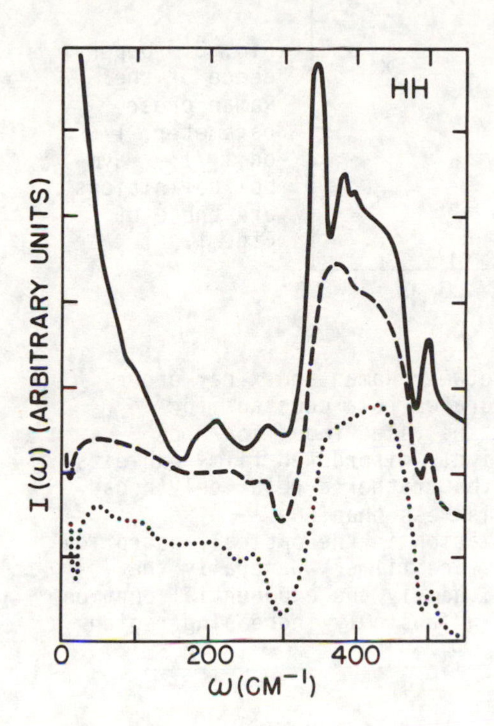

Fig. 2. Variation of HH Raman spectra of a-P with deposition.
Solid: T_s = 177°C, P_{Ar} = 60 μ,
Dashed: T_s = 0°C, P_{Ar} = 60 μ,
Dotted: T_s = -2°C, P_{Ar} = 10 μ.

As can be seen in Fig. 2 films deposited under different conditions indicate a number of detailed shape changes, although certain overall similarities occur in terms of major band positions, widths and minima. Qualitatively, the enhancement of the peak at ~350 cm^{-1} with argon pressure and T_s or T_a is consistent with the increase in structural order as observed in the x-ray data. While it is desirable to obtain a measure of order from the Raman spectra it is difficult to deconvolute spectra where "special" vibrational modes are apparently changing within a fixed band width. As a first approximation, an order parameter was determined from the moment, $\bar{\omega}$, of the major band between 300 - 550 cm^{-1}. This procedure does not rely on a detailed knowledge of the changes occurring within a vibrational band but is sensitive to their shift to lower frequency.[11] In Fig. 3 the inverse moment, which functions as a vibrational order parameter, is plotted vs. I_1/I_2 for the 10 μ and 60 μ films. The results

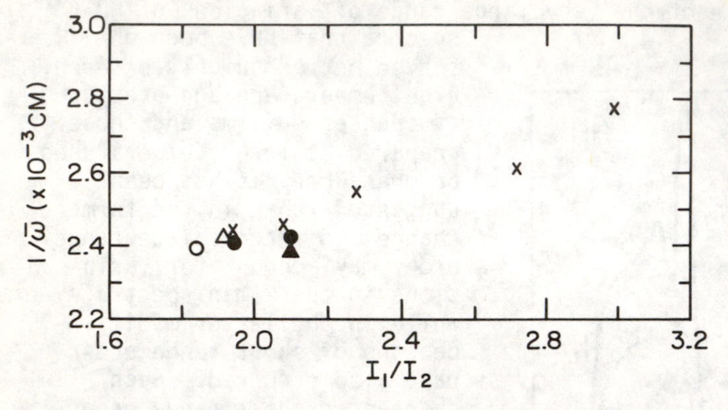

Fig. 3. Dependence of the Raman order parameter, $1/\bar{\omega}$ on I_1/I_2. Symbol definitions are those of Fig. 1.

indicate that a correlation exists between Raman and x-ray order parameters. It appears that $\bar{\omega}^{-1}$ saturates to a constant for $I_1/I_2 < 1.8$ which is consistent with the value found for a dc sputtered film made at ~90 K. This highly disordered film[3] exhibits little well-defined Raman structure, but rather a relatively broad two-band spectrum whose I_1/I_2 value is less than one.[12]

Figures 4a,b illustrate the variation in the optical absorption coefficient for the low and high pressure films. Generally the spectra have several similar regions, namely the exponential portion ranging from 2×10^2 - 3×10^3 cm^{-1} and a slowly increasing region

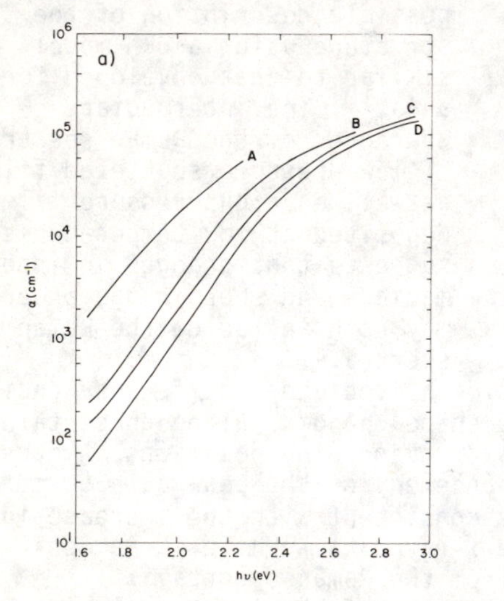

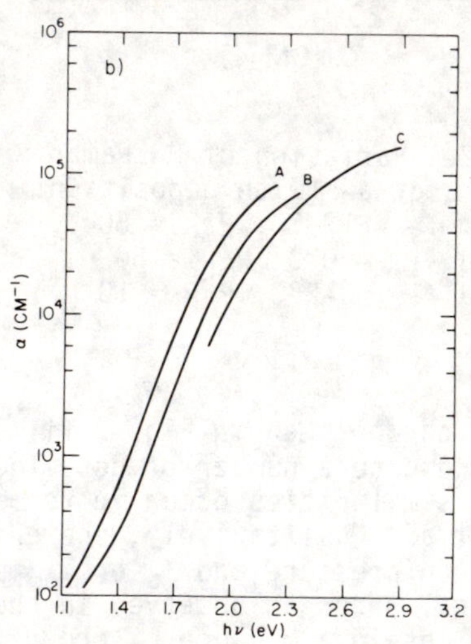

Fig. 4. Optical absorption coefficient versus $h\nu$. a) $P_{Ar} = 60$ μ, $T_A = 8°C$, $T_B = 64°C$, $T_C = 126°C$, $T_D = 180°C$; b) $P_{Ar} = 10$ μ; $T_A = -2°C$, $T_B = 84°C$, $T_C = 127°C$. Note scale differences in a) and b).

at higher values (5×10^4 - 2×10^5 cm^{-1}). The former region (where $\Delta\alpha \simeq \pm 20$ cm^{-1}) was used to determine the value of the Urbach slope whereas the optical band gap was obtained for the latter by Tauc's method with values of $\alpha > 5 \times 10^4$ cm^{-1}. In Fig. 5 the optical gap, E_0, is

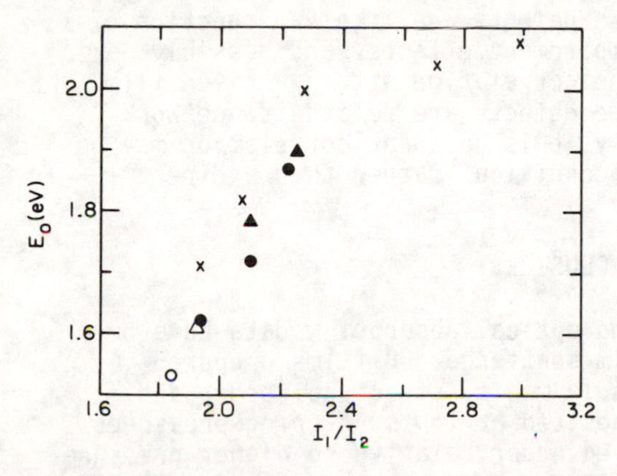

Fig. 5. Variation of the optical gap with a low angle diffraction intensity in a-P. Symbol definitions are those of Fig. 1.

plotted vs. I_1/I_2. For values of $I_1/I_2 < 2.6$ a correlation exists implying that E_0 is related to structural order. As I_1/I_2 is a measure of intermediate range order, which itself may grow as the short range order increases, E_0 may be a function of both SRO and IRO. For values of $I_1/I_2 > 2.6$ the optical gap increases very slowly suggesting a weak dependence on IRO. This weaker dependence of E_0 on I_1/I_2 is reasonable given the estimated structural correlation range ($>$ 11 Å)[13] and the limited range of interactions expected to influence the electronic states. Since E_0 *does not* correlate with $1/\Delta k$, this suggests that within a layer-like model E_0 is determined by intralayer effects. This is physically plausible and emphasizes the anisotropic character of optical absorption for a-P which parallels that of crystalline two-dimensional systems.

Figure 6 illustrates the variation of the inverse Urbach slope ($1/\Gamma$) with optical gap. For a given film there appears to be

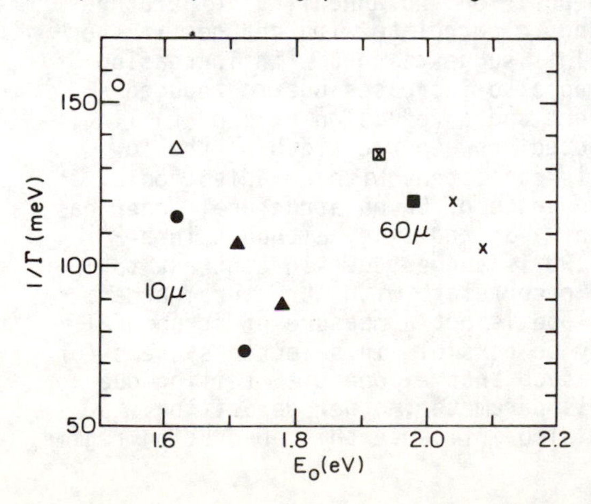

Fig. 6. Comparison of the Urbach slope, $1/\Gamma$ with the optical gap, E_0 in P_{Ar} = 10 μ and 60 μ films. Symbol definitions are those of Fig. 1. The squares represent a 60 μ film deposited at 72°C, ⊠ and subsequently annealed at 200°C, ■ .

a correlation between E_0 and $1/\Gamma$ but there are wide variations in E_0 for a fixed $1/\Gamma$, especially between the 10 μ and 60 μ films. This indicates that $1/\Gamma$ is not a useful structural order parameter as has been suggested in selected a-$Si_{1-x}H_x$ alloys.[9] Although the Urbach slope has in theory an intrinsic contribution due to localized state processes, extrinsic "defect" contributions may also influence its value. These defects are likely a function of film deposition, especially bombardment effects, and possibly structural micromorphology. The correlation within a given film may imply that a fraction of the defects are related to network structural order. Alternatively, this apparent correlation may be due to deposition or annealing conditions rather than a direct structural one.

CONCLUSIONS

Structural, vibrational and optical absorption data have been presented on thin and thick film sputtered a-P films prepared under conditions that considerably modify structural order and physical properties. Films deposited at low sputtering pressures have been found to have a reduced order relative to higher pressure films. This is a consequence of increased substrate bombardment during film deposition. These changes have been attributed to modifications of intermediate as well as short range order. The progressions observed in the x-ray diffraction and Raman spectra suggest a quasicontinuum of structural variants. The most highly disordered films studied by in situ measurements on low temperature substrates indicate Raman spectra similar to that of a continuous random network type system. This suggests a structural model which evolves from an anisotropic, local layer-like character with intermediate range order, to one without appreciable order beyond the short range. While local tube-like configurations can not be ruled out, the significant constraints imposed on the dihedral angles and ring statistics required to form tubular units, make such a model improbable for a-P.

Optical absorption measurements yield large variations in the optical gap as a function of deposition and annealing temperatures. The optical gap values are found to correlate with changes in the x-ray order parameter I_1/I_2. This suggests that with increasing structural order the optical gap also increases due to reduced fluctuations in local enviornment and a reduction of band tailing. The *absence* of a correlation noted between the width of the low angle diffraction peak and E_0 is consistent with an anisotropic, layer-like local model. A dependence of E_0 on structural order has also been noted[5,6] in studies of a-Si and a-Ge, although in a-P some additional dependency on IRO is suggested. In contrast to E_0, the Urbach slope is <u>not</u> found to correlate with structural order. This implies that the Urbach slope is not a measure of structural order in general. While it may be possible in selected systems for intrinsic effects to dominate, such that exponential tailing due to localized states determines this parameter, other deposition dependent defect processes may also influence this absorption regime.

REFERENCES

1. B. V. Shanabrook and J. S. Lannin, Phys. Rev. B24, 4771 (1981).
2. L. J. Pilione, R. J. Pomian and J. S. Lannin, Solid State Commun. 39, 933 (1981).
3. R. J. Pomian, J. S. Lannin and B. V. Shanabrook, Phys. Rev. B27, 4887 (1983).
4. J. S. Lannin, Phys. Rev. B13, 3863 (1977).
5. J. S. Lannin, L. J. Pilione, S. T. Kshirsagar, R. Messier and R. C. Ross, Phys. Rev. B26, 3506 (1982).
6. N. Maley, L. J. Pilone, S. T. Kshirsagar and J. S. Lannin, Physica 117B & 118B, 880 (1983).
7. N. B. Goodman, L. Ley and D. W. Bullett, Phys. Rev. B27, 7440 (1983).
8. H. Krebs and H. U. Gruber, Z. Naturf. 22a, 96 (1967).
9. G. D. Cody, T. Tiedje, B. Abeles, B. Brooks and Y. Goldstein, Phys. Rev. Lett. 47. 1480 (1981).
10. G. N. Greaves, E. A. Davis and J. Bordas, Phil. Mag. 34, 265 (1976).
11. Reference 1 gives an alternative Raman order parameter in which the relative intensities at 350 cm^{-1} and 440 cm^{-1} are compared.
12. B. V. Shanabrook, Ph.D. Thesis, The Pennsylvania State University, 1981 (unpublished).
13. J. S. Lannin, B. V. Shanabrook and F. Gompf, J. Non-Cryst. Solids 49, 209 (1982).

ELECTRONIC STRUCTURE OF AMORPHOUS SEMICONDUCTOR HETEROJUNCTIONS BY PHOTOEMISSION AND PHOTOABSORPTION SPECTROSCOPY

B. Abeles, I. Wagner, W. Eberhardt, J. Stöhr and H. Stasiewski
Exxon Research and Engineering Company, Annandale, N.J. 08801

F. Sette*
Brookhaven National Laboratory, Upton, NY 11973

ABSTRACT

Heterojunctions of a-Si:H/a-$Si_{1-x}C_x$:H and a-Si:H/a-SiN_x:H, deposited in situ by RF plasma CVD, were investigated by photoemission and photoabsorption using synchrotron radiation. The valence band maximum of a-Si:H coincides with that of a-Si_xC_{1-x}:H and it is 1.4 $\pm$ 0.2 eV above the VBM of a-SiN_x:H. As the a-Si:H layer thickness decreases from 800Å to 5Å, the energy bandgap, determined from the difference between the $L_{2,3}$ edge, the Si-2p binding energy and VBM, becomes progressively smaller than the optical bandgap. This suggests that excitonic effects on the $L_{2,3}$ edge are enhanced as the dimensionality is reduced from 3D to 2D.

I. INTRODUCTION

Semiconductor heterojunctions are important because of their use in compositionally modulated superlattices and in electronic devices such as LED's, lasers, solar cells, and field effect transistors. While a great deal of information is available for crystalline semiconductor heterojunctions,[1] relatively little is known in the case of amorphous systems. The information that is of interest is the relative position of the conduction bands, valence bands and core levels of the two members of the heterojunction. Another question of considerable scientific importance is how the energy bands behave in ultra thin films. It has been found that the optical band gap of hydrogenated amorphous silicon (a-Si:H) layers in a superlattice[2,3,4] formed with alternating a-SiN_x:H layers and in ultra thin films[5] increases appreciably when the layer thickness is reduced below 40Å and it was proposed that the increase in bandgap is due to quantum size effects. In this paper we present results on heterojunctions of a-Si:H with a-SiN_x:H and a-Si_{1-x} C_x:H and on size effects in a-Si:H.

II. EXPERIMENTAL

The most direct way to determine the electronic structure of heterojunctions is by photoemission spectroscopy.[1,] In this experiment (insert in Fig. 1) the top layer of the heterojunction is thin enough (5-10A) for photoelectrons to be

0094-243X/84/1200394-08 $3.00

emitted from both materials. The amorphous semiconductor films were prepared by plasma assisted chemical vapor deposition (PCVD) on polished stainless steel substrates held at 240°C on the anode of a 13.56 Hz capacitive reactor. The compositions of the gases used to make the films were: pure SiH_4 for a-Si:H; mixture 1:5 in volume SiH_4 to NH_3 for a-SiN_x:H and mixture 1:4 SiH_4 to CH_4 for a-$Si_{1-x}C_x$:H. The heterojunctions were fabricated by depositing first the bottom film and then, without interrupting the plasma, depositing the top overlayer by changing rapidly the composition of the gas in the reactor. To ensure the necessary condition for abrupt interfaces, the gas exchange time (~1 sec) was short compared to the time it took to grow a monolayer (~3 sec). The thicknesses of the layers were determined from the deposition rates measured on thick films with an accuracy of $\pm$ 10%. The concentration of H, determined by nuclear reaction with N^{15}, in specially prepared thick films were: 12 at% (a-Si:H) and 28 at% (a-SiN_x:H). The measured optical gaps, E_G, were 1.75 eV (a-Si:H) 3.9 eV (a-SiN_x:H) and 2.2 eV (a-$Si_{1-x}C_x$:H) The deposition conditions were optimized to produce a low density of defects in a-Si:H. Details of the deposition have been discussed previously.[2,6] The samples were transfered under UHV from the reactor to the measurement chamber. The photoemission measurements were made using synchrotron radiation from the storage ring SPEAR at Stanford Synchrotron Radiation Laboratory.

III. RESULTS

The photoemission spectra of the valence bands (VB) of several heterojunctions as well as of the individual films are shown in Figs. 1-4. In all cases the photon energy used was 55 eV and energies are referred to the Fermi level. The Fermi level was determined by measurement of an Au foil. Figure 2 (curve 1) shows the VB of a 7Å thick overlayer of a-SiN_x:H on top of a-Si:H film. The spectrum resembles closely that of a-SiN_x:H (Fig. 2 curve 4) with the exception of the VB top (Fig. 1) which has a broad tail due to emission from the underlaying a-Si:H film. The peak at 20 eV in Fig. 2 is due to N-2s core states. To fit the observed VB we used a linear combination of the a-Si:H and a-SiN_x:H spectra with the weighing factors and energy shifts as adjustable parameters. We used the narrow, prominent peak at 5 eV (Fig. 2) to fit the a-SiN_x:H component of the spectrum and the top of the valence band edge to fit the a-Si:H component. An excellent fit is obtained over the entire valence band with an offset between the VB maxima $\Delta E_v = 1.4 \pm 0.15$ eV. The VBM is determined by the extrapolation of the steepest descent of the leading edge of the VB. An equally good fit with $\Delta E_v = 1.4 \pm 0.2$ eV was obtained on a heterojunction consisting of a 2.5 Å a-SiN_x:H overlayer on a-Si:H.

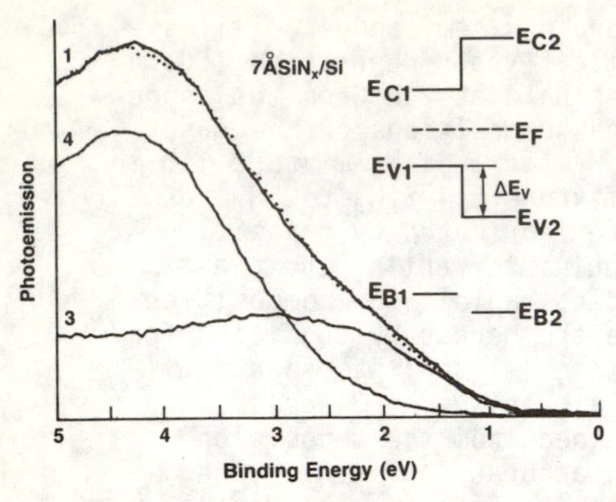

Fig. 1: Photoemission from top of valence band: (1) heterojunction 7Å a-SiN_x:H on top of a-Si:H; (3) 800Å a-Si:H, (shifted by -0.1 eV) (4) 40Å a-SiN_x:H (shifted by -0.2 eV), (....) sum of (3) and (4). Insert: heterojunction between two semiconductors; E_C and E_V are conduction and valence band energies, E_B are core energies. ΔE_V is offset between valence bands.

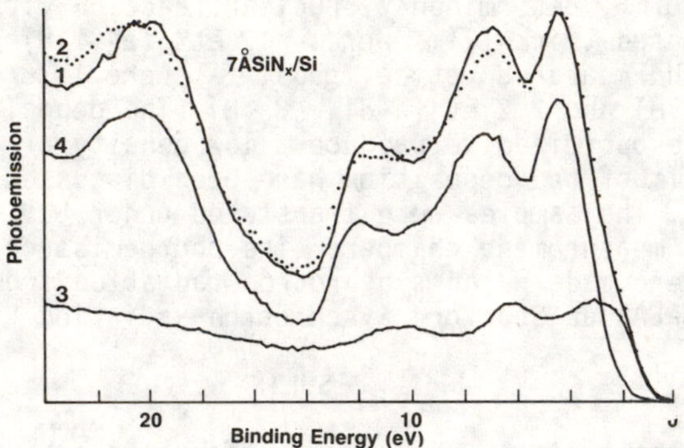

Fig. 2: Valence band photoemission: (1) heterojunction 7 Å a-SiN_x:H on top of a-Si:H. (3) 800Å a-Si:H (shifted by -0.1 eV); (4) 40Å a-SiN_x:H (shifted by -0.2 eV), (2) sum of (3) and (4).

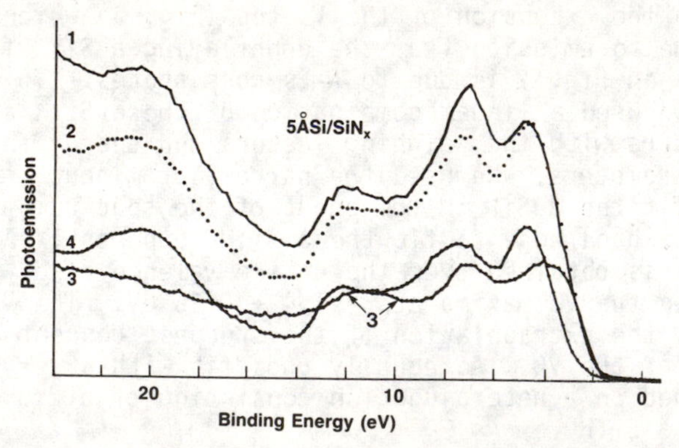

Fig. 3: Valence band photoemission: (1) heterojunction 5Å a-Si:H on top of a-SiN_x:H; (3) 800Å a-Si:H (shifted by 0.6 eV, (4) 40Å a-SiN_x:H (shifted by - 0.2 eV); (2) sum of (3) and (4).

Figure 3 shows the VB of a heterojunction in which the thin overlayer is 5Å a-Si:H and the bottom film is 40Å a-SiN_x:H. Using the same procedure as in the previous two cases we determined $\Delta E_v = 0.85 \pm 0.15$ eV. We note that in this case the fit is satisfactory only down to 5 eV. The apparent enhancement of the peak at 7 eV and at higher binding energies might be due to a H enrichment near the surface. In a heterojunction with a thicker 9Å a-Si:H overlayer on a-SiN_x:H we observed ΔE_v to be 1.4 ± 0.20 eV. Fig. 4 shows a 6Å a-Si:H overlayer on a-$Si_{1-x}C_x$:H. In this heterojunction and in the case of a 5Å a-$Si_{1-x}C_x$:H overlayer on a-Si:H we determined $\Delta E_v = 0 \pm 0.15$ eV in agreement with Evangelisti[7].

In Fig. 5 are shown the Si-2p core level spectra referred to E_F and measured at a photon energy of 115 eV in (111) c-Si (cleaned by heating and showing a clear 7 x 7 lead pattern), 800Å a-Si:H film and 9Å and 5Å a-Si:H overlayers on a-SiN_x:H. The c-Si clearly reveals the spin 1/2 and 3/2 components with a spin-orbit splitting of $\Delta = 0.6$ eV. The broadening of the peak is consistent with our instrumental resolution. The peak becomes broader in the a-Si:H films and is shifted to higher binding energies. The 5Å Si/SiN_x heterojunction shows clearly a broad peak originating from the Si-N bonds in the underlayering a-SiN_x:H film. To obtain more quantitative information on the core level binding energies we fitted the spectra with gaussian doublets:

$$\exp\left[-(E-E_{Bi})^2/2\sigma_i^2)\right] + \gamma \exp\left[-(E-E_{Bi}-\Delta)^2/2\sigma_i^2\right] \quad (1)$$

In the case of the c-Si we used one doublet (i=1) with $\gamma = 0.60$, and $\Delta = 0.6$ eV and $E_{B1} = 99.7$ eV as best fit parameters. To obtain a good fit in the case of the a-Si:H films required two gaussian doublets (i=1, 2) where for γ and Δ we used the c-Si values. The second gaussian doublet (i=2) represents the contribution due to chemically shifted core levels (Si-H and Si-N bonds). Modeling the spectrum with four gaussian doublets[8] did not significantly change the fit nor the values of E_{B1}. In the following we use E_{B1} as the reference level for the L_{23} edge and the VBM.

Figure 6 shows the $L_{2,3}$ photoabsorption edges, determined by partial electron yield spectroscopy in the same materials as in Fig. 5. The $L_{2,3}$ absorption is due to transitions from the Si-2p levels to conduction band states modified by a core hole exciton interaction.[9,10] The spin orbit splitting is resolved in all four spectra and the arrows indicate the transition energies, E_T, from the core levels centered at E_{B1}. From the fact that the L_{23} edge is well defined even in the 5Å thick layer we conclude that the interface with a-SiN_x:H is close to atomically abrupt.

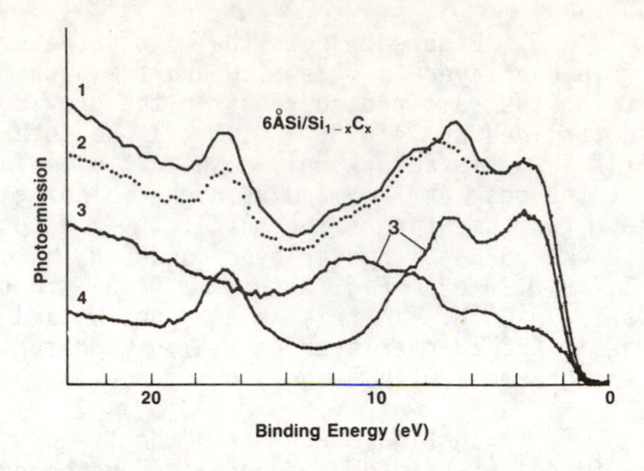

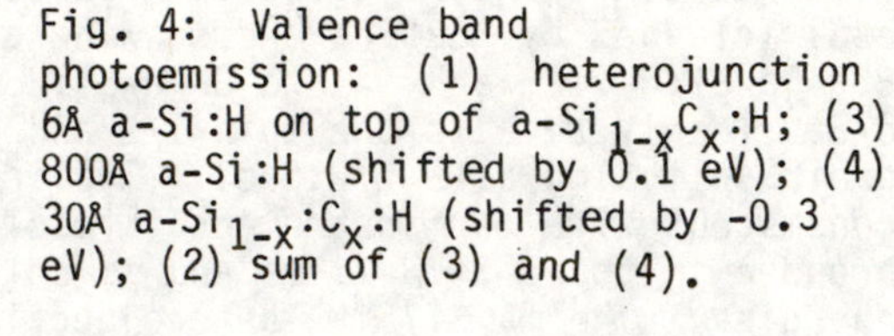

Fig. 4: Valence band photoemission: (1) heterojunction 6Å a-Si:H on top of a-$Si_{1-x}C_x$:H; (3) 800Å a-Si:H (shifted by 0.1 eV); (4) 30Å a-Si_{1-x}:C_x:H (shifted by -0.3 eV); (2) sum of (3) and (4).

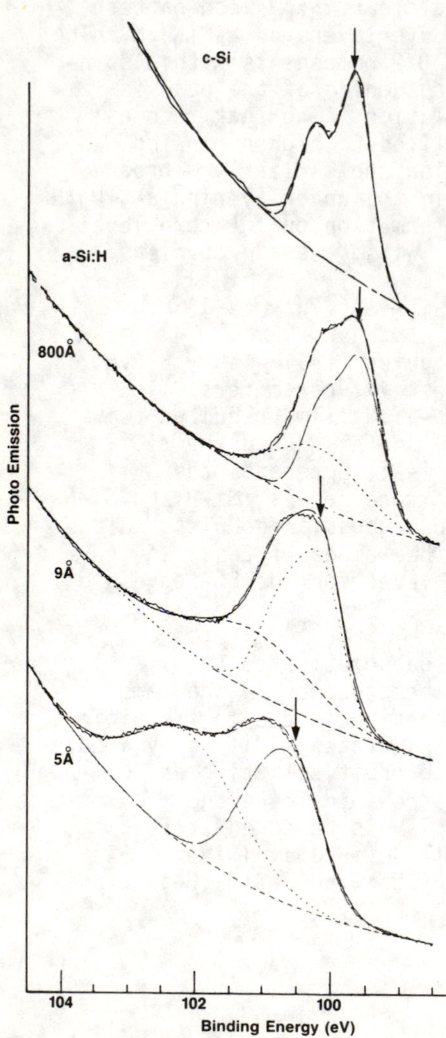

Fig. 5: Photoemission from Si-2p core states measured on c-Si, 800Å a-Si:H film and 9Å and 5Å overlayers of a-Si:H on a-SiN_x:H. Photon energy 115 eV; binding energies measured relative to E_F. The experimental data is given by the full curves. The dashed curves are the secondary electron background and fits given by Eq. (1). The arrows indicate the energy E_{B1}.

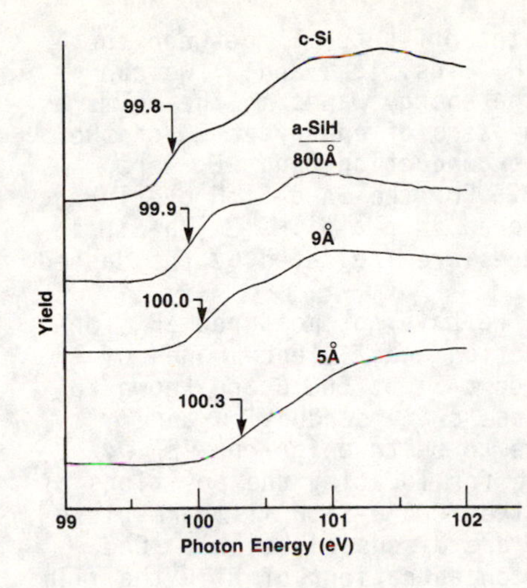

Fig. 6: $L_{2,3}$ absorption edge vs photon energy measured by partial yield, on c-Si, 800Å a-Si:H film and 9Å and 5Å overlayers of a-Si:H on a-SiN$_x$:H. The arrows indicate the transitions from E_{B1}.

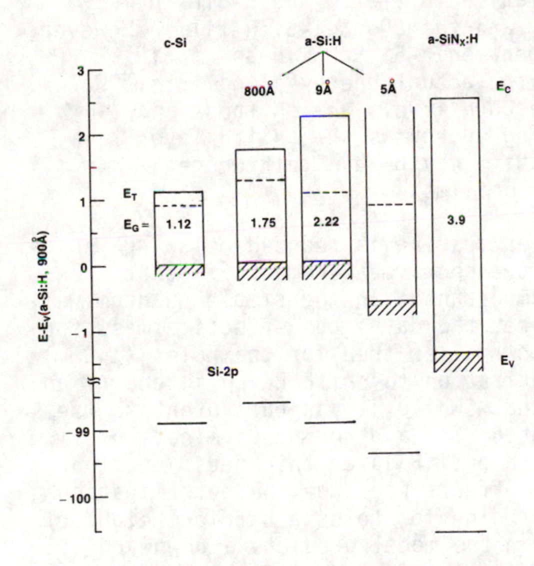

Fig. 7: Conduction band minimum, E_c, valence band maximum, E_v, $L_{2,3}$ edge energy E_T and Si-2p core levels E_{B1} of c-Si, 800Å a-Si:H film, 9Å and 5Å overlayers of a-Si:H on a-SiN$_x$:H, and a thick film of a-SiN$_x$:H. Zero of energy at VBM of 800Å a-Si:H.

IV. DISCUSSION

Using the information on the off-sets ΔE_v between the valence bands of a-Si:H and a-SiN_x:H (Figs. 1, 2 and 3) we can compare the relative positions of the energy bands in the films as shown in Fig. 7. We have placed the zero of energy at E_v of the 800Å a-Si:H film. The minima of the conduction bands E_c were placed at energies E_G above the VBM. For the 9Å a-Si:H overlayer we used the value E_G = 2.22 eV measured in a a-Si:H (8Å)/a-SiN_x:H (35Å) superlattice[2,3]. The $L_{2,3}$ edges are indicated by the dashed lines designated E_T.

In the present experiment we have not measured ΔE_v for the case of c-Si, however the conduction and valence bands of Si_3N_4 (E_G = 5 eV) deposited by CVD on c-Si at 500°C are known to be alined nearly symetrically with the c-Si conduction and valence bands.[11] We assume the same to be true for our (E_G ~4 eV) a-SiN_x:H. Using the above procedure for locating the positions of E_v, the values of ΔE_v for c-Si and the 800Å a-Si:H film are close to one another (the 9Å and 5Å films are discussed below). One would expect the same ΔE_v when the concentrations of H in the film is relatively low, as it is in our case, and when ΔE_v is determined primarily by nearest neighbor bonding at the interface and not by long range order. Another resonable scheme for setting the energy levels of c-Si relative to a-Si:H and a-SiN_x:H is to aline the core levels of c-Si and the 800 Å a-Si:H film. However, only direct measurements of between c-Si, a-Si:H and a-SiN_x:H heterojunctions will provide the actual energy alinements. We note in Fig. 7 that the Si-2p core levels are shifted progressively to higher binding energy as the a-Si:H layer thickness is reduced. This shift may be due to hydrogen enrichment (Fig. 3) in the thin films.

When the a-Si:H layer spacing is reduced below 40Å there is evidence from optical and transport measurements[2,3] that quantum size effects become important. In the simple quantum well model that was used to interpret the data[3] the effective mass for the electrons (0.2 m) was much smaller then for the holes (1.0 m) so that one would expect quantization to shift E_c up in energy and leave E_v essentially unaffected. While it was our intent to use photoemission and photoabsorption for a direct determination of the variation of E_v and E_c with a-Si:H layer thickness, we find that strong core hole exciton effects preclude the determination of E_c. We first discuss E_v. Using for holes a barrier height of 1.4 eV and effective mass 1.0 m the model predicts a downward shift in E_v of 0.1 and 0.4 eV for the 9Å and 5Å films respectively. From Fig. 7 the difference in E_v between the 800Å and 9Å thick films is 0 ± 0.3 eV. When the a-Si:H layer is reduced to 5Å, this difference becomes 0.55 ± 0.3 eV. Apart from the rather large experimental uncertainty these shifts in E_v are consistent with the model.

The steepness and the near threshold enhancement of the $L_{2,3}$ absorption edge in c-Si, a-Si and a-Si:H have been ascribed to a strong coulomb interaction between core holes and conduction electrons.[9,10] While discrete exciton lines are not resolved in the $L_{2,3}$ edge an exciton binding energy can be defined by $E_{ex} = E_C - E_T$. From Fig. 7 the values of E_{ex} are 0.46 $\pm$ 0.2 and 1.2 $\pm$ 0.2 eV for 800Å and 9Å a-Si:H films respectively. A simple explanation for the increase in E_{ex} with decreasing a-Si:H layer thickness is that the exciton energy increases when the dimensionality is reduced from 3D to 2D. In the case of the weakly bound valence hole excitons[12] the increase in E_{ex}, when the dimensionality is reduced from 3D to 2D, is a factor of 4.

In conclusion we have shown that the VB photoemission spectra of the heterojunctions behave as a superposition of the VB spectra of the individual components of the heterojunction. From this, as well as from the fact that the $L_{2,3}$ edge has a well defined a-Si:H character in a 5Å thick layer, we conclude that the interfaces are atomically abrupt. The VBM of a-Si:H is 1.4 $\pm$ 0.2 eV above that of a-SiN_x:H and coincides with the VBM of a-$Si_{1-x}C_x$:H. The core hole exciton binding energy increase with decreasing a-Si:H layer thickness is attributed to the reduction in dimensionality from 3D to 2D. We wish to thank the staff of SSRL and J. Feldhaus for valuable assistance, W. Varady and S. Constantino for help with computer programing W. Lanford for hydrogen analysis and T. Tiedje and P. Persans for helpful discussions.

* Present address, AT&T, Bell Laboratories, Murray Hill, N.J.

1. A. D. Katnani and G. Margaritondo Phys. Rev. B28, 1944 (1983).
2. B. Abeles and T. Tiedje, Phys. Rev. Lett. 51, 2003 (1983).
3. T. Tiedje, B. Abeles, P. D. Persans, B. G. Brooks and G. D. Cody, J. Non-Cryst. Solids 66, 351 (1984).
4. J. Kakalios, H. Fritzsche, N. Ibaraki and S. R. Ovshinski, J. Non-Chryst. Solids 66, 339 (1984).
5. H. Munekata and H. Kukimoto, Jap. J. of Appl. Phys. 22, L542 (1983).
6. B. Abeles, T. Tiedje, K. S. Liang, H. W. Deckman, H. C. Stasiewski, J. C. Scanlon and P. M. Eisenberger, J. Non-Cryst. Solids 66, 351 (1984).
7. F. Evangelisti, P. Fiorini, C. Giovannella, F. Patella, P. Perfetti, C. Quaresima and M. Capozi. Appl. Phys. Lett 44, 764 (1984).
8. L. Ley, J. Reichardt and R. L. Johnson, Phys. Rev. Lett., 49, 1664 (1982).
9. F. C. Brown, R. Z. Bachrach and M. Skibowski, Phys. Rev. B15, 4781 (1977).
10. J. Reichardt, L. Ley and R. L. Johnson Proceedings of 10th International Conf. on Amorphous and Liquid Semicon. p.329, North Holland Publishing Co. Amsterdam, 1983.
11. A. M. Goodman, Appl. Phys. Lett. 13, 275 (1968).
12. B. A. Vojak, N. Holonyak and W. D. Kaidig Solid State Comm. 35, 447 (1980).

PHOTOEMISSION STUDIES OF AMORPHOUS SILICON HETEROSTRUCTURES

F. Patella
Dipartimento di Fisica, II Universita´ di Roma,
Tor Vergata, Roma, Italy

F. Evangelisti and P. Fiorini
Dipartimento di Fisica, Universita´ "La Sapienza",
00185 Roma, Italy

P. Perfetti and C. Quaresima
ISM, Via E. Fermi, 00044 Frascati (Roma), Italy

M. K. Kelly, R. A. Riedel and G. Margaritondo
Department of Physics, University of Wisconsin,
Madison, WI 53706, USA

ABSTRACT

We investigated the heterostructures a-Si/a-$Si_{0.5}C_{0.5}$:H, a-Si/c-Si and a-Si/a-Si:H by photoemission techniques. In particular, we measured the band bending changes during the interface formation and this enabled us to unambiguosly determine the valence band discontinuity, ΔE_v, of each interface. The ΔE_v´s were found to be negligible for all these heterostructures. We discuss the relevance of these results to the understanding of the a-Si p-i-n solar cell behavior and of the role of hydrogen in widening the pseudo-gap of a-Si.

INTRODUCTION

Much work has been dedicated in recent years to the microscopic study of crystalline semiconductor interfaces by photoemission techniques.[1] This effort is justified both by fundamental reasons and by the widespread applications of semiconductor interfaces in solid-state electronics. Particularly important is the substantial progress made in understanding heterojunction interfaces, for example in clarifying the origin of the band discontinuities and the role of the interface states. The complete understanding of heterojunction interface parameters such as the built-in potential and the band discontinuities is a necessary first step in the explanation and improvement of the performances of the corresponding devices.

For similar reasons, interfaces involving amorphous semiconductors are of extreme interest. Very few microscopic experiments, however, have been dedicated to these fundamental systems. This interesting field of interface research is still in its first stages of development -- while applied research on amorphous semiconductors has already produced good results, e.g. in the areas of photovoltaic detection[2] and superlattices.[3]

The microscopic study of a heterojunction is performed by following the evolution of valence band and core level photoemission spectra step-by-step during the interface formation process. The interface is formed by growing in situ one semiconductor on top of the other. Due to the surface sensitivity of photoemission spectroscopy, this procedure enables one to directly determine the band discontinuities, the position of the Fermi level at the interface and the magnitude of the band bending effects. This approach has been extensively applied to the study of crystalline heterojunctions and of heterojunctions involving a crystalline substrate and an amorphous overlayer.[1] Recently, the first results were obtained on heterojunction interfaces involving two amorphous semiconductors.[4]

One particularly interesting byproduct of the research on amorphous heterojunctions is the possibility to investigate the effects of disorder and/or hydrogenation on the parameters of amorphous semiconducting materials, e.g. on the band edges and on the forbidden gaps. This is accomplished by studying pseudo-heterojunctions obtained by growing an amorphous overlayer of a given material on a crystalline or hydrogenated-amorphous substrate of the same material.

In the present work we studied two different kinds of heterostructures. First, heterojunctions between amorphous silicon (a-Si) and an amorphous hydrogenated silicon-carbon alloy (a-Si_xC_{1-x}:H). These systems are of great technological interest for the production of high-efficiency solar cells. A detailed study of the core levels of this system confirmed the preliminary conclusions of Ref. 4, i.e. that the high efficiency of these solar cells is explained by their negligible valence band discontinuity.

The second kind of heterostructures, a-Si/c-Si (c-Si = single-crystal silicon) and a-Si/a-Si:H, yielded information on the effects of disorder and hydrogenation on the valence band edge. We found that these factors do not influence the absolute position in energy of the valence band edge. Therefore, the hydrogen-induced widening of the optical gap is due to a shift of the conduction band edge.

EXPERIMENTAL

The measurements were performed at the University of Wisconsin Synchrotron Radiation Center, using its storage ring Tantalus and a grazing-incidence "Grasshopper" monochromator. The photoelectron energy distribution curves were measured with a PHI double-pass cylindrical-mirror analyzer at photon energies $h\nu = 40$ and 135 eV. The overall resolution (monochromator plus electron analyzer, FWHM) was 0.2 eV at $h\nu = 40$ eV and 0.65 eV at $h\nu = 135$ eV.

The c-Si substrates were obtained by cleaving "in situ" (at a working pressure in the low 10^{-10} Torr range) single crystal samples. The a-Si films were deposited in situ by electron bombardment. The hydrogenated silicon samples and the silicon-carbon alloys were grown by RF glow-discharge in a capacitive reactor from pure silane or from a mixture of 70% CH_4 and 30% SiH_4. One of the Si-C alloys was p-doped by adding 0.1% B_2H_6 to the discharge gas. The substrates grown by glow discharge were cleaned in situ by Ar ion sputtering. In one case hydrogen was added to argon to compensate the possible preferential removal of hydrogen from the sample during the sputtering process. No difference, however, was found between the results given by this sample and those given by the samples cleaned by sputtering in a pure argon atmosphere. The composition of the silicon-carbon alloys was determined after the sputtering by Auger spectroscopy and found to be x = 0.5. Infrared absorption measurements detected a hydrogen content of about 10%.

RESULTS AND DISCUSSION

The evolution of the valence band density of states during the formation of a a-Si/a-$Si_{0.5}C_{0.5}$:H heterostructure is shown in Fig. 1 for increasing a-Si coverages. These spectra were aligned with respect to each other to compensate for the changes in band bending during a-Si deposition. The alignment was done by bringing the leading spectral edges determined by linear extrapolation to the same position in energy. The corresponding position of the Fermi level, E_F, is shown at the right-hand-side of each curve. The main structure 6-8 eV below the top of the clean-substrate valence band, E_v, is due to the sp states of silicon and to the 2p states of carbon, together with the Si-H bonding states. Notice the rather low density of states at the top of the valence band of the clean substrate. The density of states in this reagion increases with the a-Si coverage due the 3p band of silicon. Furthermore, the a-Si coverage introduces a new Si-3s-derived feature ~ 12 eV below E_v. The top spectrum in Fig. 1 is typical of bulk a-Si. In particular, the intensity of the sp band at ~-6.5 eV is reduced with respect to c-Si.

One important implication of the spectra of Fig. 1 is that the valence band discontinuity is negligible for this heterostructure. This conclusion is affected by an uncertainty of ± 0.15 eV, and agrees with our preliminary conclusions published in Ref. 4. We also confirmed this point by analyzing the Si2p core level spectra taken at $h\nu$ = 135 eV and shown in Fig. 2. The photon energy was selected to obtain a photoelectron escape depth similar to that of the spectra of Fig. 1. The two left curves refer to the clean substrate and to a thick a-Si overlayer. Notice the larger linewidth and the energy shift of the a-$Si_{0.5}C_{0.5}$:H peak with respect to the a-Si peak. The lineshape of the a-$Si_{0.5}C_{0.5}$:H peak is due to the simultaneous presence of large amounts of C and H (10 at. %) which introduce a large compositional disorder and a variety of bonding configurations for the Si atoms. The resulting lineshape is the envelope of different Si2p peaks affected by different chemical shifts. The lineshape gradually evolves from that of a-$Si_{0.5}C_{0.5}$:H to that of a-Si for progressive a-Si coverages. The corresponding right-hand spectra of Fig. 2 were fitted by a linear superposition of the a-Si and a-$Si_{0.5}C_{0.5}$:H lineshapes with fixed energy separation, using the peak intensities as fitting parameters. The solid lines show the results of the fitting procedure and the dotted lines show the two components for each spectrum.

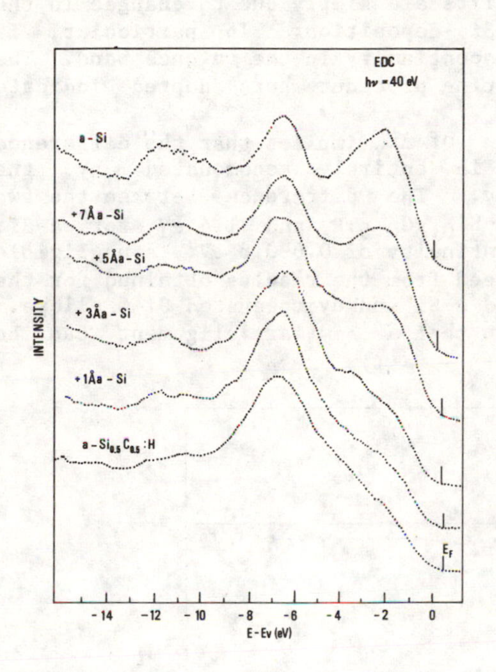

Fig. 1. Photoelectron energy distribution curves (EDC´s) of a clean a-$Si_{0.5}C_{0.5}$:H substrate and of the same substrate covered by an a-Si overlayer of increasing thickness. The horizontal scale is referred to the top of the valence band, E_v. The spectra of the a-Si-covered surfaces were shifted in energy to have E_v in the same position as the clean-substrate spectrum. The corresponding position of the Fermi level is shown by the vertical line at the right-hand-side of each spectrum.

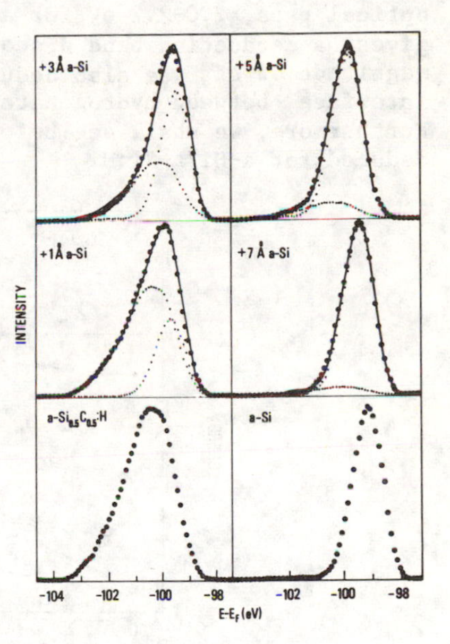

Fig. 2. Si2p spectra for clean and a-Si-covered a-$Si_{0.5}C_{0.5}$:H taken at a energy hν = 135 eV. The solid lines correspond to the results of the fitting procedure explained in the text.

Figure 3 (bottom) shows the position in energy of the two components of the fit (dashed lines) together with the overall shift of the Si2p band (solid line) as a function of the a-Si coverage. The top part of this figure shows the corresponding shift of the leading spectral edge. All the shifts are plotted in Fig. 3 keeping constant the position of E_F. Notice that the two components of the Si2p band closely follow the shift of E_v. This demonstrates that all these shifts are simply due to changes in the band bending during the a-Si deposition. In particular, it confirms that there is no discontinuity in the valence band. The good results of the simple fitting procedure here adopted indicate that the interface is abrupt.

The negligible magnitude of ΔE_v implies that the difference between the two pseudogaps is entirely accomodated by the conduction band discontinuity. The difference between the two optical gaps, 2.0-2.2 eV for a-$Si_{0.5}C_{0.5}$:H and 1.4 eV for a-Si, gives a conduction band discontinuity of 0.6-0.8 eV. A negligible magnitude of ΔE_v was also deduced from the results obtained for the interface between hydrogenated a-Si and hydrogenated Si-C alloys.[4] Furthermore, we shall see below that a similar alignment can be deduced for a-Si/a-Si:H.

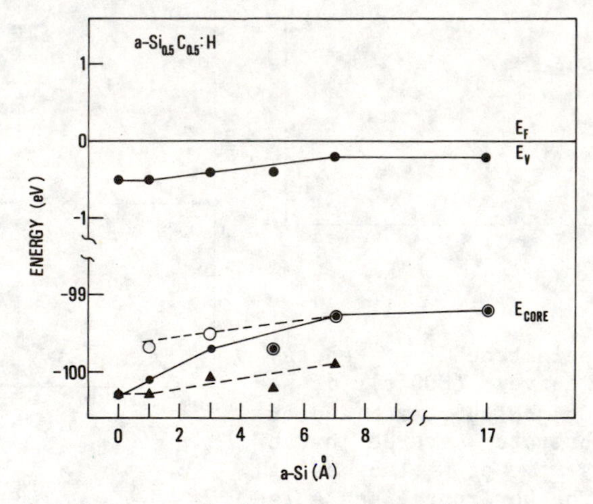

Fig. 3. Bottom: shift in energy of the centroid of the Si2p band (small dots and solid line) and of the two components of this band (large dots and triangles, dashed lines). The two components were deconvolved with the fitting procedure explained in the text. The shifts are plotted as a function of the a-Si overlayer thickness for a-Si/a-$Si_{0.5}C_{0.5}$, keeping constant the position of E_F. Top: Shift of the leading edge of the valence band spectrum.

The band alignment emerging from our experimental results is extremely interesting, since it provides a straightforward explanation for the high collection efficiency of p-i-n solar cells in which the p-doped region is a-$Si_{0.5}C_{0.5}$:H. In fact, the large conduction band discontinuity prevents the back-diffusion of the photoexcited electrons. This reduces the electron-hole recombination at the p-i interface and enhances the overall photovoltaic efficiency of the solar cell.

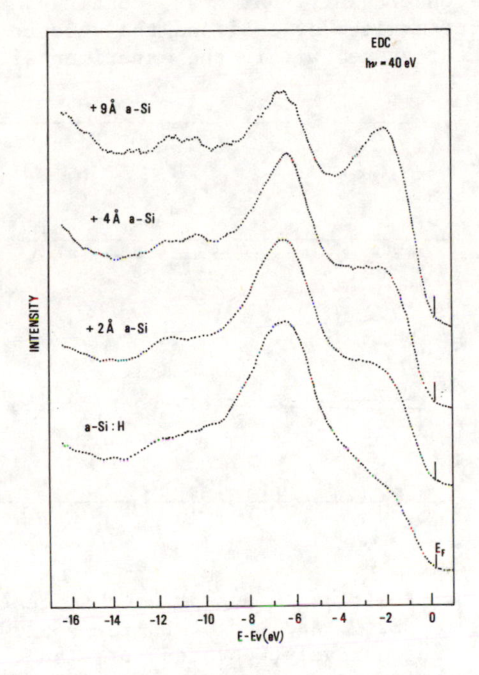

Fig. 4. Photoemission spectra of a clean a-Si:H substrate and of the same substrate progressively covered by an a-Si overlayer.

The most important implication of our experiments on a-Si/c-Si and a-Si/a-Si:H is that neither the disorder nor the hydrogenation affect the position in energy of the valence band edges. The valence band spectra of a-Si/a-Si:H are shown in Fig. 4 and will be discussed in detail in a forthcoming article.[5] We would like to emphasize here the low density of states near the top of the a-Si:H valence band compared to a-Si. This removal of states by hydrogenation reduces the optical absorption coefficient for transitions involving states in the affected energy region. The decrease in the density of states, however, does not affect the position of E_V. In fact, a detailed analysis of the leading spectral edges in Fig. 4 and of the corresponding Si2p spectra rules out a shift of E_V on going from a-Si:H to a-Si. Furthermore, Fig. 5 (triangles) shows that the a-Si deposition does not change E_V nor the position in energy of the Si2p band, i.e. it does not change the band bending.

Similar results were given by the study of the a-Si/c-Si interface as shown by Fig. 5 (dots). The data on E_V suggest a possible upward shift of 0.1 eV which is, however, comparable to the experimental uncertainty. We must conclude, therefore, that the valence band maximum of c-Si and the valence mobility-edge of a-Si concide in our system within the experimental uncertainty.

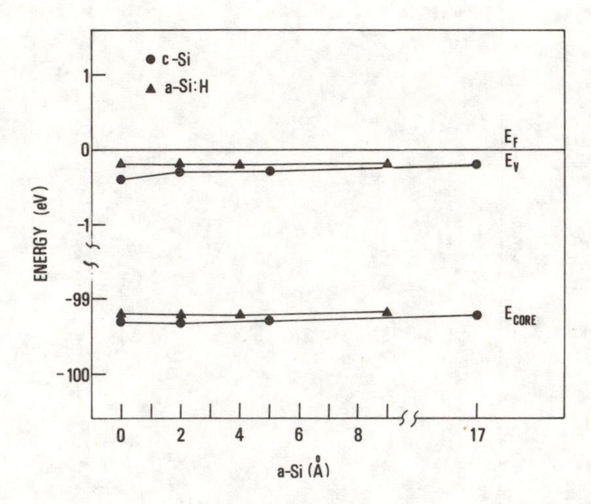

Fig. 5. Position in energy of the centroid of the Si2p band and of E_V for a-Si:H and c-Si progressively covered by an a-Si overlayer.

The fact that neither disorder nor hydrogenation change the position in energy of the top of the valence band is not surprising. Due to the similarity of the chemical properties of the component elements, the microscopic interface dipoles are likely to be negligible for these heterostructures. Therefore, the valence band discontinuity is essentially established by the absolute position in energy of the valence band edges of the components of each heterostructure.[6] The main features of the valence band are accounted for by nearest-neighbor interaction.[7] Even for high levels of hydrogenation, many Si atoms do not have a hydrogen atom among their nearest neighbors. The concentration of these Si atoms essentially unaffected by hydrogenation is equal or larger than the 10^{21} cm^{-3} level which corresponds to the leading spectral edge empirically deduced from the photoemission spectra. Similarly, due to compositional disorder a large portion of the Si atoms are surrounded by four Si atoms in a-$Si_{0.5}C_{0.5}$:H and this explains the alignment of the valence band for the a-Si/a-$Si_{0.5}C_{0.5}$:H heterostructure.

One of the most important consequences of the hydrogenation of a-Si is the change of the optical gap, which increases with the hydrogen content. This fundamental effect was attributed[8] to the removal of states from the valence band edge accompanied by a recession of the valence band edge. Our present results confirm that the hydrogenation causes a depletion of states near the top of the valence band which, in turn, is consistent with the observed change in slope of the semi-logarithmic plot of α vs. $h\nu$. (α = optical absorption coefficient). They show, however, that no recession of the top of the valence band occurs for ~ 10 at.% hydrogen concentration. This indicates that the hydrogen-induced widening of the optical gap is primarily due to a shift in energy of the conduction band edge. This effect was theoretically predicted by Economou and Papaconstantopoulos.[9]

REFERENCES

1. G. Margaritondo, Solid State Electron. 26, 499 (1983), and references therein.
2. Y. Tawada, H. Okamoto and Y. Hamakawa, Appl. Phys. Letters 39, 237 (1981).
3. B. Abeles and T. Tiedje, Phys. Rev. Letters 51, 2003 (1983).
4. F. Evangelisti, P. Fiorini, C. Giovannella, F. Patella, P. Perfetti, C. Quaresima and M. Capozi, Appl. Phys. Letters 44, 764 (1984).
5. F. Patella, F. Evangelisti, P. Fiorini, P. Perfetti, C. Quaresima and G. Margaritondo, unpublished.
6. A. D. Katnani and G. Margaritondo, Phys. Rev. B28, 1944 (1983).
7. W. Harrison: "Electronic Structure and Properties of Solids" (Freeman, San Francisco 1980).
8. B. von Roedern, L. Ley and M. Cardona, Phys. Rev. Letters 39, 1576 (1977).
9. E. N. Economou and D. A. Papaconstapoulos, Phys. Rev. B23, 2042 (1981).

AMORPHOUS SILICON HETEROJUNCTIONS STUDIED BY TRANSIENT PHOTOCONDUCTIVITY

R. A. Street and M. J. Thompson
Xerox Palo Alto Research Center, Palo Alto, CA 94304

ABSTRACT

Transient photoconductivity is used to characterize the electronic properties of a–Si:H interfaces. We show how it is possible to measure the sign and width of the surface band bending, and the density of interface states. At a silicon nitride/a–Si:H interface we find an electron accumulation layer. The interface state density is much larger when the nitride is deposited after the a–Si:H rather than vice versa. The native oxide results in an electron depletion layer, and the same is true of other oxide layers, although under some circumstances, accumulation is observed.

INTRODUCTION

The electronic properties of a semiconductor are usually modified near a surface or interface. Band bending can occur either due to the mismatch of the work functions of the two materials in contact, or because of defects or impurities at the interface. The interface of a–Si:H with oxides or nitrides are of particular interest. For example, band bending at the free surface or substrate interface could cause a conducting channel and influence the results of planar conductivity data.[1] Such effects have been observed in connection with measurements of adsorbed gases on the surface of a–Si:H.[2] The nitride interfaces are of interest in studies of multilayer structures[3] and of MIS transistors.[4]

Capacitance measurements are the most common techniques for studies of surface layers. However, these require a conducting bulk material. Hence capacitance measurements are suitable for n–type a–Si:H, but have not been successful on undoped samples. This paper describes a different technique for measuring surface layers, based on transient photoconductivity. The method is particularly suitable for undoped a–Si:H, and its application to interfaces of a–Si:H with various oxide and nitride layers are described.

EXPERIMENTAL DETAILS

The experiment uses transient photoconductivity to explore the internal electric field near the interface. Further details of the technique are described elsewhere.[5,6] Fig. 1 illustrates the experimental arrangement. The sample is in the form of a parallel plate capacitor made with semitransparent evaporated metal electrodes. Any band bending near the interface results in an internal electric field throughout the width of the surface region (shown as an electron depletion layer in Fig. 1). The experiment observes the photocurrent due to excited carriers moving in the internal field. The excitation is a 5 nsec pulse of

0094-243X/84/1200410-07 $3.00

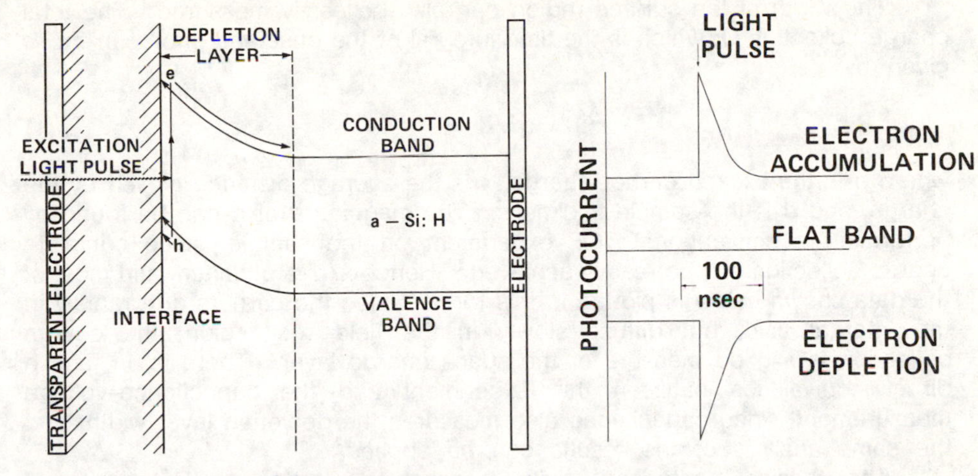

Fig. 1. A schematic diagram of the experiment using transient photoconductivity to measure band bending near an interface, showing the excitation of carriers near the interface, and their motion across the surface layer.

Fig. 2. Typical photocurrent pulses corresponding to an accumulation layer, a depletion layer, and flat bands.

light at ~2.3 eV, and is of sufficiently low intensity that the carriers do not significantly perturb the internal field. In each case studied, the interface is with a material with much higher band gap than a–Si:H, so that the excitation pulse is absorbed only in the a–Si:H layer, and within about 1000 Å of the interface. As the excited carriers move across the surface layer, they induce charge on the back electrode, just as in a conventional time–of–flight (TOF) experiment. The movement of the charge is registered as a current pulse in the external circuit. The photocurrent transient represents the motion of the charge packet, providing there is no screening of the charge by thermally generated carriers. This requirement is the same as for the TOF experiment and limits the measurements to samples of low conductivity.

The most immediate observation that can be made is of the sign of the photocurrent at zero bias. This corresponds to the sign of the internal field and tells directly the polarity of the band bending. In general the band bending can be changed and inverted by the application of a suitable bias potential. Schematic diagrams of typical transients corresponding to accumulation, depletion and flat bands are shown in Fig. 2. Since the electric field generally decreases rapidly away from the interface, the current pulse is seen as a sharp spike of duration usually less than 100 nsec. A zero response occurs when the bands are flat, and the bias voltage at which this occurs can typically be found to better than 0.1V. Note that even with flat bands there may be a weak photocurrent transient due to the diffusion of carriers. Since the diffusion is at most equivalent to a band bending of kT, the error in the flat band determination is generally small and is neglected here.

The width of the surface region can also be easily measured. The total charge collection Q, which is the time integral of the observed photocurrent, is given by

$$Q = qx/d \tag{1}$$

where q is the excited carrier charge, x is the average distance moved by the charge, and d is the sample thickness. The magnitude of q can be found by performing a conventional TOF experiment on the sample, since complete charge collection can be readily achieved.[5] Hence x/d is obtained, and most of the data shown below is plotted in this form. Since the carriers drift rapidly in the internal field, but diffuse slowly in the field-free region, the charge collection is a good measure of the space charge layer. There is, in fact, a direct equivalence between this experiment and the capacitance-voltage measurement, since capacitance also measures the depletion layer width, and the same analysis of the results can be applied.

The transient photoconductivity can also be used to profile the internal field, since the photocurrent is proportional to the local electric field acting on the charge packet. This technique has been used before in measurements of Schottky barriers,[5] but has not been used in these studies of interfaces.

RESULTS

A) Nitride Interfaces

Samples were made consisting of 5 μm of a-Si:H and 3000 Å of silicon nitride, both prepared by plasma decomposition. Pairs of samples were studied in which the nitride was deposited either before the a-Si:H (bottom nitride) or after (top nitride). An example of the charge collection as a function of bias voltage is shown in Fig. 3 for one particular pair of samples. The charge collection is given as a fraction of full collection as determined by TOF experiments, and is also converted to the layer width using the known thickness of the sample. The different results for the two interfaces show that the electronic structure depends on the order of deposition. At zero bias the sign of the photocurrent demonstrates that both structures have an electron accumulation layer at the interface. In the case of the bottom nitride a bias of only -0.1V is required to make the bands flat, and a larger negative bias voltage results in a depletion layer. The small flat band voltage and the large slope of the charge collection data are both characteristic of a low interface charge density.

In contrast, the top nitride sample requires a bias of -4.5 V to obtain flat bands. The corresponding charge at the interface is given by $C_N V_F$, where C_N is the capacitance of the nitride layer, and V_F is the flat band voltage. The magnitude of the charge is 5×10^{11} cm^{-2}, which is the total density of interface states lying between the zero bias Fermi energy, and the flat band Fermi energy. From the flat band voltage, the corresponding charge density for the

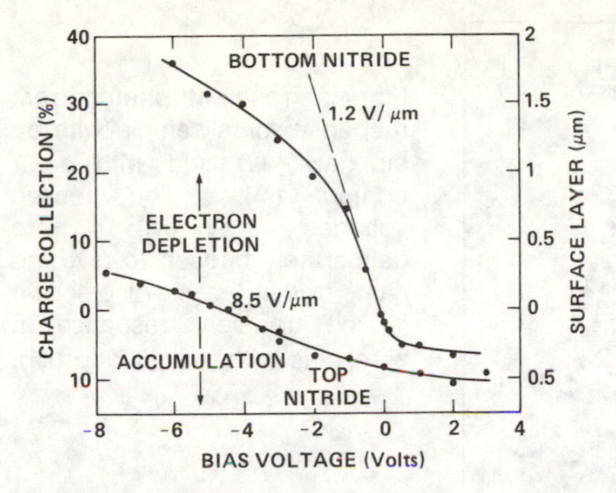

Fig. 3. Charge collection data as a function of bias voltage for the top and bottom nitride samples. The slope of the data at the flat band voltage is indicated and gives a measure of the density of interface states.

bottom nitride is evidently more than an order of magnitude lower. This difference is partly due to a smaller density of interface states, and partly to a smaller zero bias band bending. When the top nitride is biased into depletion, the charge collection increases but much more slowly than for the bottom nitride. Assuming that the band bending is dominated by interface rather than bulk states, the slope of the charge collection data where it changes sign is proportional to the density of states at the flat band Fermi energy. The results of Fig. 3 therefore show a density of states differing by almost an order of magnitude between the two interfaces.

More extensive measurements of the nitride interface are described elsewhere.[6,7] These show that the band bending at the top nitride is about 0.5 eV, that the interface states are distributed in energy, and that the states are indeed located at the interface.

B) <u>The surface oxide of a–Si:H</u>

The exposed surface of a–Si:H forms an oxide, and may contain defects, either of which could cause band bending. The main problem in using this technique to study the surface of a–Si:H is the requirement of having a top electrode. This problem can be overcome by using a thin air gap, or a dielectric spacer (mylar or polythene have been tried). Each of these gives similar results, so only the air gap data is presented. The top electrode is evaporated onto a separate substrate and brought near to the sample. The actual width of the air gap can be conveniently measured by TOF as shown in Fig. 4. The electron transit times of the air gap sample are compared with a sample from the same deposition which had an electrode evaporated directly onto it. To obtain the same transit time, the voltage has to be four times larger in the air gap sample, because much of the voltage is dropped across the air gap. Taking into account the different dielectric constants, the gap is therefore

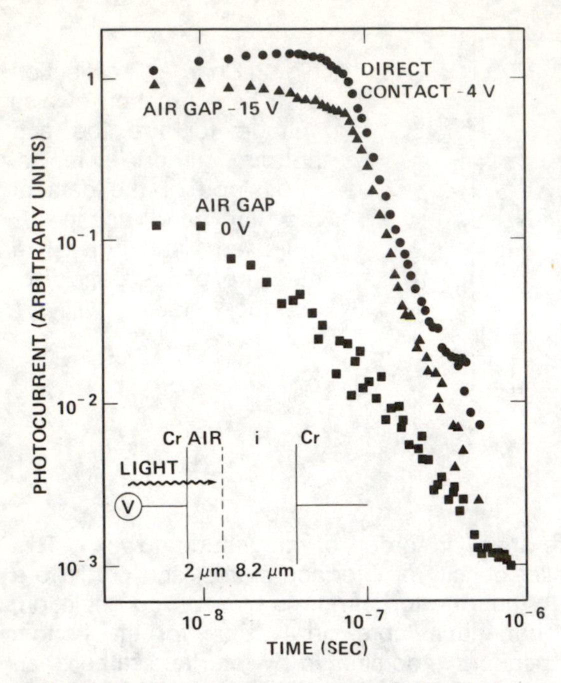

Fig. 4. Transient photocurrent response for a sample with an air gap (▲) and with a Cr contact (●). The applied voltages shown were deliberately chosen to give the same transit times. Also shown is the transient response at zero bias with the air gap (■).

deduced to have a thickness of about 2 μm. Furthermore, the fact that the transit time is relatively well defined implies that the air gap is reasonably uniform across the sample.

Fig. 4 shows the zero bias transient response of the air gap. The sign of the photocurrent demonstrates that the bands bend up at the surface to give an electron depletion layer. A measurement of charge collection gives a zero bias layer width of ~1.1 μm. For comparison, the depletion layer width for the evaporated Cr contact is 1.3 μm. The similarity of these results shows that the built–in potential must be almost the same in the two cases. Since the built–in potential of Cr is ~0.2 eV,[5] this is also the approximate value for the free surface. These results agree with recent measurements using a Kelvin probe.[8]

C) Deposited Oxide Layers

It is of interest to measure the band bending of a–Si:H in contact with other forms of silicon oxide which might correspond to a substrate material or to a gate dielectric. Fig. 5 shows some charge collection data for four different structures. Two of these are made by e–beam evaporation of SiO_2, which is known to give a silicon rich film with poor dielectric properties. The third sample uses a deposited oxide grown in an r.f. plasma Coyote reactor, and contains 6-8% nitrogen and about 5% hydrogen. The fourth interface is to a thermal oxide grown on a thin polycrystalline layer which also acts as the transparent electrode. The e–beam oxides give a fairly large zero bias depletion layer both when it is deposited before or after the a–Si:H layer. The thermal oxide is also in depletion although with a smaller band bending. In contrast the

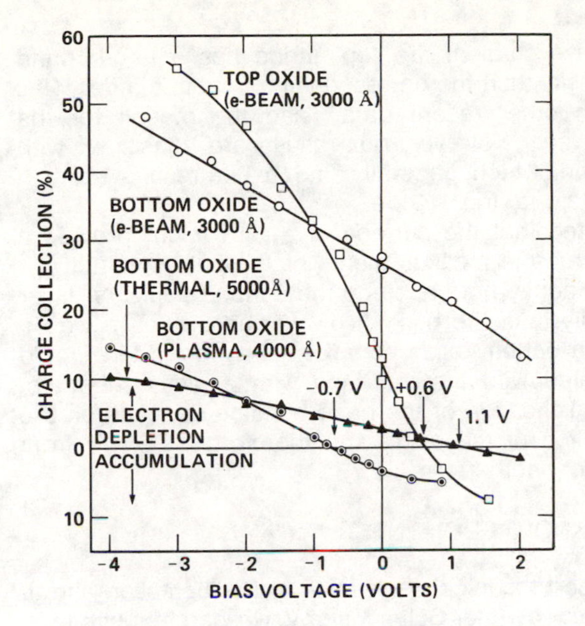

Fig. 5. Charge collection data for various oxide interfaces as a function of bias voltage. In each caes the a–Si:H sample thickness is 5μm.

plasma oxide has an accumulation layer at zero bias. For this interface, the flat band voltage was -0.7V. This value and the slope of the charge collection data indicate an interface state density intermediate between that of the top and bottom nitrides. The thermal oxide has a flat band voltage of 1.1V, and appears to have a higher interface state density than the plasma oxide.

The results of the e–beam oxides appear to show a lower density of interface states, although a larger zero bias band bending. However, when the e–beam oxide samples are measured at different bias voltages, there is a slow drift in the charge collection data. This effect is caused by charge trapping in the oxide and was particularly strong when the oxide is deposited before the a–Si:H. Under these conditions it is not possible to make reliable estimates of the interface state density. Both the plasma and thermal oxides are found to be much more stable, as are the nitride layers.

DISCUSSION

From our results it is evident that the interface of a–Si:H with silicon oxide or nitride is very sensitive to the details of the preparation of the samples. As yet we have no information as to the precise origin of the interface states and this will require structural studies of the interfaces. The variability of the results makes it seem unlikely the the interface states are due to mismatch between ideal amorphous networks. Rather the data suggest that interface impurities may be important. Specifically, the observation that of all the oxide interfaces studied, the only one to have a zero bias accumulation layer contained some nitrogen, suggests that this element is particularly important in determining the band bending.

In all the samples we have studied, the top nitride has a larger band bending and interface state density than the corresponding bottom nitride. This result appears to conflict with other recent data,[9] and the reason for the difference remains to be resolved. However, our results are consistent with measurements of MIS transistors which generally find better results using a bottom nitride gate compared to a top nitride.

It has often been suggested that the surface of a-Si:H might provide a conducting channel that dominates in measurements of the bulk conductivity. Our results show that this is evidently not so, since the surface depletion layer should have a lower conductivity than the bulk. However, in one case we do find an accumulation layer for a bottom oxide sample. Therefore, there could be a conducting layer associated with the interface with a glass substrate. From our data it is apparent that the sign of the band bending depends on the specific structure and composition of the oxide and therefore may vary from one type of glass substrate to another.

ACKNOWLEDGMENTS

We are grateful for the expert technical assistance of R. Thompson and J. Zesch. This work is supported by the Solar Energy Research Institute.

REFERENCES

1. e.g. see H. Fritzsche, Solar Energy Materials, 3, 447, (1980).
2. M. Tanelian, Phil. Mag., 45, 435, (1982).
3. B. Abeles and T. Tiedje, Phys. Rev. Lett., 51, 2003, (1983)
4. H. C. Tuan, M. J. Thompson, N. M. Johnson and R. A. Lujan, IEEE Electron Device Lett., EDL-3, 357, (1982).
5. R. A. Street, Phys. Rev., B27, 4924, (1983).
6. R. A. Street, M. J. Thompson and N. M. Johnson, Phil. Mag., in press.
7. R. A. Street and M. J. Thompson, Appl. Phys. Lett., in press.
8. B. Aker, S-Q. Peng, S-Y. Cai and H. Fritzsche, J. non-Cryst. Solids, 35 and 36, 509, (1983).
9. C. B. Roxlo, B. Abeles and T. Tiedje, Phys. Rev. Lett., 52, 1994, (1984).

PHOTOLUMINESCENCE IN AMORPHOUS SILICON/AMORPHOUS SILICON NITRIDE DOUBLE HETEROSTRUCTURES

T. Tiedje, B. Abeles, B. G. Brooks
Corporate Research Science Laboratories
Exxon Research and Engineering Company
Annandale, N.J. 08801

ABSTRACT

Photoluminescence measurements on amorphous silicon/amorphous silicon nitride multilayer structures as a function of the amorphous silicon sublayer thickness, show that the interfaces contain non-radiative recombination centers, and that both the bandgap and the localized state density near the band edges increase in the quantum size regime.

INTRODUCTION

It has recently been discovered that amorphous hydrogenated silicon (a-Si:H) and related amorphous semiconductors can be deposited in multilayer superlattice structures with uniform layers and high-quality nearly atomically-abrupt interfaces.[1] This discovery raises the question as to whether carrier confinement structures that exhibit enhanced photoluminescence, can be made from amorphous double-heterojunctions.

To investigate this question we have studied the CW photoluminescence (PL) of a series of a-Si:H/a-SiN:H superlattice materials in which the thickness of the silicon nitride (a-SiN:H) layers was held fixed at 35A and the thickness of the a-Si:H layers was varied from 8A to 1200A. The materials were prepared by plasma-assisted chemical vapor deposition from silane and ammonia as described previously.[2] The nitride material was a wide bandgap insulator with a silicon to nitrogen ratio close to unity as determined by IR absorption.

Photoemission experiments on silicon nitride prepared by high temperature chemical vapor deposition on crystalline silicon[3], and plasma deposited silicon nitride on amorphous silicon[4] both show that the bandgap discontinuity at the interface between the two materials is approximately equally divided between the conduction and valence band edges. Thus, the silicon nitride should provide a potential barrier to both electrons and holes in the a-Si:H.

THICKNESS DEPENDENCE OF PHOTOLUMINESCENCE EFFICIENCY

The PL efficiency measured at 10K is shown as a function of a-Si:H layer thickness L in Fig. 1. The samples in Fig. 1 were deposited on smooth quartz substrates and were all about 1 μm thick. The PL was excited by about 10mw of light from a Kr ion laser incident through the transparent substrate. The PL spectrum was measured with a 1/4 m monochromator and a cooled InSb detector. The uncertainty in the relative PL efficiency, indicated by the error bars in Fig. 1, is small because the samples were measured successively with no significant adjustments to the optical

0094-243X/84/1200417-08 $3.00

alignment. The absolute through-put of the optical system was obtained from a spectral calibration with a standard lamp, referenced to the collection efficiency for a known intensity of 1 μm radiation from a HeNe laser scattered off a white surface. The accuracy of the absolute calibration is estimated to be about a factor of three.

An additional complication is that the PL efficiency typically increases by a factor of 3-6 when the samples are textured so as to eliminate interference fringes in the spectra. The texturing was performed after deposition by an ion beam milling process.[5] The apparent increase in the efficiency with texturing is believed to be due to enhanced coupling of trapped optical modes inside the film, with the outside medium. The PL efficiency in Fig. 1 was obtained from measurements on untextured samples and thus represents an underestimate of the true PL efficiency.

We attribute the reduction in PL efficiency in Fig. 1 for thin a-Si:H layers to non-radiative recombination at the a-Si:H/a-SiN:H interface. By analogy with the role of dangling bonds in bulk a-Si:H,[6] we assume that any photo-excited electron-hole pair created within a critical radius R_c = 70A of a non-radiative center,[7] will recombine non-radiatively; otherwise it recombines radiatively. We make the further assumption that the non-radiative centers are all located at the interfaces, but otherwise randomly distributed.

In this case, if an electron-hole pair is created at a distance $x < R_c$ away from the interface, it will recombine radiatively as long as the disc intercepted by the capture sphere at the interface, is free of non-radiative centers. Elementary geometry shows that the area of this disc is $\pi (R_c^2 - x^2)$. Now the PL efficiency can be calculated as a function of layer thickness L.

Let the probability that an individual interface atom is a non-radiative site be p. Then the probability that m adjoining interface sites are all free of non-radiative centers is $(1-p)^m$ or exp (-mp) for large m. Since mp is the average number of non-radiative centers for an ensemble of m sites, it follows that the electron-hole pair, located at a distance x from the interface, will recombine radiatively with probability $\exp[-\pi (R_c^2 - x^2)N]$. In this expression N is the number density of non-radiative centers at the interface.

As an additional simplification we assume that the two interfaces (a-Si:H on a-Si:H or vice-versa) are not symmetric and that all of the non-radiative centers are located at one of the interfaces. Other unrelated experiments support this hypothesis that the interfaces are asymmetric.[8,9] The PL efficiency I for a layer of thickness $L < R_c$ is given by:

$$I = \frac{1}{L} \exp(-\pi R_c^2 N) \int_0^L \exp(\pi x^2 N) dx \qquad (1)$$

and for $L > R_c$:

$$I = \frac{1}{L} \left[\int_0^{R_c} \exp[\pi(x^2 - R_c^2)N] dx + (L - R_c) \right] \qquad (2)$$

The solid line in Fig. 1 is a fit of Eqs. (1) and (2) to the data with R_c = 70A from Wilson et al[7] and $N = 1.4 \times 10^{12}$ cm^{-2} as determined by the best fit. No attempt was made to fit the absolute efficiency because of the relatively large experimental uncertainties in this quantity as discussed above; instead the model curve was scaled to match the efficiency of the bulk material at large L, generally believed to be close to unity.

We conclude from the thickness dependence of the efficiency that the a-Si:H/a-SiN:H interface is a source of non-radiative recombination centers, and that at low temperatures these centers control the recombination, for a-Si:H layers less than about 100A thick. This characteristic thickness is defined by how far an electron-hole pair can tunnel in a time equal to the radiative lifetime.

THICKNESS DEPENDENCE OF OPTICAL GAP

Another characteristic length in these materials is the a-Si:H thickness at which the quantum shift in the electron energy levels becomes comparable with the band tail widths. This length is about 30A in a-Si:H as shown in Fig. 2 where we plot the optical gap ($(\alpha E)^{1/2}$ gap) as a function of L, for the same series of samples as in Fig. 1. The solid line in Fig. 2 is the calculated change in the separation between the lowest state in the conduction band and the highest state in the valence band for a one dimensional quantum well model in which $m_e^* = 1.0$, $m_h^* = 1.0$, and $U_c = U_v = 1.0$ eV. The quantum size effect gives a satisfactory explanation for the observed change in the optical gap, as illustrated by the good agreement between the solid line and the optical gap in Fig. 2, and as discussed previously.[10]

Like the optical absorption bandgap, the peak in the PL emission spectrum also shifts to higher energies and the emission spectrum broadens with decreasing layer thickness as shown in Fig. 3. However, the shift in the PL spectrum is less than half as large as the shift in the optical gap.

The PL excitation spectrum provides another method by which the bandgap can be determined. In this experiment the PL spectrum is measured as a function of excitation wavelength. A typical set of emission spectra that results is shown in Fig. 4 where PL spectra excited by five different Kr ion laser lines are shown for the same sample. Note that the emission peak shifts to lower energies as the photon energy of the excitation light is reduced. The arrows in Fig. 4 mark the peaks in PL spectra, as determined from a least squares fit of cubic polynomials to the spectra. The peak positions determined from the polynomial fits are plotted as a function of excitation energy in Fig. 5 for three representative samples. This figure (Fig. 5) shows that the emission spectrum is independent of pump wavelength at high energies, but shifts to lower energies when the energy of the excitation light is reduced below a value approximately equal to the optical gap. Similar behavior has been reported earlier for bulk a-Si:H films.[11] The excitation wavelength at which the PL spectrum begins to change can be determined more quantitatively from the intersection of linear fits to the high and low energy regimes, as illustrated in Fig. 5.

The "knee" in the graph of the PL peak position E_{pk} as a function of excitation energy E_{ex}, is a measure of the electron-hole pair energy at which the thermalization rate becomes comparable with

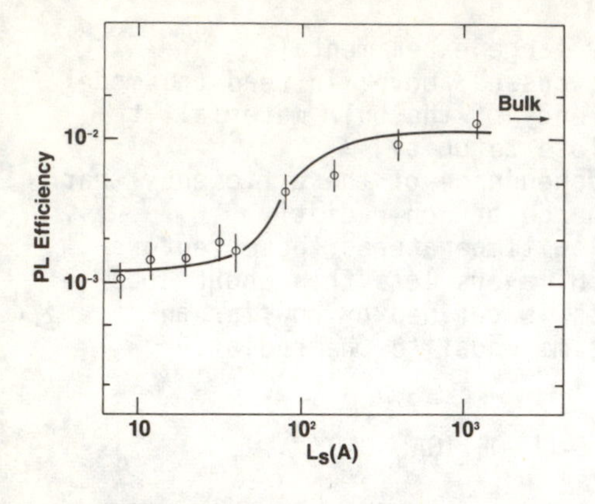

Fig. 1. Photoluminescence efficiency as a function of silicon layer thickness at 10K for silicon nitride layer thickness of 35A. The solid line is a fit to the data as discussed in the text.

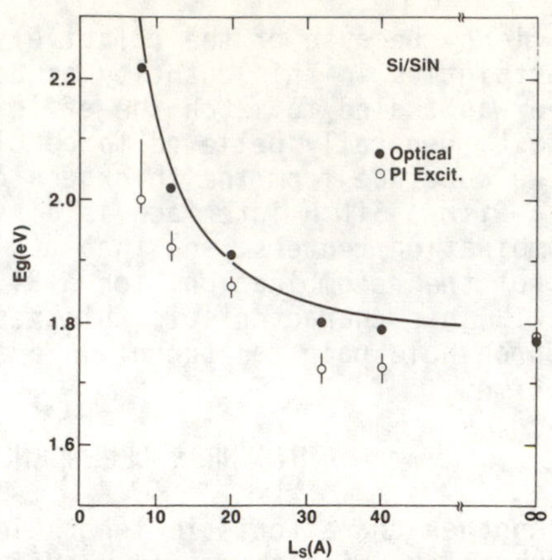

Fig. 2. Optical gap as a function of silicon layer thickness determined from linear extrapolation of $(\alpha E)^{1/2}$ vs E (o) and energy of "knee" in Epk vs Eex as shown in Fig. 5 (o). The solid line is a quantum well model with electron and hole effective masses unity, and conduction and valence band discon-tinuities of 1 ev.

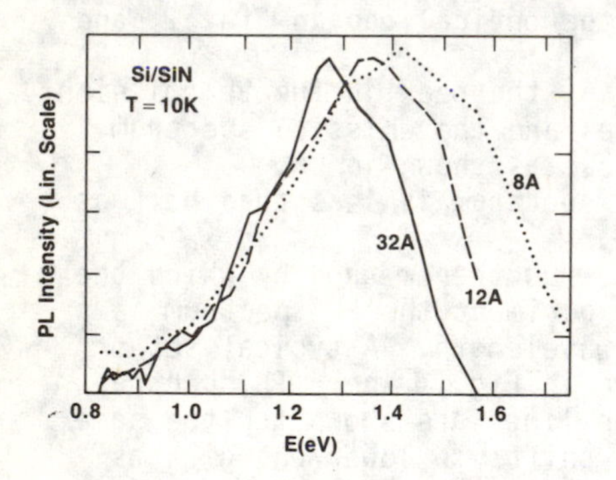

Fig. 3. Photoluminescence spectra for three films with a-Si:H layer thicknesses of 8, 12 and 32Å as indicated.

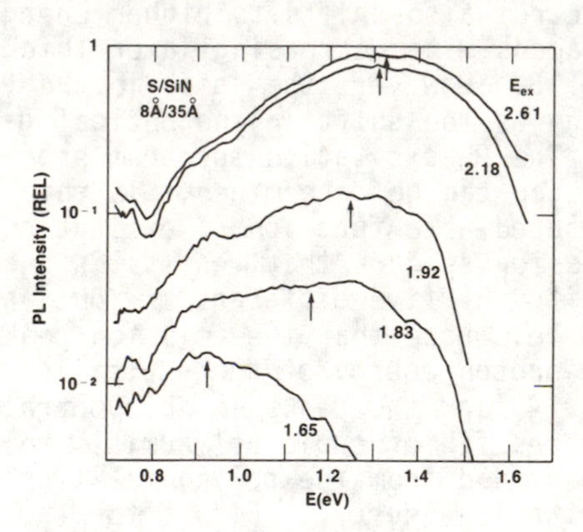

Fig. 4 Photoluminescence spectra as a function of excitation energy on a sample with 8A thick silicon layers. The arrows indicate the peak energy as determined from a least squares fit of cubic polynomials to the data, on a linear scale.

the recombination rate. Since the thermalization rate decreases very rapidly below the mobility gap, this knee is expected to be close to the mobility gap. Indeed this measure of the bandgap parallels the optical gap quite closely, as illustrated in Fig. 2 where the photon energy corresponding to the knee in the E_{pk} vs E_{ex} curves (marked by arrows for the three samples in Fig. 5) is plotted along with the optical absorption gap, as a function of L.

The slope of the low energy part of the E_{pk} vs E_{ex} graph in Fig. 5 is between 1/2 and 1. A slope of 1/2 was reported earlier for a bulk a-Si:H sample.[11] A slope of unity implies that the Stokes shift in the PL is a constant independent of where the electron-hole pairs are positioned in the band tails.

The experimental data discussed above indicates that the quantum size effect has a significant effect on the electronic energy levels for L< 30A. However, there is no corresponding increase in the PL efficiency, even though one would expect an increase in the radiative recombination rate for two dimensional electron-hole pairs relative to 3D, since the electron-hole binding energy is four times larger in 2D than in 3D.

LOCALIZED STATES NEAR THE BAND EDGES

We speculate that the explanation for why there is no enhancement of the PL in the quantum regime (L<30A) lies in the fact that the density of states tails broaden as L decreases, either because of structural disorder associated with the interfaces or because of quantum-size-effect-induced stretching out of the localized state distribution.[1] As long as the random potential associated with the structural disorder is larger than the electron-hole coulomb attraction, one might not expect changes in this attraction to have a significant effect on the radiative recombination rate, consistent with the experimental observations.

The increase in the width of the Urbach tail with decreasing L shown in Fig. 6, supports the hypothesis that the disorder broadening of the band edges increases in the quantum size regime. Further support for this hypothesis comes from the PL spectra which broaden with decreasing L as shown in Fig. 4, consistent with an increase in the width of the localized state distribution.

The small increase in the PL emission energy relative to the increase in the optical gap for L < 30A, is also consistent with a wider distribution of band-tail-like states in the thin a-Si:H layers. The electron-hole pairs will be able to penetrate deeper into the localized state distribution before radiating if these states are more numerous.

A wider distribution of localized states for small L is also consistent with the temperature dependence of the PL. The temperature dependence of the PL intensity I(T), can be fit by the expression,[12]

$$I(T) = I_o \,(\exp(T/T_o) + 1)^{-1} \qquad (3)$$

as shown in Fig. 7. Fig. 7 shows that the temperature dependence of the PL becomes progressively weaker as L is reduced below ~30A. The parameter T_o that describes the temperature dependence of the PL is

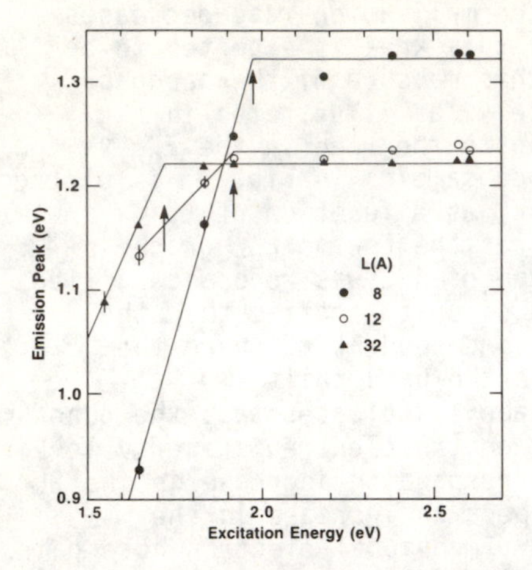

Fig. 5 Photoluminescence peak energy as a function of excitation energy, for three a-Si:H layer thicknesses as indicated. The arrows mark the intersection between linear fits to the data obtained at high and low excitation energies.

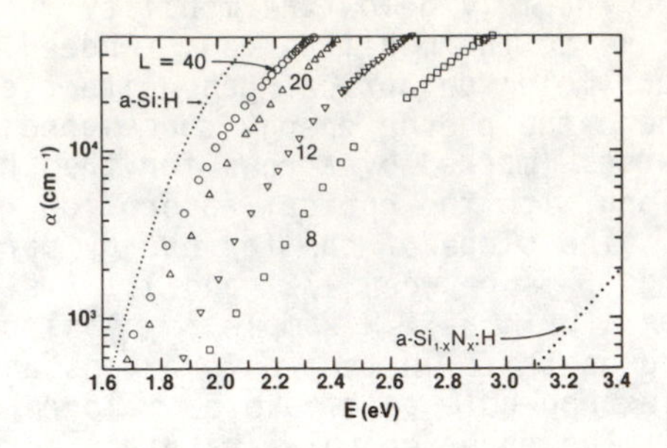

Fig. 6. Absorption as a function of photon energy for various a-Si:H layer thicknesses. Also shown is the absorption of a-Si:H and a-SiN:H films prepared under the same conditions as in the multilayer structures.

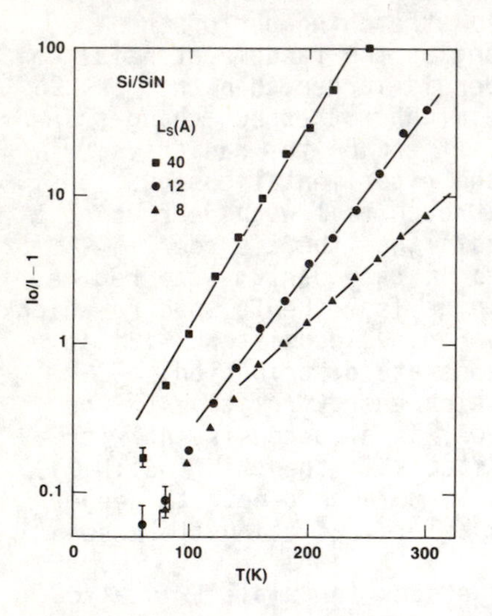

Fig. 7. Temperature dependence of the photoluminescence intensity I(T) normalized as indicated to T_o, the efficiency at 10K. The a-Si:H sublayer thicknesses are shown.

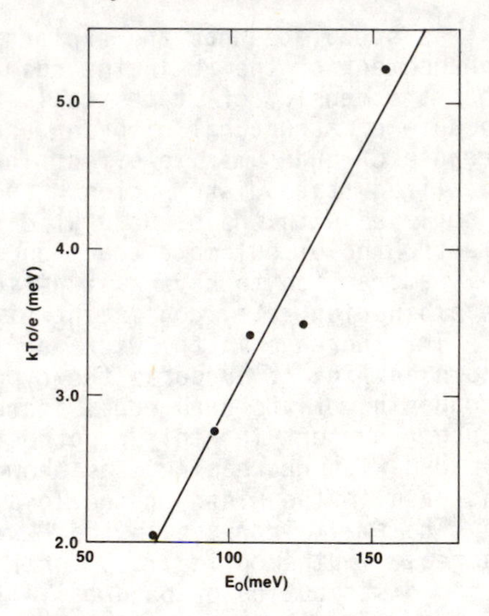

Fig. 8. Photoluminescence temperature dependence parameter T_o as a function of Urbach Slope parameter E_o for a-Si:H layer thickness varying from 8A to 40A. A non-layered 1μm film of a-Si:H is also shown. The Urbach slope parameter was determined from transmission measurements of the optical absorption.[14]

~ $T_c/10$ according to theory,[12] where T_c is the width of the distribution of energy barriers for non-radiative recombination. The width of this distribution might in turn be expected to be comparable with the width of the band tails or the width of the Urbach tail. In Fig. 8 we plot T_o as a function of the Urbach slope parameter E_o, with L as a parameter. The linear relation between E_o and T_o in Fig. 8 suggests that the reduced ability of electron-hole pairs in the thin a-Si:H layers to diffuse to non-radiative centers with increasing temperature, may be due to broadening of the localized state distribution and only indirectly due to the effects of the 2D confinement.

We note that PL data very similar to that reported here has been observed in a-SiN_x:H alloys with $x < 0.3$.[13] The similarity could mean that the phenomena have the same origin. For example the dilute a-SiN_x:H alloys may be inhomogeneous in the sense that they may contain Si rich regions surrounded by N rich regions. In the specific case of the superlattice material with the thinnest a-Si:H layers (8A), when the thickness of the interfaces are taken into account (~ 4A)[2], the size of the pure a-Si:H regions could be similar to that in a dilute alloy (X ~ 0.3 for example[13]), with randomly distributed N atoms.

CONCLUSIONS

Photoluminescence and optical absorption measurements on a series of a-Si:H/a-SiN:H superlattice materials with different a-Si:H layer thicknesses, show that the interfaces are a source of non-radiative recombination, and that the disorder broadening of the band edges increases together with the optical gap in the quantum size regime. The enhanced room temperature photoluminescence in thin silicon/silicon nitride multilayers can be explained by enhanced localization associated with a wider distribution of localized states and only indirectly by carrier confinement in double heterostructures. More work needs to be done to confirm the generality of these results with regard to different sample preparation conditions.

ACKNOWLEDGEMENTS

We thank G. D. Cody for the optical absorption data, H. W. Deckman and J. Dunsmuir for texturing the samples, and R. Carius and C. B. Roxlo for helpful discussions.

REFERENCES

1. B. Abeles and T Tiedje, Phys. Rev. Lett. 51, 2003 (1983).
2. B. Abeles, T. Tiedje, K. S. Liang, H. W. Deckman, H. E. Stasiewski, J. C. Scanlon and P. M. Eisenberger, Proc. Int. Top. Conf. on Transport and Defects in Amorphous Semiconductors, Bloomfield Hills, MI, (to be published 1984).
3. D. J. DiMaria, P. C. Arnett, Appl. Phys. Lett. 26, 711 (1975).
4. B. Abeles, I. Wagner, W. Eberhardt, J. Stohr, H. E. Stasiewski (this proceedings).

5. H. W. Deckman, J. H. Dunsmuir Appl. Phys. Lett. 41, 377 (1982).
6. R. A. Street, J. C. Knights, D. K. Biegelsen Phys. Rev. B 18, 1880 (1978).
7. B. A. Wilson, A. M. Sergent, J. P. Harbison Bull. Am. Phys. Soc. 29, 508 (1984).
8. J. Mort, F. Jansen, S. Grammatica, M. Morgan, I. Chen, J. Appl. Phys. 55, 3197 (1984).
9. C. B. Roxlo, B. Abeles, T. Tiedje, Phys. Rev. Lett. 52, 1994 (1984).
10. T. Tiedje, B. Abeles, P. D. Persans, B. G. Brooks, G. D. Cody, ibid 2.
11. J. Shah, A. Pinczuk, F. B. Alexander, B. G. Bagley, Sol. State Commun. 42, 717 (1982).
12. G. S. Higashi, M. Kastner, J. Phys. C 12, L821 (1979).
13. R. Carius, K. Jahn, W. Siebert, W. Fuhs, Journal of Luminescence (to be published 1984).
14. C. B. Roxlo, B. Abeles, C. R. Wronski, G. D. Cody, T. Tiedje Solid State Comm., 47, 985 (1983).

DOPING MODULATED AMORPHOUS SEMICONDUCTORS

J. Kakalios*, H. Fritzsche, and K. L. Narasimhan†
The University of Chicago, Chicago, IL 60637

ABSTRACT

Amorphous semiconductor doping superlattices consisting of alternating layers of n- type and p- type doped r.f. glow discharge deposited hydrogenated silicon have been prepared using a closed two-chamber system. After very brief light exposure we observed large excess conductivities that persisted for many days but could be removed by annealing above 450°K. These are explained in terms of Döhler's model of charge separation by the internal fields. The photoluminescence peak of the npnp.... films is blue-shifted with respect to the peak positions of the n-type and p-type films.

INTRODUCTION

Interesting new materials can be synthesized by depositing alternating thin layers of different materials. Crystalline heterojunctions as well as p-n junction superlattices have been studied experimentally and theoretically for a number of years[1]. A very large choice of materials becomes available when the superlattices are made out of amorphous films because there is then no need to match lattice constants at the interfaces. Moreover, amorphous superlattices can be deposited on flexible and curved substrates, which led to their use as optical elements for soft X-rays[2]. More recently several groups have investigated amorphous semiconductors that consist of many alternating thin layers of two materials having different energy gaps[2-5]. The compositionally modulated films are the amorphous analogs to the crystalline compositional superlattices.

In this paper we report on the preparation and properties of doping modulated amorphous semiconductors; i.e. a periodic sequence of n- and p- doped layers of hydrogenated amorphous silicon (a-Si:H). These are the amorphous analogs of nipi crystals for which interesting new effects were predicted, such as very long electron-hole recombination lifetimes and tunable electronic and optical properties[1].

EXPERIMENTAL DETAILS

The doping modulated a-Si:H films were prepared in a system that consists of two separate r.f. plasma deposition chambers as sketched in Fig. 1. The sample substrate S is mounted on a flat face of a stainless steel ball that is tightly held by copper gaskets but rotatable from the outside. Each plasma chamber has its separate gas

* Bell Labs Ph.D. Scholar
† Present address: Tata Institute, Bombay-400005, India.

0094-243X/84/1200425-08 $3.00

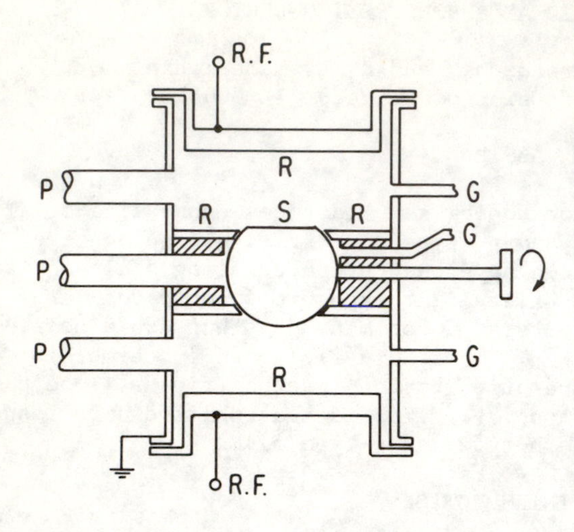

Fig. 1. Sketch of multilayer deposition chamber, S=sample, R=reference samples, G=gas inlet, P=pump line.

supply G and exit pump line P. Cross contamination between the chambers is further reduced by a constant flow of inert gas through the differential chamber that surrounds the ball. The pressure of the inert gas exceeds those of the two plasma chambers. Besides the substrate S onto which alternating layers of n- and p- doped a-Si:H are grown, there are stationary substrates R in both chambers. These enable one to grow reference samples of just n- and p- layers under the same conditions as for the npnp .... film on substrate S. All substrates, those on the grounded plates as well as on the plates connected to the r.f. source, are heated to 500°K.

We used 100 ppm PH_3 in SiH_4 and 100 ppm B_2H_6 in SiH_4 in the n- and p- chambers, respectively. The plasma is extinguished before the ball is rotated to the other chamber. The same apparatus was used[2] to grow

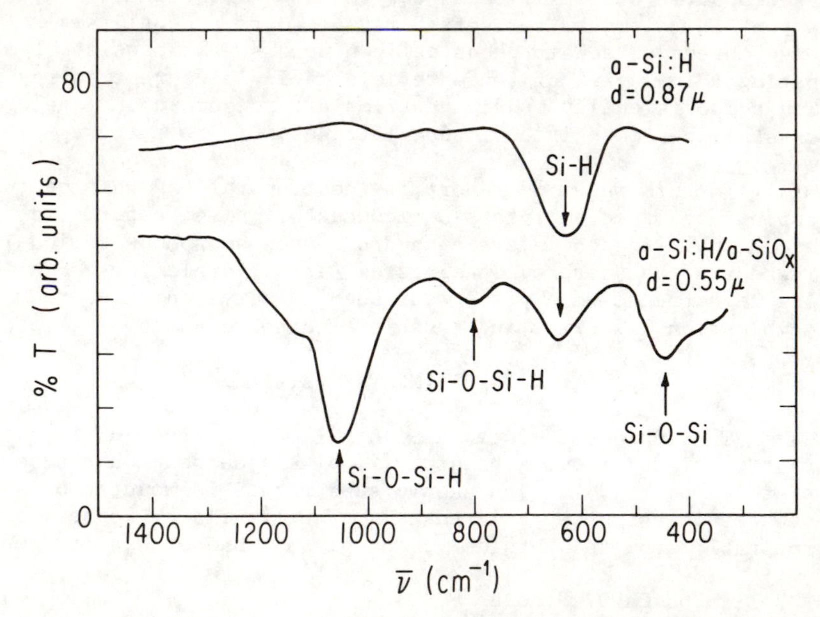

Fig. 2. Infrared spectra of a-Si:H/SiO_x sample and an a-Si:H reference. The latter does not show oxygen contamination.

compositionally modulated multilayers consisting of undoped a-Si:H and a-SiO_x. In this case the lower chamber was used for an oxygen plasma for growing the alternate layers of SiO_x. This preparation sequence enabled us to test the isolation quality of the differential chamber. Figure 2 compares the infrared absorption spectra of the multilayer sample S and of a reference sample R. There is no trace of a Si-O absorption band in the reference spectrum. The absence of cross contamination can also be checked by spectroscopic analysis of the silane plasma.

PHOTOLUMINESCENCE

Part of the 7059 glass substrate S and of reference substrates R had been roughened with #600 mesh SiC powder. These areas were used for photoluminescence (PL) studies in order to avoid interference fringes and to enhance the PL light collection. The latter is quite small in the case of smooth films because of total internal reflections. Good reproducibility of the roughening procedure permit meaningful comparison of the PL intensity of different samples. The PL spectra were measured at 78°K with an excitation intensity of $2X10^{18}$ photons/cm^2s of $h\nu$=2.41 eV. Figure 3 shows the PL spectra of a npnp .... layer containing eight 500 Å thick n-layers and eight 500 Å thick p- layers. Its peak is blue-shifted by 0.08 eV and the PL intensity is decreased by a factor 4 compared to our standard undoped films. The PL of the n-type and p-type reference samples is slightly red-shifted as expected for doped samples. After 15 min laser light exposure at 78°K we observed a PL fatigue of 7,11,19, and 10 percent in the undoped, n-type, p-type, and npnp... film, respectively. This fatigue is probably due to photo-induced creation of dangling bond defects that can be removed by annealing above 470°K.[7] We cannot explain the PL blue-shift of the doping modulated sample. The internal fields should be screened out at the high excitation intensity used. Furthermore, all the samples mentioned above had the same optical gap E_o=1.81 eV.

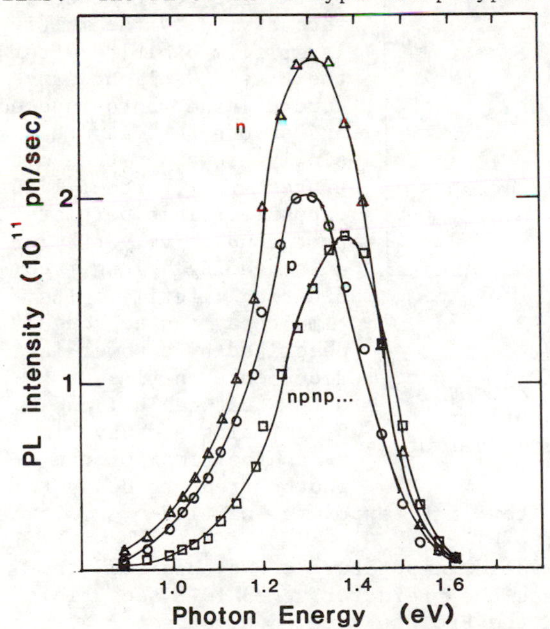

Fig. 3. Photoluminescence spectra at 78°K doping multilayer (blue-shifted) and n-type and p-type reference samples; d_n=d_p= 500 Å.

The multilayer samples were scratched with a diamond scribe before applying carbon paint to assure that ohmic contacts were made to all layers. The electrodes were separated by 0.1 cm and were 0.5 cm long. The in-plane sheet conductance was measured after annealing the sample at 450°K for 1h in an oil-free vacuum. The effects of adsorbates and of prior light exposure are eliminated in this annealed state A. Starting from state A a multilayer film having 11 n-layers and 10 p- layers, each 275 Å thick, was exposed repeatedly for 20s to 50 mW/cm^2 heat-filtered light ($h\nu > 1.5eV$) from a tungsten-halogen lamp at 300°K. The result is shown in Fig. 4 where the total film thickness was used for calculating the conductivity. Figure 5 shows for comparison the results obtained in the same manner with the n-type and p-type reference samples that were prepared with this multilayer film. One observes some remarkable differences: first, the photoconductivity σ_p of the multilayer film is about a factor two larger than σ_p of the n-type reference; secondly, only the multilayer film exhibits after the brief 20s light exposure a large photo-induced excess conductivity that slowly grows and then saturates after further light exposure. This persistent photoconductivity (PPC) has an exceedingly long recombination lifetime. Its time decay can be roughly described by a power-law dependence on time, $t^{-\alpha}$, with α between 0.1 and 0.2. By defining a decay time τ_R as the time for the photocurrent to decay to half the value it had 4 min after light exposure, we find

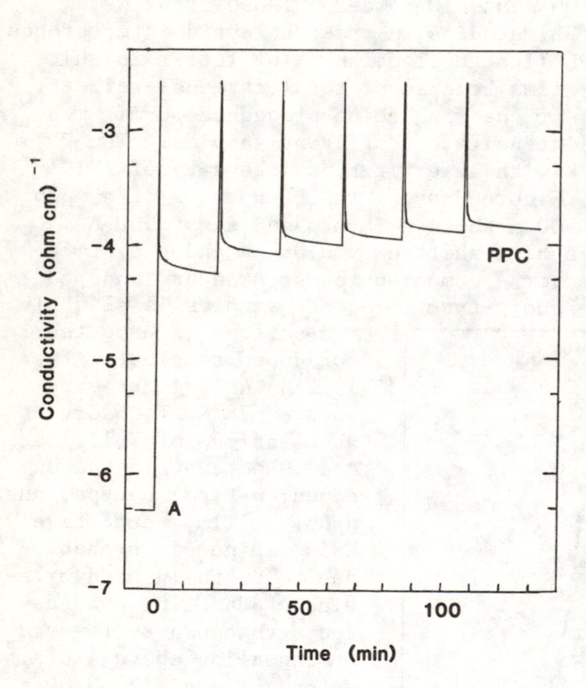

Fig. 4. Effect of six 20s light exposures on conductivity of doping multilayer at 300°K, $d_n=d_p=275$ Å.

$$\tau_R = \tau_{RO} \exp (E_R/kT) \quad (1)$$

by measuring the decay at elevated temperatures. We obtained[6] the activation energy $E_R=0.5$ eV and the prefactor $\tau_{RO}=3X10^{-5}$ sec. This means that after 10h at 300°K the PPC has decayed by a factor 4.

Figure 6 shows σ_p and the PPC effect of a doping modulated a-Si:H film that has 6 n-layers and 5 p-layers each 500 Å thick. The

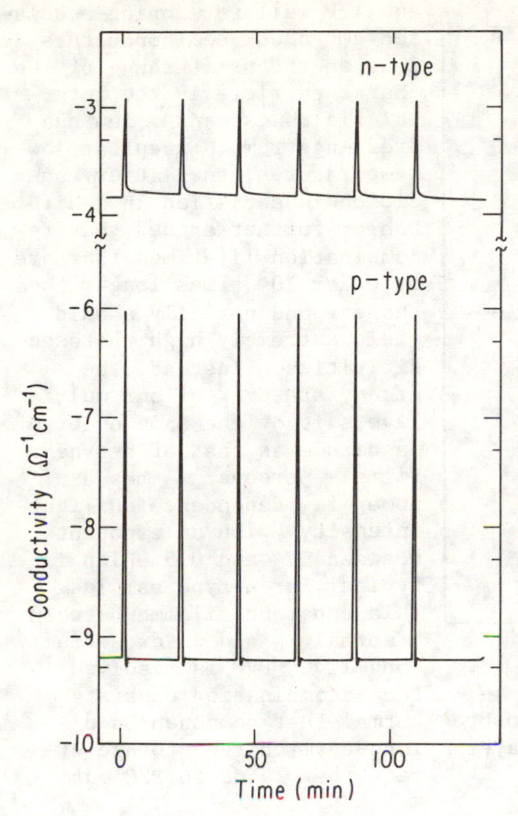

Fig. 5. Effect of six 20s light exposures on conductivity of n-type and p-type reference samples.

exposure time was 15 sec and the light intensity was identical to that used for Fig. 4. The steady state value of σ_p was 10^{-2} ohm^{-1} cm^{-1} compared to $7.2x10^{-3}$ ohm^{-1} cm^{-1} of the n-type reference sample.

Figure 7 shows the annealing curves of the PPC effect for the same sample. At moderate temperatures the PPC curves are sufficiently stable for determining an approximate value of the conductivity activation energy E_a. Figure 7 was taken after a 15s light exposure followed by a 17h relaxation period. E_a as well as σ_o of the PPC state are greatly reduced when annealing is started from the saturated PPC state. Above 450°K the PPC quickly anneals. It is interesting to note that the annealed state A is approached from below at the highest temperatures. This suggests that we are dealing with two light-induced effects that anneal at different temperatures. The second may in this case be the Staebler-Wronski effect[7] even though the samples received a light exposure that is about 100 times less than normally used for creating a substantial number of Staebler-Wronski defects. We reported elsewhere[6] that we were unable to quench the PPC by infrared light of energy, $h\nu<1.1$ eV. On the contrary, this low energy light enhanced the PPC when the PPC was not already saturated. We also observed that the steady state σ_p produced by $h\nu<1.1$ eV light was more than 5 times larger when the PPC state was illuminated instead of the annealed state A.

DISCUSSION

Döhler[1] predicted that crystalline doping superlattices should exhibit persistent photoconductivities and exceedingly long recombination life-times. He argued that the photo-excited electron-hole pairs will become spatially separated by the internal p-n junction fields. Recombination is then possible only by tunneling or by thermal excitation over the potential barriers. These arguments hold

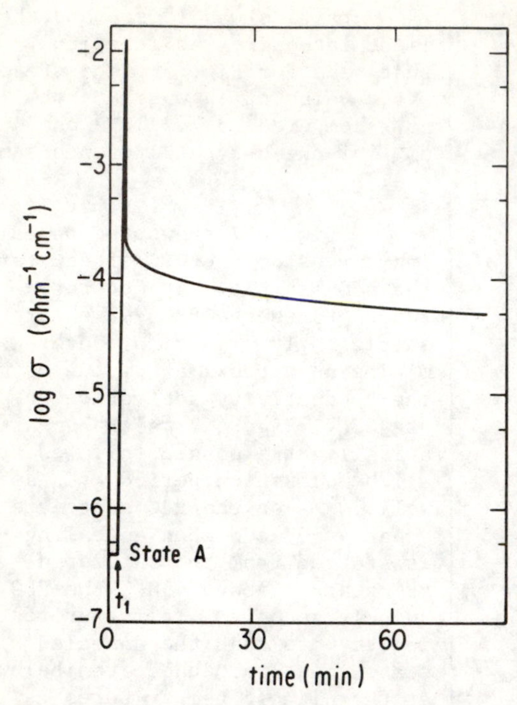

Fig. 6. Effect of 15s light exposure on conductivity of doping multilayer at 300°K; d_n=d_p=500 Å.

equally well for doping-modulated amorphous semiconductors as long as the drift range of the charge carriers in the internal fields exceeds a few hundred Angstrom. Except at low temperatures, the latter condition is satisfied in a-Si:H. Döhler further argued that recombination lifetimes that are of order 10^8 times longer than those found normally should yield extremely high photosensitivities. Instead, the steady state σ_p of our multilayers is of the same order of magnitude as that of n-type films. Moreover σ_p has a power-law dependence on light intensity[6] with an exponent between 0.4 and 0.5 which is typical for n-type samples. This apparent dilemma between a normal σ_p and an exceedingly long-lived PPC is resolved if the recombination consists of a normal fast component and a long-term charge storage effect that leads to PPC but that is self-limiting. The separation of charges by the internal fields is naturally self-limiting since these separated charges tend to neutralize the intrinsic space charge modulation and hence the internal fields. Yet internal barriers of some height, the value of which depends on temperature, are needed to retard recombination. This determines the saturation value of the PPC in this model. A normal fast recombination process is always present because a fraction of the material is effectively field-free. At higher light intensities the fields are screened out and the slow recombination process disappears altogether. Hence both relaxation processes will depend on the position of the quasi-Fermi levels as well as on the local fields and doping levels.

The annealing curve shown in Fig. 7 is strikingly similar to those observed by Aker et al.[8] after illuminating p-type a-Si:H films that had a surface oxide layer. Those authors explained their observation of a large photo-induced excess conductivity by an increase of the negative charge in the oxide. This charge separation is similar to what we believe occurs in our doping multilayers. Figure 8 shows however that there are significant differences in the build-up rates for the excess conductance. The light intensity was essentially the same for both experiments. The excess conductances were measured

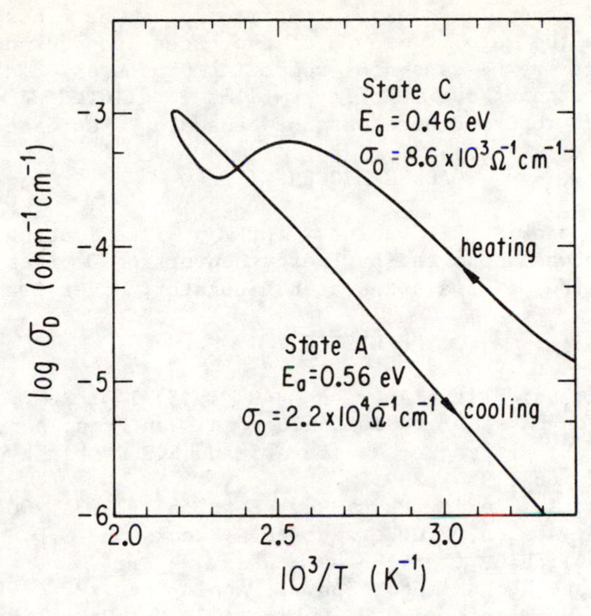

Fig. 7. Annealing of persistent photoconductivity of $d_n = d_p = 500$ Å multilayer film.

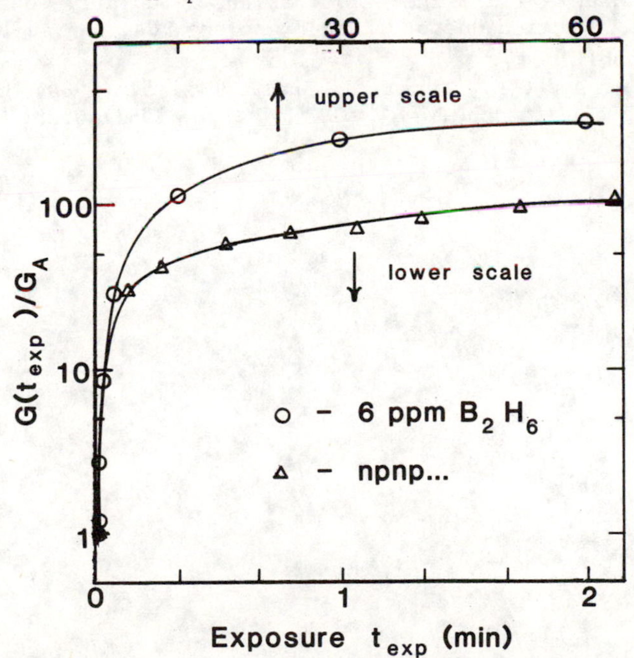

Fig. 8. Increase of excess conductivity at 300°K with exposure time of $d_n = d_p = 500$Å multilayer and p-type doped a-Si:H.

as a function of accumulated exposure time t_{exp}. One finds a factor 30 difference in the rate of growth of the excess conductance. This difference though may be caused by competing processes of different magnitudes. By etching our multilayer films in diluted HF we made sure that the PPC reported here was not caused by a surface oxide.

ACKNOWLEDGEMENT

We wish to thank S. R. Ovshinsky for many illuminating discussions. The work was supported by Energy Conversion Devices, NSF DMR 9008225, and by the Materials Research Laboratory under Grant 79-24007.

REFERENCES

1. G. H. Dohler, Scientific American 249 (1983) 144; Prof. of the 17th Intl. Conf. on Physics of Semiconductors, San Francisco,CA,8/1984.
2. J. Kakalios, H. Fritzsche, N. Ibaraki and S.R.Ovshinsky, J. Non-Cryst. Solids 66 (1984) 339.
3. B. Abeles and T. Tiedje, Phys. Rev. Lett. 51 (1983) 2003.
4. T. Tiedje, B. Abeles. P.D.Persans, B.G.Brooks and G.D.Cody, J. Non-Cryst. Solids 66 (1984) 345.
5. H. Munekata and H. Kukimoto, Jpn. J. Appl. Phys. 22 (1983) 6544; H. Munekata, M. Mizuta, and H. Kukimoto,J. Non-Cryst. Solids 59-60 (1983) 345.
6. J. Kakalios and H. Fritzsche, Proc. of the 17th Intl. Conf. on the Physics of Semiconductors, San Francisco, CA, Aug. 1984; and to be published.
7. D.L. Staebler and C.R.Wronski, J. Appl. Phys. 51 (1980) 3262.
8. B. Aker and H. Fritzsche, J. Appl. Phys. 54 (1983) 6628.

ELECTROABSORPTION MEASUREMENTS OF INTERFACE CHARGES IN a-Si:H/a-SiN_x:H SUPERLATTICES

C. B. Roxlo, T. Tiedje and B. Abeles
Exxon Research and Engineering Company
Corporate Research Laboratories
Annandale, New Jersey 08801

ABSTRACT

Electroabsorption spectroscopy shows that large (4×10^5 V/cm) electric fields exist in a-Si:H/a-SiN_x:H superlattice materials. These fields are due to interface charges which are assymmetrically distributed, showing that interface properties strongly depend upon the order of deposition. The charge density observed is compared with that obtained from optical, resistivity and photoluminescence measurements.

Amorphous semiconductor interfaces are important in a wide variety of semiconductor devices, including solar cells and thin film transistors. The introduction of amorphous semiconductor superlattices[1] has provided an opportunity to study interfacial properties, because of the high concentration of interfaces. We have reported[2] electroabsorption measurements which show that the interface charge in a-Si:H/a-SiN_x:H superlattices depend strongly on the order of deposition. The interface where silicon is deposited onto silicon nitride has a charge density up to 6×10^{12} cm^{-2} higher than the opposite interface. In this paper we compare these results with other measurements, including optical absorption, resistivity, and photoluminescence. Photothermal Deflection Spectroscopy (PDS) measurements show that the defect concentration in unlayered a-SiN_x:H is much lower than that in the superlattices. This and other data indicate that the electronically active defects in superlattice materials are concentrated at the interfaces. Comparison of the results gives more information about the charge distribution at these interfaces.

The samples used here were prepared using techniques described earlier.[1] They consisted of alternating layers of amorphous silicon and amorphous silicon nitride, with sublayer thicknesses varying from 8Å to 1200Å.

In crystalline materials electroabsorption and electroreflection spectroscopies are often used to determine critical points in the band structure.[3] In amorphous semiconductors, the spectra do not show sharp structure but instead reach a broad maximum at energies near the bandgap, decreasing proportional to the absorption coefficient below the gap.[4] Because the electroabsorption signal is proportional to the square of the electric field, it can be used to determine internal, or built-in, fields in a sample subjected to

0094-243X/84/1200433-08 $3.00

both alternating and constant fields. In this case, as we will show below, the signal goes to zero when the applied dc field cancels out the built-in field. Nonomura et. al.[5] have used this technique to determine the built-in potential of n-i-p amorphous silicon solar cells.

We have used electroabsorption to measure fields in the silicon sublayers of superlattice structures.[2] In this case, it is important that the applied fields, especially the ac field, be dropped uniformly across the sample. Because of the nitride layers, these superlattices have very high ($\sim 10^{12}$ Ω-cm) perpendicular resistivities. This yields a dielectric relaxation time (~ 1 sec) which is much slower than the modulation frequencies used (1 KHz). In addition, we do not expect that sufficient charge movement will occur within a single silicon sublayer to significantly alter the fields, because the depletion width of bulk a-Si:H is several thousand Ångstroms, greater than the thickest silicon sublayers used here.

Experimentally, the built-in fields measured were changed by less than 10% when the modulation frequency was increased to 100 KHz and the dc voltage was applied as a pulse of only 3 ms duration. Experiments done at a sample temperature of 50 K, where all charges are virtually immobile, also showed little change. These experiments prove that the applied fields are uniform.

Thus we can make use of the boundary condition that the component of the applied electric displacement $D=\epsilon E$ normal to the layers is constant. The field within the sample is the sum of applied and built-in fields:

$$D(x) = D_{dc} + D_{ac} \cos(\omega t) + D_{bi}(x) \tag{1}$$

Here D_{dc} and D_{ac} are the constant and modulated components of the applied field, and ω is the modulation frequency. D_{bi} is the built-in field which depends on distance x normal to the layers. Because of the local isotropy of an amorphous material, symmetry requires that the change in absorption, $\Delta\alpha$, vary as the square of the field

$$\Delta\alpha(x) = \alpha_0(x)\, K(x)\, D^2(x) \tag{2}$$

where α_0 is the zero-field absorption coefficient and K is an electro-optic coefficient. Squaring the field in Eq. (1) yields several terms; a lock-in amplifier is used experimentally to isolate that term which varies at the frequency ω:

$$D^2|_{\omega} = 2\,(D_{bi} + D_{dc})D_{ac} \tag{3}$$

Because of the high bandgap of a-SiN_x:H, the nitride layers do not contribute to the electroabsorption signal. In addition we assume that all silicon sublayers are identical so the electroabsorption signal observed in transmission equals the spatial average of $\Delta\alpha(x)$ over a single silicon sublayer:

$$\Delta\alpha_\omega = (2\alpha_{os}K_s D_{ac}/L_s)\int_0^{L_s}(D_{bi}(x) + D_{dc})\,dx \qquad (4)$$

Here L_s is the thickness of a single silicon sublayer; α_{os} and K_s are constants of a-Si:H. It can be seen from this equation that $\Delta\alpha_\omega$ goes to zero when the applied dc field cancels out the built-in field. Expressed in terms of the applied voltages V_{ac} and V_{dc}, this is

$$\Delta\alpha_\omega \sim V_{ac}\ [\phi_S - \frac{V_{dc}}{N}\ (1 + \frac{L_N\varepsilon_S}{L_S\varepsilon_N})^{-1}\ ] \qquad (5)$$

Here ϕ_S is the built-in electric potential across a single silicon layer. Positive ϕ_S describes a field which points away from the substrate. The dielectric constants are taken as $\varepsilon_S = 12\ \varepsilon_o$ for the silicon sublayers and $\varepsilon_N = 7.5\ \varepsilon_o$ for the nitride.

Figure 1 shows the signal $\Delta\alpha_\omega$ plotted against V_{dc} with V_{ac} as a parameter, for a sample with sublayer thicknesses $L_S = 12$ Å and $L_N = 35$Å, and N=332 layer pairs. In this experiment a He-Ne laser at 1.96 eV was used as a light source to provide excellent signal to noise. This data shows that Eq. (5) is obeyed to a very high degree of accuracy. The points for the three V_{ac} values were least-squares fitted to obtain three lines which intersect at $\Delta\alpha_\omega$=0 as shown in the inset. The slope of the lines was proportional to the ac voltage within 0.3%. The intercept of $V_{bi} = 79.58 \pm .02$ V determine the single layer built-in potential as

$$\begin{aligned}\phi_S &= (V_{bi}/N)\ (1 + L_N\varepsilon_S/L_S\varepsilon_N)^{-1} \\ &= 42\ \text{meV}\end{aligned} \qquad (6)$$

This potential implies a built-in electric field of 4×10^5 V/cm in the silicon sublayers. Because the material as a whole has no net field of this magnitude, this field must be opposed by an inverse field in the nitride. The electroabsorption technique measures only fields which are asymmetrical in this way. Such fields can only be caused by interface charges which are asymmetrically distributed. The sign of the charges relative to the direction of growth is known from the sign of the built-in voltage. The simplest such charge distribution is that where a positive charge σ present on the interface where silicon is deposited onto silicon nitride is balanced by a negative charge on the opposite interface. For this charge distribution, simple electrostatics gives the silicon sublayer potential as

$$\phi_S = \sigma e L_S L_N/(L_S\varepsilon_N + L_N\varepsilon_S) \qquad (7)$$

For the material of Figure 1, this gives a surface charge of $\sigma = 2.8 \times 10^{12}\ \text{cm}^{-2}$ at each interface, with the sign of the charge alternating with the order of deposition. This is the charge distribution which maximizes the built-in potential for a given amount of charge. In a distribution which has some charge placed symmetrically or which is more evenly distributed spatially, more charge would be required. Location of all charge at one interface would require 6×10^{12} cm^{-2}.

Figure 2 shows the built-in potential ϕ_S for materials with a number of different sublayer thicknesses. ϕ_S is relatively independent of the number of layer pairs, which varied for these samples from 10 to 700. The ~25% scatter in the data is presumably due to slightly different deposition conditions; measurements on different areas of the same sample yield results which agree to within 5%. The potential generally increases with the nitride thickness L_N, although this dependence is difficult to characterize with this limited selection of samples. The dependence on silicon thickness L_S is given by the black dots for L_N = 35Å. ϕ_S is greatest in the L_S = 12-160Å range and decreases by a factor of 3 for thicker L_S. The interface charge distribution given above in Eq. (7) would predict ϕ_S increasing linearly with L_S and then saturating above 60Å, contrary to the data.

We have shown[2] that the dependence given in Figure 2 can be explained using a slightly more detailed charge distribution. In this model, a positive interface charge of 6×10^{12} cm^{-2} at the interface where silicon is deposited onto silicon nitride is balanced by adjacent electrons in the silicon. These electrons are exponentially distributed over the 20Å near the interface.

The structure and interatomic spacing of a-SiN_x:H differs substantially from that of a-Si:H. This results in considerable strain at the interface which we expect to be relieved through the introduction of defects. We attribute the striking dependence of interface charge on order of deposition to charges located at such defects. These defects may be distributed asymmetrically because it is easier to introduce defects into material as it is growing than it is into material which has already been deposited.

Optical measurements also show that these charges are concentrated at the interfaces. The charge density derived above would be equivalent to a bulk defect density of over 10^{19} cm^{-3} if distributed uniformly throughout the layers. It is well known that the electrically active defect density in bulk a-Si:H is much lower than this. To determine the density of defects intrinsic to bulk a-SiN_x:H, we have performed Photothermal Deflection Spectroscopy measurements on unlayered nitride of a composition representative of that used in the superlattices (Figure 3). This spectrum shows an optical bandgap of 3.2 eV and an exponential edge which extends throughout the entire visible region, continuing to the $\alpha = 3$ cm^{-1} level at 1.2 eV. This edge is much broader (291 meV) than that observed in a-Si:H (60 meV). This might be due to defects near the valence band maximum as proposed by Robertson and Powell.[6] Absorption from these defects might be broadened into an exponential edge by the high level of disorder present in this material. The very low level of absorption present at 1.2 eV can be used to determine an upper bound to the density of midgap defects. Photoemission measurements[7] show that the Fermi level in this material is 2.4 eV above the valence band, so that defects immediately below the Fermi level would absorb light at photon energies above 0.8 eV. Assuming

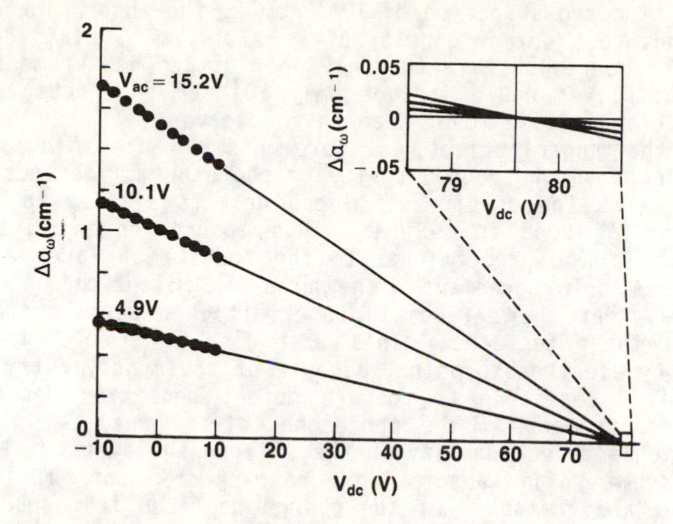

Fig. 1: Electroabsorption signal vs. V_{dc} for three values of the ac voltage.

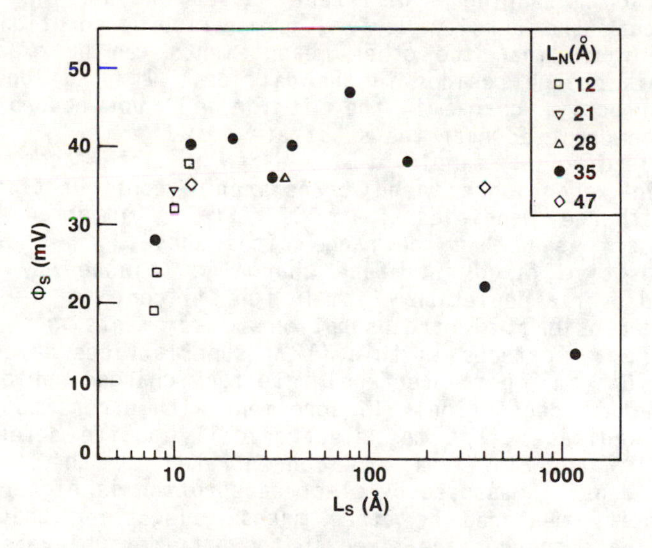

Fig. 2: Built-in potential across a single silicon sublayer ϕ_S for materials with silicon sublayer thicknesses L_S and nitride thicknesses L_N.

an absorption cross-section of 10^{-16} cm^{-2}, the absorption at 1.2 eV corresponds to a defect density of 3 x 10^{16} cm^{-2}. This density is too low to be responsible for the superlattice measurements; a 35Å a-SiN_x:H sublayer would include only 10^{10} cm^{-2} defects, more than an order of magnitude lower than that observed.

In the superlattices, measurements of the midgap optical absorption using photoconductivity or photothermal deflection spectroscopy gives the density of midgap defects similar to the well-known dangling bond in a-Si:H. This absorption (at 1.1 eV) has been shown[8] to be proportional to the density of layers, as shown in Figure 4. In agreement with the a-SiN_x:H optical measurement, this shows that the defects in superlattice materials are concentrated at the interfaces. This must also be true of the charges measured by electroabsorption, because of their assymmetry.

It is interesting to compare defect densities determined by various measurements. The photoconductivity results in Figure 4 give a midgap defect density in the silicon sublayers of 1.4 x 10^{11} cm^{-2} (assuming an absorption cross-section of 10^{-16} cm^{-2}). Similarly, measurements[8] of the charge depleted from superlattices near the substrate yield 2 x 10^{11} cm^{-2} charged states per layer. Photoluminescence efficiency measurements[9] show that the density of non-radiative recombination centers is 8 x 10^{11} cm^{-2} per interface.

The wide variation in these measurements, from 1.4 x 10^{11} cm^{-2} to 3 x 10^{12} cm^{-2} per interface, may partly reflect the fact that they are each measuring a different subset of interface defects. The fact that the density derived from electroabsorption is considerably higher than the other measurements can be explained if most of this charge resides in the nitride. If positioned in the bandgap correctly, charge in the nitride will not absorb light at 1.1 eV or be depleted near the substrate.

Band-bending at single a-Si:H/a-SiN_x:H interfaces has been observed in xerographic[10] and transistor[11] configurations. The interface charge densities (5 x 10^{11} cm^{-2}) inferred from the measurements[11] is within the range cited above. In these structures a positive fixed interface charge is balanced by electrons distributed over a depletion width in the silicon.

In conclusion, electroabsorption measurements show that a-Si:H/a-SiN_x:H interfaces in thin (12Å) superlattices have substantial (3 x 10^{12} cm^{-2} per interface) interface charges, which depends on the order of deposition. In agreement with other measurements, this data indicates that the electronically active defects in a-Si:H/a-SiN_x:H superlattices are concentrated at the interfaces. The charge density measured by electroabsorption is higher than the defect density measured by other means, suggesting that the majority of these charges are fixed in the nitride sublayers near the interfaces.

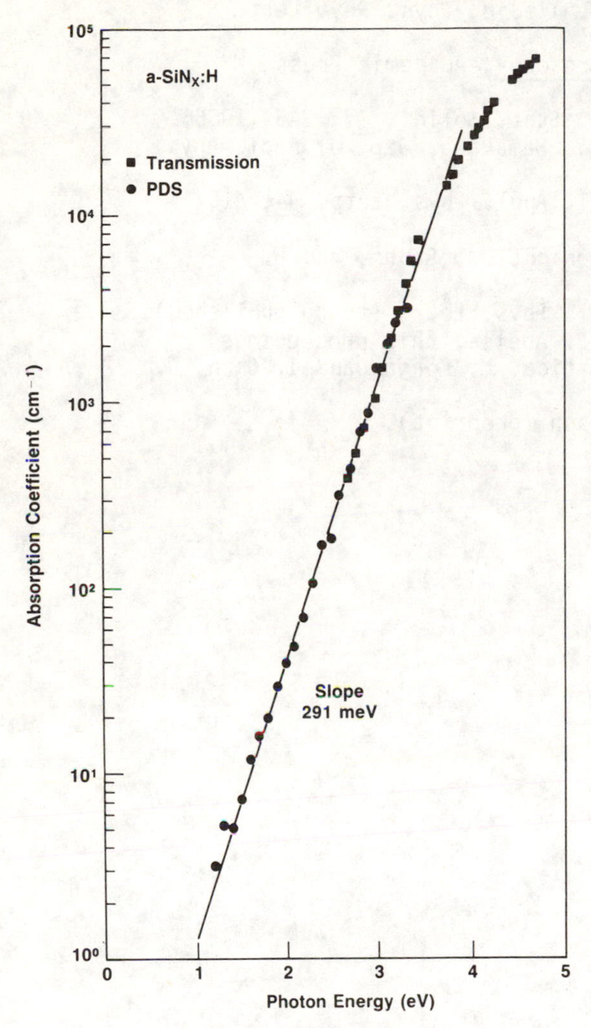

Fig. 3: Optical absorption spectrum of a-SiN$_x$:H determined using Photothermal Deflection Spectroscopy (PDS) and Transmission measurements.

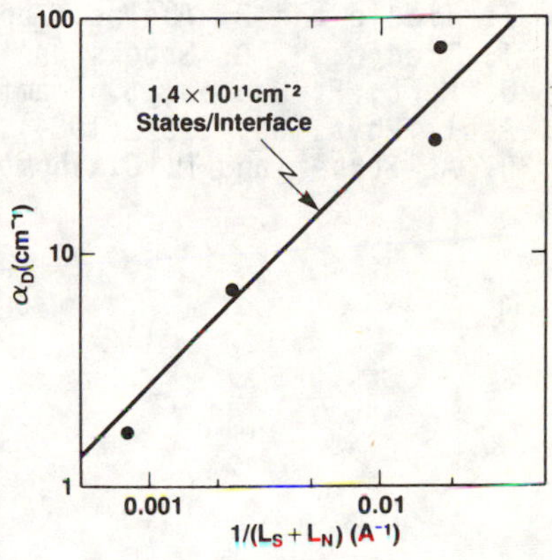

Fig. 4: Density of silicon defects determined from the photoconductivity spectrum α_D at 1.1 eV, plotted against the density of layers. The line gives the behavior expected from materials with 1.4×10^{11} defect states per interface and an absorption cross-section of 10^{-16} cm^2.

REFERENCES

1. B. Abeles and T. Tiedje, Phys. Rev. Lett. 51, 2003 (1983).
2. C. B. Roxlo, B. Abeles and T. Tiedje, Phys. Rev. Lett. 52, 1994 (1984).
3. M. Cardona, Modulation Spectroscopy, Academic Press, NY (1969).
4. J. Stuke and G. Weiser, Phys. Stat. Solidi, 17, 343 (1966).
5. S. Nonomura, H. Okamoto and Y. Hamakawa, Jap. J. Appl. Phys. 21, L464 (1982).
6. J. Robertson and M. J. Powell, Appl. Phys. Lett. 44, 415 (1984).
7. B. Abeles, I. Wagner, W. Eberhardt, J. Stöhr, and H. Stasiewski, this proceedings.
8. T. Tiedje and B. Abeles, Appl. Phys. Lett. (to be published).
9. T. Tiedje, B. G. Brooks and B. Abeles, this proceedings.
10. J. Mort, F. Jansen, S. Grammatica, M. Morgan and I. Chen, J. Appl. Phys. 55, 3197 (1984).
11. R. A. Street and M. J. Thompson (preprint).

OPTICAL AND RAMAN INVESTIGATION OF AMORPHOUS POLYPHOSPHIDES

D. Olego, R. Schachter, J. Baumann, M. Kuck and S. Gersten
Stauffer Chemical Company, Eastern Research Center,
Elmsford, N.Y. 10523

ABSTRACT

The intrinsic properties of amorphous MP_{15} polyphosphides (M = alkali metal) have been investigated. Attention is focused on thin films of KP_{15} grown at different substrate temperatures. Results of Raman, photoluminescence and transmission spectroscopy are presented. The structure of the MP_{15} polyphosphides consists of parallel unidimensional P tubes of pentagonal cross-section. The order parameters of the structure are the intratube and intertube correlations. Changes in the order parameters with substrate temperature have been determined and correlated with the optical properties. It is shown that a continuous transition from the ordered to the disordered structure takes place for MP_{15} polyphosphides.

INTRODUCTION

In this publication we report about vibrational and opto-electronic properties of KP_{15} thin films measured by Raman, photoluminescence and transmission spectroscopy. The KP_{15} thin films, typical representatives of the amorphous MP_{15} polyphosphides (M = alkali metal), are novel non-tetrahedrally bonded semiconducting materials.[1] They also constitute a very interesting model system in which to investigate changes in range order in the amorphous network when more than one order parameter is involved.[2]

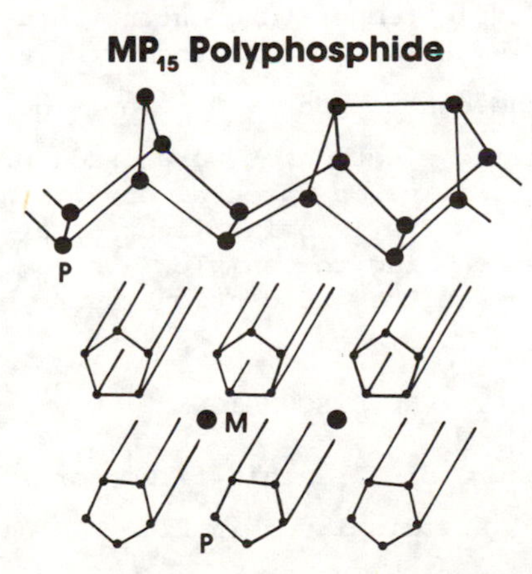

Fig. 1 View of the pentagonal P tube and of the MP_{15} crystalline structure

The disordered network of the KP_{15} thin films is derived from the low symmetry MP_{15} crystalline structure. The MP_{15} compounds are triclinic crystals with the P atoms arranged in the form of infinite tubes with pentagonal cross-section.[3] The tubes are aligned parallel to each other with bonding provided by the metal atoms M and by van der Waals forces. Figure 1 shows the structure of one tube and the arrangement of tubes in the MP_{15} crystals. The relevant order parameters (long - or intermediate range) of the MP_{15} structures are the intertube and intratube correlations (among tubes and within a tube).

The degree of order in the network of the KP_{15} thin films can be altered by growing the films at various substrate temperatures T_S.[1,2] We have studied the changes in the order parameters with T_S by following the behavior of Raman peaks corresponding to intertube and intratube vibrational modes and their correlation with photoluminescence and transmission measurements. It can be concluded from these investigations that for the MP_{15} systems a continuous transition takes place from the ordered structure to the disordered network.

EXPERIMENTAL

The KP_{15} thin films were grown on glass substrates by vapor transport methods in a sealed quartz ampoule from separate K and P sources with T_S as a variable parameter. The temperature of the K source was $\sim$ 620K and that of the P source $\sim$ 770K. The thicknesses of the films were in the range 1 to 5 μm. The compositions of the films were determined by means of X-ray fluorescence. The films were also characterized by X-ray diffraction. For $T_S \lesssim$ 600K the films are amorphous to X-rays. When $T_S >$ 600K the films are polycrystalline.

The Raman and photoluminescence measurements were taken in the back-scattering geometry with the samples attached to a cold finger of a closed cycle cryostat. The sample temperature during these measurements was $\sim$ 12K. The Raman spectra were excited with laser light at 6000 Å, and the photoluminescence spectra were obtained with light at 5260 Å. Photon counting electronics were used to detect the scattered light. The transmission spectra were taken with the samples at room temperature. The incoming light of a Xe-lamp was monochromatized before reaching the sample. The transmitted intensity was normalized to the incident one and detected by a photomultiplier.

RAMAN SCATTERING: RESULTS AND DISCUSSIONS

Typical first order Raman spectra of KP_{15} thin films for different substrate temperatures T_S are displayed in Fig. 2. The spectra of Fig. 2 show the evolution of the Raman modes when progressive disorder is introduced in the network of the thin films with decreasing T_S. This evolution will be analyzed in detail for

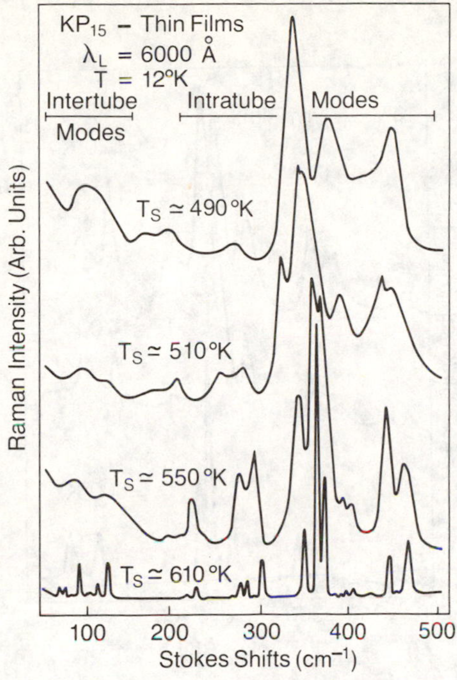

Fig. 2 Raman spectra of KP_{15} thin films for different substrate temperatures

three temperature ranges: $T_S >$ 600K, $520 \lesssim T_S \lesssim 600$K and $T_S \lesssim 510$K.

When $T_S >$ 600K the Raman spectra of the KP_{15} thin films show a large number of very sharp lines. The spectrum of Fig. 2 for $T_S \simeq$ 610K is a typical example. The frequencies and linewidths of the peaks are indistinguishable from those measured in the spectrum of single crystal KP_{15}.[2,4] The Raman peaks in the frequency range up to 140 cm^{-1} correspond to intertube vibrations. These are low frequency modes because they involve vibrations of pentagonal tubes against each other (very large masses) and weak restoring forces. Above 200 cm^{-1} the Raman spectra consist of intratube modes. They correspond to vibrations that propagate along the covalently bonded P atoms of a tube. The intrinsic properties of the intratube and intertube modes of MP_{15} crystals have been discussed in detail in ref. 4. The large number of Raman modes are expected because the MP_{15} unit cell has 32 atoms. The results of Fig. 2 for polarized and depolarized spectra are the same, an effect attributed to the fact that Γ_1 is the only symmetry element of the MP_{15} structure.

The sharp intratube and intertube modes measured for $T_S >$ 600K indicate the presence of long range correlations in the microscopic structure of the films. The existence of a well ordered structure is confirmed by X-ray diffraction patterns that gave indications of polycrystallinity. The crystallization temperature of KP_{15} can be estimated to be 610 $\pm$ 10K from the Raman and X-ray diffraction results. This value is very close to two-thirds of 910 $\pm$ 10K, which is the melting temperature of KP_{15} measured by the authors.

Disorder is introduced in the structure of the films for T_S below the crystallization temperature. The spectra of Fig. 2

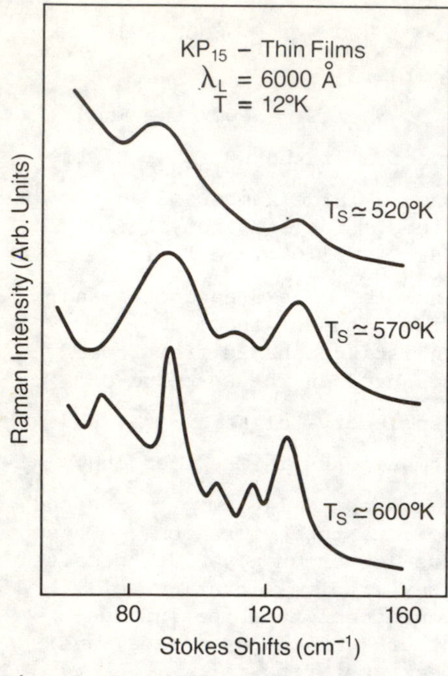

Fig. 3 Intertube modes of KP_{15} thin films in the substrate temperature range $600 \gtrsim T_S \gtrsim 520K$

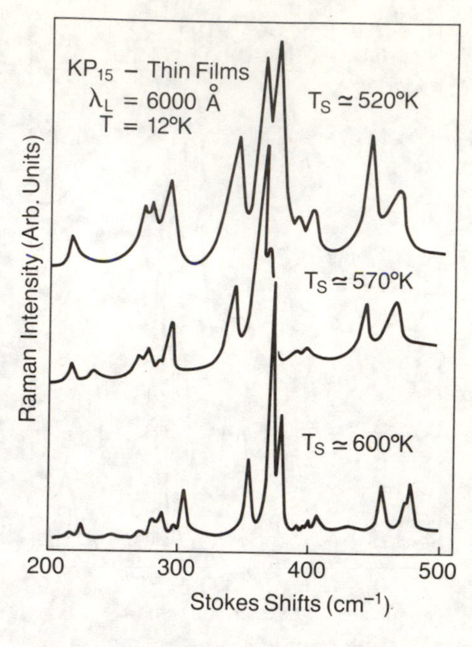

Fig. 4 Intratube modes of KP_{15} thin films in the substrate temperature range $600 \gtrsim T_S \gtrsim 520K$

indicate a progressive broadening of the Raman lines with decreasing T_S. The evolution of the intratube and intertube modes in the temperature range $600 > T_S > 520K$ is exemplified in Figs. 3 and 4. Long range correlations among tubes are lost in this range of T_S as indicated by broad bands replacing the sharp intertube peaks in Fig. 3. However, long range order within a tube is further preserved. The intratube modes can still be recognized in Fig. 4. This behavior is reminiscent of the common situation of disordered molecular crystals when the molecules retain their identities. For the MP_{15} polyphosphides, the tubes play the role of very large molecules.[4]

With decreasing T_S the intratube modes are shifted to lower frequencies and somewhat broadened as observed in Fig. 4, because of the uncertainty in the wave vector component related to the lack of intertube periodicity. The measured red shift and broadening of the intratube modes of the film grown at $T_S \simeq 520K$ are both ~ 10 cm^{-1}. The maximum dispersion of the intratube modes of single crystal KP_{15} is also ∿ 10 cm^{-1} measured with second order Raman scattering.[4] These observations point out that the uncertainty in the intertube wave vector component for $T_S \simeq 520K$ is on the order of the magnitude of the wave vector component itself. Therefore, it can be concluded that the intertube correlations are of intermediate range with a coherence length on the order of the distance between two tubes bridged by the metal atom in the MP_{15} structure.[3,4] Also attributable to the breakdown of intertube periodicity are the changes of relative intensities of some intratube modes of Fig. 4.

The long range correlations among the tubes are lost first because the weaker bondings in the MP_{15} structure are those provided by the intertube forces. The disparity in the strengths of the intertube and intratube bondings is reflected in the mode Grüneisen parameters, which have been determined for the KP_{15} crystals.[4] The Grüneisen parameters of the intertube modes are one order of magnitude larger than the corresponding parameters of the intratube modes.[4]

Progressive loss of intratube correlations takes place in the network of the films grown with $T_S < 520K$. Further broadenings, red shifts and changes of relative intensities of the intratube modes are observed in the Raman spectra shown in Fig. 2 for $T_S \simeq$ 510 and 490K. The lineshape of the spectrum of the film with $T_S \simeq$ 490K consists of broad but easily distinguishable bands which indicate the presence of intermediate range intertube and intratube correlations. Except for the band at around 100 cm^{-1} the other structures of the Raman lineshape of the film grown at $T_S \simeq 490K$ correspond to broadened modes of the fivefold P tubes. The richness of the spectrum of the disordered films reflects the large density of phonon states of the nearly dispersionless intratube modes of the MP_{15} systems. The lack of dispersion results from the low symmetry of the crystalline structure involved and the large number of atoms per unit cell.[4]

The spectrum of Fig. 2 for $T_S \simeq 490K$ is indistinguishable from the Raman spectra of the amorphous red P.[2,5] Therefore, the behavior of the Raman modes of Fig. 2 leads us to conclude that the phosphorus tubes shown in Fig. 1 are also the structural units of the amorphous red P network. Similar conclusions have been reached by investigating the vibrational properties of Hittorf's P and other P allotropes that have structures based on pentagonal tubes.[2] The nature of the network of amorphous red P inferred from the Raman investigations confirms the results of X-ray diffraction and photoemission spectroscopy.[6,7]

The transition from the ordered structure to the disordered network followed with the Raman spectra of Fig. 2 is a continuous one. It proceeds by first losing long range intertube correlations followed by the loss of intratube long range order. To the best of our knowledge, a continuous transformation of the microscopic structure has not been reported before for a semiconducting system. This transition is possible for the MP_{15} materials because of the low symmetry of the structure and of the different strengths of the bonding forces.[3,4]

PHOTOLUMINESCENCE AND TRANSMISSION SPECTROSCOPY: RESULTS AND DISCUSSIONS

The optical gaps of the thin films of KP_{15} for different T_S have been determined by means of photoluminescence and transmission spectroscopy. Typical experimental results are shown in Figs. 5 and 6. The photoluminescence lineshapes of Fig. 5 consist of two bands whose relative intensities depend on T_S. In the recombination spectra of KP_{15} single crystal, the stronger band at 1.8 eV corresponds to transitions across the energy gap and the weaker band at ∿ 1.45 eV is due to recombinations involving midgap defect states.[1,4]

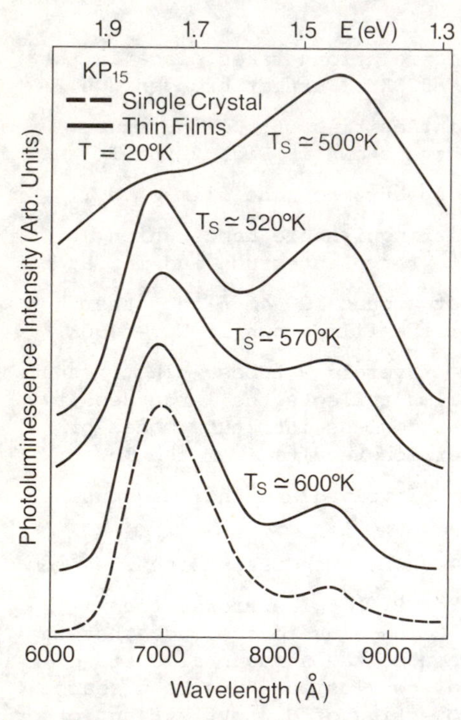

Fig. 5 Photoluminescence spectra of single crystal and thin films of KP_{15} for different substrate temperatures

Radiative recombinations at ∿ 1.8 eV are also measured in the photoluminescence spectra of KP_{15} thin films grown with T_S > 520K as shown by the data of Fig. 5. These observations imply that the optical gaps of the films are at the same energy as in the crystal, when intratube correlations are of long range order.[1] The same conclusions are obtained by measuring the transmission edges at room temperature. The band gap energies determined by this method of the films with T_S > 520K (Fig.7) and of the single crystal (Fig. 1 ref. 1) agree within the experimental error.

The retention of the spectral position of the energy gap in the film results from the intratube nature of the electronic states at the band edges.[4] This situation is comparable to the behavior of the intratube vibrational modes shown in Fig. 4.

The lack of long range intertube correlations for $T_S > 520K$ results in a strong dependence of the intensity of the photoluminescence bands at ∿ 1.8 eV with T_S. Between $T_S \simeq 620K$ and $T_S \simeq 520K$ the intensity decreases monotonically by a factor of ∿ 100. Other effects of the wave vector non-conservation are the broadening with decreasing T_S of the photoluminescence bands and transmission curves of Figs. 6 and 7 when compared with the single crystal spectra. It can also be noted in Fig. 5 that the relative intensity of the line at ∿ 1.45 eV increases with decreasing T_S indicating a progressively larger number of defect states with disorder.

When intertube and intratube intermediate range correlations are present ($T_S < 510K$), the photoluminescence spectra of the thin films display a broad band peaking at around 1.45 eV. The properties of this photoluminescence band are similar to those reported for the recombination spectra of amorphous red P.[8] The optical gap determined by transmission in Fig. 6 is ∿ 1.9 eV. This value is approximately 0.1 eV smaller than the optical gap determined[9] for amorphous red P. We conjecture that the presence of K in the amorphous red P - like network is responsible for the red shift of the optical gap.

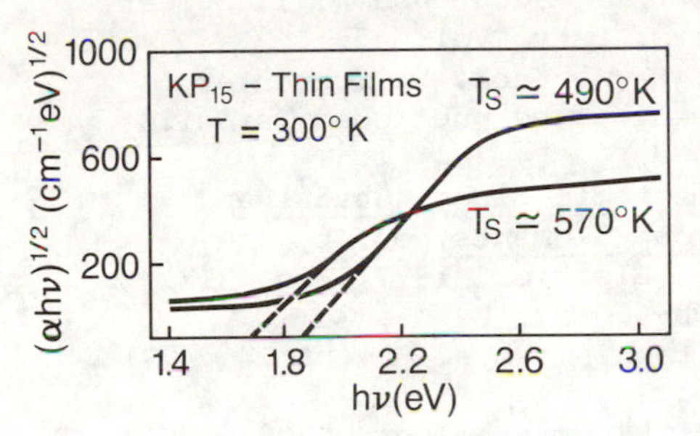

Fig. 6 Transmission spectra of thin films of KP_{15}

CONCLUSIONS

Raman, photoluminescence and transmission spectroscopy have been applied to investigate the vibrational and optical properties of amorphous KP_{15} polyphosphides. The behavior of these properties as a function of the growth parameters (in the present study of substrate temperature) has been correlated with the degree of microscopic order. We have shown that two order parameters are needed to characterize the amorphous MP_{15} network. The order parameters are the intratube and intertube correlations.

For the MP_{15} systems, a continuous transition takes place from the ordered structure to the disordered network. The transition proceeds by first losing long range intertube correlations followed by the loss of intratube long range order. The vibrational and electronic states of intratube nature retain their identities when long range intratube correlations are present.

In the presence of intermediate range intertube and intratube order the vibrational and optical properties of the MP_{15} polyphosphides are quite similar to those of amorphous red phosphorus. These similarities lead us to conclude that fivefold P tubes are the building blocks of the network of amorphous red phosphorus.

ACKNOWLEDGMENT

Discussions with C. Michel, L. Polgar and P. Raccah are gratefully acknowledged. We would like to thank S. Adler for his comments about the manuscript.

REFERENCES

1. R. Schachter, C. G. Michel, M. A. Kuck, J. A. Baumann, D. J. Olego, L. G. Polgar, P. M. Raccah and W. E. Spicer, to be published in Appl. Phys. Lett. Vol. 45.
2. D. J. Olego, J. A. Baumann, M. A. Kuck, R. Schachter, C. G. Michel and P. M. Raccah, to be published in Solid State Comm.
3. H. G. von Schnering, in Homoatomic Rings, Chains and Macromolecules of Main - Group Elements, ed. by A. L. Rheingold (Amsterdam: Elsevier, 1977), pg. 317.
4. D. J. Olego, submitted to Phys. Rev. B.
5. B. V. Shanabrook and J. S. Lannin, Phys. Rev. B 24, 4771 (1981).
6. H. von Thurn and H. Krebs, Acta Crystallogr. B 25, 125 (1969).
7. N. B. Goodman, L. Ley and D. N. Bullett, Phys. Rev. B 27, 7440 (1983).
8. P. B. Kirby and E. A. Davis, J. Non-Cryst. Solids 35 - 36, 945 (1980).
9. L.J. Pilione, R.J. Pomian and J.S. Lannin, Solid State Commun. 39, 933 (1981).

PRESSURE INDUCED STRUCTURAL CHANGE OF CLUSTERS IN CHALCOGENIDE GLASSES

Kazuo MURASE
Department of Physics, Osaka University, 1-1 Machikaneyama, Toyonaka 560, Japan

and

Toshiaki FUKUNAGA
Oki Electric Industry Co. Research Laboratory
550-5 Higashiasakawa, Hachioji, Tokyo 193, Japan

ABSTRACT

It is found that the intensity of the companion A_1 band (A_1^c) relative to that of the A_1 band decreases with increasing pressure in g-$Ge_{1-x}Se_x$. In g-$GeSe_2$, a simple extrapolation of the intensity ratio gives a vanishing of the A_1^c band at 48 kbar. The effect is related to a morphological transition for lack of a quasi-planer correlation at high pressures. In sulfur rich glasses, the three characteristic modes, 153, 219 and 475 cm^{-1}, of the S_8 ring disappear at 20-27 kbar, depending upon x and the laser light intensity. It is suggested that sulfur chains in g-$Ge_{1-x}S_x$ are like broken S_8 rings in contrast with selenium chains in g-$Ge_{1-x}Se_x$ having long spatial extent. We have observed a softening of the SnS_4-A_1 mode at high pressures in g-$Ge_{0.6}Sn_{0.4}S_3$ where tin atoms are not incorporated into sulfur clusters but into the $GeS_{4/2}$ clusters. The glass forming tendency of the (Ge,Sn) chalcogenide glass system is discussed, based on our high pressure works and in relation to a significant composition x=0.8.

INTRODUCTION

The (Ge,Sn) chalcogenide glasses are favorable to investigating the physics of the glass, since the bonds are mainly covalent with weak ionicity differences between unlike atoms.[1] The (Ge,Sn) chalcogenide glasses are known to be micro-heterogeneous with different kinds of low dimensional molecular clusters. The morphology of the medium range order (MRO) has been successfully studied by vibrational[2] and Mössbauer spectra etc..[3] For the Ge(Se,or S)$_{4/2}$ cluster, a tentative, but heuristic model has been proposed, which is called an outrigger raft (OR).[4] Calculated vibrational density of states (VDOS) of the OR and modified OR's well reproduce many important features of the infrared and Raman spectra of the stoichiometric glasses, (Ge,Sn)(Se,or S)$_2$.[2,5] It is recognized that MRO of clusters is topologically high with partially broken chemical orders in the stoichiometric glass, especially in $GeSe_2$.[1,6,7] In the Raman spectra (at 50K), the 247 cm^{-1} peak of g-$GeSe_2$ and the 443 cm^{-1} peak of g-$GeSe_2$ have been identified as an chalcogen-chalcogen stretching mode (CS),[2] probably at the outrigger. The peaks at 145 cm^{-1} in g-GeS_2 and 205 cm^{-1} in g-GeS_2 are identified as the corresponding bending mode (CB).[5]

0094-243X/84/1200449-08 $3.00

It becomes realistic to investigate interaction among clusters as well as structure, spatial extent, strain and dimensionality of the clusters.[8,9] Pressure dependent Raman measurements are made at room temperature with a diamond anvil cell.

$Ge(Se,or\ S)_{4/2}$ CLUSTERS

Though each of the $Ge(Se,or\ S)_{4/2}$ clusters is surrounded by chalcogen clusters in a wide composition range, the $Ge(Se,or\ S)_{4/2}$ clusters play a leading part in the infrared response spectra. Complementary informations about the clusters are given by Raman spectra.[10] In g-$GeSe_2$, the 203 cm^{-1} peak at 50K has been ascribed to the symmetric breathing A_1 mode of the $GeSe_4$ tetrahedron. The high energy side band at 219 cm^{-1} is called the companion A_1 line (A_1^c) which has similar polarization character to the A_1 band. The A_1^c band gives us important suggestion for the interaction among clusters. We have inferred that the A_1^c is due to an out-of-phase stretching motion of two Ge-Se bonds coupled with a bond bending mode at the outrigger Se-Se. Pressure dependent Raman spectra in g-$Ge_{1-x}Se_x$ are shown in Figure 1(a), (b) and (c). The intensity of the A_1^c band relative to that of the A_1 band decreases with increasing either selenium composition or pressure as shown in Figure 1 and 2(a). In near stoichiometric glasses, the 178 cm^{-1} mode is observed, which is identified as an A_{1g} mode of an ethane like molecule $Ge_2(Se_{1/2})_6$.[10] The A_{1g} mode seems to disappear with increasing pressure. If the A_1^c band really comes from the outrigger Se-Se, the simultaneous decrease of the A_1^c and A_{1g} bands indicate that the chemically disordered Se-Se and Ge-Ge bonds are broken with increasing pressure and then coalescence of the $GeSe_{4/2}$ and $Ge_2(Se_{1/2})_6$ clusters proceed. A simple extrapolation of the intensity ratio in g-$GeSe_2$ gives a vanishing of the A_1^c band at 48 kbar. It has been reported that the x-ray first sharp diffraction peak (FSDP) is absent at most beyond 50 kbar and it almost recovers after releasing pressure.[11] If the origin of the FSDP is due to a quasi-planer correlation,[6] the high pressure decrease of the chemically disordered bonds is accompanied by a morphological transition with disappearance of the layered structure. In the selenium rich glasses, where the ethane like molecules are absent, the $GeSe_{4/2}$ clusters are connected with selenium chains with increasing x. With increasing pressure, remaining Se-Se bonds at outriggers may become increasing unstable and bond rearrangements give rise to polymerization between the $GeSe_{4/2}$ clusters and the selenium chains.

In Figure 2(b), A_1 peak positions are plotted as functions of pressure in g-$Ge_{1-x}Se_x$. The pressure coefficient becomes very small near the stoichiometric composition, while in g-$Ge_{1-x}S_x$, the GeS_4-A_1 peak position changes little with composition x and the pressure dependence is similar to that of g-GeS_2.[8] In order to investigate strain and stability of g-$Ge(Se,or\ S)_2$, we will compare the pressure peak shift of the A_1 and A_1^c bands with those of A-like mode in the layered $GeSe_2$ and GeS_2 crystals. Typical Raman spectra of the crystals at different pressures are shown in Figure 3(a) and

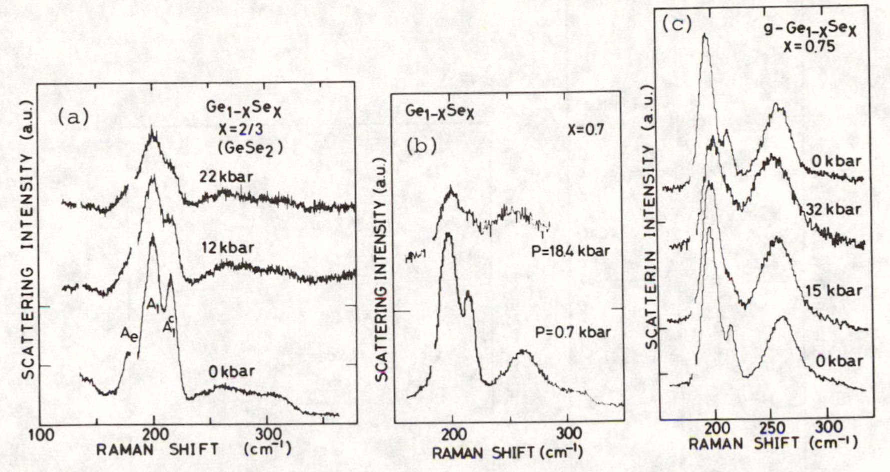

Fig. 1 Pressure dependent Raman spectra of g-$Ge_{1-x}Se_x$ at room temperature excited by the 6328A line of He-Ne laser; (a) x=2/3, (b) x=0.7 and (c) x=0.75. The A_1^c and A_e (ethane) band decrease with pressure.

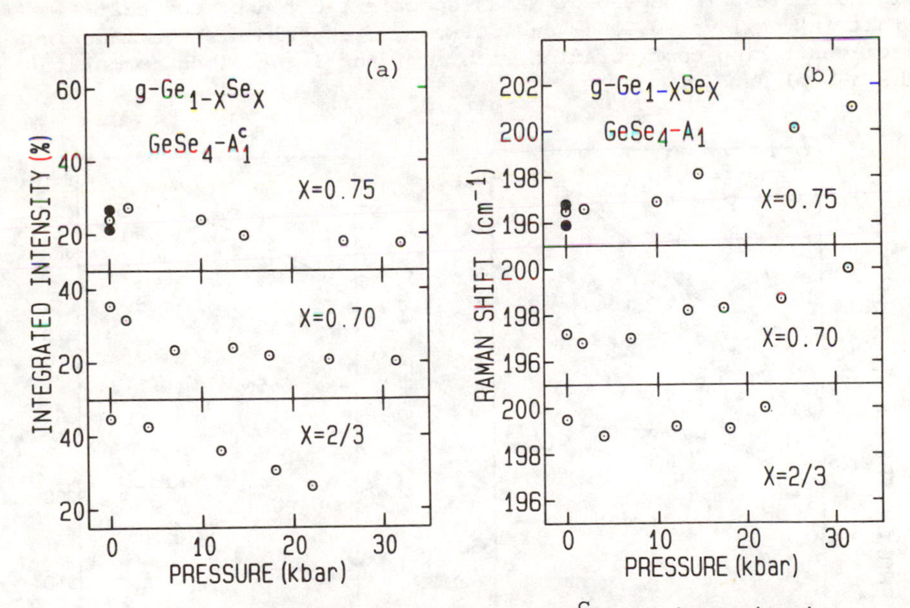

Fig. 2 Pressure dependence of; (a) the A_1^c band intensity in g-$Ge_{1-x}Se_x$ divided by the corresponding A_1 band intensity, where the black circle is after compression; (b) A_1 peak positions. The pressure coefficient becomes very small at the stoichiometric composition.

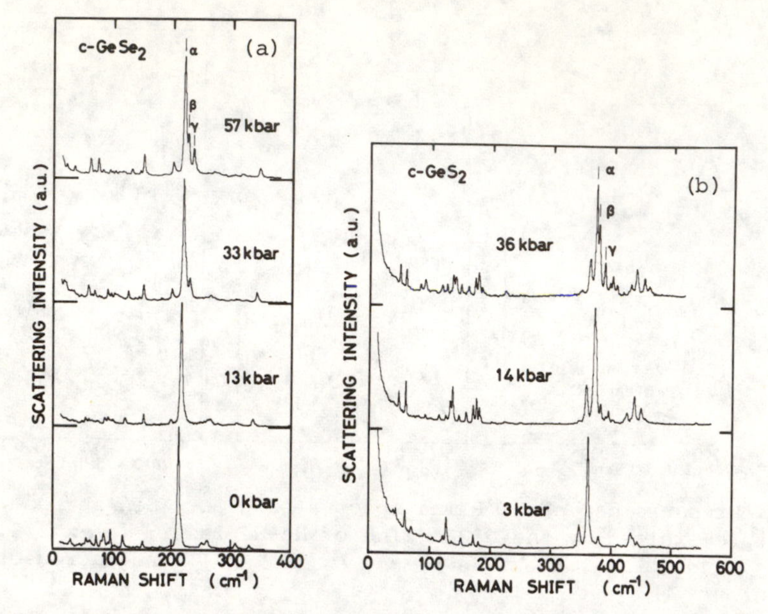

Fig. 3. Pressure dependent Raman spectra excited by the 6328Å light in the layered $GeSe_2$ and GeS_2. The experiments are made in a back-scattering configuration with incident light along normal to the a-b plane.

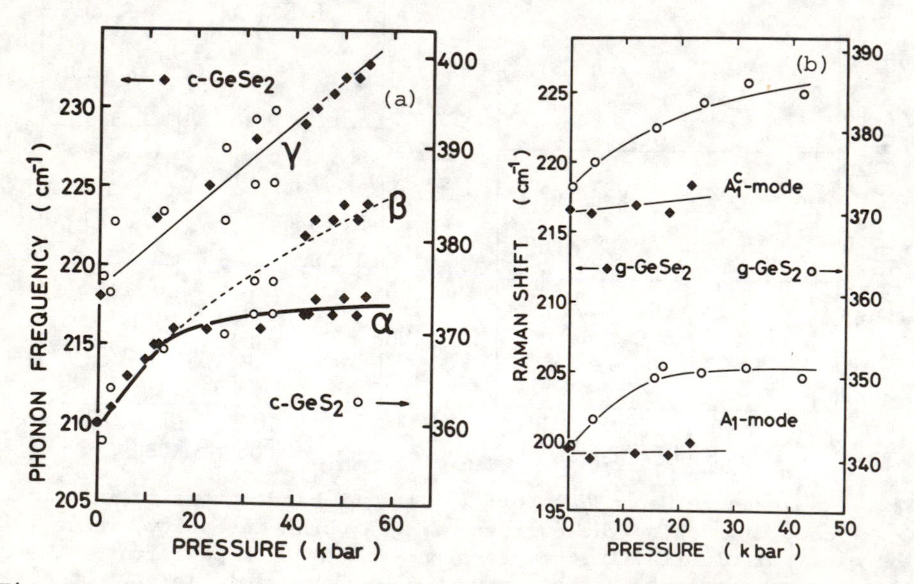

Fig. 4. A-like Raman peaks as a function of pressures,(a) in layered crystals, $GeSe_2$ and GeS_2 and (b) the stoichiometric glass. A bold line is drawn along the dominant A -peak (α).

(b). The dominant peaks at 210 cm^{-1} (c-$GeSe_2$) and 360 cm^{-1} (c-GeS_2) at the atmospheric pressure are due to A-like modes. Around the dominant peak, a few subsidary peaks are seen and are noticeable in c-GeS_2. In Figure 4(a), these peak positions are plotted as functions of pressure.[7,9] The scales of the ordinate are changed so as to fit the 210 cm^{-1} to the 360 cm^{-1} (c-GeS_2). The A-like peak positions of both crystals behave very similarly with increasing pressure. The pressure coefficient of the dominant peaks becomes very small beyond 15 kbar. Pressure dependences of the A_1 and A_1^c modes in g-$GeSe_2$ and g-GeS_2 are shown in Figure 4(b). It should be noted that the pressure behavior of the GeS_4-A_1 band of g-GeS_2 resembles that of the 360 cm^{-1} peak of c-GeS_2. On the other hand, the pressure dependence of the $GeSe_4$-A_1 band of g-$GeSe_2$ is very small even below 15 kbar. It is natural to compare the small dependence with that of the 210 cm^{-1} peak of c-$GeSe_2$ at more than 15 kbar. We have estimated that the $GeSe_{4/2}$ clusters are internally stressed with 20 kbar even without external pressure, based on the composition and pressure dependence of the A_1 band [Figure 2(b)].[2]

In g-$Ge_{1-x}Se_x$, the A_1 peak energy measured by Raman scattering at the atmospheric pressure increases linearly with (1-x) and begins to increase superlinearly at x=0.8. The onset composition x=0.8 accords with the most favorable glass forming composition x as is predicted by the topological theory of Phillips[12] and Döhler.[13] The large internal stresses at $x<0.8$ seems to be understandable due to the overconstraint in the mean field sense. With increasing selenium compositions, the internal strain will be relaxed by the selenium chains surrounding the $GeSe_{4/2}$ cluster, where the softening of the Se-A_1 with x due to inter-chain interactions may cooperate in the relaxation. We have concluded that the lateral extension of the $GeSe_{4/2}$ cluster is larger than that of the $GeS_{4/2}$, since the high frequency F_2 band (258 cm^{-1}) in the infrared spectra is very sharp and the Raman CS band is relatively weak.

In g-$Ge_{1-x}S_x$, the $GeS_{4/2}$-A_1 position is practically constant with x. Instead, the band width increases at $x<0.8$. It is probable that the $GeS_{4/2}$ cluster is chain-like and more flexible and possibly of larger disorder. The surrounding chalcogen clusters have different morphology the S_8 ring is dominant at $x>0.8$. We will discuss more about the sulfur clusters in the next section.

CUTTING OF S_8 RING

In sulfur rich glasses, the three characteristic modes at 153, 219 and 475 cm^{-1} due to the S_8 ring disappear at high pressure[2], as shown in Figure 5(a) and (b). The phenomenon may be called a pressure induced photo-structural change, where the rate of disappearance depends on laser wavelengths and intensities. The integrated intensity of the S_8 band normalized by the $GeS_{4/2}$-A_1 band intensity decreases with increasing pressure as shown in Figure 6(a), in parallel with increase of the sulfur chain mode intensity [Figure 6(b)]. A method of decomposition of the high frequency band into the ring and chain components is shown elsewhere.[8] After re-

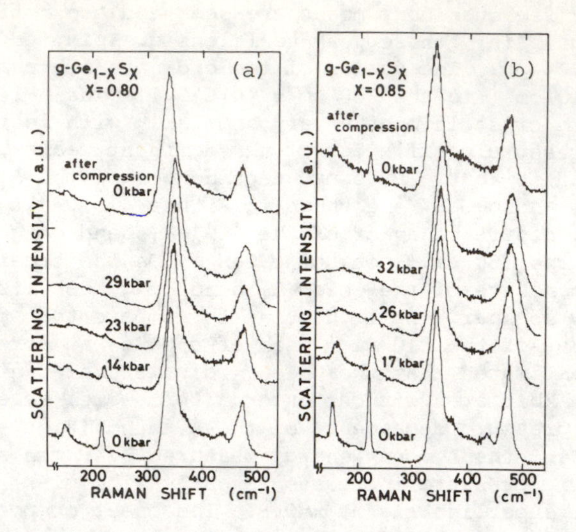

Fig. 5. Pressure dependence of Raman spectra excited by the 6328A light with power levels less than 10 mW in g-$Ge_{1-x}S_x$; (a) at x=0.80 and (b) at x=0.85. The characteristic spectra at 153, 219 and 475 cm^{-1} due to S_8 rings disappear at high pressures.

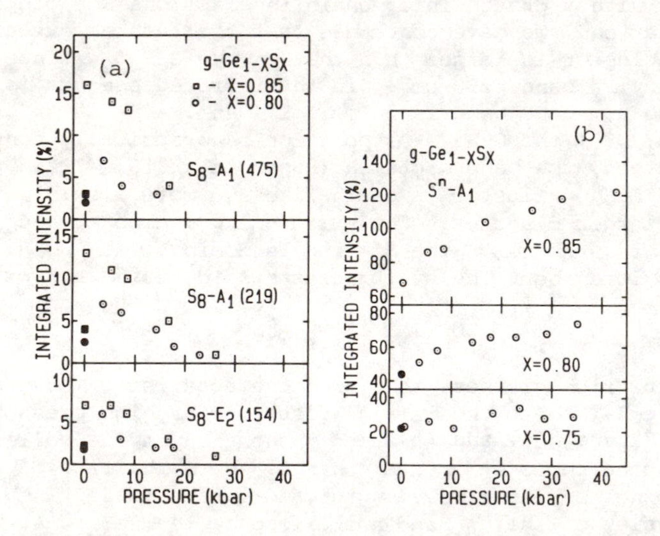

Fig. 6. Pressure dependence of Raman intensities ;(a) due to S_8 rings. The zero pressure spectral positions are shown in the parentheses. (b) due to sulfur chains S^n (475 cm^{-1}). Closed circles and squares are for after compressions.

leasing pressure, about one third of the S_8 band intensity recovered with some broadening.

It is known that the pressure coefficient of the S^n-A_1 mode is positive in contrast with that of the Se^n-A_1 mode.[8] It should be noted that the three S_8 band positions all increase with pressure. The S^n-A_1 peak position at the atmospheric pressure increases at $x<0.8$. These trends suggest to us that the sulfur chain is a cut S_8 ring in shape and is fairly strained near the stoichiometric compositions. Pressure dependences of the optical gap also suggest the presence of sulfur chains made of broken S_8 rings in g-$Ge_{1-x}S_x$.[8]

Sn-CONTAINED-GLASSES

It has been clarified that the tin atoms are tetrahedrally bonded with chalcogen atoms in the chalcogen rich glasses. We have suggested that they are incorporated into $GeSe_{4/2}$ or $GeS_{4/2}$ clusters rather than into chalcogen chains or rings. Insistent two-mode type behaviors of the high energy F_2 and the A_1 modes have been observed and are reproduced by VDOS calculations for a modified OR. Typical pressure-Raman spectra of a tin contained sulfur rich glass are shown in Figure 7(a).[9] The peak positions and the line widths of the SnS_4-A_1, GeS_4-A_1 and S^n-A_1 bands are shown in Figure 7(b) and (c). The softening of the SnS_4-A_1 is observed beyond 10 kbar in parallel with a line-broadening. This is a clear evidence of an instability of the tetrahedral bonding of the tin atom, which is unstable in the crystalline Sn(Se,or S)$_2$. The CdI_2 structure is stable therein.

In g-$(Ge_{1-y}Sn_y)_{1-x}Se_x$, the rate of the LO-TO separation of the high frequency $SnSe_{4/2}$-F_2 modes increases with decreasing selenium composition x and levels off at $x<0.8$.[14] At $x>0.8$, the glass forming tin composition range[7] is wide, $0<y<0.9$, and the boundary com-

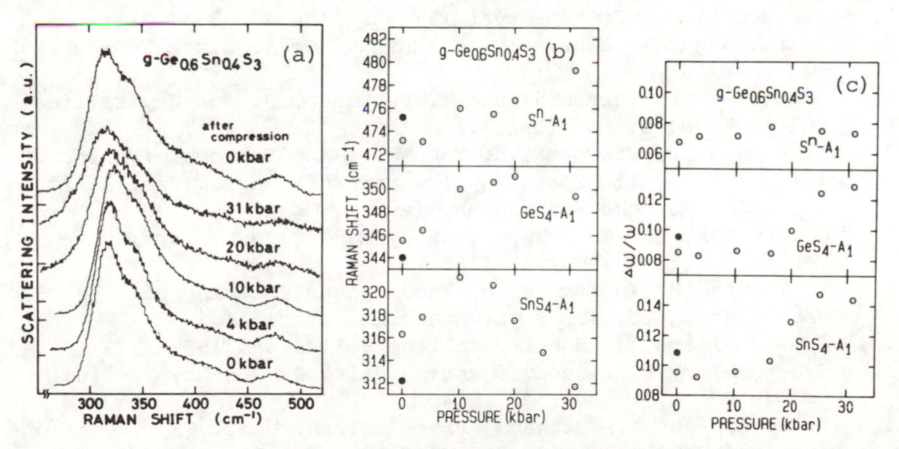

Fig. 7. Pressure dependence of (a) Raman spectra in a sulfur rich tin contained glass and (b), Gaussian resolved peak positions. Softening of the SnS_4-A_1 band is apparent. The band width is broaden at high pressures.

position y_b is nearly constant (=0.9), suggesting that tin atoms are mostly incorporated into $GeSe_{4/2}$ clusters. The boundary composition y_b begins to decrease near x=0.8 and becomes very small at x=2/3. The trend of the exclusion of tin atoms is probably due to increasingly large internal strain near the stoichiometric composition x=2/3 together with relatively large ionicity of $SnSe_{4/2}$ and ionic radius of the tin atoms. It should be pointed out that a realization of a wide composition range g-(Ge,Sn)(Se,or S)$_2$ by force with a rapidly quenching method with small samples may be due to an extremely large internal strain,[15] which allows an instability of the tetrahedral bonding as is observed in Figure 9.

ACKNOWLEDGEMENTS

The authors would like to thank K. Yakushiji, T. Yoshimi and I. Yunoki for their help in the experiments and numerical calculations. This work was supported in part by the Grant-in-aid for Scientific Research from the Ministry of Education, Science and Calture.

REFERENCES

1. P. Tronc, M. Bensoussan, A. Brevac and C. Sebenne, Phys. Rev. B8, 5947 (1973).
2. K. Murase, T. Fukunaga, Y. Tanaka, K. Yakushiji and I. Yunoki, Physica 117B & 118B, 962 (1983).
3. P. Boolchand, J. Grothaus, W. J. Bresser and P. Surayn, Phys. Rev. B25, 2925 (1982).
4. P.M. Brindenbaugh, G.P. Espinasa, J.E. Griffiths, J.C. Phillips and J.P. Remeika, Phys. Rev. B20, 4140 (1979).
5. K. Murase, T. Fukunaga, K. Yakushiji, T. Yoshimi and I. Yunoki, J. Non-Crystalline Solids 59 & 60, 883 (1983).
6. J.C. Phillips, J. Non-Crystalline Solids 43, 37 (1981).
7. T. Fukunaga, Y. Tanaka and K. Murase, Solid State Communs. 42, 513 (1982).
8. K. Murase, K. Yakushiji and T. Fukunaga, J. Non-Crystalline Solids 59 & 60, 855 (1983).
9. K. Murase, T. Fukunaga, K. Yakushiji and T. Yoshimi, Proc. Int. Sympo. Solid State Physics under Pressure, Jan. 18-21, 1984, Izu Nagaoka Spa, Japan, (ed. B. Okai)
10. G. Lucovsky and T.M. Hayes, ed by M.H. Brodsky (Springer-Verlag, Berlin, 1979) p. 215.
11. K. Tamura, H. Tsutsu and H. Endo, Annual Meeting of Phys. Soc. Japan, March 27, 1983; private communications.
12. J.C. Phillips, J. Non-Crystalline Solids 34, 153 (1979).
13. G.H. Döhler, R. Dandoloff and H. Bilz, J. Non-Crystalline Solids 42, 87 (1980).
14. K. Murase and T. Fukunaga, Proc. Intern. Conf. Phys. Semiconductors, San Francisco, CA, USA, August 6-10, 1984.
15. M. Stevens, J. Grothaus and P. Boolchand, Solid State Communs. 47, 199 (1983).

REVERSIBLE AND METASTABLE CHANGES IN THE RAMAN SPECTRUM OF GeS_2 GLASS INDUCED BY COMPRESSION

B.A. Weinstein and M.L. Slade
Xerox Webster Research Center-114, Webster, N.Y. 14580

We report cryobaric Raman measurements for amorphous (a-) GeS_2 at 13K to 56 kbar. Hydrostaticity in a diamond-anvil cell is maintained by an argon pressure-transmitting medium. Pressure-induced peak shifts are fractionally small and reversible. However, peak broadenings are large (~50-100%) and metastably retained, along with a newly emergent peak at 486cm^{-1}, after pressure release and dark-annealing. We find evidence for pressure-assisted quasicrystallization at P=56 kbar. These results support a molecular model in which pressure squeezes out free (intermolecular) volume. We estimate that internal stress-fluctuations caused by random intercluster coupling do not exceed 2 kbar at P=0 (1 atm.). The metastable broadening is attributed to retained densification ~20%. The 486cm^{-1} peak, assigned to sulfur chain and/or ring modes, indicates S-rich (presumably compensated by Ge-rich) regions formed under pressure by fairly stable atomic rearrangements.

INTRODUCTION

Raman scattering has proven to be an effective probe of structure in a-GeS_2 and similar Ge-chalcogenide glasses.[1] It was determined early on that the stoichiometric composition was largely chemically ordered, and that the basic structural unit was a $GeS_{4/2}$ tetrahedron loosely coupled into a covalent network.[2,3] The quasi-isolation of this unit stemmed from the near 90° Ge-S-Ge bond angle and weak sulfur-centered bond-bending forces. This led to highly polarized molecular-solid-like Raman spectra in the bond-stretching regime. In particular the strongest peak at 343cm^{-1} of A_1 symmetry is due to symmetric breathing of the $GeS_{4/2}$ tetrahedra. The anomalously rapid composition dependence of the sharp polarized A_1-companion-line, A_{1C}, provided evidence of larger molecular units thought to be rings.[4] Subsequently it was argued that the A_1-A_{1C} splitting arose from S–S dimers terminating the edge of large partially-polymerized raft-like molecules.[5] Contrary to previous supposition,[3] these molecules (resembling fragments of crystalline GeS_2 layers) have < 3d network-topology. Convincing evidence for the broken chemical ordering implied by this model has emerged from Mossbauer studies.[6] Recent Raman experiments have shown that quasi- and micro-crystallization, involving coalescence of molecular clusters, can be photo-induced in bulk a-GeS_xSe_{1-x} alloys.[7] The atomic rearrangements during this process can be reversed (back to the glassy state) by dark-annealing well below the glass transition.

The advantage of pressure for studying structure in molecular glasses is that it selectively probes weak bonds irregardless of disorder.[8] Pressure-Raman studies at room temperature were previously reported for a-GeS_2,[8] and a-$Ge_x(S, Se)_{1-x}$ alloys.[9] The present work reports the first cryobaric Raman measurements on a chalcogenide glass, in which we are able to study pressure-induced broadening, uncomplicated by thermal effects.

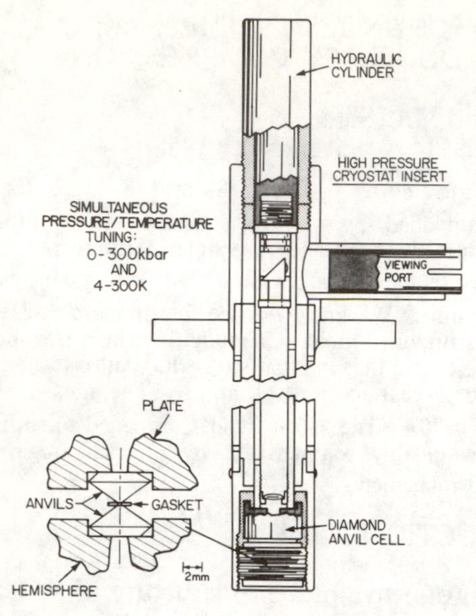

Fig. 1 Cryogenic hydrostatic-pressure apparatus used in this experiment. Cut-away drawing of gasketed Bridgman-anvil configuration used in the diamond-anvil cell shown in insert.

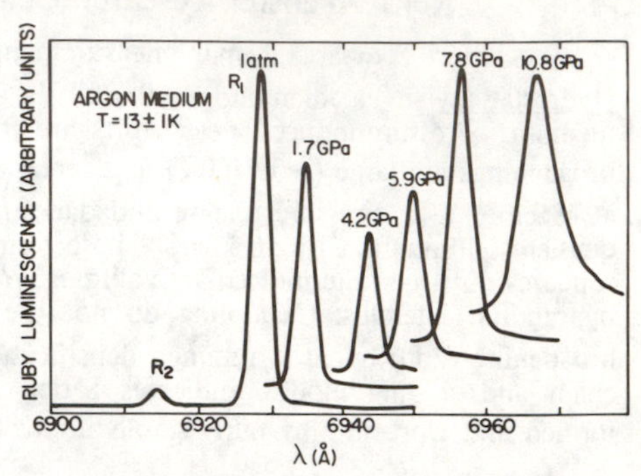

Fig. 2 Typical set of ruby (R_1-line) pressure calibration measurements to 108 kbar at 13K. 1 GPa = 10 kbar.

CRYOBARIC APPARATUS AND EXPERIMENT

The cryostat insert shown in Fig. 1 illustrates the high-pressure and optical aspects of the present apparatus. A Be-Cu (RC-40) diamond-anvil cell (DAC), consisting of a piston in cylinder arrangement with NBS-type[10] diamond mounts, threads into the outer tension sustaining tube. The inner tube provides compression to the DAC piston when activated by the room temperature hydraulic cylinder at the top. Both tubes are 1/8" wall 304 stainless steel. Optical access is through the bottom cryostat window and bottom (cylinder mounted) diamond, or through the internal microscope arrangement (lenses and 90° prism) and the top (piston mounted) diamond. The cryostat insert sits in a He-gas flow variable temperature (4-300K with 1K precision) optical dewar. The apparatus is designed for 300 kbar operation using room temperature parameters for 304 s.s.

A standard Inconel gasket[10] with ~250μm diameter sample chamber is used. The sample and ruby pressure calibrant are loaded into the gasket chamber, but sealing is initially prevented by a soft indium spacer between the piston and cylinder diamond mounts. The cell is then mounted and cooled. Argon is introduced into the apparatus and liquified at $83 < T < 87K$ until it floods the sample area. Force is now applied, the DAC is sealed trapping solid argon as the pressure medium, and the remaining untrapped argon is heated to vapor >87K and evacuated. Measurements may now begin.

Figure 2 shows a typical series of ruby pressure measurements up to 108 kbar (10.8 GPa) at 13K. The room temperature ruby scale is used since studies show little scale variation down to cryogenic temperatures.[11] The narrow line-width shows that the solid argon medium is quite hydrostatic. The broadening at 108 kbar is caused by the ruby (thicker than the sample) being pinched between the diamonds.

The Raman spectrum of a-GeS_2 and the Ruby luminescence were excited by ~10mW (at the sample) 6471 Å Kr^+-laser light focused to a ~50 μm spot. The standard back-scattering configuration[12] was used. The collection optics had an effective f/4 efficiency determined by the DAC geometry. The ability to view the sample in transmission (by means of the cryostat x50 microscope, Fig. 1) or in reflection (by a monochromator slit-periscope) greatly aided in alignment, and allowed inspection for laser damage. A computer-interfaced Spex double monochromator, equipped with photon counting (RCA C31034A) detection, analyzed and recorded the spectra. Polarized scattering was not studied. At each pressure three runs were multiplexed. The ruby R_1-line was recorded before and after each run-set. Pressure did not vary by more than 5 kbar between the two measurements. The temperature was held at 13±1K, as measured by a diode sensor in good mechanical contact with the Be-Cu DAC.

Several factors prevented us from knowing the exact power density incident on the sample. These include defocusing inside the front diamond and the diamond reflectivity, both of which depend on the highly uniaxial strain pattern inside the Bridgman-anvil configured diamonds. Visible monitoring of the sample during the experiment showed no laser damage at any pressure. In addition, except at the highest pressure (56 kbar), the incident flux was insufficient to produce any quasicrystallization (QC), as evidenced by the absence of sharp low-frequency Raman structure.[7]

Bulk glassy GeS_2 was synthesized from the high-purity elements in evacuated silica ampoules by standard rocking furnace techniques. Raman spectra measured at P=0 indicated that stoichiometry was maintained within 1%. DAC samples ~75x50x20 μm were prepared by chipping the glass, and selecting only those chips showing no internal cracks and otherwise good clarity. No other mechanical or chemical treatment was applied.

RESULTS

The effect of pressure on the Raman spectrum of a-GeS_2 at 300K[8] and at 13K is displayed in Figs. 3 and 4. Each spectrum is the sum of three runs at the

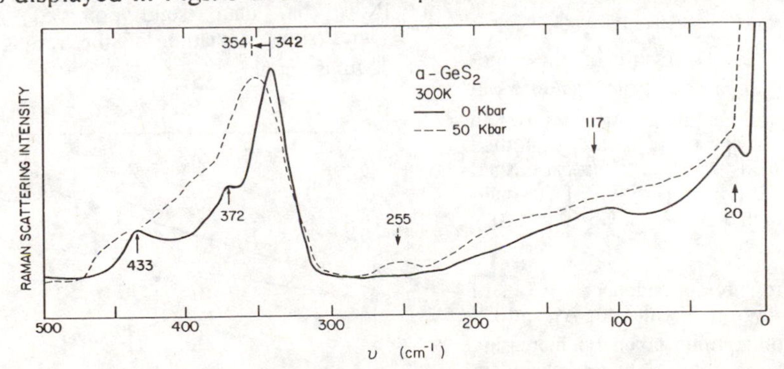

Fig. 3 Effect of pressure on the Raman spectrum of a-GeS_2 at room temperature (after Ref. 8).

labeled pressure, smoothed by a 12-point Savitsky-Golay routine. We find that at P=0 the spectral definition is somewhat improved at 13K over room temperature in the high frequency bond-stretching band of primary interest here (300-500cm^{-1}). For example, the minimum between A_1 and A_{1C} is more pronounced, and the A_1 peak is ~7% narrower.

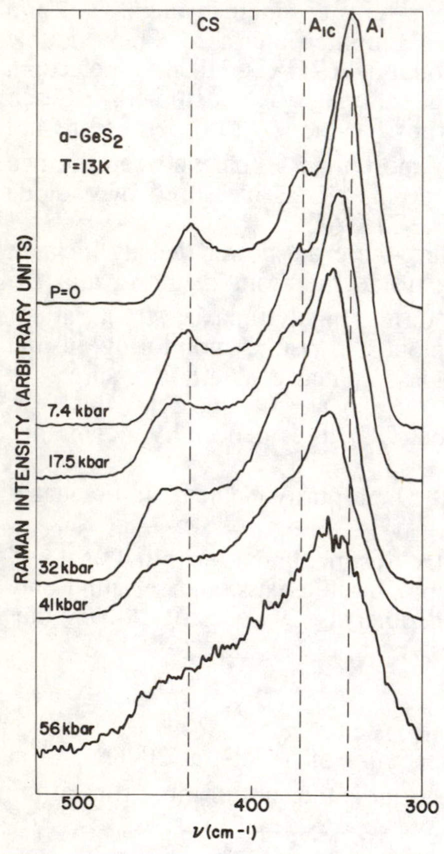

Fig. 4 Effect of pressure on the bond-stretching Raman spectrum of a-GeS_2 at 13K. Experimental parameters given in text. Each spectrum is the smoothed sum of 3 measurements. Vertical dashed lines mark the P=0 position of the main features labeled A_1, A_{1C}, CS (see text).

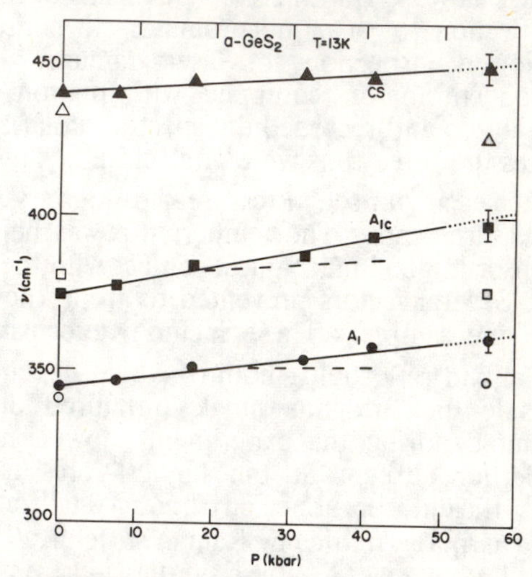

Fig. 5 Pressure-dependence of peak frequencies for A_1, A_{1C}, CS. Data for P<56 kbar and dotted points at 56 kbar determined from a 4-Gaussian fit to the Fig. 4 spectra (each the sum of 3 runs); solid points at 56 kbar obtained from a 5-Gaussian fit to the first 56 kbar run (see text). Solid points taken on pressure increase; open P=0 points taken after pressure release from 56 kbar and dark-annealing at P=0, 100K for ~8 hrs. Solid lines are least square fits to P<56 kbar data (solid points); Dashed lines show deviation of the results of Murase et al.[9]

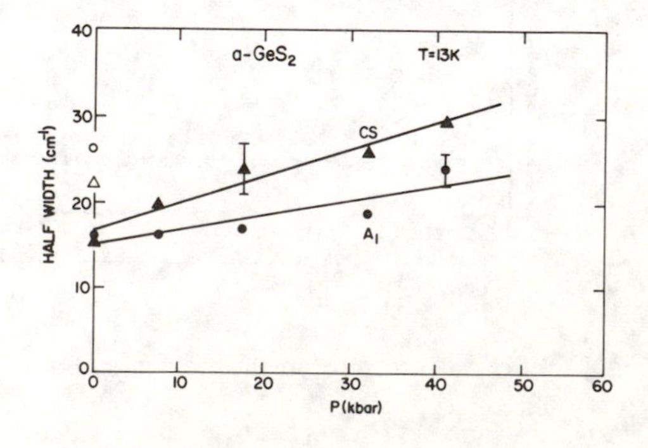

Fig. 6 Pressure-dependence of Gaussian (1/e) half-width for A_1 and CS peaks. Solid points taken on increasing pressure; 56 kbar data is excluded. Open points at P=0 as in Fig. 5. Solid lines are best fits to solid points.

It is clear that the general effect of pressure is to up-shift and broaden the bond-stretching band. The frequency increase is fractionally small, but the broadening is fractionally large. To demonstrate this more quantitatively, the sum of four (except for the solid points at 56 kbar, see below) Gaussian peaks was fit to the 13K data using a least-square minimization procedure. Referring to Fig. 4, these peaks simulate A_1, A_{1C}, CS (recently attributed to a chalcogen-stretch mode at sulfur dimers[9,13]), and a filler peak between CS and A_{1C}. Figures 5 and 6 show the peak shifts (for A_1, A_{1C}, CS) and broadenings (for A_1, CS) determined from these fits. Whereas the shifts are ~5% in 50 kbar, the peaks broaden by ~50% in the same span. Table I lists the pressure coefficients and Gruneisen parameters obtained from the best linear fit (solid lines) to the data for $P < 56$ kbar in Figs. 5 and 6. The 300K results of Murase et al.[9] for the peak positions agree with the present findings at low pressure, but deviate above 20-30 kbar, as shown by the dashed lines in Figs. 5 and 6. These authors also noted pressure-induced broadening but did not quantify their results.

TABLE I. Measured pressure coefficients and Gruneisen γ's (using bulk modulus $B=135$ kbar[8]) for prominent bond-stretching peaks ν and their half-widths σ in a-GeS_2 at 13K.

Peak	$\nu(P=0)$ cm^{-1}	$d\nu/dP$ $cm^{-1}/kbar$	γ_ν $B(d\ell n\nu/dP)$	$\sigma(P=0)$ cm^{-1}	$d\sigma/dP$ $cm^{-1}/kbar$	γ_σ $B(d\ell n\sigma/dP)$
A_1	343	0.33	0.13	15	0.18	1.6
A_{1C}	374	0.45	0.16	--	--	--
CS	439	0.19	0.06	17	0.31	2.5

At the highest pressure measured, qualitative spectral differences were observed that indicate structural changes in addition to simple densification. The spectrum degraded with time, such that the third 56 kbar run showed substantial broadening to lower energy, and overall loss of definition and intensity with respect to the first run at that pressure. This degradation is apparent in Fig. 4, even for the three runs summed together, and in Fig. 5, where at 56 kbar dual values resulted from separate treatment of the first run (solid points) and the summed runs (dotted points). The dotted points correspond to the <u>same</u> 4-peak fit used for $P < 56$ kbar, but the solid points were obtained by adding a 5th Gaussian peak at ~345cm^{-1} to account for the asymmetric low-energy broadening of A_1. Because of this ambiguity, the solid-line fits in Fig. 5 were calculated using only $P < 56$ kbar data; the dotted extension of these lines to the solid points at 56 kbar supports our modified fitting procedure. This ambiguity also leads us to exclude the 56 kbar half-width data in Fig. 6. In addition to the above spectral changes, at least 6 sharp low-frequency peaks (at 62, 80, 110, 122, 135, 157cm^{-1}) were observed in the 3rd 56 kbar run, but not at lower pressure.

We find even at low-temperature that the peak shifts are reversible within experimental error when pressure is released from 56 kbar. However the broadening is not. This is illustrated in Fig. 7, and also by the double-valued points at P=0 in Figs. 5 and 6 (solid for before pressurization, open for after). The FINAL P=0 spectrum in Fig. 7 was measured at 13K after dark-annealing the sample overnight at P=0, T=100K. The qualitative changes at P=56 kbar just discussed also reverse for the most part, as evidenced by the disappearance of the low-frequency peaks, and the normal "FINAL" shape (save for broadening) in Fig. 7. However, a weak new peak appears in the "FINAL" spectrum at $486cm^{-1}$, above the normal cutoff for optical phonons in a-GeS_2. This peak is evidence for a microscopic structural change at high pressure that has been metastably retained at P=0. The new peak went undetected at 56 kbar because it was masked by the up-shifted spectrum.

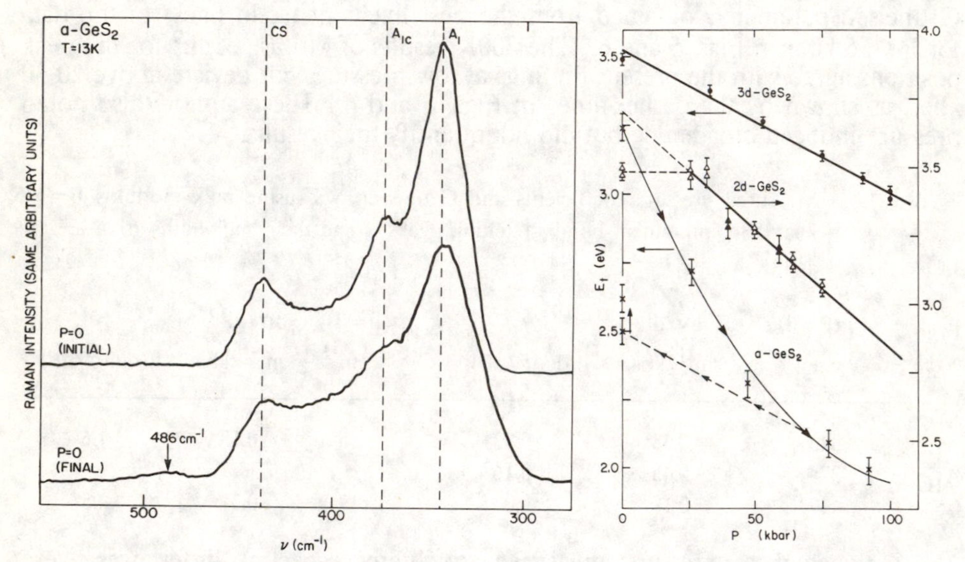

Fig. 7 Raman spectra of a-GeS_2 at 13K measured before pressurization, and after pressure release from 56 kbar and dark-annealing at P=0, T=100K for ~8 hrs. Vertical dashed lines mark initial positions of A_1, A_{1C}, CS. Note new peak at $486cm^{-1}$ and asymmetric low-energy broadening of A_1 in FINAL spectrum.

Fig. 8 Bandgap pressure-dependence of the glassy and the two crystalline forms of GeS_2 at 300K (after Ref. 8). Dashed line indicates hysteresis of bandgap shift for a-GeS_2 upon pressure release from 90 kbar. This hysteresis is attributed to a retained density excess.

DISCUSSION

We regard the major effect of compression in a-GeS_2 to be the induced broadening. It reflects enhanced sampling of structural disorder by the normal modes through increases in local strain-fluctuations. These increases are roughly 50-100% in 50 kbar (Fig. 6). As noted, the peak shifts are fractionally much smaller. This dichotomy is illustrated by the Gruneisen parameters in Table 1 pertaining to peak position and width. Whereas the former are ~0.1, typical of internal modes in a molecular solid, the latter are ~2, which is characteristic of external modes.[12] Hence, the interactions that determine peak position vary with pressure like intramolecular forces, but those that determine peak width vary like forces governing the frequency of intermolecular phonons. This supports the view that a-GeS_2 is not a rigid 3d-network glass, but consists of distinct molecular units. Our mechanism for the observed broadening is, then, pressure-enhanced intermolecular coupling as the free (viz. intermolecular) volume is squeezed out.

The measured peak shifts enable us to estimate the magnitude of the local internal strain-fluctuations in a-GeS_2. For example, at P=0 the ~15cm^{-1} half-width of the A_1-peak corresponds to ~50 kbar. However, it is a mistake to think that this large internal pressure exists in a-GeS_2 because of the molecular nature of this glass. Since the ratio of high-frequency to low-frequency phonon modes is ~6 (force constant ratio of 36), the actual internal pressure needed to produce sufficient local strain to explain the broadening is only 1-2 kbar. It is clear from Fig. 6 that for an external pressure of 50 kbar, the local fluctuations in internal pressure still do not exceed 4 kbar. On this basis we disagree with the assertion by Murase et al.[9] that internal clusters in a-$GeSe_2$ are stressed by ~20 kbar at P=0. Furthermore, within the molecular-cluster picture[5] adopted in Ref. 9, the equally narrow line-widths of both A_1 and A_{1C} in a-$GeSe_2$ at P=0 also argue against large internal stress. Otherwise the requisite cluster-cluster interactions would broaden A_{1C} much more than A_1, because A_{1C} vibrations are localized on cluster surfaces. By the same reasoning the accelerated broadening of CS over A_1 in Fig. 6 lends credence to the assignment of CS by Murase et al.[9] to cluster-edge localized dimer-stretch vibrations.[13] Unfortunately, because of the small A_1 - A_{1c} splitting, it is not clear in either a-GeS_2 or a-$GeSe_2$[9] which of these two peaks broadens faster with pressure.

We attribute the apparent reduction in peak frequency at 56 kbar (dotted points in Fig. 5), and other time-dependent spectral changes at this pressure to photo-induced QC, similar to effects seen at P=0 in a-GeS_xSe_{1-x} alloys.[7] This interpretation is further supported by the low-frequency peaks observed in the 3rd 56 kbar run, and by the reversion of the entire spectrum (save for broadening and the 486cm^{-1} peak) back to its prepressurized form after dark-annealing at P=0, T=100K (Fig. 7).[7] It is likely that the present QC is pressure-assisted, since it did not occur for P < 56 kbar for the same incident flux. At least part of this pressure-assist derives from the "red-shift" of the bandgap, shown in Fig. 8, which increases the absorbed photon density at higher pressure. The remainder is undoubtedly due to squeezing out of free volume. We suggest that the observations in Ref. 9 of reduced peak frequencies in a-GeS_2 for P > 20-25 kbar (dashed lines in Fig. 5), and of small pressure coefficients for A_1 and A_{1C} in a-$GeSe_2$ may be related to earlier onset of QC. Clearly, the conditions and mechanism for pressure-assisted QC in Ge-chalcogenide glasses bear further study.

We turn now to metastable effects. The density excess upon returning from 56 kbar to P=0 can be estimated from the bandgap hysteresis in Fig. 8. Using a constant compressibility of 7.4x10^{-3} kbar^{-1}, [8] we obtain an excess of ~20%. Our finding that peak-shift is reversible within experimental error (see Fig. 5), but peak-broadening is not (Fig. 6) shows that the 20% density increase affects the average local internal strain very little, but substantially increases the local strain-fluctuations. This again supports a model in which pressure squeezes out the free volume between distinct molecular units or clusters, thereby metastably increasing the cluster-cluster coupling below room temperature. A portion of the retained broadening, especially in the low-energy wing of A_1 (see Figs. 4 and 6), may also be due to incompletely annealed QC.

The detailed nature of the atomic rearrangements accompanying pressure-assisted QC is not apparent at this time. However, the new peak at 486cm^{-1} (Fig. 7) provides a possible clue. We assign this peak to symmetric S−S chain and/or ring modes since such modes occur at similar frequencies in sulfur rich a-Ge_xS_{1-x} alloys.[3,9] It follows that microscopic regions of our sample have become S-rich, and must then be compensated by neighboring Ge-rich regions to maintain stoichiometry. The molecular entities in such regions should contain some Ge−Ge bonds. The 5th peak, necessary to explain the asymmetric low-energy broadening of A_1 at 56 kbar, could be due to scattering from Ge-rich regions. Figure 7 shows that this asymmetric broadening, though weaker than at 56 kbar, is retained along with the 486cm^{-1} peak after pressure release. The term metastable has been used to describe the 486cm^{-1} peak. However, the pressure-assisted atomic rearrangements associated with this peak appear to be quite stable, since the 486cm^{-1} peak remained after 1 week dark-annealing at 300K.

The authors are indebted to J.C. Mikkelsen, Jr. for supplying the a-GeS_2 samples, and to R. Zallen for discussions on network topology.

REFERENCES

1. G. Lucovsky, T.M. Hayes, Amorphous Semiconductors, Topics in Applied Physics, Vol. 36, ed. by M.H. Brodsky (Springer-Verlag, Berlin, 1979), p.215.
2. P. Tronc, M. Bensoussan, A. Brenac, Phys. Rev. B8, 5947 (1973).
3. G. Lucovsky, F.L. Galeener, R.C. Keezer, R.H. Geils, H.A. Six, Phys. Rev. B10, 5134 (1974).
4. R.J. Nemanich, S.A. Solin, G. Lucovsky, Solid St. Commun. 21, 273 (1977).
5. P.M. Bridenbaugh, G.P. Espinosa, J.E. Griffiths, J.C. Phillips, J.P. Remeika, Phys. Rev. B20, 4140 (1979); J.C. Phillips, C.A. Beevers, S.E.B. Gould, Phys. Rev. B21, 5724 (1980).
6. W.J. Bresser, P. Boolchand, P. Suranyi, J.P. deNeufville, Phys. Rev. Lett. 46, 1689 (1981); P. Boolchand, J. Grothaus, J.C. Phillips, Solid State Commun. 45, 183 (1983).
7. J.E. Griffiths, G.P. Espinosa, J.P. Remeika, J.C. Phillips, Phys. Rev. B25, 1272 (1982); Phys. Rev. B28, 4444 (1983).
8. R. Zallen, B.A. Weinstein, M.L. Slade, J. dePhysique 42, C4-241 (1981); B.A. Weinstein, R. Zallen, M.L. Slade, J.C. Mikkelsen, Phys. Rev. B25, 781 (1982).
9. K. Murase, T. Fukunaga, Y. Tanaka, K. Yakushiji, I. Yunoki, Physica 118B, 962 (1983); K. Murase, K. Yakushiji, T. Fukunaga, J. of Non-Crystalline Solids 59&60, 855 (1983); K. Murase, T. Fukunaga, K. Yakushiji, T. Yoshimi, I. Yunoki, op. cit. p.883; see also K. Murase, T. Fukunaga, in this volume.
10. G.J. Piermarini, S. Block, Rev. Sci. Instrum. 46, 973 (1975).
11. R.A. Noack, W.B. Holzapfel in: High-Pressure Science and Technology, ed. by K.D. Timmerhaus, M.S. Barber (Plenum, NY, 1979), Vol. I, p. 748.
12. B.A. Weinstein, R. Zallen: Light Scattering in Solids IV, Topics in Applied Physics, Vol.54, eds. M. Cardona, G. Guntherodt (Springer, Berlin, 1984) p. 465.
13. J.A. Aronovitz, J.R. Banavar, M.A. Marcus, J.C. Phillips, Phys. Rev. B28, 4454 (1983); for an earlier assignment see G. Lucovsky, Wong, JNCS.

RAMAN SCATTERING, LASER ANNEALING AND PRESSURE-OPTICAL STUDIES OF ION BEAM DEPOSITED AMORPHOUS CARBON FILMS

S. K. Hark, M. A. Machonkin, F. Jansen, M. L. Slade, B. A. Weinstein
Xerox Webster Research Center, Webster, New York 14580

ABSTRACT

Optical studies of amorphous carbon films, deposited at liquid nitrogen and room temperature, by ion beam sputtering of a graphite target onto several crystalline and amorphous substrates, are reported. Room temperature Raman spectra of hydrogenated and unhydrogenated films are presented. Spectra of laser annealed films and carbon black composed mainly of microcrystals of graphite are also included. A comparison of these spectra with recent model studies is discussed. In addition the effect of pressure on the absorption edge tail was measured. The results are consistent with a model of amorphous carbon in which there is a network structure of both three-fold and four-fold coordinated atoms.

INTRODUCTION

Hydrogenated amorphous carbon (a–C:H) films, prepared under suitable deposition conditions, have been reported to show unusual physical and chemical properties, e.g. hardness[1,2,3], high electric resistance[4,5,6], transparency in the infrared[3] and inertness towards corrosive chemicals[7]. These combined properties have led to the suggestion of their usefulness as a protective coatings of optical components. Hard a–C:H films, sometimes being referred to as diamond-like or i–carbon[8] films, may be prepared in a number of ways. R.f. plasma decomposition of hydrocarbon gases and ion beam sputtering are probably two of the more widely used methods. The properties of the resulting films can differ widely, because, as for amorphous silicon and germanium, they depend critically upon the preparation conditions. The variations in the properties are attributed to the differences in the local structure of the films.

Carbon exists in two allotropic forms, namely, graphite and diamond. The graphitic form, in which three-fold coordinated carbon atoms form a layered structure, is a soft semimetal and the more stable form at ordinary temperature and pressure. The metastable diamond form which has a four-fold coordinated network structure is a hard insulator. The different degrees of mixing of the three and four-fold coordinated carbon is believed to be responsible for the wide range of properties exhibited by a–C:H films[4,8].

An interesting and important question is how the preparation conditions of a-C:H films affect the properties, especially the structure, of the film.

Traditional structural probes for crystalline materials, such as x-ray, neutron and electron diffraction, when applied to amorphous materials, often yield less precise structural information in the form of radial distribution functions (RDF). RDF is not sensitive to the fine details of the structure. When they are applied to a−C:H films, the difficulty in distinguishing the graphitic and diamond C−C nearest neighbour distance (1.42Å for graphite, 1.54Å for diamond) makes diffraction techniques even more inconclusive[9,10]. To supplement and complement diffraction studies, other structurally sensitive techniques are desirable. In this report, we have used Raman scattering and pressure-optical effects to characterize a−C:H films prepared by ion beam deposition techniques. Raman scattering has been known to be a sensitive structural probe and has been used extensively to study amorphous materials[11,12]. Effects of hydrostatic pressure on optical properties have also been used to test the network dimensionality of amorphous materials [23,24].

EXPERIMENTAL DETAILS

All but one of the a−C:H films studied were prepared by ion beam deposition onto temperature controlled substrates. The one sample used for optical absorption edge measurement under pressure was made by r.f. cathodic glow discharge of CH_4. The ion beam was composed of mixtures of high purity hydrogen and argon gas. The geometry of the system was arranged in such a way that both the target and the substrate could be simultaneously bombarded by ions from the ion beam[25]. The substrate could be shadowed from direct bombardment by partially blocking the ion beam. The range of sputtering ion energies varied from 1000 to 1350eV. The substrates were cooled by either liquid nitrogen or room temperature water. Both crystalline and non-crystalline substrates were used. The former category included Si wafers, α-quartz, diamond, highly oriented pyrolytic graphite and the latter a−Si:H on glass, evaporated Cu, Al and Ni, 7059 glass and fused silica. During deposition, the beam current was typically 50 mA, the chamber pressure was 3×10^{-4} Torr. The deposition rate depended on whether the sample was shadowed or resputtered. Typical rates were 12 and 4 Å/min for shadowed and resputtered samples respectively. The resulting films, except those unhydrogenated, were pale yellow to brownish in color and have good adhesion to the substrates. Unhydrogenated films tended to crack or even to peel off when thicker than about 1 μm. The range of thicknesses of the samples varied from 0.1 to 2.5 μm. The extreme film hardness was demonstrated by scratch tests with a mechanical stylus. No visible traces could be found on ion beam deposited films but the films deposited by glow discharge decomposition of CH_4 were found to be softer[25].

The properties of a−C:H are critically dependent upon preparation conditions. Some of the properties of the a−C:H films which were used in this work, are listed in Table I. The details of the characterization will be published elsewhere[25] and will not be discussed here.

Table I. Optical properties of amorphous carbon films

	Hydrogenated	Unhydrogenated
Refractive index, n, at 280 nm	1.8~2.4	2.4
Optical absorption, α, at 3eV	$\sim 5\times10^4$ cm^{-1}	2×10^5 cm^{-1}
Optical gap, $\Delta\varepsilon_{opt}$	1.1eV	0.6eV

Refractive indices, n, at 280nm were determined by measuring the near normal reflectivity, r. At this wavelength, all the carbon films are highly absorbing and the reflectivities are then independent of the substrates. The reflectivity is approximately related to n by $r=(1-n)^2/(1+n)^2$, since κ, the extinction coefficient, is always less than 0.2 . Measurements (at P=0 Kbar) of the optical absorption, α, were made on samples deposited on quartz substrates from 1 to 3 eV, using a Cary 1413 spectrophotometer. Optical gaps were determined from extrapolating the absorption data to the abscissa in the $(\alpha h\nu)^{1/2}$ versus $h\nu$ plots. Hydrogen contents were determined by SIMS and by Galbraith Lab using a pyrolysis method. The two methods agree with each other to within a few percent.

Raman scattering was performed in the backscattering configuration using a point focussed 5145Å Ar^+ ion laser incident beam and a Spex 1401 double monochromator. Typical laser power at the sample was ~ 30mW. It was found that with laser powers exceeding 100mV, annealing of the sample occurs. The onset of annealing is influenced by the ability of the substrate to conduct heat away from the sample. For example, films on metals could withstand over 200 mW without being annealed or burned. For comparison, the Raman spectrum of pelletized carbon black was also obtained. The average particle size of the carbon black used was 20nm.

To measure the optical absorption under pressure, a small piece of plasma deposited film peeled off from the substrate and a chip of ruby pressure calibrant were placed in a diamond anvil high pressure cell. A mixture of methanol and ethanol was used as the hydrostatic pressure transmitting fluid. The experimental details for this absorption measurement have been described elsewhere.[24]

RESULTS AND DISCUSSION

Raman spectra were obtained of as-deposited amorphous carbon films prepared under various deposition conditions. It was found that in general the substrate material and its crystallinity or lack thereof have no effect on the spectra. The voltage of the ion beam deposition system, in the range studied, also has no effect. Weissmantel et al.[13] have suggested the importance of secondary ion impact on the formation of hard films. We found that the shadowed films which were not bombarded by argon ions and resputtered films show essentially the same measured properties. Raman spectra of films deposited onto substrates cooled by water and liquid nitrogen show qualitatively similar features. This is different from what was observed by Wada et al.[14] who found the Raman spectra of their films to be sensitive to substrate temperature. Jones and Stewart[15] and Meyerson and Smith[16] also found that the electrical conductivity and optical absorption of their hydrogenated amorphous carbon films strongly depended on the substrate temperature. We note that the two substrate temperatures used are considerably lower than the lowest substrate temperatures in Ref. 14-16.

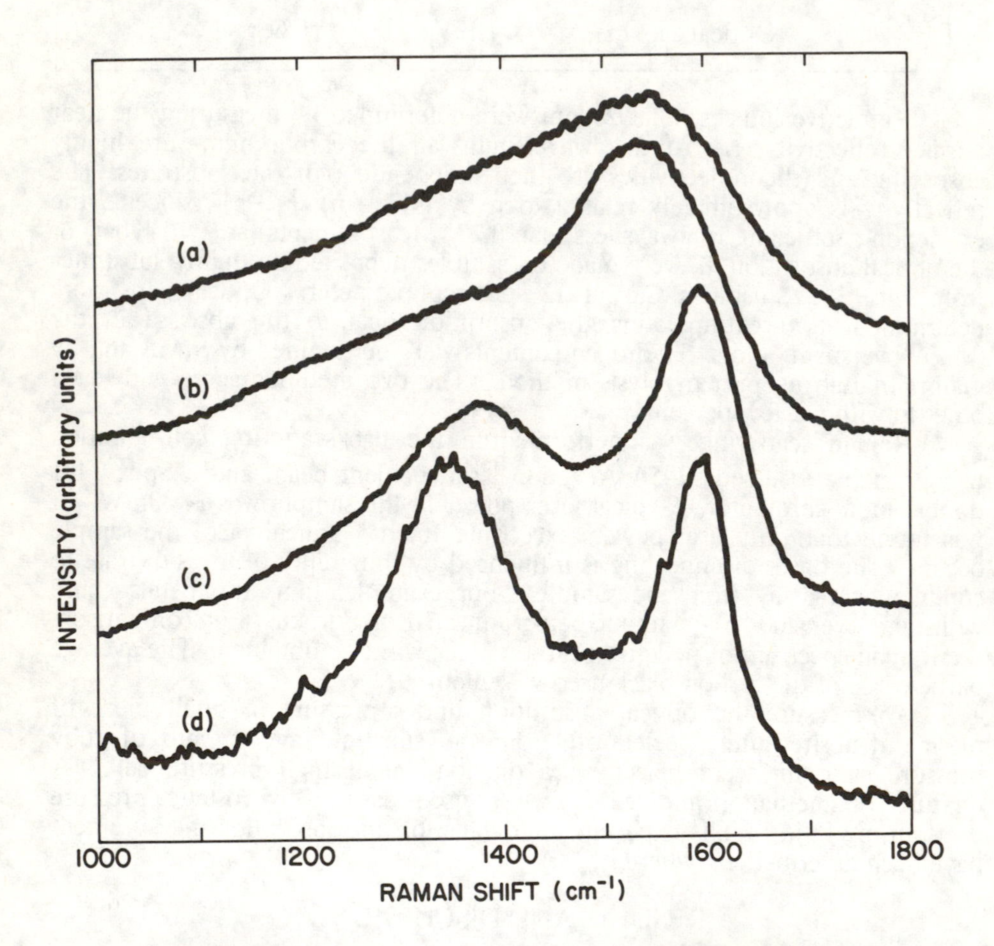

Fig.1. Raman spectra of (a) unhydrogenated amorphous carbon prepared by ion beam sputtering of graphite (b) a$-$C:H containing 40 at% hydrogen. (c) annealed a$-$C:H and (d) pelletized carbon black, average particle size 20nm.

Fig.1 shows the Raman spectra of variously prepared amorphous carbon films and of carbon black in the 1000 to 1800 cm^{-1} range. The Raman spectra of several forms of graphite have been studied by Tuinstra and Koenig[17] and recently by Lespade et al.[18] Tuinstra and Koenig noted that, while single crystal graphite exhibits only a sharp Raman line at 1575 cm^{-1} all other graphitic materials give a line at 1355 cm^{-1} in addition to the omnipresent line near 1575cm^{-1} They also established, with the aid of x-ray diffraction, an inversely linear relationship between the intensity ratio I_{1355}/I_{1575} of the 1355cm^{-1} to 1575cm^{-1} peaks and the graphitic in-plane correlation range, L_a. Solin and Kobliska [19] have pointed out that this inverse relationship is linear only over the range of L_a>40Å, for L_a<40Å saturation effects occur. Applying this relationship to the Raman spectra of carbon black (Fig. 1d) gives L_a< 40Å. Using phonon density of states for an infinite two-dimensional layer, and a model expressing the breakdown in the wave-vector selection rule in microcrystalline graphite and difference in photon-phonon coupling constants of different phonon branches, Lespade et al. have satisfactorily reproduced Raman spectra of carbon materials in various degrees of graphitization.

An important feature of the Raman spectra of carbon materials made from microcrystallites of graphite, and a feature emphasized by the model of Lespade et al. is that there are always two well resolved peaks, even when L_a is only 16Å. Fig.1a shows the spectrum of our amorphous carbon without hydrogen. There is a single, asymmetric peak at 1550cm^{-1}. This indicates that amorphous carbon is structurally different from a system composed entirely of microcrystalline graphite. Recently, Beeman et al.[20] have built large cluster models with different percentages of three- and four- fold coordinated carbon atoms . They also calculated the corresponding Raman spectra. They found that, when the tetrahedrally bonded atoms are uniformly distributed throughout the model, the effect is to modify the vibrational spectrum of the model as a whole, rather than to introduce spectral features typical of scattering from "diamond molecules". They also proposed that the structure of amorphous carbon probably consists of three-fold coordinated planar regions with occasional four-fold coordinated atoms acting as cross linking sites among these planes. Based upon their model, they suggest that the percentage of tetrahedrally coordinated atoms in amorphous carbon is probably less than 10%.

Fig.1b shows the Raman spectrum of a−C:H. The hydrogen content in these hydrogenated films ranges from 30 to 40 at %. These figures are comparable with films deposited by decomposition of hydrocarbon gases[6]. Meyerson and Smith[5] have suggested that the amount of hydrogen depends upon the substrate temperature which, in turn, determines the properties of the a−C:H films. We have not found a systematic variation of hydrogen content with substrate temperature, in agreement with the results of Jones and Stewart.[6] Furthermore, our Raman spectra of films containing more than 30 at % hydrogen are all very similar to each other and, when broadening is taken into account, are similar to that of unhydrogenated carbon. Similar results have been observed between ion beam deposited unhydrogenated amorphous carbon and r.f.-deposited carbon which naturally contains hydrogen by Dillon et al.[21] Line-width narrowing has also been observed in a−Si upon hydrogenation.[22] In a-C:H, hydrogen probably plays the same role, as it does in a−Si:H--viz. increasing the local order, in addition to passivation of dangling bonds.

Upon annealing of a−C:H by the laser, we observed that 1580 cm^{-1} peak shifts to 1590cm^{-1} and that another peak grows in strength at around 1370cm^{-1}. Fig.1c shows the Raman spectrum of an annealed film. It is similar to the spectra, obtained by Wada et al.[14] and Dillon et al.[21] of amorphous carbon annealed at 500°C. The changes in the spectrum are ascribed to the growth of graphitic domains. For samples thin enough to allow us to measure the Stoke's/anti-Stoke's ratio of the substrate material, it is estimated that the temperature due to laser heating is about 120°C. Since it is in intimate contact with the substrate, we feel that the temperature of the film will not be much higher. Our ability to anneal the film at seemingly low temperatures can also be related to the increase in local order due to hydrogenation. This observation is consistent with a similar effect observed for the "recrystallization" temperature of a−Si as observed by Tsu et al.[22] From comparison of the spectra in Figs. 1c and 1d, we note that the annealed a−C:H film is still considerably different from a system of graphite microcrystallites. In particular, the peaks are much too wide for crystallites of dimension, L_a~80Å, the correlation range suggested by the intensity ratio using the rule of Tuinstra and Koenig.

Owing to the short wavelength cut-off of our high pressure microspectrometer system for wavelength less than 3500Å, and the large gap (~3.5eV) of our plasma deposited films,we were only able to observe the tail of the absorption edge in the diamond-cell even at P=0 Kbar. Nevertheless, we found that the band-tail "blue" shifted toward high energy when pressures up to 40 Kbar were applied, indicating that the external compression was transferred in large part to the covalent nearest neighbour bonds. This trend further supports the prescence of at least a partially 3d-connected covalent network, with substantial numbers of tetrahedrally bonded atoms acting as cross-linking centres[23,24].

CONCLUSION

It is concluded from the Raman scattering meausrements on a large variety of well characterized, hydrogenated and unhydrogenated ion beam deposited films and presure-optical absorption studies, that neither the microcrystalline graphite nor random 2-d network model is compatible with the structure of amorphous carbon. Instead, the data are qualitatively consistent with a network model consisting of both three-fold and four-fold coordinated carbon atoms, in agreement with other recent Raman studies. The role of hydrogenation is to increase the local order in addition to passivation of dangling bonds.

ACKNOWLEDGEMENT

We would like to thank D. Beeman for sending us their paper before publication.

REFERENCES

1. K. Bewilogua, D. Dietrich, L. Pagel, C. Schurer and C. Weissmantel, Surface Sci. 86, 308 (1979).
2. Toshio Mori and Yoshikatsu Namba, J. Vac. Sci. Technol. A, 1, 23, 1983.
3. B. Dischler, A. Bubenzer and P. Koidl, Appl. Phys. Lett 42, 636, 1983.
4. D. A. Anderson, Phil. Mag., 35, 17, 1977.
5. B. Meyerson and F. W. Smith, J. Non-Cryst. Solids 35&36, 435, 1980.
6. D. I. Jones & A. D. Stewart J. de Physique C4, 42, 1085, 1981, Phil. Mag. B 46, 423, 1982.
7. E. G. Spencer, P. H. Schmidt, D. C. Joy and F. J. Sausalone, App. Phys. Lett. 29, 118 (1976).
8. J. Fink, T. Muller Heinzerling, J. Pfluger, A. Bubenzer, P. Koidl and G. Crecelius, Solid State Comm. 47, 687, 1983.
9. D. F. R. Mildner and J. M. Carpenter, J. Non Cryst. Solids, 47, 391, 1982.
10. C. A. Majid, J. Non Cryst. Solids 51, 137, 1983.
11. S. A. Solin, Structure and Excitations of Amorphous Solids, AIP conference proceedings, Williamsburg, Va 1976, ed by G. Lucovsky and F. L. Galeener, p. 205.
12. M. H. Brodsky, Light Scattering in Solids ed. M. Cardona Springer-Verlag, NY, 1975.
13. C. Weissmantel, K. Bewilogua, D. Dietrich, H. J. Erler, H. J. Hirmeberg, S. Klose, W. Nowick and G. Reisse, Thin Solid Films 22, 19 (1980).
14. N. Wada, P. J. Gaczi and S. A. Solin, J. of Non Cryst. Solids 35 & 36, 543, 1980.
15. D. I. Jones and A. D. Stewart, Phil. Mag. B 46, 423, 1982.
16. B. Meyerson, and F. W. Smith, J. Non Cryst. Solids 35 & 36, 435, 1980.
17. F. Tuinstra and J. L. Koenig, J. Chem. Phys. 53, 1126, 1970.
18. P. Lespade, R. Al-Jishi and M. S. Dresselhaus, Carbon 20, 427, 1982.
19. S. A. Solin and R. J. Kobliska, proceedings of 5th Int. Conf. on Amorphous and Liq. Semiconductors, 1251, edited by J. Stuke and W. Brenig (Taylor and Francis, London) 1974.
20. D. Beeman, J. Silverman, R. Lynds and M. R. Anderson to be published in Phys. Rev. B.
21. R. O. Dillon, J. A. Woollam and V. Katkanant Phy. Rev. B 29, 3482, 1984.
22. R. Tsu, J. Gonzalez-Hernandez, J. Doehler and S. R. Ovshinsky, Solid State Comm. 46, 79, 1983.
23. B. A. Weinstein, R. Zallen, M. L. Slade and A. deLozanne, Phys. Rev. B 24, 4652, 1981.

24. B. A. Weinstein, R. Zallen, M. L. Slade and J. C. Mikkelsen, Phys. Rev. B 25, 781, 1982.
25. F.Jansen, M.A. Machonkin, S. Kaplan and S. K. Hark, J. Vac. Sci. Technol., to be published.

SPIN POLARIZED PHOTOEMISSION AS A SENSITIVE TOOL TO STUDY STRUCTURAL SURFACE PHASE TRANSITIONS

F. Meier and D. Pescia
Laboratorium für Festkörperphysik, ETH Hönggerberg, CH-8093 Zürich, Switzerland

ABSTRACT

The photoelectron spin polarization created by optical spin orientation is completely determined by the symmetry of the electronic states involved in the optical transition. Therefore, this type of spin polarized photoemission experiment is sensitive to structural changes of surfaces.

The production of spin polarized photoelectrons from an unpolarized ground state by means of optical excitation (optical spin orientation) is a well-established technique by now [1]. The most essential feature of optical spin orientation is that the final state polarization in a given transition depends only on the symmetry of the states involved. The details of the radial parts of the wave functions are unimportant. The circumstance that symmetry alone determines the polarization of a transition - sign as well as magnitude - makes optical spin orientation especially useful for the investigation of band hybridization [2], identification of orbitals [3], or enhanced resolution in photoelectron spectroscopy [4].

The sensitivity of the electron polarization with respect to the symmetry of the wave functions opens the possibility to investigate structural changes of a solid by optical spin orientation. It should be noted that photoemission is generally not a very suitable tool for this purpose: angle-integrated, energy-resolved photoelectron spectra of solid and liquid metals are generally almost identical [5]. The dielectric constants measured by reflection spectroscopy are very similar, the only difference being some structure due to critical point interband transitions of the crystalline solid. This structure is, however, superimposed on a large signal background [6]. In this paper an experiment is reported which shows the response of the photoelectron spin polarization obtained by optical spin orientation to a structural change of the

0094-243X/84/1200473-05 $3.00

system. It is large enough to nourish optimism that structural phase transformations can be studied successfully in the future by spin polarized photoemission.

In the present experiment the spin polarization obtained from crystalline and amorphous germanium is compared. This system deserves special interest because the atomic nearest-neighbor arrangement in both phases is tetrahedral. If the polarization depends only on short range order on the scale of nearest neighbor atomic distances, the polarization spectra of crystalline and amorphous germanium were expected to be quite similar.

The process of optical spin orientation requires the use of circularly polarized light. In Fig. 1 transitions

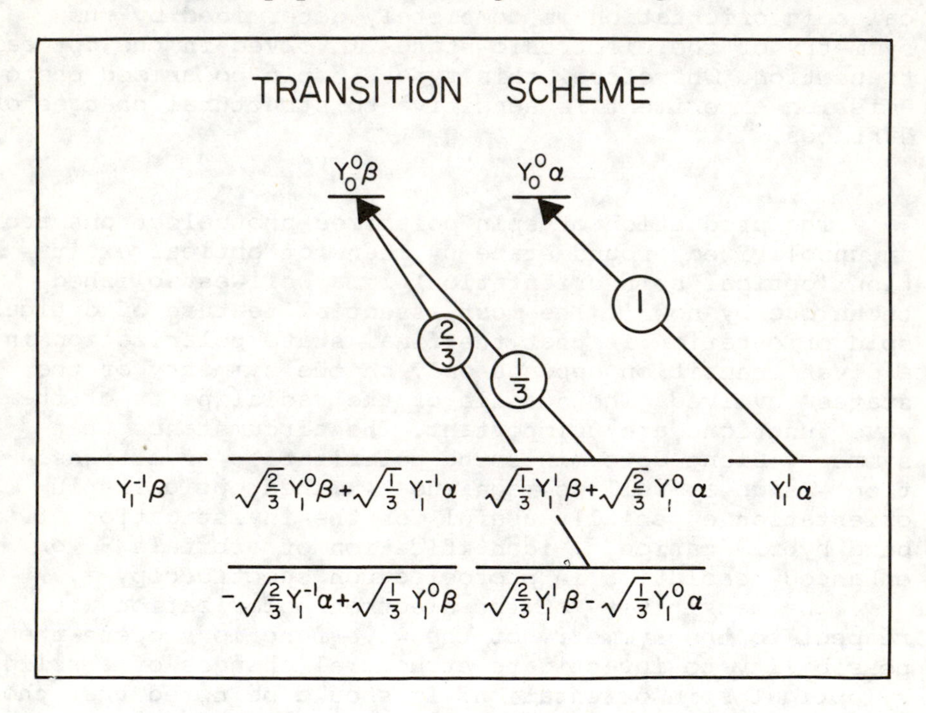

Fig. 1. Transitions by circularly polarized light ($\Delta m = -1$) at the Γ-point of Ge from the topmost valence bands to the lowest conduction band. The two initial state valence bands are separated by the spin-orbit splitting. The relative intensities of the transitions are given by the encircled numbers.

induced at the Γ-point of crystalline Ge by right-hand circularly polarized light are indicated. The symmetries of the wave functions are expressed in terms of spherical harmonics and spin functions to permit the use of atomic selection rules. For a more rigorous treatment based on group theory see Ref.(1). Inspection of Fig. 1 shows that the transitions from the upper valence band result in a polarization

$$P = \frac{N\uparrow - N\downarrow}{N\uparrow + N\downarrow} = \frac{1 - 1/3}{1 + 1/3} = 50\%$$

The measured spectrum $P(h\nu)$, $1.6 < h\nu < 4$ eV, is shown in Fig. 2a. Although at photothreshold the transitions are not exactly at Γ (the gap at Γ is 0.9 eV) but only nearby the polarization still equals 50%; for an explanation see Ref.(2). Some more structure due to transitions from the spin-orbit split valence band (at 2.4 eV) and due to transitions to an upper conduction band (at 3.1 eV) are also resolved.

All the polarized transitions of Fig. 2a occur at or close to Γ. In this part of the Brillouin zone the bands of Ge are rather isotropic. Therefore the polarization spectrum is independent of the crystal orientation. This has been explicitly checked by measuring (100), (110), (111), and (211) faces and a polycrystalline surface. Consequently, if the polarization in amorphous Ge were determined by nearest neighbor tetrahedra, it should not be influenced by their random orientation.

Amorphous germanium was prepared by evaporation on a molybdenum substrate held at room temperature or slightly above. By high resolution electron diffraction the amorphous nature of the sample was checked [7]: no indication of crystallite formation was found.

The optical orientation experiment on amorphous Ge yielded the result shown in Fig. 2b: the spin polarization is not only reduced with respect to crystalline Ge but entirely wiped out. The amorphous film was annealed at 250°C such that crystallites of roughly 100 - 1000 Å were formed. Then the polarization spectrum of Fig. 2a was obtained again.

The conclusion from the experiment is twofold: first, it is not the geometry of nearest neighbors which determines the polarization in an optical spin orientation experiment but a larger region. Secondly, there is a maximum change of the polarization between crystalline and amorphous germanium giving rise to the expectation to use spin-polarized photoemission for the investigation of more subtle surface structural phase transitions generally.

A probable cause for the P = O result is the vanishing of the spin-orbit splitting of infinitely extended electron states in amorphous material [8]. Then, because the light operator acts only on the orbital motion of the

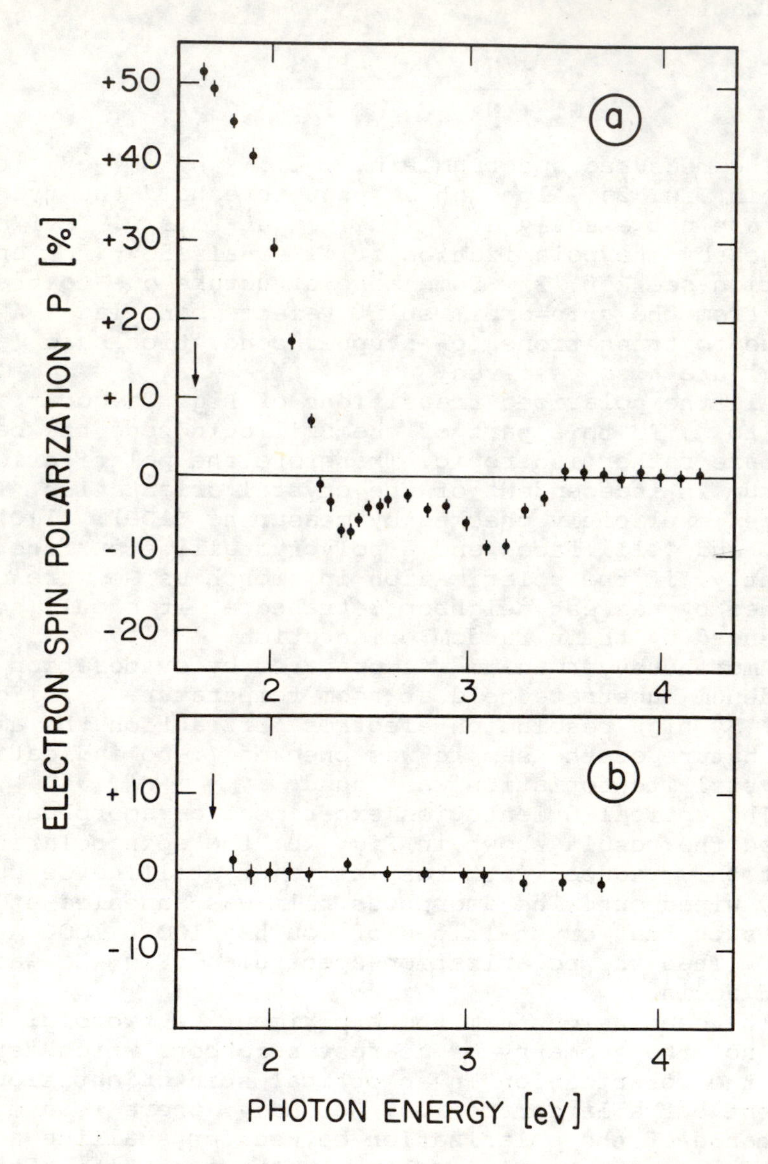

Fig. 2 a: Spectrum of spin polarization of a Ge single crystal. b: Spectrum of spin polarization of amorphous germanium. The arrows indicate the photothresholds obtained by cesiation of the surface.

electron the spin polarization is always zero. As an example, one may consider the transition of Fig. 1 for zero spin-orbit splitting. The initial state becomes 6-fold degenerate and the total polarization of the 3 simultaneously excited transitions is $P = (1-1/3-2/3)/(1+1/3+2/3) = 0$.

The additional information to be gained from the experiment is that a region of the size of the optical excitation in a photoemission experiment is already large enough to make the spin-orbit splitting vanish in amorphous Ge. The determination of the critical size of an agglomerate of Ge-atoms where its crystalline properties begin to appear is a challenging problem. Its solution should answer the question at which crystallite size the wave functions become those of the infinitely extended solid. Here a long-standing problem of small particle physics [9] may get an useful hint from an unexpected angle.

We thank H.C. Siegmann for his continuous encouragement and M. Baumberger for his help during the measurements. The financial support by the Schweizerische Nationalfonds is gratefully acknowledged.

REFERENCES

1. "Optical Orientation" edited by F. Meier and B. Zakharchenya (North-Holland Publ. Comp., Amsterdam) to be published 1984.
2. R. Allenspach, F. Meier, and D. Pescia, Phys. Rev. Lett. 51, 2148 (1983).
3. F. Meier and D. Pescia, Phys. Rev. Lett. 47, 374 (1981).
4. F. Meier, R. Allenspach, G.L. Bona, and D. Pescia, Proc. Int. Conf. Physics of Semiconductors, (San Francisco, August 6-10, 1984), to be published.
5. F. Greuter, PhD thesis, Eidgenössische Technische Hochschule Zürich, Report No. 6468, 1979 unpublished; F. Greuter and P. Oelhafen, Z. für Physik B34, 123 (1979).
6. N.H. March, Can J. Chem. 55, 2165 (1977); J.N. Hodgson "The Optical Properties of Liquid Metals" in "Liquid Metals", Editor S.Z. Beer (Marcel Dekker Inc., N.Y. 1972), p. 331 ff.
7. We thank H.-U. Nissen, ETH Zürich, for having made these studies.
8. J.M. Ziman, J. Phys. C: Solid State Physics 4, 3129 (1971).
9. J.A.A.J. Perenboom, P. Wyder, and F. Meier, Physics Repts. 78, 175 (1981).

REVIEW OF METASTABLE DEFECTS IN a-Si:H

H. Fritzsche
James Franck Institute and Department of Physics
The University of Chicago, Chicago, IL 60637

ABSTRACT

The latest results on the Staebler-Wronski effect will be discussed in the light of present theories for the creation of metastable defects in hydrogenated amorphous silicon (a-Si:H).

INTRODUCTION

More than six years ago Staebler and Wronski[1] discovered metastable changes in the properties of hydrogenated amorphous silicon (a-Si:H) resulting from prolonged exposure to band gap light. These changes can be reversed by annealing between 160 and 200°C. This effect is one of the most important and challenging problems in the field of amorphous semiconductors.

Detailed accounts of the various manifestations of the Staebler-Wronski (SW) effect can be found in some recent reviews of this subject[2-4]. Here I mention only a few of the essential observations in order to highlight some controversies in present interpretations and to explore experimental tests that may lead us to a better understanding of this phenomenon.

DEFECT IDENTIFICATION AND CREATION

The SW defects are silicon dangling bonds that have a positive effective correlation energy. They appear to be indistinguishable from the residual dangling bond defects found in annealed a-Si:H films. The ionization energy of the negatively charged defect state is 0.85 ± 0.05 eV and that of the neutral dangling bond state is 1.2 ± 0.05 eV. The latter is identified by the g-value, g=2.0055, and the ESR line width of its spin.

The SW- defects are produced by electron-hole recombination processes following excitation for instance by light[1], X-rays, electron bombardment, or carrier injection[5]. The creation process does not seem to be thermally activated since it proceeds with nearly unchanged efficiency at 78°K and 300°K in some samples. In others the rate is reduced by several orders of magnitude at low temperatures. The latter samples show also a large drop in photoconductivity with decreasing temperature.

The initial rate of defect creation is approximately 10^{-5} defects/recombination event. The rate decreases with time, and ultimately one needs about $10^{25}cm^{-3}$ recombination events for $10^{17}cm^{-3}$ defects. One reason for the very low creation efficiency is the following. We know from spin-dependent photoconductivity measurements that the recombination via dangling bond defects is the dominant recombination channel[6]. Since these recombinations

0094-243X/84/1200478-09 $3.00

obviously cannot produce dangling bonds, another less efficient channel such as nonradiative recombination between tail states of the conduction and valence bands must be responsible. The creation efficiency of defects is fairly independent of photon energy down to $h\nu=1.2$ eV in undoped a-Si:H[7].

The total concentration of SW-defects rarely exceeds $10^{17}cm^{-3}$. The few exceptions are related to samples that contain unusually large hydrogen contents. For instance, Yoshida et al.[8] found up to $6X10^{19}cm^{-3}$ new spins in films prepared from disilane that contain about 40 at.% hydrogen.

Earlier experiments by Carlson et al.[9] and Zhang et al.[10] suggested that the SW-effect is accompanied by reversible changes in the Si-H vibration spectra. If this were true, then the $10^{17}cm^{-3}$ SW-defects might be only a small by-product of more massive changes in the structure of the material. At this conference it was reported that normal films do not exhibit changes in the Si-H vibration spectra[11] and that there is no change in the slope of the Urbach tail of optical absorption[12]. We therefore associate the SW-effect with the production of about 10^{17} metastable defects.

KINETICS OF DEFECT CREATION

Several studies have shown that the Staebler-Wronski effect does not obey the rule of photographic reciprocity[13]. This means it does not simply depend on the product of light intensity I and exposure time t (which is proportional to the total number of recombination processes). Stutzmann et al.[14] measured the kinetics of the increase in dangling bond defect density N_s over a range of I and t. This can be done by measuring directly the spin density N_s or by following the time dependence of the photoconductivity σ_p. The latter is proportional to the drift mobility and the total number of trapped electrons n or holes p, respectively. Attributing the SW-effect to tail→tail recombination processes Stutzmann et al. described the increase in N_s as follows:

$$\frac{dN_s}{dt} = C_{sw}\, b\, np \tag{1}$$

where b is the nonradiative tail to tail recombination and C_{sw} is the fraction that leads to metastable SW-defects. As mentioned above the dominant recombination proceeds through the N_s dangling bonds. When σ_p is proportional to the electron-hole generation rate G, then both n and p are proportional to G/N_s. Substituting this result into Eq.(1) leads to

$$\frac{dN_s}{dt} \propto C_{sw} b \frac{G^2}{N_s{}^2} \tag{2}$$

This equation expresses the self-limiting nature of this model since the creation rate dN_s/dt decreases as $1/N_s{}^2$.

Integration yields

$$N_s(t)^3 - N_s(o)^3 \propto C_{sw} b\, G^2 t \tag{3}$$

For longer exposure times, the term $N_s(o)^3$ can be neglected and

$$N_s(t) \propto G^{2/3} t^{1/3} \quad \text{for } t>>o \tag{4}$$

The initial slope of the logarithmic decrease of the photoconductivity is

$$\frac{1}{\sigma_p} \frac{d\sigma_p}{dt} = - C_{sw}\, b\, G^2/N_s(o) \tag{5}$$

Using a sample that obeyed monomolecular recombination, $\sigma_p \propto G$, Stutzmann et al. verified the specific dependencies on G and t expressed by Eqs.(4) and (5). Unfortunately, monomolecular recombination is rarely observed in a-Si:H at room temperature; instead, $\sigma_p \propto G^{\gamma}$ with $0.5 < \gamma < 0.9$. This probably arises from the fact that the recombination rate via N_s centers is limited by hole diffusion through localized valence band tail states[6]. This diffusion process depends in turn on the position of the trap quasi-Fermi level and hence on light intensity. The specific forms of Eqs.(4) and (5) for different values of γ have not yet been worked out. Nevertheless, Stutzmann et al.'s work illustrates the importance of studying the kinetics for identifying the processes leading to the SW-effect.

ANNEALING AND REVERSIBILITY

The SW-defects can only be removed by annealing. Attempts to quench the defects by infrared irradiation have failed. The decrease of N_s at a fixed elevated temperature neither follows an exponential decay with time, $\exp(-t/\tau)$, nor a bimolecular recombination kinetics[15,16]. One may define a recovery time τ_R during which a change of a certain physical quantity is recovered by half during the course of annealing. Activation energies of τ_R between 0.5 and 1.5 eV can be observed depending on the temperature at which the SW-defects were created. Lower values correspond to lower creation temperatures. Moreover a prior anneal at T_1 decreases the anneal rate at $T_2>T_1$[15]. These observations suggest that one is dealing with a broad spectrum of barrier heights that stabilize the SW-defects. This spectrum can of course extend to lower energies when the centers are created at lower temperatures. It would be interesting to find out whether high barrier height defects can be created with similar efficiency at low T as at high T or with $h\nu=1.2$ eV photons as with 2 eV photons. If the efficiencies turn out to be very different then it is possible to create selectively SW-defects of different stability.

The original properties of a-Si:H are essentially restored by a 30 min anneal at 200°C. This means that there is an upper limit to the stability of the SW defects. In a few instances films submitted to repeated cycles of light exposure and annealing have shown some lack of reversibility[17] that may be interpreted as a gradual increase in residual defect centers. Since these experiments extend over long time periods however, one cannot exclude the possibility that surface oxidation or other changes unrelated to the SW-effect were causing these irreversibilities. Nevertheless, the degree of reversibility is an important piece of information that one has to consider in searching for the origin of this effect.

ROLE OF IMPURITIES

Some early work indicated that the SW-effect is related to impurities[2,3]. A much greater deterioration was observed when a-Si:H solar cells were made of material in which oxygen and nitrogen were introduced through an air leak into the plasma chamber. A variety of impurities are present in ordinary a-Si:H films at concentration levels 100-1000 times larger than that of SW-defects.

This question has finally been resolved by Tsai et al.[18] who prepared very pure a-Si:H films in a bakeable high vacuum system using the highest quality silane. These precautions reduced the concentration of the most common contaminants by two or three orders of magnitude. These authors found that the concentration of SW-defects detected by ESR remained essentially unchanged as the concentration of nitrogen was increased from 10^{17} to $10^{20}cm^{-3}$ and that of oxygen from 10^{18} to $10^{20}cm^{-3}$. A further increase led to an enhancement of SW-defects by a factor of 4. Impurity concentrations $>10^{20}cm^{-3}$ correspond however to the alloy regime where the structure and the hydrogen concentration of the material are changing. This alloy regime was studied by Morimoto et al.[19] who found not only more SW-defects but also a larger number of dangling bond defects in the annealed state.

Some impurities such as boron modify the structure and the hydrogen concentration[20] when they are present at densities as low as 10^{16} or $10^{17}cm^{-3}$. These impurities may therefore affect the creation of SW-defects indirectly without being responsible or essential for the effect.

WEAK BONDS

Dersch, Stuke and Schweitzer[21] suggested that the first step in creating SW-defects is the breaking of weak Si-Si bonds. This corresponds to the nonradiative tail-tail state recombination process discussed by Stutzmann et al.[14]. Weak or strained bonds will contribute to the tail states because of their small energy splitting of the bonding and antibonding states. An electron raised to the antibonding state can lower its energy by breaking the bond of the two resulting neutral dangling bond states T_3^o which lie below the mean energy level of the electron-hole pair.

The two dangling bonds facing each other are unstable and are prone to reform the bond. This must be prevented by some stabilizing mechanism. Dersch et al.[21] suggest that a neighboring hydrogen atom changes its position, thereby separating the two dangling bonds. In order to observe an ESR line that is not broadened by exchange interactions, the authors originally estimated that a separation of at least 10 Å is needed. This would require at least two hydrogen bond transpositions. More recent estimates[22] decrease the necessary separation to 5 Å, so that only one hydrogen atom has to change its bond position. This model requires that a hydrogen atom is close to the weak bond. However, a hydrogen atom bonded to silicon decreases the local strain because the local covalent connectivity is lowered from 4 to 3. One therefore should expect that the probability of finding a hydrogen next to a weak bond is less than the statistical value derived from the hydrogen concentration[23]. This may be another factor reducing the creation efficiency of SW-defects.

It is sometimes argued that the energy, that is released when an electron and a hole recombine from their respective tail states, helps switching the nearby hydrogen into the stabilizing position at one of the newly created dangling bonds. However, the energy balance is much less favorable because even a weak bond represents a lower energy state than two broken bonds. The energy difference is at least the energy separation ($\sim$1.3±0.1 eV) between a conduction band tail state and a valence band tail state. Only this energy and not more is released in a nonradiative recombination process between tail states. Therefore the original excited state A* (a weak bond with an excited electron-hole pair in tail states) and the final state B (two dangling bonds with a H-atom moved into a blocking position) have essentailly the same energy when E_F lies near the gap center. The annealed state A (a weak bond with a H-bond nearby) lies lower by about 1.3 eV, yet it cannot be reached from state B unless annealing overcomes a barrier of about 1 eV. It is difficult to see how the transition from A* to B can proceed without a noticeable activation barrier even though state A* lies considerably below the top of the barrier that is needed to stabilize B.

In n-type doped a-Si:H the state B is decreased relative to its energy in intrinsic material by 2Δ where Δ is the energy gained by transferring an electron from the Fermi energy E_F to a SW-defect which in turn becomes negatively charged. This D^- state lies about 0.85 eV below the electron mobility edge E_C so that $\Delta \lesssim 0.6$ eV. In p-type a-Si-H an equivalent lowering of state B by 2Δ occurs because each SW-center will release an electron to an empty state near E_F in the lower half of the gap. It would be interesting to investigate whether the average annealing barrier between states B and A is affected by doping. Nevertheless as state B moves down by 2Δ as a result of doping, it is easier to envision a smooth transition from A* to B. This model predicts that the rate of producing SW-defects increases with doping. However, doping also increases the dangling bond defect density in state A and hence decreases the branching ratio between the SW- and the dominant recombination channel. Nevertheless, the effect of the Fermi-level position on the SW-defect creation rate should be explored.

DEFECTS WITH NEGATIVE EFFECTIVE CORRELATION ENERGY

The model proposed by Adler[24] for the formation of metastable SW-defects predicts the same state B as the weak bond model: an increased concentration of Si dangling bonds that have a positive effective correlation energy U. Since the two models arrive at the same final state, they explain all aspects of the SW-effect with equal success. The major differences between the weak bond model and Adler's model are the initial or annealed state A and the kinetics leading to the degraded state B.

Adler argues that a negatively charged Si atom behaves chemically like a phosphorus atom and thus should be 3-fold coordinated with p^3 orbitals at angles of about 100°. Similarly a positively charged Si is chemically equivalent to Al that also prefers 3-fold coordination with nearly planar sp^2 bonds forming angles of 120°. Such 3-fold charged Si atoms relieve local strains and should therefore occur naturally - perhaps at concentrations of $10^{17}cm^{-3}$. The $Si^-(p^3)$ form hole traps close to E_V, the valence band mobility edge, and the $Si^+(sp^2)$ form electron traps close to E_C. If the covalent bond angle distortions could easily be overcome, the $Si^-(p^3)$, $Si^o(sp^3)$, and $Si^+(sp^2)$ configurations would form a negative U triplet similar to the valence alternation centers in chalcogenide glasses: interconversion is possible and the first ionization energy is larger than the second one. However, the necessary change in bond angles is more difficult within a surrounding tetrahedral network. Hence there is a potential barrier hindering the interconversion

$$2\ Si^o(sp^3) \rightleftarrows Si^+(sp^2) + Si^-(p^3) \qquad (6)$$

The height of these potential barriers will be lower when one or two of ligand atoms is hydrogen or when the 3-fold coordinated Si atom is at an internal surface. It is high when the Si atom in question is constrained by tetrahedrally bonded Si. As a consequence and in contrast to (-U) centers in chalcogenide glasses, the (-U) centers $Si^-(p^3)$ and $Si^+(sp^2)$ can co-exist with $Si^o(sp^3)$. The latter can therefore change their charge state and act as positive U-centers without suffering interconversion. Conversion into the diamagnetic negative U-centers, following the right-hand arrow in Eq.(6) can occur by annealing at elevated temperatures except for some very constrained 3-fold (sp^3) centers that remain. These are the residual dangling bond centers of state A.

The creation of SW-defects from the pool of diamagnetic $Si^-(p^3)$ and $Si^+(sp^2)$ centers occurs according to this model by trapping of holes and electrons respectively. After capturing a conduction band electron the then neutral $Si^o(sp^2)$ changes bond angles and moves to the position of $Si^o(sp^3)$ below midgap. About 1 eV is thereby released to the environment. This energy should be sufficient to overcome the barrier against bond angle changes. Similar arguments apply to $Si^-(p^3)$ centers that capture a hole. The energy released at either center depends again on E_F and hence on doping. The saturation limit of the SW-effect is in this model given by the finite number of convertible (-U) centers.

This work was supported by NSF Grant DMR 8009225.

The fact that Adler's model deals with strain-relieving 3-fold coordinated Si ions that should naturally be present in this material is a very attractive feature. It also explains why the SW-effect is strong in films having high H-contents: this facilitates bond angle changes. There are however some difficulties, (a) the large capture cross sections of the charged (-U) centers should yield a very efficient conversion rate; (b) the (-U) centers should pin E_F until the doping level exceeds their concentration; (c) the creation of SW-defects is governed by trapping, hence the rate should be proportional to n and p separately instead of the product np that governs tail to tail recombination.

Let us respond to these objections. (a) The dwell time of trapped carriers in these centers can be extremely short because a high capture rate implies a large thermal release rate[25]. This may yield the observed inefficiency of the SW-defect creation process. (b) As soon as the concentration of dopants exceeds that of (-U) centers E_F becomes unpinned. This should happen near a few ppm. There is no clear evidence against an unpinning threshold. (c) Indeed, the creation kinetics appears to be different in the two model, however, there is also evidence for metastable defects created by trapping[26]. Perhaps there is more than one SW-process.

THE ROLE OF HYDROGEN

The SW-effect has been observed so far only in hydrogenated amorphous silicon. If it were possible to produce amorphous Si without hydrogen but with a low density of gap states and of dangling bond defects then one could test whether hydrogen is essential for the effect.

At least one should be able to determine whether the SW defects occur predominantly in a-Si:H regions where the Si-H bonds are dilute and randomly dispersed or in the clustered hydrogen-rich regions[27]. One would expect that the former regions contain more strained bonds whereas the latter give greater freedom for bond angle changes.

Other forms of hydrogen are H_2 molecules trapped in voids[28] and atomic hydrogen. The molecules amount to a few at.%,but it is unlikely that they are playing any role in the SW-effect. Atomic hydrogen would be ideal for blocking the healing of a broken bond but only a very low concentration of H has been detected in films that contain a lot of oxygen[29].

DIFFERENT TYPES OF DEFECTS

Evidence for different kinds of SW-defects has been reported recently. They differ in the way they affect the recombination lifetime or the subbandgap absorption near $h\nu=1.2$ eV, in their annealing kinetics,and the efficiency of creation by light exposure at different temperatures[30].At this conference,Krühler et al.[26] distinguished metastable trapping (T) and recombination (R) centers that are produced at different rates depending on whether the a-Si:H film is subjected to injection of electrons or of holes or to electron-hole recombinations. Moreover, calculations by Lucovsky et al.[31] showed that the energy of the dangling bond state increases as the electronegativity of one ligand atom of the Si in question increases from Si or H to C, N, and O. It is also likely that the energies of SW-defects in the H-dilute and in the H-clustered phases are different as well as their effect on the electronic and optical properties.

CONCLUSIONS

We are now in the position of designing quantitative experiments that permit us to test the different creation and annealing kinetics predicted by the two theories. It will be very valuable to know how the creation and annealing rates depend on the position of the Fermi level. This might best be studied by creating accumulation and depletion regions in a field effect geometry. Preliminary experiments show some surprising results[18]. It would be interesting to study the temperature dependence of the creation rate down to helium temperatures.

It is generally assumed that the defects created by electron bombardment are the same as the SW-defects. Very much can be learned by exploring whether there are differences in the creation and annealing kinetics as a function of temperature and doping. Many more defects can be produced by electron bombardment than the usual limit experienced by prolonged illumination. Hence these defects cannot be Adler's (-U) charged centers. If they are related to weak bonds one should find a sharpening of the conduction band tail due to the depletion of weak bonds.

Without intending to favor one model over the other, I wish to compare some numbers relevant to the weak bond model. About $10^{25}cm^{-3}$ recombination processes are needed to create $10^{17}cm^{-3}$ SW-defects in a-Si:H. If we assume that weak bonds contribute states to the conduction band tail we estimate that their concentration does not exceed $10^{19}cm^{-3}$. Only 1/10 of these will have a H-atom nearby to block recombination when the weak bond breaks. The number will be smaller when Pankove's argument holds that strain-relieving H will not likely be close to a weak bond. Therefore, the conversion of weak bonds with a neighboring H-bond into SW-defects must occur with an efficiency of 10 percent or higher. The major cause for the inefficiency of the SW-effect is the very small branching ratio between the nonradiative tail state recombination and recombination via dangling bonds.

What can be done to stop the SW-effect? One probably cannot alter the recombination branching ratio. Actually our attempts to prepare better material having fewer dangling bonds will enhance the probability for tail state recombinations. It appears that the best course of action is to relieve strain in the material by incorporating atoms of low coordination that do not switch their bond position such as hydrogen does. Fluorine[32], chlorine, and perhaps selenium seem to be promising candidates.

REFERENCES

1. D. L. Staebler and C. R. Wronski, Appl. Phys. Lett. 31, 292 (1977); J. Appl. Phys. 51, 3262 (1980).
2. D. E. Carlson, Solar Energy Mat. 8, 129 (1982).
3. J. I. Pankove, Solar Energy Mat. 8, 141 (1982).
4. C. R. Wronski in "Semiconductors and Semimetals" Vol. 21 Part C, edited by J.I.Pankove (Academic Press, N.Y. 1984).
5. D. L. Staebler, R. Crandall and R. Williams, Appl. Phys. Lett. 39, 733 (1981); S. Guha, J. Yang, W. Czubatyj , S. J. Hudgens and M. Hack, Appl. Phys. Lett. 42, 588 (1983).
6. H. Dersch, L. Schweitzer and J. Stuke, Phys. Rev. B28, 4678 (1983).
7. C. C. Tsai, private communication.
8. M. Yoshida, K. Morigaki, I. Hirabayashi, H. Ohta, A. Amamou and S. Nitta, this volume; I. Hirabayashi, K. Morigaki and M. Yoshida, Solar Energy Mat. 8, 153 (1982).
9. D. E. Carlson, A. R. Moore, D. J. Szostak, B. Goldstein, R. W. Smith, P. J. Zanzucchi and W. R. Frenchu, Solar Cells 9,19 (1983).
10. P. X. Zhang, C. L. Tan, Q. R. Zhu and S. Q. Peng, J. Non-Cryst. Solids 59-60, 417 (1983).
11. H. Fritzsche, J. Kakalios and D. Bernstein, this volume.
12. D. Han and H. Fritzsche, this volume.
13. S. Guha, Appl. Phys. Lett. in press.
14. M. Stutzmann, W. B. Jackson and C. C. Tsai, Appl. Phys. Lett. in press; and this volume.
15. M. Stutzmann, C. C. Tsai and W. B. Jackson, Proc. MRS-Europe Conf. Strasbourg (1984).
16. C. Lee, W. D. Ohlsen, P. C. Taylor, H. S. Ullal and G. P. Caesar, this volume.
17. S. Nitta, Y. Takahashi and M. Noda, J. de Phys. 42, C4-403 (1981).
18. C. C. Tsai, M. Stutzmann and W. B. Jackson, this volume.
19. A. Morimoto, H. Yokomichi, T. Atoji, M. Kumeda, I. Watanabe and T. Shimizu, this volume.
20. K. Chen and H. Fritzsche, Solar Energy Mat. 8, 205 (1982).
21. H. Dersch, J. Stuke and J. Beichler, Appl. Phys. Lett. 38, 456 (1981).
22. M. Stutzmann, private communication.
23. J. I. Pankove, private communication.
24. D. Adler, J. de Phys. 42, C4-3 (1981).
25. M. Silver, private communication.
26. W. Krühler, H. Pfleiderer, R. Plättner and W. Stetter, this volume.
27. J. A. Reimer, R. W. Vaughan and J. Knights, Phys. Rev. Lett. 44, 193 (1980); Phys. Rev. B23, 2567 (1981).
28. Y. J. Chabal, C. K. N. Patel and J. B. Harbison, Proc. 17th Intl. Conf. on the Physics of Semiconductors, August 1984.
29. W. M. Pontuschka, W. E. Carlos, P. C. Taylor and R. W. Griffith, Phys. Rev. B25, 4362 (1982).
30. D. Han and H. Fritzsche, J. Non-Cryst. Solids 59-60,397 (1983).
31. G. Lucovsky and S. Y. Lin, this volume.
32. S. R. Ovshinsky and A. Madan, Nature 276, 482 (1978).

Author Index

Abeles, B......394, 417, 433
Abkowitz, M. 117
Adler, D..............70,197
Adriaenssens, G.J.........110
Allen, J.W...............341
Amamou, A................141
Asano, A.................318
Atoji, T.................221
Austin, I.G.........189,288

Baumann, J...............441
Bernstein, D.............229
Bhat, P.K...........189,288
Biegelsen, D.K............32
Bishop, S.G...............86
Biter, W.J...............170
Boulitrop, F.............178
Brodsky, M.H..............24
Brooks, B.G.........349,417

Carius, R................125
Carlson, D.E.............234
Catalano, A..............234
Cavenett, B.C.............48
Ceasar, G.P..............205
Chen, G..................266
Chopra, K.L..............189
Cody, G.D................349
Cohen, J.D................16
Cohen, M.H...............371
Collins, R.W.............170
Culbertson, J.C..........157

D'Aiello, R.V............234
Dickson, C.R.............234

Eberhardt, W.............394
Economou, E.N............371
Evangelisti, F...........402

Fiorini, P................402
Foley, G.M.T.............117
Freitas, J.A., Jr..........86
Fritzsche, H..229,296,425,478
Fuhs, W..................125
Fukunaga, T..............449

Gelatos, A.V..............16
Gersten, S...............441
Guha, S...................40

Hack, M...................40
Han, D...................296
Harbison, J.P........16, 149
Hark, S.K................465
Haruki, H................318
Hauser, J.J..............102
Hirabayashi, I.........8,141
Hirsch, M.D..............133
Homewood, K.P.............48
Hosokawa, Y. 250

Iida, S..................258
Inuishi, Y...............333
Inushima, T...............24
Itoh, M. 250

Jackson, W.B....213,242, 341
Janai, M.................364
Jang, J..................280
Jansen, F................465
Johnson, N.M..............32

Kakalios, J.........229,425
Kamiyama, M..............318
Kanicki, J................24
Kastner, M.A.......78,94,183
Kelly, M.K...............402
Klein, P.B...............157
Kocka, J.................272
Kosarev, A...............272
Krühler, W...............311
Kuck, M..................441
Kumagai, N. 333
Kumeda, M................221
Kuwano, Y................303

Lanford, W...............349
Lannin, J.S..............386
LeComber, P.G.............48
Lee, Charles. 205
Lee, Choochon............280
Lin, S.Y..................55
Liu, H-n. 1
Lucovsky, G...............55

Machonkin, M.A. 465
Margaritondo, G..........402
Maris, H.J...............102
Markovics, J.M...........117
Meier, F.................473

Author Index Continued

Mescheder, U..............356
Michiel, H..............110
Monroe, D..............94
Morigaki, K..............8,141
Morimoto, A..............221
Murase, K..............449
Murayama, K..............326

Nakamura, N..............303
Narasimhan, K.L..............425
Ninomiya, T..............326
Nishikuni, M..............303
Nishiura, M..............318
Nitta, S..............141

Oh, S.-J..............341
Ohlsen, W.D..............205
Ohnishi, M..............303
Ohta, H..............141
Ohtaki, T..............258
Okuno, T..............250
Okushi, H..............250
Olego, D..............441
Olshaker, M..............364
Orlowski, T.E..............163
Oswald, R.S..............234

Palumbo, A.C..............117
Patella, F..............402
Perfetti, P..............402
Persans, P.D..............349
Pescia, D..............473
Pfleiderer, H..............311
Pfost, D..............1
Pilione, L.J..............386
Pomian, R.J..............386
Pratt, B..............364
Plättner, R..............311

Quaresima, C..............402

Riedel, R.A..............402
Robertson, J..............63
Robins, L.H..............183
Roxlo, C.B..............433
Ruppert, A.F..............349

Sakai, H..............318
Schachter, R..............441
Scheffler, M..............379
Scher, H..............163
Schweitzer, L..............379
Searle, T.M..............189,288
Sergent, A.M..............149
Serino, R.J..............24
Seki, T..............258
Sette, F..............394
Shimizu, T..............221,266
Shirafuji, J..............333
Shur, M..............40
Silver, M..............197
Slade, M.L..............457,465
Soukoulis, C.M..............371
Spear, W.E..............48
Stasiewski, H..............394
Stathis, J.H..............78
Stetter, W..............311
Štika, O..............272
Stöhr, J..............394
Strait, J..............102
Street, R.A..............410
Strom, U..............86,157
Stuchlík, J..............272
Stutzmann, M..............213,242
Suzuki, H..............326

Tada, T..............326
Takahama, T..............303
Tanaka, K..............8, 250
Tauc, J..............1, 102
Taylor, P.C..............205
Thompson, M.J..............410
Thomsen, C..............102
Tiedje, T..............417, 433
Tříska, A..............272
Tsai, C.C..............213,242,341
Tsuda, S..............303

Uchida, Y..............318
Ullal, H.S..............205

van Berkel, C..............48
Vaněček, M..............272
Vanier, P.E..............133
Vardeny, Z..............1,102,364
Varmazis, C..............133

Wagner, I..............394
Wang, H..............266
Wang, Y..............266

Watanabe, I...............221
Watanabe, K...............303
Weil, R...................364
Weinstein, B.A..163,457, 465
Weiser, G.................356
Wilson, B.A...............149
Wolf, S.A.................157

Xu, X.....................266

Yamasaki, S.............8,250
Yokomichi, H..............221
Yoshida, M................141

Zhang, F..................266

AIP Conference Proceedings

		L.C. Number	ISBN
No.1	Feedback and Dynamic Control of Plasmas	70-141596	0-88318-100-2
No.2	Particles and Fields - 1971 (Rochester)	71-184662	0-88318-101-0
No.3	Thermal Expansion - 1971 (Corning)	72-76970	0-88318-102-9
No.4	Superconductivity in d-and f-Band Metals (Rochester, 1971)	74-18879	0-88318-103-7
No.5	Magnetism and Magnetic Materials - 1971 (2 parts) (Chicago)	59-2468	0-88318-104-5
No.6	Particle Physics (Irvine, 1971)	72-81239	0-88318-105-3
No.7	Exploring the History of Nuclear Physics	72-81883	0-88318-106-1
No.8	Experimental Meson Spectroscopy - 1972	72-88226	0-88318-107-X
No.9	Cyclotrons - 1972 (Vancouver)	72-92798	0-88318-108-8
No.10	Magnetism and Magnetic Materials - 1972	72-623469	0-88318-109-6
No.11	Transport Phenomena - 1973 (Brown University Conference)	73-80682	0-88318-110-X
No.12	Experiments on High Energy Particle Collisions - 1973 (Vanderbilt Conference)	73-81705	0-88318-111-8
No.13	π-π Scattering - 1973 (Tallahassee Conference)	73-81704	0-88318-112-6
No.14	Particles and Fields - 1973 (APS/DPF Berkeley)	73-91923	0-88318-113-4
No.15	High Energy Collisions - 1973 (Stony Brook)	73-92324	0-88318-114-2
No.16	Causality and Physical Theories (Wayne State University, 1973)	73-93420	0-88318-115-0
No.17	Thermal Expansion - 1973 (lake of the Ozarks)	73-94415	0-88318-116-9
No.18	Magnetism and Magnetic Materials - 1973 (2 parts) (Boston)	59-2468	0-88318-117-7
No.19	Physics and the Energy Problem - 1974 (APS Chicago)	73-94416	0-88318-118-5
No.20	Tetrahedrally Bonded Amorphous Semiconductors (Yorktown Heights, 1974)	74-80145	0-88318-119-3
No.21	Experimental Meson Spectroscopy - I974 (Boston)	74-82628	0-88318-120-7
No.22	Neutrinos - 1974 (Philadelphia)	74-82413	0-88318-121-5
No.23	Particles and Fields - 1974 (APS/DPF Williamsburg)	74-27575	0-88318-122-3
No.24	Magnetism and Magnetic Materials - 1974 (20th Annual Conference, San Francisco)	75-2647	0-88318-123-1
No.25	Efficient Use of Energy (The APS Studies on the Technical Aspects of the More Efficient Use of Energy)	75-18227	0-88318-124-X

AIP Conference Proceedings

No.	Title	LC Number	ISBN
No.26	High-Energy Physics and Nuclear Structure - 1975 (Santa Fe and Los Alamos)	75-26411	0-88318-125-8
No.27	Topics in Statistical Mechanics and Biophysics: A Memorial to Julius L. Jackson (Wayne State University, 1975)	75-36309	0-88318-126-6
No.28	Physics and Our World: A Symposium in Honor of Victor F. Weisskopf (M.I.T., 1974)	76-7207	0-88318-127-4
No.29	Magnetism and Magnetic Materials - 1975 (21st Annual Conference, Philadelphia)	76-10931	0-88318-128-2
No.30	Particle Searches and Discoveries - 1976 (Vanderbilt Conference)	76-19949	0-88318-129-0
No.31	Structure and Excitations of Amorphous Solids (Williamsburg, VA., 1976)	76-22279	0-88318-130-4
No.32	Materials Technology - 1976 (APS New York Meeting)	76-27967	0-88318-131-2
No.33	Meson-Nuclear Physics - 1976 (Carnegie-Mellon Conference)	76-26811	0-88318-132-0
No.34	Magnetism and Magnetic Materials - 1976 (Joint MMM-Intermag Conference, Pittsburgh)	76-47106	0-88318-133-9
No.35	High Energy Physics with Polarized Beams and Targets (Argonne, 1976)	76-50181	0-88318-134-7
No.36	Momentum Wave Functions - 1976 (Indiana University)	77-82145	0-88318-135-5
No.37	Weak Interaction Physics - 1977 (Indiana University)	77-83344	0-88318-136-3
No.38	Workshop on New Directions in Mossbauer Spectroscopy (Argonne, 1977)	77-90635	0-88318-137-1
No.39	Physics Careers, Employment and Education (Penn State, 1977)	77-94053	0-88318-138-X
No.40	Electrical Transport and Optical Properties of Inhomogeneous Media (Ohio State University, 1977)	78-54319	0-88318-139-8
No.41	Nucleon-Nucleon Interactions - 1977 (Vancouver)	78-54249	0-88318-140-1
No.42	Higher Energy Polarized Proton Beams (Ann Arbor, 1977)	78-55682	0-88318-141-X
No.43	Particles and Fields - 1977 (APS/DPF, Argonne)	78-55683	0-88318-142-8
No.44	Future Trends in Superconductive Electronics (Charlottesville, 1978)	77-9240	0-88318-143-6
No.45	New Results in High Energy Physics - 1978 (Vanderbilt Conference)	78-67196	0-88318-144-4
No.46	Topics in Nonlinear Dynamics (La Jolla Institute)	78-057870	0-88318-145-2
No.47	Clustering Aspects of Nuclear Structure and Nuclear Reactions (Winnepeg, 1978)	78-64942	0-88318-146-0
No.48	Current Trends in the Theory of Fields (Tallahassee, 1978)	78-72948	0-88318-147-9
No.49	Cosmic Rays and Particle Physics - 1978 (Bartol Conference)	79-50489	0-88318-148-7

AIP Conference Proceedings

No.50	Laser-Solid Interactions and Laser Processing - 1978 (Boston)	79-51564	0-88318-149-5
No.51	High Energy Physics with Polarized Beams and Polarized Targets (Argonne, 1978)	79-64565	0-88318-150-9
No.52	Long-Distance Neutrino Detection - 1978 (C.L. Cowan Memorial Symposium)	79-52078	0-88318-151-7
No.53	Modulated Structures - 1979 (Kailua Kona, Hawaii)	79-53846	0-88318-152-5
No.54	Meson-Nuclear Physics - 1979 (Houston)	79-53978	0-88318-153-3
No.55	Quantum Chromodynamics (La Jolla, 1978)	79-54969	0-88318-154-1
No.56	Particle Acceleration Mechanisms in Astrophysics (La Jolla, 1979)	79-55844	0-88318-155-X
No. 57	Nonlinear Dynamics and the Beam-Beam Interaction (Brookhaven, 1979)	79-57341	0-88318-156-8
No. 58	Inhomogeneous Superconductors - 1979 (Berkeley Springs, W.V.)	79-57620	0-88318-157-6
No. 59	Particles and Fields - 1979 (APS/DPF Montreal)	80-66631	0-88318-158-4
No. 60	History of the ZGS (Argonne, 1979)	80-67694	0-88318-159-2
No. 61	Aspects of the Kinetics and Dynamics of Surface Reactions (La Jolla Institute, 1979)	80-68004	0-88318-160-6
No. 62	High Energy e^+e^- Interactions (Vanderbilt , 1980)	80-53377	0-88318-161-4
No. 63	Supernovae Spectra (La Jolla, 1980)	80-70019	0-88318-162-2
No. 64	Laboratory EXAFS Facilities - 1980 (Univ. of Washington)	80-70579	0-88318-163-0
No. 65	Optics in Four Dimensions - 1980 (ICO, Ensenada)	80-70771	0-88318-164-9
No. 66	Physics in the Automotive Industry - 1980 (APS/AAPT Topical Conference)	80-70987	0-88318-165-7
No. 67	Experimental Meson Spectroscopy - 1980 (Sixth International Conference , Brookhaven)	80-71123	0-88318-166-5
No. 68	High Energy Physics - 1980 (XX International Conference, Madison)	81-65032	0-88318-167-3
No. 69	Polarization Phenomena in Nuclear Physics - 1980 (Fifth International Symposium, Santa Fe)	81-65107	0-88318-168-1
No. 70	Chemistry and Physics of Coal Utilization - 1980 (APS, Morgantown)	81-65106	0-88318-169-X
No. 71	Group Theory and its Applications in Physics - 1980 (Latin American School of Physics, Mexico City)	81-66132	0-88318-170-3
No. 72	Weak Interactions as a Probe of Unification (Virginia Polytechnic Institute - 1980)	81-67184	0-88318-171-1
No. 73	Tetrahedrally Bonded Amorphous Semiconductors (Carefree, Arizona, 1981)	81-67419	0-88318-172-X
No.74	Perturbative Quantum Chromodynamics (Tallahassee. 1981)	81-70372	0-88318-173-8

No. 75	Low Energy X-ray Diagnostics-1981 (Monterey)	81-69841	0-88318-174-6
No. 76	Nonlinear Properties of Internal Waves (La Jolla Institute, 1981)	81-71062	0-88318-175-4
No. 77	Gamma Ray Transients and Related Astrophysical Phenomena (La Jolla Institute, 1981)	81-71543	0-88318-176-2
No. 78	Shock Waves in Condensed Matter - 1981 (Menlo Park)	82-70014	0-88318-177-0
No. 79	Pion Production and Absorption in Nuclei - 1981 (Indiana University Cyclotron Facility)	82-70678	0-88318-178-9
No. 80	Polarized Proton Ion Sources (Ann Arbor, 1981)	82-71025	0-88318-179-7
No. 81	Particles and Fields - 1981: Testing the Standard Model (APS/DPF, Santa Cruz)	82-71156	0-88318-180-0
No. 82	Interpretation of Climate and Photochemical Models, Ozone and Temperature Measurements (La Jolla Institute, 1981)	82-071345	0-88318-181-9
No. 83	The Galactic Center (Cal. Inst. of Tech., 1982)	82-071635	0-88318-182-7
No. 84	Physics in the Steel Industry (APS.AISI, Lehigh University, 1981)	82-072033	0-88318-183-5
No. 85	Proton-Antiproton Collider Physics - 1981 (Madison, Wisconsin)	82-072141	0-88318-184-3
No. 86	Momentum Wave Functions - 1982 (Adelaide, Australia)	82-072375	0-88318-185-1
No. 87	Physics of High Energy Particle Accelerators (Fermilab Summer School, 1981)	82-072421	0-88318-186-X
No. 88	Mathematical Methods in Hydrodynamics and Integrability in Dynamical Systems (La Jolla Institute, 1981)	82-072462	0-88318-187-8
No. 89	Neutron Scattering - 1981 (Argonne National Laboratory)	82-073094	0-88318-188-6
No. 90	Laser Techniques for Extreme Ultraviolt Spectroscopy (Boulder, 1982)	82-073205	0-88318-189-4
No. 91	Laser Acceleration of Particles (Los Alamos, 1982)	82-073361	0-88318-190-8
No. 92	The State of Particle Accelerators and High Energy Physics(Fermilab, 1981)	82-073861	0-88318-191-6
No. 93	Novel Results in Particle Physics (Vanderbilt, 1982)	82-73954	0-88318-192-4
No. 94	X-Ray and Atomic Inner-Shell Physics-1982 (International Conference, U. of Oregon)	82-74075	0-88318-193-2
No. 95	High Energy Spin Physics - 1982 (Brookhaven National Laboratory)	83-70154	0-88318-194-0
No. 96	Science Underground (Los Alamos, 1982)	83-70377	0-88318-195-9

No. 97	The Interaction Between Medium Energy Nucleons in Nuclei-1982 (Indiana University)	83-70649	0-88318-196-7
No. 98	Particles and Fields - 1982 (APS/DPF University of Maryland)	83-70807	0-88318-197-5
No. 99	Neutrino Mass and Gauge Structure of Weak Interactions (Telemark, 1982)	83-71072	0-88318-198-3
No. 100	Excimer Lasers - 1983 (OSA, Lake Tahoe, Nevada)	83-71437	0-88318-199-1
No. 101	Positron-Electron Pairs in Astrophysics (Goddard Space Flight Center, 1983)	83-71926	0-88318-200-9
No. 102	Intense Medium Energy Sources of Strangeness (UC-Santa Cruz, 1983)	83-72261	0-88318-201-7
No. 103	Quantum Fluids and Solids - 1983 (Sanibel Island, Florida)	83-72440	0-88318-202-5
No. 104	Physics,Technology and the Nuclear Arms Race (APS Baltimore-1983)	83-72533	0-88318-203-3
No. 105	Physics of High Energy Particle Accelerators (SLAC Summer School, 1982)	83-72986	0-88318-304-8
No. 106	Predictability of Fluid Motions (La Jolla Institute, 1983)	83-73641	0-88318-305-6
No. 107	Physics and Chemistry of Porous Media (Schlumberger-Doll Research, 1983)	83-73640	0-88318- 306-4
No. 108	The Time Projection Chamber (TRIUMF, Vancouver, 1983)	83-83445	0-88318-307-2
No. 109	Random Walks and Their Applications in the Physical and Biological Sciences (NBS/La Jolla Institute, 1982)	84-70208	0-88318-308-0
No. 110	Hadron Substructure in Nuclear Physics (Indiana University, 1983)	84-70165	0-88318-309-9
No. 111	Production and Neutralization of Negative Ions and Beams (3rd Int'l Symposium, Brookhaven, 1983)	84-70379	0-88318-310-2
No. 112	Particles and Fields-1983 (APS/DPF, Blacksburg, VA)	84-70378	0-88318-311-0
No. 113	Experimental Meson Spectroscopy - 1983 (Seventh International Conference, Brookhaven)	84-70910	0-88318-312-9
No. 114	Low Energy Tests of Conservation Laws in Particle Physics (Blacksburg, VA, 1983)	84-71157	0-88318-313-7
No. 115	High Energy Transients in Astrophysics (Santa Cruz, CA, 1983)	84-71205	0-88318-314-5
No. 116	Problems in Unification and Supergravity (La Jolla Institute, 1983)	84-71246	0-88318-315-3
No. 117	Polarized Proton Ion Sources (TRIUMF, Vancouver, 1983)	84-71235	0-88318-316-1

No. 118	Free Electron Generation of Extreme Ultraviolet Coherent Radiation (Brookhaven/OSA,1983)	84-71539	0-88318-317-X
No. 119	Laser Techniques in the Extreme Ultraviolet (OSA, Boulder, Colorado, 1984)	84-72128	0-88318-318-8
No. 120	Optical Effects in Amorphous Semiconductors (Snowbird, Utah, 1984)	84-72419	0-88318-319-6